Application of Soil Physics in Environmental Analyses

Progress in Soil Science

Aims and Scope

Progress in Soil Science series aims to publish books that contain novel approaches in soil science in its broadest sense – books should focus on true progress in a particular area of the soil science discipline. The scope of the series is to publish books that enhance the understanding of the functioning and diversity of soils in all parts of the globe. The series includes multidisciplinary approaches to soil studies and welcomes contributions of all soil science subdisciplines such as: soil genesis, geography and classification, soil chemistry, soil physics, soil biology, soil mineralogy, soil fertility and plant nutrition, soil and water conservation, pedometrics, digital soil mapping, proximal soil sensing, soils and land use change, global soil change, natural resources and the environment.

For further volumes:
http://www.springer.com/series/8746

Wenceslau Geraldes Teixeira
Marcos Bacis Ceddia
Marta Vasconcelos Ottoni
Guilheme Kangussu Donnagema
Editors

Application of Soil Physics in Environmental Analyses

Measuring, Modelling and Data Integration

Springer

Editors
Wenceslau Geraldes Teixeira
Guilheme Kangussu Donnagema
Soil Physics
Embrapa Soils
Rio de Janeiro, RJ, Brazil

Marcos Bacis Ceddia
Soils Department
Rural Federal University of Rio de Janeiro
Seropédica, RJ, Brazil

Marta Vasconcelos Ottoni
Department of Hydrology
Companhia de Pesquisa de Recursos Minerais (CPRM)
Geological Survey of Brazil
Rio de Janeiro, RJ, Brazil

ISBN 978-3-319-06012-5 ISBN 978-3-319-06013-2 (eBook)
DOI 10.1007/978-3-319-06013-2
Springer Cham Heidelberg New York Dordrecht London

Library of Congress Control Number: 2014942554

Printed on acid-free paper

Springer is part of Springer Science+Business Media (www.springer.com)

Foreword

The singular importance of soil as the Earth's life-supporting system is increasingly recognized. The focus on soils becomes a global phenomenon, and the United Nations General Assembly has declared 2015 as the International Year of Soils. Various disciplines supply and use knowledge about soils, and soil physics traditionally provides essential information about soil physical properties and processes and characterizes soil as a complex medium where a multitude of chemical and biological processes act and interact. Agricultural practices and ecological trends both affect and are affected by soil physical properties. Ecological services of soils as well as the reaction of soil to extreme events cannot be understood and managed without soil physics analysis.

It is the time of opportunity for soil physics. The demand on understanding the mechanisms of process level, understanding of flow, transport, storage, and transformation processes in soils is increasing at various scales. The existing empirical knowledge appears to be not sufficient since it was accumulated in climate conditions and management practices that are changing. Tropical regions and currently warming polar regions include vast areas of interest for agriculture, where little is known about the physical properties of soils. Agriculture in those regions is devoid of means to predict management-related trends. At larger scales, understanding mechanisms of physical processes in soils is needed to evaluate the potential of carbon sequestration, to address food security issues, and to quantify the soil controls of the climate.

New prospects in soil physics open as novel measurement techniques such as wireless soil moisture sensor networks, air-borne and ground-based remote sensing, and hydrogeophysical measurement techniques are developing and become available. Clearly, we have now entered the "big data" era in the field of soil water content sensing with data sets often comprising millions of measurements and providing unique opportunities to study soil water dynamics. This creates substantial challenges in managing, sharing, analyzing, and visualizing soil moisture data. The development of smart adaptive data collection and integration strategies presents an exciting avenue to explore.

Soil physics already benefits from the current trend toward multidisciplinary research. The fast-developing discipline of hydropedology emphasizes the relationship between soil structure and its hydrologic functioning as one of the cornerstones. This provides the much needed conceptual base for the interpretation and presentation of soil water dynamics in complex landscapes. The structure-function relationships in soil physics are equally important at finer scales, where computer tomography creates the opportunity for understanding the habitat and functioning of soil microbial communities, and at coarser scales, where the structure of soil cover controls and is affected by the spatial and temporal plant community and land use structure. Progress in sister disciplines will undoubtedly create new opportunities for interdisciplinary enrichment in soil science, and soil physics in particular. It is important to monitor this progress and seek interactions.

A strong monitoring tradition in soil physics has immensely helped to understand physical processes in soils. It is realized nowadays that these benefits will be magnified if the monitoring of soil physical variables will be coupled with the monitoring of chemical, biological, and plant-related variables. A recent scientific breakthrough related to the isotope analysis of water coming to, stored in, and leaving soil is one successful example. Soil physics results find active use in research and management of agricultural systems. The concurrent monitoring and modeling of plant and soil variables and interactions appears of utmost significance since the response of soil-plant systems to droughts and inundations gains importance, and adaptation strategies of even the most widely studied plants remain only partially understood. The combination of monitoring and modeling via data assimilation has a potential to provide new data about the adequacy of existing models.

The demand on soil physical information has exposed several areas where knowledge needs to be substantially improved. Soils are intrinsically heterogeneous, and some heterogeneities, such as macropores, albeit rare, control the ability of soil to store and conduct water soil physical processes at field scale. It is a good time to reflect on how these heterogeneities are characterized and what information about them is of maximum use to assist in management decisions. Heterogeneity causes variability, and the efficient techniques of the characterization of soil physical variability remain the object of scientific pursuit.

A critical look is needed at the products that soil physics provides for most widely used applications. Soil physical quality evaluation, for example, is invaluable for screening purposes and sustainability projections. It is important to ask whether the current metrics of soil physical quality are suited for these purposes and whether there are data and ideas to develop their modern replacements. Pedotransfer functions, that is, empirical relationships between process parameters and easily measured soil properties, find broad applications in modeling projects. Many pedotransfer functions have been developed, and an analysis seems to be in order to derive guidelines on their individual or combined use. Regional pedotransfer development projects have been useful, and more needs to be done in this field. Soil physics traditionally translates the process level understanding into mechanistic models, thus achieving knowledge packaging and dissemination. One research need consists in eliminating the discrepancy between measuring water flow parameters in equilibrium conditions and using these parameters to simulate nonequilibrium flow and transport.

Measuring, modeling, and data integration create the application of soil physics in research areas impacting global issues where soil physics (and soil science) can make important contributions, such as characterizing plant–soil interactions in the changing hydrologic cycle and food production and valuing and preserving the soil-associated ecosystem services. Soil physics as a relatively young discipline will greatly benefit from taking on these big challenges.

Beltsville, MD, USA Yakov Pachepksy

Preface

The green economy and the recognition of the importance to preserve soil and water have been discussed in many interdisciplinary forums in the world, which invokes the necessity to better characterize the soil and understand the behavior of water in this medium. Soil scientists are an integral part of interdisciplinary studies about the interactions in the soil-plant-atmosphere continuum, soil management practices, and indicators of soil quality. The higher frequency of natural disasters, such as landslides and thunderstorms, addresses the importance to integrate soil characteristics in predictive models. This book seeks to integrate the efforts of scientists to meet this demand. Soil physics research has grown considerably, specially in the use of innovative sensors; software, soil database, algorithms, and modeling techniques have been introduced into the soil-water relationship and environmental monitoring. These advances are thoroughly dispersed in articles and conference proceedings. In this volume, the authors will bring together the effectiveness of many new field and lab sensors and examine the current state-of-the-art in modeling and data analysis. It also includes innovative approaches and case studies in tropical soils. Future directions in soil physics research are given by key researchers in this discipline. The topics tackled by the chapters were divided into five parts that include: **Part I – Integrating data in soil physics** to discuss the potential to use data sets and modeling methodologies in major environmental problems. It is proposed to re-establish the knowledge chain, linking tacit knowledge to cutting-edge science. Moreover, the use of field soil data in flow phenomena, or hydropedology, is discussed, and some examples are provided. **Part II – Data analysis in soil physics and pedotranfer functions** presents the analysis of data in state-space and geoestatistical approaches. State-space has been used in the last decades to describe spatial processes of transport- or biomass-related state variables. The opportunities to use this technique for analyzing experiments are addressed, and examples of use in tropical soils are given. The development and use of pedotransfer functions (PTF) for soil hydraulic parameters and functions are exemplified for a Brazilian database. The significance of predictors is discussed, and many PTF equations for tropical soil hydraulic parameters and functions are presented. **Part III – Different approaches to characterize soil physical quality indicators** has focused on new techniques used to characterize and to interpret soil physical parameters, with case studies in a semiarid region and in the Amazon. Techniques for mapping and monitoring the quality of agricultural soils are in large demand for the society. The use of a friendly algorithm for this purpose is demonstrated for many soil physical properties in an agroecological farm, which can be used in precision agriculture. The challenge of assessing soil physical quality is discussed using ascending complexity classification from the simplest to the more complex indicators. A preliminary proposal of a structural classification system of soil pore space similar to that of textural classification is proposed. Several horticulture crops cultivated in the Brazilian semiarid region have root systems capable of reaching deeper soil layers and are drought-tolerant. Therefore, deficit irrigation is discussed and considered as an opportunity for reduced water application. **Part IV Sensors and monitoring in soil physics** centers the discussion in equipment and sampling techniques for monitoring soil physical parameters. Using monitoring data to periodically correct modeling results consists of replacing simulated values with values that are closer to the measured ones. This operation is called data assimilation (DA), and it is discussed as a possibility to improve conceptual models of water fluxes integrating developments in PTF, temporal stability of soil water patterns, and innovative soil moisture. Looking inside an undisturbed soil sample was considered impossible a century ago. Nevertheless, technological advances of imaging techniques such as X-ray computed tomography (CT) provide powerful means for visualization of the soil interior to evaluate pore topological properties and structure and study dynamic processes in a noninvasive way. Advantages and limitations of the technique are demonstrated. Determination of accurate, unsaturated hydraulic conductivity functions may be required to solve many questions in agriculture and environmental research. Examples of field measurements of unsaturated hydraulic functions using tension infiltrometers and the use of scaling theory and piecewise continuous regression to deal with soil variability are given. Interpretations of the results to compare different land use systems in the Amazon are pointed out. The data sets of soil moisture and soil water storage are a necessary input in many hydrological and climate circulation models. A comparison with in situ and remotely sensed data of soil moisture and limitations in using these data for hydrological modeling are discussed.

Part V Creating databases and models applied to soil physics discusses to the solving of practical environmental problems. Alternative approaches for modeling water flow and solute transport in soil-stressing flow and transport processes in the vadose zone between the soil surface and the groundwater table are given and exemplified. A review of multicomponent solute transport models and examples of their use in agricultural and environmental applications are presented. Unsaturated soil hydraulic properties for their applicability in models simulating unsaturated water transport are reviewed. Problems of common parameterizations of hydraulic functions in the critical regions near and at saturation and recent progress with respect to including film and corner flow components in the hydraulic conductivity function are addressed. The phenomenon of dynamic nonequilibrium in soil water flow is discussed as the need of a paradigm change in the modeling of soil water transport. Finally, the availability and the modeling of root water uptake are discussed with more advanced approaches than the simple concepts of the field capacity and permanent wilting point.

The last two chapters of the book discuss the challenges to the development of a soil database. The chapter "Experiences and Challenges in Developing European Soil Hydrological Databases" presents the phases of development of the Hydraulic Properties of European Soils (HYPRES) database. It is considered a successful project, and the derived PTFs have been widely used. It not only discusses the technical aspects employed to create a collaborative soil hydrological database, but the lessons learnt and the agreement needed for the upkeep of the database are also stressed. In many parts of the book, it is pointed out that the use of PTF derived in temperate soils for solving questions in tropical soils can lead to erroneous results. Therefore, the last chapter makes an overview of hydrophysical data on Brazilian soils available for pedotransfer function development. Some statistical analysis and the structure of metadata on soil hydrophysical information are provided. The results of this study suggest that there are excellent prospects for water retention PTF development at the Brazilian national scale. The challenge in the development of hydrophysical databases for PTF application is also stressed.

Early versions of most of these chapters were presented at the Second Brazilian Soil Physics Meeting held in Rio de Janeiro, May 6–10th, 2013.

We want to thank the reviewers of the chapters for their valuable time and scientific expertise. Many authors of this book also helped us to review chapters. In particular, we wish to thank Boris Faybishenko, Gustavo Vasques, João Herbert Moreira Viana, José Ronaldo de Macedo, Marcel Schaap, Pedro Luiz de Freitas, and Waldir Carvalho. We are much indebted to Maiara Santos, who proofread the references in all chapters.

Thanks also go to Maryse Elliott, senior editor at Springer, who motivated us to edit this volume, and Melanie van Overbeek, the publishing assistant who answered all questions through the publication process.

Finally we thank our families and friends for supporting us during this time.

February 2014
Rio de Janeiro, Brazil

Wenceslau G. Teixeira
Marcos B. Ceddia
Marta V. Ottoni
Guilheme K. Donnagema

Contents

Part III Different Approaches for the Characterization of Soil Physical Quality Indicators

Part IV Sensors and Monitoring in Soil Physics

Contributors

Leandro Sanzi Aquino Agronomy Post-Graduate Program, Faculty of Agronomy, Federal University of Pelotas, Capão do Leão, Brazil

Alexandre Hugo Cezar Barros Brazilian Agricultural Research Corporation (EMBRAPA) – National Centre of Soil Research, Recife, PE, Brazil

Luis Henrique Bassoi Brazilian Agricultural Research Corporation (EMBRAPA) – Embrapa Semiárido, Petrolina, PE, Brazil

Johan Bouma Soils Department, Wageningen University, Wageningen, The Netherlands

Ary Bruand Institut des Sciences de la Terre d'Orléans (ISTO) UMR6113, Université d'Orléans, CNRS/INSU, Université François Rabelais – Tours, Orléans, France

Marcos Bacis Ceddia Soils Department, Rural Federal University of Rio de Janeiro, Seropédica, RJ, Brazil

Sílvio Crestana Embrapa Agricultural Instrumentation, Brazilian Agricultural Research Corporation-EMBRAPA, São Carlos, SP, Brazil

Guilheme Kangussu Donnagema Soil Physics, Embrapa Soils, Rio de Janeiro, RJ, Brazil

Alvaro Pires da Silva Departamento de Ciência do Solo, Escola Superior de Agricultura Luiz de Queiroz - Universidade de São Paulo, Piracicaba, São Paulo, Brazil

Euzebio Medrado da Silva Empresa Brasileira de Pesquisa Agropecuária – Embrapa Cerrados, Planaltina, DF, Brazil

Afonso Augusto Magalhães de Araujo Departamento de Recursos Hídricos e Meio Ambiente, Escola Politécnica de Engenharia, Universidade Federal do Rio de Janeiro, Rio de Janeiro, RJ, Brazil

Lígia Maria Nascimento de Araujo Laboratório de Recursos Hídricos e Meio Ambiente (LABH2O), Programa de Engenharia Civil, Instituto Alberto Luiz Coimbra de Pós-Graduação e Pesquisa em Engenharia – COPPE, Universidade Federal do Rio de Janeiro, Campus Ilha do Fundão – Cidade Universitária, Rio de Janeiro, RJ, Brazil

Agência Nacional de Águas – ANA, Setor Policial Sul – SPS, Brasília, DF, Brazil

Cláudia Liane Rodrigues de Lima Soil Science Department, Faculty of Agronomy, Federal University of Pelotas, Capão do Leão, RS, Brazil

Rosane Ferreira de Oliveira Department of Mathematics, Federal Rural University of Rio de Janeiro (UFRRJ), Seropédica, RJ, Brazil

Rafael Carneiro Di Bello Tribunal de Contas da União – TCU, Setor de Administração Federal Sul – SAFS, Brasília, DF, Brazil

Efstathios Diamantopoulos Institute for Geoecology, Soil Science and Soil Physics, Technische Universität Braunschweig, Braunschweig, Germany

Durval Dourado-Neto Crop Production Department, ESALQ/USP, Piracicaba, SP, Brazil

Wolfgang Durner Institute for Geoecology, Soil Science and Soil Physics, Technische Universität Braunschweig, Braunschweig, Germany

Neyde Fabíola Balarezo Giarola Departamento de Ciência do Solo e Engenharia Agrícola, Universidade Estadual de Ponta Grossa, Ponta Grossa, PR, Brazil

Célia Regina Grego Embrapa Satellite Monitoring, Campinas, São Paulo, Brazil

Rachel Muylaert Locks Guimarães Departamento de Agronomia, Universidade Tecnológica Federal do Paraná, Pato Branco, PR, Brazil

Roland Hiederer European Commission, Joint Research Centre, Institute for Environment and Sustainability, Ispra, Italy

Bernd Huwe Soil Physics Department, University of Bayreuth, Bayreuth, Germany

Sascha C. Iden Institute for Geoecology, Soil Science and Soil Physics, Technische Universität Braunschweig, Braunschweig, Germany

Diederik Jacques Institute for Environment, Health and Safety, Belgian Nuclear Research Centre (SCK•CEN), Mol, Belgium

Vilson Antônio Klein Faculdade de Agronomia e Medicina Veterinária, Universidade de Passo Fundo, Passo Fundo, RS, Brazil

Paulo Renato Orlandi Lasso Embrapa Agricultural Instrumentation, Brazilian Agricultural Research Corporation-EMBRAPA, São Carlos, SP, Brazil

Bertrand Leterme Institute for Environment, Health and Safety, Belgian Nuclear Research Centre (SCK•CEN), Mol, Belgium

Allan Lilly The James Hutton Institute, Aberdeen, Scotland, UK

Henry Lin Department of Ecosystem Science and Management, The Pennsylvania State University, University Park, PA, USA

Robélio Leandro Marchão Empresa Brasileira de Pesquisa Agropecuária – Embrapa Cerrados, Planaltina, DF, Brazil

Maria Leonor Ribeiro Casimiro Lopes-Assad Department of Natural Resources and Environmental Protection, Federal University of São Carlos (UFSCar), Araras, SP, Brazil

Jean Dalmo Marques Federal Institute of Education, Science and Technology of the Amazonas – IFAM, Campus Center of Manaus, Manaus, Brazil

Gonzalo Martinez Department of Natural Resources and Agroecology, Andalusian Institute of Agricultural Research and Training, Fisheries, Food and Organic Production (IFAPA), Seville, Spain

Daniel Medeiros Moreira Laboratório de Recursos Hídricos e Meio Ambiente (LABH2O), Programa de Engenharia Civil, Instituto Alberto Luiz Coimbra de Pós-Graduação e Pesquisa em Engenharia – COPPE, Universidade Federal do Rio de Janeiro, Campus Ilha do Fundão – Cidade Universitária, Rio de Janeiro, RJ, Brazil

Divisão de Hidrologia Aplicada, Serviço Geológico do Brasil, CPRM, Rio de Janeiro, RJ, Brazil

Carolina Naveira-Cotta Department of Mechanical Engineering, POLI&COPPE, Federal University of Rio de Janeiro, Rio de Janeiro, RJ, Brazil

Attila Nemes Division of Soil and Environment, Bioforsk, Norway

Marta Vasconcelos Ottoni Department of Hydrology, Companhia de Pesquisa de Recursos Minerais (CPRM), Geological Survey of Brazil, Rio de Janeiro, RJ, Brazil

Laboratório de Recursos Hídricos e Meio Ambiente (LABH2O), Programa de Engenharia Civil, Instituto Alberto Luiz Coimbra de Pós-Graduação e Pesquisa em Engenharia – COPPE, Universidade Federal do Rio de Janeiro, Rio de Janeiro, RJ, Brazil

Theophilo Benedicto Ottoni Filho Department of Water Resources and Environment, Federal University of Rio de Janeiro, Rio de Janeiro, RJ, Brazil

Yakov Pachepsky Environmental Microbial and Food Safety Laboratory, USDA-ARS Beltsville Agricultural Research Center, Beltsville, MD, USA

Feng Pan Energy and Geoscience Institute, The University of Utah, Salt Lake City, UT, USA

Letiane Helwig Penning Soil and Water Management and Conservation Post-Graduate Program, Faculty of Agronomy, Federal University of Pelotas, Capão do Leão, RS, Brazil

Elizabeth M. Pontedeiro Department of Nuclear Engineering, POLI&COPPE, Federal University of Rio de Janeiro, Rio de Janeiro, RJ, Brazil

Tiago B. Ramos CEER-Biosystems Engineering, Institute of Agronomy, Technical University of Lisbon, Lisbon, Portugal

Amir Raoof Department of Earth Sciences, Utrecht University, Utrecht, The Netherlands

Klaus Reichardt Soil Physics Laboratory, CENA/USP, Piracicaba, SP, Brazil

Willian Daniel Reynolds Greenhouse and Processing Crops Research Centre, Agriculture and Agri-Food Canada, Harrow, ON, Canada

Otto Corrêa Rotunno Filho Laboratório de Recursos Hídricos e Meio Ambiente (LABH2O), Programa de Engenharia Civil, Instituto Alberto Luiz Coimbra de Pós-Graduação e Pesquisa em Engenharia – COPPE, Universidade Federal do Rio de Janeiro, Campus Ilha do Fundão – Cidade Universitária, Rio de Janeiro, RJ, Brazil

Glenio Guimarães Santos Escola de Agronomia e Engenharia de Alimentos – Universidade Federal de Goiás, Goiânia, Goiás, Brazil

Benedikt Scharnagl Institute for Geoecology, Soil Science and Soil Physics, Technische Universität Braunschweig, Braunschweig, Germany

Götz Schroth Sustainable Agriculture Department, Rainforest Alliance, Wageningen, The Netherlands

Jiří Šimůnek Department of Environmental Sciences, University of California, Riverside, CA, USA

Todd H. Skaggs U.S. Salinity Laboratory, USDA-ARS, Riverside, CA, USA

Wenceslau Geraldes Teixeira Soil Physics, Embrapa Soils, Rio de Janeiro, RJ, Brazil

Luís Carlos Timm Rural Engineering Department, Faculty of Agronomy, Federal University of Pelotas, Capão do Leão, RS, Brazil

George Clarke Topp Eastern Cereal and Oilssed Research Centre, Agriculture and Agri-Food Canada, Ottawa, ON, Canada

Cássio Antônio Tormena Departamento de Agronomia, Universidade Estadual de Maringá, Maringá, PR, Brazil

Markus Tuller Department of Soil, Water and Environmental Science, The University of Arizona, Tucson, AZ, USA

Martinus Th. van Genuchten Department of Mechanical Engineering, POLI&COPPE, Universidade Federal do Rio de Janeiro, Rio de Janeiro, RJ, Brazil

Department of Earth Sciences, Utrecht University, Utrecht, The Netherlands

Quirijn de Jong van Lier Center for Nuclear Energy in Agriculture, University of São Paulo (CENA/USP), São Paulo, Brazil

Carlos Alberto Alves Varella Engineering Department, Federal Rural University of Rio de Janeiro (UFRRJ), Seropédica, RJ, Brazil

Carlos Manoel Pedro Vaz Embrapa Agricultural Instrumentation, Brazilian Agricultural Research Corporation-EMBRAPA, São Carlos, SP, Brazil

Sergio Drumond Ventura Department of Mathematics, Federal Rural University of Rio de Janeiro (UFRRJ), Seropédica, RJ, Brazil

Sidney Rosa Vieira Agronomy Institute (IAC), Campinas, São Paulo, Brazil

André Luis Oliveira Villela Technical College of the Federal Rural University of Rio de Janeiro (CTUR-UFRRJ), Seropédica, RJ, Brazil

Ole Wendroth Department of Plant and Soil Sciences, University of Kentucky, Lexington, KY, USA

J. Henk M. Wösten Alterra – Wageningen University and Research Centre, Wageningen, The Netherlands

Adilson Elias Xavier Programa de Engenharia Sistemas e Computação, Instituto Alberto Luiz Coimbra de Pós-Graduação e Pesquisa em Engenharia – COPPE, Universidade Federal do Rio de Janeiro, Rio de Janeiro, RJ, Brazil

Luciano Nóbrega Rodrigues Xavier Centro de Pesquisa de Energia Elétrica (CEPEL), Avenida Horácio de Macedo, 354, Cidade Universitária, Rio de Janeiro, RJ, Brazil

Yang Yang Department of Plant and Soil Sciences, University of Kentucky, Lexington, KY, USA

Haoliang Yu Department of Ecosystem Science and Management, The Pennsylvania State University, University Park, PA, USA

College of Water Resources and Civil Engineering, China Agricultural University, Beijing, China

Weihua Zhang Department of Ecosystem Science and Management, The Pennsylvania State University, University Park, PA, USA

College of Resources and Environment, Southwest University, Chongqing, China

Part I
Integrating Data in Soil Physics

Chapter 1
Knowledge Chains, Linking Cutting-Edge Research on Flow Phenomena in Soils with Tacit Knowledge from Field Research

Johan Bouma

Abstract The impressive set of measuring, monitoring and modeling methodologies of soil physics is currently not being used to its full potential in inter- and transdisciplinary studies on major environmental problems. Suggestions are made to: (i) re-establish the knowledge chain, linking tacit knowledge to cutting edge science in both directions; (ii) make the scattered field of soil science research more visible to outsiders by focusing research on the seven soil functions, as defined in the Soil Protection Strategy of the Commission of the European Communities; (iii) better integrate the various subdisciplines of soil science where particular attention is suggested for hydropedology: soil physicists working together with pedologists; (iv) show the relevance of soil research by demonstrating its crucial input when defining strategic concepts, such as the Planetary Boundaries of Rockstrom; (iv) focus on specific case studies demonstrating the central role that soil science can play in studying and solving "wicked" environmental problems, and (v) be aware that considering business opportunities, as in the two case studies, can provide essential support to the scientific research effort.

Keywords Sustainable development • Land evaluation • Soil erosion • Soil physics • Precision agriculture • Soil functions

1.1 Introduction

Impressive results of soil physical research, including water, solute and gas transport, energy balances and thermal regimes have recently been summarized in the Handbook of Soil Sciences (Huang et al. 2012). A long list of operational methods for soil physical analyses were earlier presented by the Soil Science Society of

J. Bouma (✉)
Soils Department, Wageningen University, The Netherlands
e-mail: Johan.bouma@planet.nl

W.G. Teixeira et al. (eds.), *Application of Soil Physics in Environmental Analyses: Measuring, Modelling and Data Integration*, Progress in Soil Science,
DOI 10.1007/978-3-319-06013-2_1,

America (Dane and Topp 2002) and several methods have been added since that review. Minasny et al. (2013) provide a state-of-the-art review of Pedometrics in terms of variability, sampling and modeling methodologies. Many successful applications of soil physical theory have been presented in literature. The issue to be raised in this chapter does therefore not relate to the quality and applicability of soil physical theory and methodology (which we consider to be beyond doubt) but to a concern as to whether or not this vast expertise is being applied to its full potential when studying the major environmental issues of today. There is reason for doubt whether this is indeed occurring.

Rockstrom et al. (2009) have identified nine so-called Planetary Boundaries defining a safe operating space for humanity. Boundaries are defined by: biodiversity, climate, nitrogen, land use, fresh water, poison, aerosols, acidification of the oceans and ozone. Soils play a key role in affecting at least seven of the nine boundaries and this is hardly reflected in current interdisciplinary programs characterizing the major environmental issues, associated with the planetary boundaries. For example, climate change modeling does mention the importance of land use change and considers soil carbon,- moisture contents and – temperatures in a biogeochemical context, but does not link this with soil types which is surprising as different soil types have quite different characteristics and dynamics in terms of these three features (Denman et al. 2007). Soils of the world contain 1,500 Gt of C; plants only 560 Gt, while most C is in the oceans (38,000 Gt). Soil respiration is estimated to be equal to plant respiration at 60 Gt yr^{-1}, while land use change contributes only 0.9 Gt Cyr^{-1} (GLOBE carbon cycle project: info@globecarboncycle.unh.ed). When considering the C-cycle, the importance of C contents in and fluxes from the soil and the relative low estimated values due to land-use change have to be taken into account. When hydrologists model the water regimes of watersheds, their consideration of soils is often minor or even absent (e.g. Bouma et al. 2011a; Droogers and Bouma 2014). Ecologists define soil biodiversity but tend to focus on typology of organisms without defining physical and chemical soil processes that are likely to affect both the occurrence and functioning of observed species (e.g Bonkowski et al. 2012). Agronomists are quite successful in exploring the potential of the land to grow crops but, again, their consideration of soils is limited. An example is the successful program Decision Support System for Agrotechnology Transfer – DSSAT (Jones et al. 2002), that has a sophisticated plant production module but characterizes soils in terms of the "available water" and "field capacity" concepts of the 1930s. Soil physics can do better than that.

There is an obvious gap between what is known in soil science and what is being applied in the interdisciplinary context that prevails when studying the major environmental issues of today. Bouma (2010) has described this in terms of the knowledge paradox: why is so much of our knowledge not being used in the real world?

Somewhat limited attention to soil science expertise is not restricted to the interdisciplinary scientific arena but can also be observed in policy circles. For example, the recent, impressive report from United Nations Environment

Programme (UNEP) (2011) on the need for a Green Economy does not mention the role of soils. Another report from the Consultative Group on International Agricultural Research (CGIAR) on food security (Beddington et al. 2011) does not mention soils neither in its key messages nor in its action points.

Within soil science soil physics has a strong identity of its own and soil biologists and chemists tend to focus on their own specialty as well, although excellent integrated research is done on interfacial interactions on micro level. In fact, on microlevel interdisciplinary interaction is excellent and an example for activities at macrolevel, discussed here (e.g. Huang and Violante 2012; Xu and Sparks 2013). Pedologists take a broad view of soil science, studying natural soils as they occur in the landscape. They describe physical, chemical and biological properties of soils in their soil survey reports but these have – except for chemical and mineralogical soil profile data- a qualitative, descriptive character that is valuable but inadequate to feed modern simulation models (e.g. West and Wilding 2012).

Focusing on soil physics, this chapter will discuss the gap between "what is known" and "what is being used" as an opportunity, analysing the way soil physical research is practiced and the way that developments in society increasingly require the research community to analyse and possibly revise its attitude towards and interaction with society both in terms of various stakeholder groups and the political arena. This issue, obviously, extends far beyond soil physics and soil science but its implications are of relevance to these disciplines.

The objectives of this chapter are, therefore, to explore: (i) the role of soil science as a partner in interdisciplinary studies focused on major environmental issues that define whether or not Planetary Boundaries are or will be exceeded; (ii) the particular role of soil physics within soil science, and (iii) ways in which application of this expertise can be improved in future, considering new demands of the information society.

1.2 Soil Science as a Partner in Interdisciplinary Environmental Studies

1.2.1 The Effects of Not Being Part of the Interdisciplinary Research Team

Realizing that exploratory simulation models play a central role in climate, food, energy, hydrological and biodiversity studies, all of which directly related to the Planetary Boundaries, it cannot be denied that, as discussed above, many of these interdisciplinary models either ignore the soil or assume the presence of some "standard soil" everywhere. Be that as it may, when modelling major environmental problems many factors are involved and any comprehensive model has to strongly schematize and simplify a highly complex reality. Researchers have a natural tendency to resist simplification of their own subdisciplines and when soil scientists don't take part in such modeling exercises nor in planning committees (which they often don't) they may find that their soil component of the model is being simplified. This is the situation. Interdisciplinary aspects of soil science have been discussed in ten subchapters in the Handbook of Soil Science (e.g. Levy 2012) with a strong focus on soils. This provides valuable information but does not address the question as to what soils input is needed for the interdisciplinary simulation models, discussed above, that focus on major planetary issues such as climate, hydrology, food production, biodiversity etc. Discussing soil erosion and soil conservation is, for example, very important and relevant but the next step addressing the question as to what this means for the planetary issues should be taken as well and this is often not done.

The tendency to simplify soil input into models and for soil scientists of not being involved is enforced by the wide availability of soil data in accessible databases and published pedotransfer functions, the latter relating by regression analysis basic available soil data (such as texture, bulk density and %organic matter) to physical and chemical parameters needed for the models (Bouma 1989; Pachepsky and Rawls 2004). Such simplified representations of soils and use of pedotransfer functions can easily be handled by non-soil scientists and may create the impression that soil scientists are no longer required (e.g. Bouma 2010). In short, soil scientists would be well advised to not assume beforehand that their expertise is indispensible and that they have to make a serious effort to show that it is. The tendency of research is to aim for cutting-edge techniques and methods. To run models demonstrating the effects of excluding one's own disciplinary data and showing that results would be better when such data are included, is a highly unusual but necessary procedure to prove the value of one's particular discipline. Before continuing this analysis, attention will be paid to the basic question as to the role of knowledge in the information age. The phenomenal development of information technology during the last decade has fundamentally changed the relation between science and society and this should be taken into account when discussing future developments in any scientific discipline.

1.2.2 Impact of the Information Revolution

Internet is providing easy access to all kinds of information. The modern citizen of the twenty-first century is well informed both at technical and political level, confronting researchers with the need to better explain their procedures and results (e.g. In 't Veld 2010). The time that researchers considered themselves to be the exclusive owners of information that they were gracefully willing to share with ignorant stakeholders is over, if it ever existed.

Of course, the need to reach out to practitioners and policy makers has been well recognized long ago and Wenger et al. (2002) proposed the widely used term *Communities of Practice* to illustrate scientists working together with stakeholders.

Bouma (2005) proposed the additional term *Communities of Scientific Practice* indicating that, in his opinion, the science community was (and is) not well prepared to face interaction with the real world in the twenty-first century. Just presenting research results or plans is not enough anymore, however brilliant they may be from a scientific point of view. In a review of a major Dutch research effort on sustainable agriculture, Bouma et al. (2011b) concluded that projects were only considered to be successful by stakeholders when continued to completion and this took major efforts during several years, including the activities of so-called "knowledge brokers", members of the scientific community forming a link between the science community and various stakeholders, such as entrepreneurs, Non Governmental Organisations (NGO's), Governmental Agencies and citizen groups. Knowledge brokers, who are capable to inject the right type of knowledge to the right kind of people at the right time (and in the right way!), can be seen as representing: "Extension 2.0" (e.g. Bouma et al. 2011b). Rather than traditional soil extension officers, who transmitted research knowledge to farmers in a linear process, knowledge brokers take part in joint learning of a team in which they can play a leading role depending on their input. If they just babble, they will be kicked out. They are, in our opinion, important members of investigative teams of the future (and need to be paid accordingly). In their study, Bouma et al. (2011b) used the concept of *connected value development*, indicating that quite different opinions and value judgements of various stakeholders and policy makers have to be somehow harmonized, illustrating the cumbersome, time-consuming procedures that turned out to be necessary.

Modern environmental problems are:"wicked", sharing a number of characteristics: (i) there is no definite formulation of the problem nor a single, straightforward "solution"; (ii) solutions are never true or false but only better or worse; there are only alternative "options"; (iii) there is no immediate test of any given option; (iv) every problem is unique and every attempt to derive options is significant; (v) the explanation ("Framing") of the problem often already suggests the proposed solution as seen by those who frame the problem. Pure objectivity is difficult to reach (Rittel and Webber 1973). Introduction of the concept of sustainable development in the 1980s has added the additional criterium to express each option in terms of its economic, social and environmental consequences. This is useful as one-sided and unbalanced options, often those only considering economic considerations, are thus exposed.

Rather than just design a system and let others worry about its realization, as was often the case in traditional research, systems are designed here in a cooperative effort by co-design involving the various stakeholders and paying attention to various institutional and legal problems that often block realization of plans in the real world. At the same time, basic research is needed to face problems being encountered, and Bouma et al. (2011b) made a plea, therefore, for closing the so-called knowledge chain, linking empirical, qualitative knowledge of practitioners and general experts with cutting edge scientific knowledge. The knowledge chain concept will therefore be discussed next.

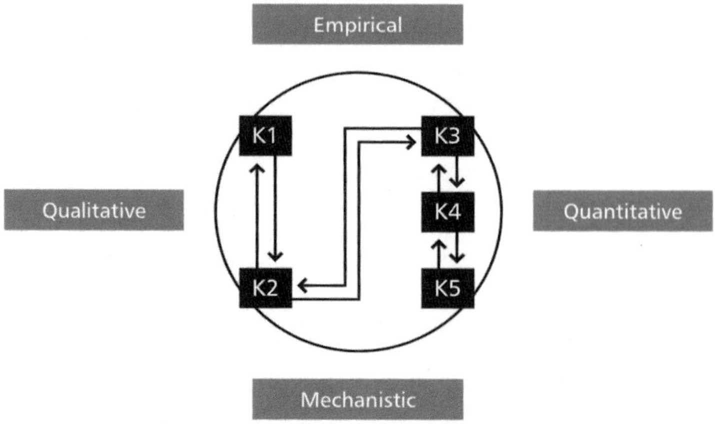

Fig. 1.1 Knowledge diagram, defining five types of knowledge as a function of four criteria, as indicated. The *K1 to K5* and the *K5 to K1* lines represent knowledge chains, as discussed in the text

1.2.3 The Knowledge Chain: Linking Different Types of Soil Knowledge

As many other scientific disciplines, soil science tends to be rather self-centred. Most papers presented at international conferences and published in international refereed journals focus on cutting-edge results obtained with modern techniques and models and there is no question as to whether or not this represents excellent research. It does. If, however, our problem is communication with other professions and the outside world, as mentioned above, we need to at least raise the question as to how we might possibly re-connect soil science with society. The knowledge diagram in Fig. 1.1 (Bouma et al. 2008) may be helpful in this context, defining five types of knowledge (K1–K5) in terms of two ranges of characteristics: from qualitative to quantitative and from empirical to mechanistic.

K1 represents user expertise, sometimes referred to as *tacit knowledge*. K2 is broad expert knowledge from practitioners or applied scientists. The still essentially empirical character of this knowledge is supported by a better understanding of the underlying processes and may still be considered *tacit in* nature. K3–K5 represents increasingly specific scientific knowledge. K3 represents e.g. empirical statistical studies. An example are widely used pedotransfer functions, discussed above. The underlying physical processes are increasingly expressed in quantitative terms in K4 by simple, operational models. An example is the application of the hydrological SWAT model in the Green Water Credits study in Kenya (Bouma et al. 2011a; Hunink et al. 2012, 2013; Kauffman et al. 2014). Finally, K5 represents cutting-edge science, a favourite and almost exclusive entry point for publishing in international scientific journals. An arbitrary example is the work of Murphy et al. (2009) on modelling soil moisture conditions applying Digital Terrain Models. Bouma and Droogers (1999) calculated the moisture supply capacity of a prime agricultural soil in the Netherlands, comparing the results of K3, K4 and K5 methods. They reported significant differences, allowing choices to be made among the methods, also considering a steep rise in costs from K3 to K5. Sometimes, questions can be solved with K2 or K3 expertise and the not unusual tendency of researchers to jump right away to the K4 or K5 level (that can result in publishable results) may represent overkill.

Two types of arrows are shown in Fig. 1.1: from K1 to K5 indicating use of tacit knowledge when planning research and from K5 to K1 indicating the flow of knowledge from research to practitioners, experts and politicians. The latter used to be considered in terms of extension services when dealing with farmers but they have been privatized in many countries and have become less effective. As discussed above, modern communication occurs more in terms of joint learning with involvement of knowledge brokers (Bouma et al. 2011b).

The problem we see now is that the chains in the knowledge diagram are broken. K4–K5, and particularly the latter, is disconnected from the chain and K4–K5 is thus hardly fed by user's expertise and questions while results of K4–K5 research are not communicated down the chain. This would appear to be a major reason why communication of soil scientists with users of soil information is problematic at this time. We are too self centred and have to get: "*out-of-the-box*" and, for a start, reconnecting the knowledge chain. Originally, the knowledge chain was strong in soil science as soil surveyors walked over the land as they prepared their maps, talking to and learning from land users, thus providing inspiration to more detailed studies by soil physicists, – chemists and – biologists. As soil surveys have been completed, his form of interaction has ended. Of course, field experiments are still being made by soil physicists, – chemists and – biologists, but many of those are more science oriented with fewer direct contacts with land users, while many studies are restricted to the laboratory where information is widely available through the internet.

Some suggestions to again close the knowledge chain are discussed in the next section.

1.2.4 How to Get Our Soil-Science Act Together?

Suggestions have already been made elsewhere to improve interactions with colleague scientists in interdisciplinary programs on the one hand and with stakeholders and policy makers in a transdisciplinary context on the other (Bouma 2010; Bouma et al. 2011b). First, a focus on key contributions may clarify potential soil science input in interdisciplinary programs because the soil science literature currently covers a wide array of separate topics that are difficult to interpret for outsiders. The seven soil functions, defined by Commission of the European Communities – CEC (2006), provide an attractive focal point: (1) biomass production, including agriculture and forestry; (2) storing, filtering and transforming nutrients, substances and water; (3) biodiversity, such as habitats, species and genes;

(4) physical and cultural environment for humans and human activities; (5) source of raw materials; (6) acting as carbon pool; and (7) archive of geological and archaeological heritage. The condition of land in a given area can now be defined in terms of the seven functions, requiring standards in terms of indicators allowing comparisons among soils. Also, thresholds are valuable to define ecological "tipping points" beyond which any function is irreversibly changed. Soil functions can also be used to define "soil quality", in terms of: "the ability to perform" any of the functions (Bouma 2010). Bouma (2002) defined a soil quality measure based on function 1. So far, soil quality is still an undefined, elusive term while water and air quality are well defined. Also, "ecosystem services" follow from the characterizations of the soil functions, as, when performed, every function can be seen as providing a soil-related ecosystem-service to society. Ideally, all functions should be "satisfied" in a large area, such as a country, but in smaller areas emphasis can be on individual functions when conditions for that function are particularly suitable. This can be an excellent basis for spatial land-use planning, aiming for the optimal use of land in small and large areas.

Second, incorporation of the knowledge chain in research, as discussed in the previous section, may facilitate interaction with stakeholders and policy makers. Rather than jump right away into any project, all too often from a K5 perspective, research could start with K1 asking local stakeholders about their opinions as to what is known and needed. Next, expert advise (K2) is solicited and then research can start from K3 onwards until adequate information is obtained. This "step-by-step" approach (Bouma 1997) engages stakeholders and policy makers, utilizes their input and presents the researcher as somebody who does not aspire to do more than is really necessary to solve the problem at hand. The case studies in Bouma et al. (2011b) further illustrate this procedure that has proved quite valuable to us.

We do not claim that the two procedures described here will solve all problems, nor do we, of course, exclude the possibility that other procedures can be quite effective as well. We intend to only emphasize the need to be more interactive in the field of soil science.

Finally, the suggestion was made to specifically define operational "toolkits" for each of the soil functions to be applied when characterizing soil functions at different K-levels. This would allow soil scientists to react quickly when questions arise (e.g. Bouma and McBratney 2013).

1.3 The Key Role of Soil Physics in the Soil Science Discipline

An integrated approach covering soil physics, – chemistry, – biology and pedology is needed to adequately characterize the soil functions. Each subdiscipline on its own cannot deliver the potential contribution of soil science that is possible with an integrated approach. There would, however, appear to be a logical sequence of applying the various subdisciplines. A soil, wherever it occurs on earth, only comes to life when there is input of water and energy, the dynamics of which are defined by physical laws. Next, adding chemistry defines the important (micro)nutrient status of the soil, resulting in conditions for particular types of biota to thrive. Pedology has a special niche in this context as it defines the occurrence of different types of soils in different parts of landscapes, as expressed on soil maps. It also describes morphological soil features reflecting physical, chemical or biological processes that are or have been occurring, including soil structure that has major effects on water flow patterns in soils. While physical, chemical and biological soil processes are most often studied within a given soil profile, applying a K4–K5 focus, pedology has a much wider landscape dimension with a K2 character. This forms a barrier for internal communication. Still, examples have been presented elsewhere showing that combining soil physical theory with pedological information, as expressed in Hydropedology (Bouma 2006, 2012) can be profitable to improve the characterization of flow phenomena in soils. These examples will not be repeated here.

1.4 Case Studies Demonstrating the Role of Soil Science Participating in Interdisciplinary Studies on Wicked Environmental Problems

1.4.1 Introduction

Rather than present conceptual discussions and theoretical analyses of the role of soil science in interdisciplinary environmental studies, presenting specific case studies demonstrating this role is much more convincing (e.g. Bouma and McBratney 2013). The case study approach has been quite successful for Harvard Business School and the soil science profession would be well advised to follow this approach, focusing on the soil functions and covering the entire knowledge chain to involve stakeholders and policy makers. A major inter- and transdisciplinary study in the Netherlands on sustainable agriculture also followed this approach by analysing a number of case studies as they developed over a period of sometimes more than 10 years (Bouma et al. 2011b). Case studies are also valuable for teaching soil science, where innovative approaches are necessary (e.g. Hopmans 2007; Field et al. 2011). Two case studies will be analysed with reference to source publications. One deals with simulating water regimes and soil erosion in a Kenyan watershed coupled with possible payment to farmers for ecosystem services. The other deals with precision agriculture on farm level in the Netherlands. Both deal with input of soil data when developing innovative forms of soil management (e.g. Droogers and Bouma 1997).

1.4.2 *Achieving Soil Conservation by Engaging Stakeholders: Green Water as a Guiding Principle in the Tana Region, Kenya*

Experience shows that just presenting studies, however brilliant they may be, about the negative effects of soil degradation does not by itself lead to changes in practices. The "Green Water Credit" (GWC) concept was therefore designed to promote soil conservation by initiating direct payments to farmers that follow effective soil conservation practices, by water and electricity companies (e.g Kauffman et al. 2014). "Green Water" refers to water in the unsaturated zone in the soil above the watertable that is transpired by plant growth; blue water is free surface water or groundwater and brown water carries soil downslope by erosion. Ideally, rainfall will infiltrate into the soil at the surface and flow downwards through unsaturated soil until it reaches the groundwater (or bedrock when no groundwater is present). Thus, the groundwater is replenished which is good because water tables drop quickly in many areas due to excessive water extraction for human use. Soil degradation, such as compaction and crusting of surface soil, can result in a strong reduction of water infiltration rates and to surface runoff in sloping terrain, resulting in erosion of soil, followed by deposition downslope or in basins used to store water for either direct urban use or for electricity generation. Also, production of crops is lower when runoff of water and erosion occurs, carrying soil and nutrients, leading to lower water contents in the soil and lower production in dry periods.

Water and electricity companies are interested in avoiding silting up of their reservoirs and they are therefore in favor of soil conservation measures by farmers, thereby replenishing their groundwater supply and the longevity of their storage basins. They are, however, not used to pay for those conservation measures and they are not prepared to do so either. The GWC program intends to convince the business partners that such payments represent a good business opportunity as long as, of course, payments are lower than benefits achieved over a period of years. It takes hard data to convince the business partners!

The Tana River basin in Kenya (Fig. 1.2) was selected as a test area (Hunink et al. 2011) to demonstrate the feasibility of the GWC concept, as the catchment is facing severe challenges to meet increasing water demands due to poverty and population growth (Githui et al. 2009). In the past century farmers occupied the Tana upstream land area, altering the water balance of the basin by clearing of natural vegetation for agriculture, reducing downstream water availability, increasing soil erosion and reservoir sedimentation. Water downstream is used for hydropower, irrigation, industrial and domestic use in the capital Nairobi, creating a high downstream dependency on upstream soil and water management that can enhance water availability through improved upstream land management interventions at individual plots, at watershed and landscape scales. The main crops are maize, coffee and tea, but also flowers, vegetables and fruits are produced for the international market. Livestock farming systems occur in the lower reaches of the watershed into the drier areas. Downstream, five major reservoirs have been built for

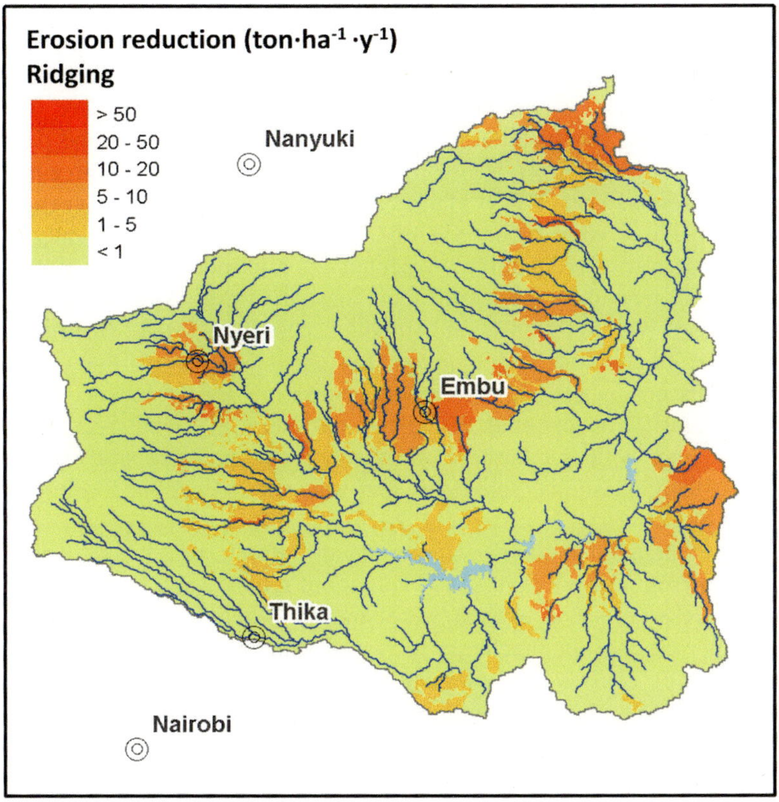

Fig. 1.2 Spatial distribution of relative erosion reduction for the ridging scenario, based on yearly averages for the Upper Tana Basin, Kenya. Similar maps were produced for the other conservation measures, also in terms of their effect of increasing Green Water

hydropower and flow regulation. Two smaller upstream reservoirs are used for water supply of Nairobi. Siltation of the reservoirs and thus loss of storage capacity is one of the main threats being faced. Population growth in recent years has caused a steady increase in water and electricity demand, both upstream and downstream, rural and urban. Recent severe droughts made it necessary to ration water and electricity. Inhabitants of rural areas have noticed hydrological changes in the basin in terms of permanent tributaries becoming seasonal and much lower water levels in water wells.

The GWC program started with discussions with farmers and business partners from the water and electricity companies, engaging them, at first at K2 level and later, as discussions progressed, at ever higher K levels, emphasizing costs versus benefits to be achieved by introducing effective conservation methods. This continuing engagement was crucial to create a joint learning environment and mutual trust and represents the step-by-step approach mentioned above. A research program was devised that could answer questions being raised such as: (i) which conservation methods are likely to produce the best results and how reliable are estimates being provided? (ii) Are farmers going to do what they promise, and (iii) what are the associated benefit/cost relationships? Only quantitative analyses can provide satisfactory answers. A K4 methodology was therefore designed, using an innovative combination of existing modeling procedures and databases. In summary, this involved: (i) transforming the available soil map into a map with Hydraulic Response Units (HRU's) used for simulations (e.g. Wosten et al. 2013); (ii) running the hydrologic model SWAT (Neitsch et al. 2002) for not only existing conditions but also for conditions expected after introducing certain conservation measures, defined by the World Overview of Conservation Approaches and Technologies – WOCAT (2007) and Hunink et al. (2012, 2013); (iii) running the model Water Evaluation And Planning System – WEAP for the various soil conservation scenario's to estimate benefit/cost ratio's for the various soil conservation measures (Droogers et al. 2011). Several successive workshops were organized to interact with the farmers and the business partners and research was focused on their questions, comments and suggestions without loosing sight of the overall objectives (Kauffman et al. 2014).

Results of detailed modeling studies on the effects of 11 soil conservation measures on crop transpiration, soil evaporation, groundwater recharge, sediment inflow into reservoirs and erosion were summarized by Kauffman et al. 2014. For example, groundwater recharge increased from 2 % to 23 % depending on various types of conservation measures taken and sediment inflows decreased from 1 % to 21 %; both values being highly significant for the water and electricity companies. An example of results obtained is shown in Fig. 1.2 for the calculated reduction of erosion when applying the conservation measure of ridging, as compared with the baseline study. Similar maps were prepared for the effects of other conservation measures. Economic benefit/cost ratios for the most effective conservation measures were an attractive 7 and 6, respectively. Currently, attempts are made to create an operational system which, however, faces institutional and bureaucratic hurdles (Kauffman et al. 2014).

A focus on the soil functions worked well in this program when studying reduction of erosion. Four functions were considered: biomass production (function 1) is increased as fertile surface soil is maintained and water infiltration is enhanced; higher infiltration of water allows its filtration and purification as it moves downwards (function 2), biodiversity is maintained as soil is not removed (function 3) and the C content can potentially increase as was demonstrated in a separate study (Batjes 2012) (function 6). The effects on the other functions is less obvious, although preservation of existing landscapes and maintaining viable forms of agriculture are culturally important everywhere. In turn the study addresses five of the Planetary Boundaries of Rockstrom et al. (2009): biodiversity, climate (by C adsorption in soil), nitrogen (by avoiding removal of topsoil by erosion, even though fertilization levels are low), land use and fresh water availability. The latter two boundaries are most prominently addressed. This way, erosion studies are not considered as a topic in itself, drawing insufficient attention of outsiders, but are: (i) interpreted in terms of universal soil functions; (ii) related to major environmental issues recognized in literature and (iii) linked with business propositions. This approach can only be helpful to illustrate the importance of soils.

The: "Green Water Credits" (GWC) program, with a focus on land use, was initiated by pedologists of ISRIC- World Soils information and they guided this program to completion. Pedologists have a tradition to work with land users when preparing their soil maps. Extending this cooperative approach to the business community is a logical step. However, K2 information provided by classical soil maps is inadequate to answer questions being raised. The link with hydrologists of Futurewater in Wageningen, with their modeling expertise, has been crucial to develop this program, requiring at least K4 knowledge. As such, the GWC study is a good illustration of Hydropedology at work. A key element of this program was the "translation" of the many units of the soil map and their patterns into a smaller number of Hydrological Response Units (HRU's): an example of the functional interpretation of soil maps, advocated by Bouma (1989) and others.

Questions have been raised about the use of soil maps because the definitions of map units and their: "representative soil profiles" as well as their delineations are based on K2 knowledge with no indication of accuracy. Still, relations between landscape characteristics and soil types is evident everywhere and small scale maps (e.g. 1:100.000 and smaller) use those relationships. As long as the level of interpretations does not extend beyond the reliability of the underlying soil map data, this procedure is acceptable. However, not being satisfied with this procedure, Sanchez et al. (2009) proposed the Global Soil Map program, using a set of modern methods defining soil parameters of direct relevance for users, such as %clay; %Carbon, pH, Cation Exchange Capacity (CEC), Electrical Conductivity (EC) and bulk density, to be presented for 90×90 m grids all over the world. This ambitious program is in progress and cannot be judged as yet. Sanchez et al. (2009) correctly notice lack of soil maps in many areas of the world and the Tana-GWC study was fortunate to have access to a 1:1.000.000 scale soil map which was used to "translate" map units into HRU's. As is, we consider this to be an acceptable procedure at this time as there are no operational alternatives.

1.4.3 Characterizing Soil Processes as a Basis for Precision Agriculture

Surface- and groundwater pollution with agrochemicals is a major environmental problem in many areas of the world. As pollution is due to applying more agrochemicals than the plants need at any given time it also means that farmers spend more money than they have to. And agrochemicals increase rapidly in price. Precision agriculture aims at fine-tuning application of agrochemicals (and water) to the growing crop, matching demand as a function of time, thereby avoiding pollution as a side effect. Modern remote – and proximal sensing systems have been designed that can observe the nutrient and water status of plants, allowing customized application of agrochemicals and water. This is fine but when plants show nutrient or water stress, they already suffer and it would be better to arrange applications before the stress occurs. Sensing techniques are not helpful here. To enable this, modeling real-time soil solute processes and plant growth can play an important role as will be illustrated for a Dutch case study. This approach is timely because of increasing emphasis in literature on using remote- and proximal sensing in precision agriculture and on using GPS-controlled yield registering machines, bypassing the soil.

Precision agriculture in terms of smart application of agrochemicals was studied on a 110 ha arable farm in the Netherlands in close interaction with the farmer concerned. Details of the study were reported by van Alphen and Stoorvogel (2000a, b, 2002) and Bouma et al. (2012). The following observations are relevant in the context of this chapter:

(i) Large scale soil maps (1:10.000 and larger), defining soil types, were not relevant in this context because attention focused on nitrogen use efficiency of plants and on pesticide adsorption by soil, not on the occurrence of soil types as a carrier of information. A geostatistical analysis was therefore made to assess local soil variability and this determined the dimensions of an optimal grid pattern for soil observations. For each of the 180 points being sampled, calculations were made with the K4 Water and Agrochemicals in Soil and Vadose Environment, (WAVE) simulation model of relevant functional characteristics for this particular location and farming system, when considering the N regime: (a) water stress in a dry year; (b) N stress in a wet year; (c) N leaching from the rootzone in a wet year, and (d) residual N content at harvest in a wet year (which is susceptible to leaching during the winter season). The model was fed with weather data for 30 years, a standard fertilization regime was introduced and pedotransfer functions were used to derive transport parameters for the model. Calculations were also made for each separate observation point to assess leaching risks of a number of different pesticides used in the Netherlands.

(ii) Mapped patterns were derived for each of the functional characteristics by merging the point data into four management units with significantly different behavior in terms of N dynamics (Fig. 1.3). For pesticide leaching, point data for each pesticide were interpolated to maps, expressing specific leaching risks.

(iii) The WAVE model, simulating water and N dynamics in soil as well as plant growth, was run in real time, using daily input data for precipitation, temperature and radiation. Rooting depths of plants being grown were observed a number of times and served as input of the model. The model calculated the amount of N within the rootzone as the growing season progressed. When this reached a critical level, defined as supply for a period of two growing days, a signal was given that fertilization was necessary. Figure 1.3 shows three and four fertilization moments thus derived for two management units. They are different for the different management units and they are quite different from predictions by extension services that are based on highly generalized data for different soils and years.

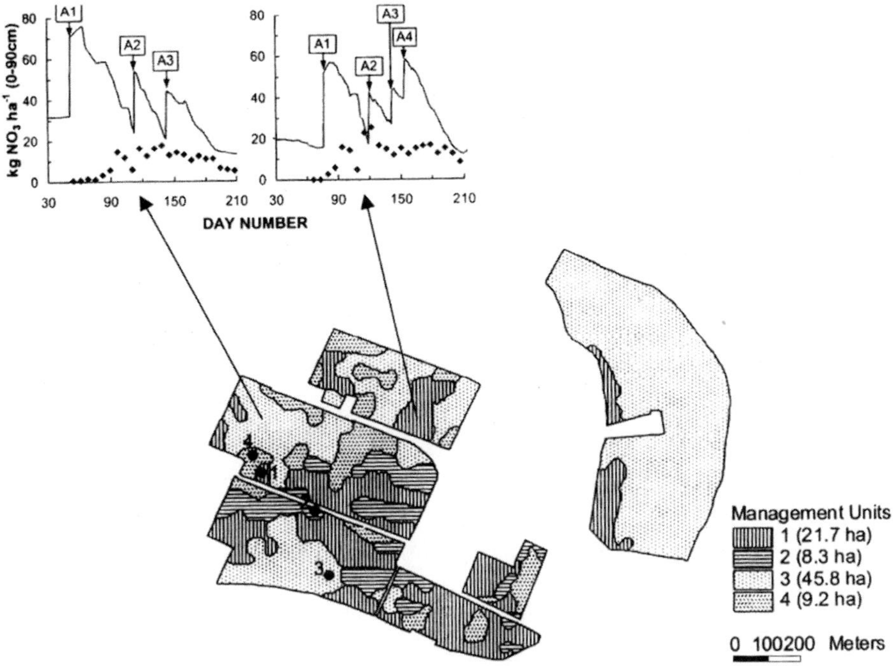

Fig. 1.3 Four management units, derived by interpolation from the functional characterization of 180 borings. The two upper figures show three and four dates when fertilization was recommended for two management units at times when the N content of the rootzone reached a critically low level. The *drawn lines* show the N content in the rootzone and the *dots* indicate daily uptake of N by the crop, all calculated with the WAVE model considering soils to a depth of 90 cm

The procedures described were only possible because of the rapid development of information technology and computer calculating powers. Using these powers produces an entirely different and innovative approach to dynamic soil characterization that is important for practical applications. Using data obtained in this study, the farmer saved 30 % on his fertilizer cost and a same amount when applying pesticides, as he could use relatively cheap types of pesticides in soils that were less susceptible to leaching. The survey of the farm cost 20.000 euro, which was acceptable to the farmer as a strategic investment. A major market for such modern surveys at farm level is emerging as part of "Good Agricultural Practices" that increasingly are considered by regulatory agencies as a means to handle environmental regulations at farm level.

The suggestion by Sanchez et al. (2009) to use small grid points of 30 × 30 m to characterize farms and fields while still only estimating the limited set of parameters as in the small-scale studies, is ill advised. The WAVE simulations considered dynamic soil and solute processes in soils also taking into account important subsoil properties. Just assembling a few static topsoil parameters in 30 × 30 m grids would be a rather sterile, meaningless exercise in the context of precision agriculture and other applications on the field level.

In this study the clear focus was on soil functions 1 (biomass production) and 2 (storing, filtering and transformation of nutrients, in this case N). In terms of Planetary Boundaries, the work was highly relevant for four boundaries: land use, nitrogen, fresh water and poison. As in the first case study, soil scientists were in charge of the work and this study also focused on: (i) business interests of users; (ii) research focused on specific soil functions and: (iii) showing a broader context of the work in terms of its relevance for planetary boundaries studies. Thus, the importance of soils when studying environmental problems is more effectively illustrated than when producing standard soil surveys or isolated simulations of soil processes.

1.5 Conclusions

1. Soil physical literature contains an impressive amount of modern measurement and monitoring techniques and models. These are indispensible to characterize dynamic soil behavior. Unfortunately, this resource is not as yet being applied to its full potential in the international environmental research arena. A plea is made for cooperation between soil physics and pedology in Hydropedology, followed by integration of soil chemical aspects thus resulting in defining the living environment for soil biota. This integration of subdisciplines of soil science is necessary to allow soil science to play its role in interdisciplinary studies on the major "wicked" environmental problems of today, that define whether or not the nine Planetary Boundaries, as distinguished by Rockstrom et al. (2009), are being exceeded.
2. Soil science would benefit by focusing its currently rather scattered research activities, thus becoming more attractive as a partner in interdisciplinary studies. One suggestion is a focus on the seven soil functions, as defined by Commission of the European Communities in 2006 in their Soil Protection Strategy. Development of an operational toolkit for each of the functions would be useful to allow quick responses to problems being distinguished.
3. Knowledge chains, linking tacit knowledge to cutting-edge research in both directions, used to be one of the basic strengths of the soil science profession. No more, as cutting-edge research tends to follow its own agenda and tacit knowledge of stakeholders is increasingly being inspired by many new sources of knowledge on the internet. Re-establishing the knowledge chain is needed to again engage stakeholders, including those in the business community, and inspire cutting-edge research while communicating research results more effectively to the real world with its many environmental problems.
4. Case studies are a suitable vehicle to illustrate the importance of contributions by soil science to studying major environmental problems. The two presented case studies on watershed and farm level demonstrate the central role that soil science can play in both initiating and executing interdisciplinary studies. Here, soil physics plays a central role that can be enhanced by integrating it with

pedological expertise in Hydropedology. Also, the case studies show that consideration of business opportunities provides a major thrust to the scientific endeavour.

References

Batjes NH (2012) Projected changes in soil organic carbon stocks upon adoption of recommended soil and water conservation practices in the Upper Tana River catchment, Kenya. Land Degrad Dev. doi:10.1002/ldr.2141

Beddington J, Asaduzzaman M, Fernandez A, Clark M, Guillou M, Jahn M, Erda L, Mamo T, Van Bo N, Nobre CA, Scholes R, Sharma R, Wakhungu J (2011) Achieving food security in the face of climate change: summary for policy makers from the commission on sustainable agriculture and climate change. CGIAR Research Program on Climate Change, Agriculture and Food Security (CCAFS), Copenhagen, Available online at: www.ccafs.cgiar.org/commission

Bonkowski et al (2012) Soil fauna. In: Huang PM et al (eds) The handbook of soil science. First volume. Part IV: soil biology and biochemistry: soil biology in its second golden age. 25. Handbook of soil sciences. Properties and processes, 2nd edn. CRC Press, Taylor and Francis, Boca Raton

Bouma J (1989) Using soil survey data for quantitative land evaluation. In: Stewart BA (ed) Advances in soil science, vol 9. Springer, New York, pp 177–213

Bouma J (1997) Role of quantitative approaches in soil science when interacting with stakeholders. Geoderma 78:1–12

Bouma J (2002) Land quality indicators of sustainable land management across scales. Agric Ecosyst Environ 88(2):129–136

Bouma J (2005) Soil scientists in a changing world. Adv Agron 88:67–96

Bouma J (2006) Hydropedology as a powerful tool for environmental policy research. Geoderma 131:275–287

Bouma J (2010) Implications of the knowledge paradox for soil science. Adv Agron 106:143–171. Academic Press, USA

Bouma J (2012) Hydropedology as a powerful tool for environmental policy and regulations: toward sustainable land use, management and planning. In: Lin H (ed) Hydropedology: synergistic integration of soil science and hydrology. Academic/Elsevier B.V., Amsterdam, pp 483–512

Bouma J, Droogers P (1999) Comparing different methods for estimating the soil moisture supply capacity of a soil series subjected to different types of management. Geoderma 92:185–197

Bouma J, McBratney AB (2013) Framing soils as an actor when dealing with wicked environmental problems. Geoderma 200–201:130–139

Bouma J, de Vos JA, Sonneveld MPW, Heuvelink GBM, Stoorvogel JJ (2008) The role of scientists in multiscale land use analysis: lessons learned from Dutch communities of practice. Adv Agron 97:177–239

Bouma J, Droogers P, Sonneveld MPW, Ritsema CJ, Hunink JE, Immerzeel WW, Kauffman S (2011a) Hydropedological insights when considering catchment classification. Hydrol Earth Syst Sci 15:1909–1919

Bouma J, Van Altvorst AC, Eweg R, Smeets PJAM, Van Latesteijn HC (2011b) The role of knowledge when studying innovation and the associated wicked sustainability problems in agriculture. Adv Agron 113:285–314. Academic Press, USA

Bouma J, Stoorvogel JJ, Sonneveld WMP (2012) Land evaluation for landscape units. In: Huang PM, Li Y, Summer M (eds) Handbook of soil science, 2nd edn. CRC Press, Boca Raton/London/New York. Chapter 34, pp 34–1 to 34–22

Commission of the European Communities (CEC) (2006) Communication from the Commission to the Council, the European Parliament, the European Economic and Social Committee, the European Parliament and the Committee of Regions. Thematic strategy for soil protection (COM 2006.231.)

Dane JH, Topp C (eds) (2002) Methods of soil analysis. Part 4. Physical methods, Soil Science Society of America Book Series 5. Soil Science Society of America (SSSA), Madison

Denman KL, Brasseur G et al (2007) Couplings between changes in the climate system and biogeochemistry. In: Solomon S, Qin D, Manning M, Chen Z, Marquis M, Averyt KB, Tignor M, Miller HL (eds) Climate change 2007: the physical science basis. Contribution of working group I to the fourth assessment report of the Intergovernmental Panel on Climate Change. Cambridge University Press, Cambridge, UK/New York

Droogers P, Bouma J (1997) Soil survey input in exploratory modeling of sustainable soil management practices. Soil Sci Soc Am J 61:1704–1710. doi:10.1080/07900627.2014.903771

Droogers P, Bouma J (2014) Simulation modeling for water governance in basins. Int J Water Res Dev http://dx.doi.org/10.1080/07900627.2014.903771

Droogers P, Hunink J, Kauffman S, Van Lynden G (2011) Costs and benefits of land management options in the Upper Tana, Kenya: using the Water Evaluation And Planning system – WEAP, Green water credits report 14. ISRIC –World Soil Information, Wageningen, 98 p

Field DJ, Koppi AJ, Jarret LE, Abbott LA, Cattle SR, Grant CD, Mc Bratney AB, Menzies NW, Waetherly AJ (2011) Soil science teaching principles. Geoderma 167–168:9–14

Githui F, Mutua F, Bauwens W (2009) Estimating the impacts of land-cover change on runoff using the soil and water assessment tool (SWAT): case study of Nzoia catchment, Kenya. Hydrol Sci J 54(5):899–908

Hopmans JW (2007) A plea to reform soil science education. Soil Sci Soc Am J 71:639–640

Huang PM, Violante A (2012) Soil physical, chemical and biological interfacial interactions. In: Huang PM et al (eds) Handbook of soil sciences. Properties and processes, 2nd edn. Part I (second volume). CRC Press, Taylor and Francis, Boca Raton, pp 1–10

Huang PM, Li Y, Summer ME (eds) (2012) Handbook of soil sciences. Properties and processes, 2nd edn. Taylor and Francis/CRC Press, Boca Raton

Hunink JE, Immerzeel WW, Droogers P, Kauffman S, van Lynden G (2011) Green and blue water resources for the Upper Tana catchment, Kenya – Soil-water management scenarios using the soil and water assessment tool, Green water credits report 10. ISRIC, Wageningen

Hunink JE, Droogers P, Kauffman S, Mwaniki BM, Bouma J (2012) Quantitative simulation tools to analyze up- and downstream interactions of soil and water conservation measures: supporting policy making in the Green Water Credits program of Kenya. J Environ Manage 111:187–194

Hunink JE, Niadas IA, Antonaropoulos P, Droogers P, de Vente J (2013) Targeting of intervention areas to reduce reservoir sedimentation in the Tana catchment (Kenya) using SWAT. Hydrol Sci J 58(3):1–15

In ‘t Veld R (ed) (2010) Knowledge democracy. Consequences for science, politics and media. Springer, Heidelberg/Dordrecht/London/New York

Jones JW, Hoogenboom G, Porter CH, Boote KJ, Batchelor WD, Hunt LA, Wilkens PW, Singh U, Gijsman AJ, Ritchie JT (2002) The DSSAT cropping system modelling. Elsevier Publ. Co., Amsterdam

Kauffman JH, Droogers P, Hunink JE, Mwaniki B, Muchena FN, Gicheru PT, Bindraban PS, Onduru D, Cleveringa R, Bouma J (2014) Green water credits – Exploiting its potential to enhance ecosystem services by reducing soil erosion in the Upper Tana basin, Kenya. Int J Biodivers Sci Ecosyst Serv Manage http://dx.doi.org/10.1080/21513732.2014.890670

Levy GJ (2012) Interdisciplinary aspects of soil science. In: Huang PM et al (eds) Handbook of soil sciences. Properties and processes, 2nd edn. Part III (first volume). CRC Press, Taylor and Francis, Boca Raton, pp 17–26

Minasny B, Whelan BM, Triantafilis J, Mc Bratney AB (2013) Pedometrics research in the vadose zone- review and perspectives. Vadose Zone J. doi:10.2136/vzj/2012.0141

Murphy PNC, Ogilvie J, Arp P (2009) Topographic modelling of soil moisture conditions. Eur J Soil Sci 60:94–109

Neitsch SL, Arnold JG, Kiniry JR, Williams JR, King KW (2002) Soil and Water Assessment Tool (SWAT). Theoretical documentation, version 2000. Taxes Water Resources Institute, College Station

Pachepsky YA, Rawls WJ (eds) (2004) Development of pedotransfer functions in soil hydrology, Developments in soil science 30. Elsevier, Amsterdam

Rittel HWJ, Webber MM (1973) Dilemmas in a general theory of planning. Policy Sci 4(2):155–169

Rockstrom J, Steffen W, Noone K, Persson A, Stuart Chapin F III, Lambin EF, Lenton TM, Scheffer M, Folke C, Schellnhuber HJ, Nykvist B, de Wit CA, Hughes T, vander Leeuw S, Rodhe H, Sorlin S, Snyder PK, Costanza R, Svedin U, Falkenmark M, Karlberg L, Corell LW, Fabry VJ, Hansen J, Walker B, Liverman D, Richardson K, Crutzen P, Foley JA (2009) A safe operating space for humanity. Nature 461:472–475

Sanchez PA, Ahamed S, Carre F, Hartemink AE, Hempel J, Huising J, Lagacherie P, McBratney AB, McKenzie NJ, Mendonca-Santos MI, Minashy B, Montanarella L, Okoth P, Palm CA, Sachs JD, Shepherd KD, Vagen TG, Vanlauwe B, Walsh MG, Winowiecki LA, Zhang GL (2009) Digital soil map of the world. Science 325:680–681

United Nations Environment Programme – UNEP (2011) Towards a green economy: pathways to sustainable development and poverty eradication – A synthesis for policy makers. www.unep.org/greeneconomy

van Alphen BJ, Stoorvogel JJ (2000a) A functional approach to soil characterization in support of precision agriculture. Soil Sci Soc Am J 64:1706–1713

van Alphen BJ, Stoorvogel JJ (2000b) A methodology for precision nitrogen fertilization in high-input farming systems. Precis Agric 2:319–332

van Alphen BJ, Stoorvogel JJ (2002) Effects of soil variability and weather conditions on pesticide leaching: a farm evaluation. J Environ Qual 31:797–805

Wenger E, Mc Dermott R, Snyder WM (2002) Cultivating communities of practice – a guide to managing knowledge. Harvard Business School Press, Boston

West LT, Wilding LP (2012) Pedology. In: Huang PM et al (eds) Handbook of soil sciences. Properties and processes, 2nd edn. First Volume, Part V. CRC Press, Taylor and Francis, Boca Raton, pp 29–39

WOCAT (2007). In: Liniger H, Critchley W (eds) Where the land is greener. Case studies and analysis of soil and water conservation initiatives worldwide. WOCAT, Bern

Wosten JHM, Verzandvoort SJE, Leenaars JGB, Hoogland T, Wesseling JG (2013) Soil hydraulic information for river basin studies in semi-arid regions. Geoderma 195–196:79–86

Xu J, Sparks DL (2013) Molecular environmental soil science. Progress in soil science book series. Springer, Heidelberg/Dordrecht/London/New York

Chapter 2
Hydropedology: Linking Dynamic Soil Properties with Soil Survey Data

Henry Lin, Weihua Zhang, and Haoliang Yu

Abstract Hydropedology is an intertwined branch of soil science and hydrology that provides a useful framework for connecting dynamics soil properties with soil survey data. Five categories of pedotransfer functions (PTFs) are illustrated in this chapter for estimating soil organic carbon, bulk density, and K_{sat} based on (1) fundamental relationships, (2) essential soil variables, (3) class variables, (4) topo- or geo-rectifications, and (5) climate or land use adjustments. Some outlooks are then discussed for enhancing PTFs developments and applications, including some guidelines for developing a dynamic soil properties database alongside traditional soil survey database as well as hydropedoinformatics for integrated soil-landscape analysis.

Keywords Soil physics • Hydrology • Pedology • Pedotransfer functions

---

H. Lin (✉)
Department of Ecosystem Science and Management, The Pennsylvania State University, University Park, PA 16802, USA
e-mail: henrylin@psu.edu

W. Zhang
Department of Ecosystem Science and Management, The Pennsylvania State University, University Park, PA 16802, USA

College of Resources and Environment, Southwest University, Chongqing 400716, China

H. Yu
Department of Ecosystem Science and Management, The Pennsylvania State University, University Park, PA 16802, USA

College of Water Resources and Civil Engineering, China Agricultural University, Beijing 100083, China

W.G. Teixeira et al. (eds.), *Application of Soil Physics in Environmental Analyses: Measuring, Modelling and Data Integration*, Progress in Soil Science, DOI 10.1007/978-3-319-06013-2_2,

2.1 Introduction

In soil science and hydrology, it is widely recognized that considerable gaps exist between what we have (e.g., National Cooperative Soil Survey databases in the U.S.) and what we need (e.g., soil hydraulic parameters needed in simulation models). The lack of sufficient field data of soil hydraulic properties often limits the application of flow and transport models. However, direct field measurements of soil hydraulic properties remain complex, time-consuming, and costly (Mualem 1986; Bouma 1989). In addition, significant spatial and temporal variability demands a large number of measurements that are often prohibitive in terms of time and money (van Genuchten et al. 1999a). This has prompted efforts to indirectly estimate soil hydraulic properties using more readily available data often found in soil surveys (such as particle-size distribution, bulk density, organic matter content, among others). Such indirect methods, generally referred to as pedotransfer functions (PTFs) (Bouma and van Lanen 1987), have been attempted widely in the last three decades (e.g., Vereecken et al. 1990; Tietje and Hennings 1996; Batjes 1996; van Genuchten et al. 1999a; Lin et al. 1999; Wosten et al. 2001; Pachepsky and Rawls 2004).

Compared to other methods of estimating soil hydraulic parameters (e.g., pore-size distribution models and inverse methods), PTFs are inexpensive, easy to derive and use, and in many practical cases provide needed estimation of missing hydraulic parameters (e.g., Verhagen and Bouma 1998; van Genuchten et al. 1999b; Wosten et al. 2001). Besides conventional regression or functional analyses, new techniques have been increasingly used to develop PTFs, such as neural networks (e.g., Schaap et al. 2001), group methods of data handling (e.g., Pachepsky and Rawls 1999), and classification and regression trees (e.g., McKenzie and Jacquier 1997; Jalabert et al. 2010). In general, the newly developed PTFs integrate more information (both quantitative and qualitative) in the estimation (e.g., Martin et al. 2009; Sequeira et al. 2014). There are also a growing number of large soils databases (national or international) that have been explored for constructing PTFs, such as the NCSS databases (USDA-NRCS 1994, 1995), UNSODA (Leij et al. 1996), WISE (Batjes 1996), HYPRES (Lilly 1997), ISRIC-WISE (Batjes 2008), HWSD (FAO/IIASA/ISRIC/ISS-CAS/JRC 2012), among others.

While various degrees of success have been achieved with different PTFs (Wosten et al. 2001; Pachepsky and Rawls 2004), limitations of existing PTFs remain. There is a tendency among users to apply PTFs developed from one area to another without careful examination of the suitability of the PTFs in the new area. This is because all empirical PTFs are constrained by the original dataset used to derive the functions. However, the vast majority of existing PTFs are completely empirical, and limited efforts have been put into systematic probing of *underlying mechanisms* for the existence of such PTFs. Furthermore, existing PTFs have not yet fully incorporated soil structural and landscape information. As such, the accuracy, reliability, and utility of existing PTFs have been limited. In the meantime, soil survey databases developed over the last century have been underused for environmental analysis. Thus, better dissemination, interpretation, application, and enhancement of soil survey databases are needed.

The objectives of this chapter are three-fold: (1) To suggest hydropedology as a useful framework for bridging the gaps between dynamic soil properties and soil survey databases; (2) To illustrate five types of PTFs to estimate dynamic soil properties (organic carbon, bulk density, and saturated hydraulic conductivity or K_{sat}); and (3) To provide some future outlooks on PTFs and related issues.

2.2 Hydropedology as a Framework for Connecting Dynamics Soil Properties with Soil Surveys

2.2.1 Concept of Hydropedology and Its Unique Features

Hydropedology is an intertwined branch of soil science and hydrology that studies interactive pedologic and hydrologic processes and properties across space and time (Fig. 2.1). Two fundamental questions of hydropedology are (Lin et al. 2008; Lin 2012):

(1) *How do soil architecture and the associated distribution of soils over the landscape exert a first-order control on hydrologic processes (and related biogeochemical dynamics and ecological functions)?*
(2) *How do hydrologic processes (and the associated transport of energy and mass) influence soil genesis, evolution, variability, and function across space and time?*

A unique feature of hydropedology lies in its encapsulation of the co-evolution of fast (water) and slow (soil) changes in multiphase soil systems, where fast and cyclic *soil functioning processes* (SFPs) involve mostly liquids, gases, and biota (in which circulating water is a key), giving rise to dynamic soil properties; whereas slow and irreversible *specific pedogenic processes* (SPPs) involve predominantly solids, largely linked to soil survey databases (Lin 2011). Each SPP is characterized by a set of solid-phase pedogenic features formed over hundreds to thousands or more years, while SFPs are dominated by diurnal, seasonal, and annual changes (Targulian 2005). Many SFPs and related cycles are not completely closed, and many input–output fluxes are not necessarily balanced in the open, dissipative soil system (Rode 1947; Targulian and Goryachkin 2004; Lin 2011). Such non-closed cycles and off-balanced fluxes of SFPs generate residual solid-phase products in soil profiles over time. Each single cycle may generate a micro amount of transformed or newly-formed solid products, which may hardly be detected; but being produced repeatedly over long time these micro amounts gradually accumulate into macro amounts that are detectable morphologically or analytically (Targulian and Goryachkin 2004). Such residual products generated in SPPs will feedback to alter SFPs, mostly in a gradual manner, but some could lead to threshold changes in soils (Lin 2011).

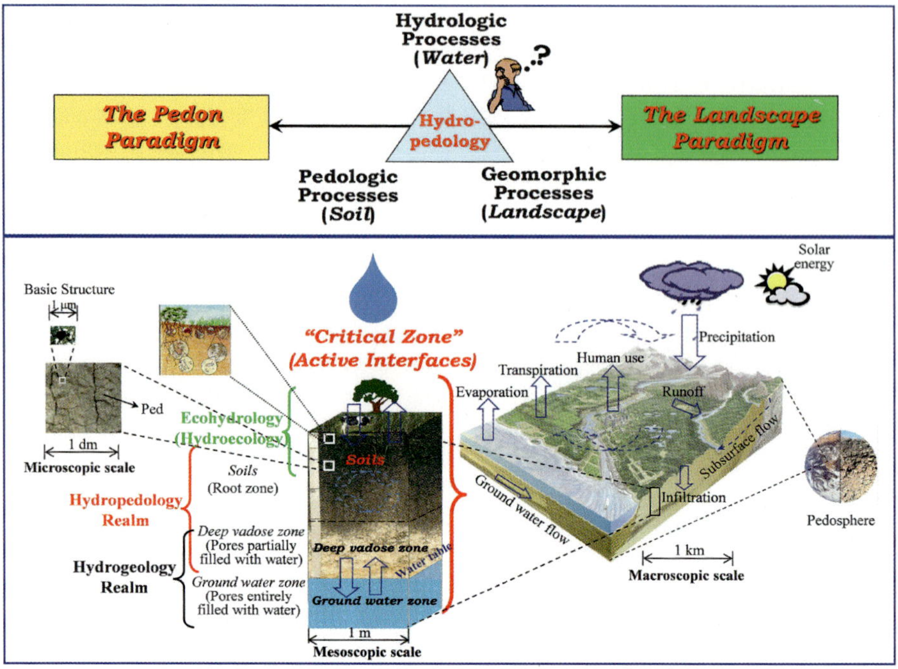

Fig. 2.1 Hydropedology connects the pedon and landscape paradigms through linking phenomena occurring at the microscopic (e.g., pores and aggregates) to mesoscopic (e.g., horizons and pedons) and macroscopic (e.g., catenas and catchments), and to megascopic (e.g., regional and global) scales. In the vertical dimension, hydropedology, combined with hydrogeology, promotes integrated systems approach to study the interactions of water with solid earth (i.e., soils, rocks, and anything in between) beneath the Earth's surface. Above the ground, hydropedology interfaces with ecohydrology and hydrometeorology to understand the feedback mechanisms between climate, vegetation, and soils. In temporal dimension, hydropedology deals with both short- and long-term changes of the landscape-soil-hydrology-ecosystem relationships, including the use of pedogenesis and soil micro- and macromorphology as signatures of soil change and soil hydrology in the past and current conditions (Modified from Lin 2003)

Another important feature of hydropedology is its emphasis on soil architecture (a broadened concept of soil structure) and the soil-landscape relationship (Lin 2012). All processes in the soil are bound to an underlying structural control that exhibits a hierarchical organization (Vogel and Roth 2003; Lin et al. 2006). *Soil structural unit* (SSU) refers to soil architecture within a pedon (including soil particles, pores, fabrics, aggregates, horizons and their sequence, and other organizational features in a soil profile), while *soil-landscape unit* (SLU) refers to soil architecture in the landscape (such as catenae, soilscapes, soil sequences, and soil zones) (Lin 2012). The linkage between soil structural and landscape units represents the soil-landscape relationship that has been traditionally studied by pedologists, including block diagrams, cross sections, and soil maps that are used to portrait the distribution patterns of soil types in relation to landforms and geology formations (e.g., Hole 1976; Soil Survey Staff 1993; Buol et al. 2001). Lin et al. (2005) suggested to add hydrologic information (e.g., water table dynamics, water flow paths, hydric soils, and restrictive soil layers) to enhance the use of the classical soil-landscape relationship information, which could significantly improve the value of soil survey products and provide more useful framework for understanding the landscape-soil-water relationship in different geographic regions for hydrologic pedotransfer functions and modeling.

A third important feature of hydropedology is its systems approach to understanding complex soil processes across scales. Jenny (1941) has noted, "*The goal of soil geographer is the assemblage of soil knowledge in the form of a map. In contrast, the goal of the 'functionalist' is the assemblage of soil knowledge in the form of a curve or an equation... Clearly, it is the union of the geographic and the functional method that provides the most effective means of pedological research.*" Such a union of soil maps and soil functions, based on process-based understanding, is what hydropedology embraces. Through identifying myriads of soil patterns and understanding how heterogeneity occurs at different scales, hydropedology can help de-mystify the seemly mind-boggling variability of field soils. For example, two general controls of soil variability may be identified: (1) regional controls associated with soil-landscape units, e.g., zonal soil patterns expressed over large areas that are the results of climatic and vegetative gradients; and (2) local controls associated soil structural units, i.e., factors that vary over short distances, such as topography and parent material. The hierarchical organization of soils and landscapes can then be used to discern deterministic and stochastic components at each scale, with the deterministic portion being dictated by the underlying soil architectural control at that scale (such as soil profile layers or catenae) and the stochastic portion being randomized processes (such as rainfall inputs or biological activities). As an example, water flow at the microscale is controlled by capillarity and laminar flow through individual soil pores and around peds; as scale increases, flow often becomes controlled by impeding layers in soil profiles, then accumulation of water downslope is dictated by soil catena and associated bedrock; then flow is routed through a stream network, the formation of which is controlled by soil-landscape relationships (Lin et al. 2006).

2.2.2 Soil Survey and Its Classical Databases

There are five orders of soil surveys in the U.S., ranging from the 1st order for detailed mapping (minimum delineation size $\leq$1 ha, 1:15,840 or larger cartographic scale, mapping units mostly consociations of phases of soil series) to the 5th order for very general mapping (minimum delineation size 252–4,000 ha, 1:250,000 or smaller cartographic scale, mapping units largely associations consisting of two or more dissimilar soil components) (Soil Survey Staff 1993; Buol et al. 2001). The orders of soil surveys are intended to assist the identification of operational procedures for the conduct of a soil survey, and to indicate general levels of quality control that affect the kind and precision of subsequent interpretations and predictions.

The Soil Survey Staff in the U.S. have developed three soil geographic databases (USDA-NRCS 1994, 1995):

- *The SSURGO (Soil Survey Geographic Database)*: This is the most detailed database at the county level, with a mapping scale ranging from 1:12,000 to 1:63,360, along with an associated laboratory characterization database of soil physical, chemical, and mineralogical properties and a soil profile description database;
- *The STATSGO (State Soil Geographic Database)*: This is compiled through generalizing the SSURGO at the state level, with a mapping scale of 1:250,000;
- *The NATSGO (National Soil Geographic Database)*: This is country level very general soil mapping database at a scale of 1:7,500,000, which is consistent with Major Land Resource Area (MLRA) boundaries.

A number of soil properties can be extracted from the NATSGO, STATSGO, and SSURGO databases. For each soil layer in a STATSGO component or a SSURGO soil series, 28 individual soil properties are provided, including texture, permeability, available water content, bulk density, and organic matter content. The lumping of spatial variability in soil properties represented by map units in these three soil databases increases from the SSURGO to the STATSGO and to the NATSGO. With the SSURGO, a map unit may consist of a dominant soil series, while in the STATSGO, a map unit may consist of a soil association made up of as many as 21 soil series. Hydrologic and biogeochemical models that derive soil input parameters from these databases will be able to better represent soil spatial variability with the SSURGO than the STATSGO or the NATSGO databases. But even models that rely on the SSURGO soil databases often underestimate the true extent of spatial variability in soil properties. This is because each SSURGO map unit can consist of up to three soil series, and the variability within and across these soil series, or inclusion within map units, is not represented.

It is important to recognize that traditional soil surveys in the U.S. (and in many other countries) have been designed for general land use planning purposes at the county or regional levels. With growing demands on site-specific applications and high resolution soils data, local level or high-intensity soil mapping has become more in demand. Thus, other means of mapping soils, such as the application of geostatistics (e.g., Heuvelink and Webster 2001), the use of proximal or remote sensing (e.g., McKenzie and Ryan 1999), and spatial modeling through terrain and other geospatial analyses (e.g., Burrough and McDonnell 1998), have been increasingly used, leading to the current booming in digital soil mapping (Zhu et al. 2001; McBratney et al. 2003; Lagacherie et al. 2007). However, it is important to keep in mind that field experience (including soft data and tacit knowledge), ground-truthing, and process-based understanding have been the hallmark of pedology (e.g., Daniels 1988; Arnold and Wilding 1991; Nordt et al. 1991; Schaetzl and Anderson 2005; Lin et al. 2005). Goryachkin (2005) noted that the loss of rich field experience would be a serious threat to pedology as a fundamental science.

Overall, soil survey databases provide a wealth of information that could potentially be utilized in the development of PTFs, hydrologic grouping or classification of soils, and testing of hydrologic models. However, traditional soil survey databases do not contain much information on dynamic soil properties required for deriving reliable PTFs (except perhaps for specific retention points such as wilting point or field capacity). Most data in traditional soil survey databases have been collected at a window in time. Hence, to enhance the value of classical soil survey databases and to facilitate the integration of pedology and hydrology in the framework of hydropedology, concerted efforts are needed to develop new and systematic PTFs. We believe that hydropedology can help lay the foundation for constructing fundamental PTFs and for bridging traditional soil surveys with dynamic soil properties.

2.2.3 Dynamic Soil Properties and Use-Dependent Properties

Dynamic soil properties are those that change with either natural disturbances and cycles (such as seasonal and diurnal changes) or human land use and management (such as tillage and irrigation). They are important for characterizing soil functions in ecosystem services and for predicting soil behaviors under anthropogenic impacts. The time scale emphasized here is biological or human time scales (from less than a day to over decades), rather than geological time scales (over thousands of years or more). Thus, soil properties may be classified into dynamic vs. relatively static properties, with surface soils having more dynamic properties while subsoils having more relatively static properties.

Grossman et al. (2001) suggested use-dependent properties as those soil properties that show change and respond to soil use and management (such as soil organic matter levels and aggregate stability), and use-invariant properties as those soil properties inherent from natural soil-forming processes that show little change over time and are not affected much by soil use and management (such as soil mineralogy and particle size). Use-dependent properties are particularly evident in surface soils, whereas use-invariant soil properties are more common in subsoils. Because of this, soil classifications have traditionally emphasized the use of subsoil properties that are less susceptible to human alternations.

With increasing emphasis on human impacts and land management practices, dynamic soil properties have received elevated attention in recent years (Tugel et al. 2005). Anthropogenic influences on soils can result in distinct characteristics that can be used to classify and model naturally-formed soils under different land management scenarios. The concepts of "genoform" (for genetically-defined soil series) and "phenoform" (for soil types resulting from a particular form of management in a given genoform), as proposed by Droogers and Bouma (1997), can facilitate the incorporation of management effects into pedologic and hydrologic characterizations of human-altered soils and could potentially enhance PTFs that involve soil series and land use classifications as carriers of soils information (e.g., Pulleman et al. 2000; Sonneveld et al. 2002). The distinction between major soil management types (phenoform) within the same soil series (genoform) can separate the morphogenetic properties used in soil taxonomic units from near-surface temporally dynamic properties used in PTFs and in cartographic units delineating management driven effects.

Pedology traditionally has focused on natural processes that do not reflect effects of short-term soil management. This was done on purpose to avoid frequent change of classifications of a given soil following different types of soil management. However, numerous studies have now shown that land use/management is the biggest and fastest driver in altering soils. As Richter and Mobley (2009) summarized well, "*More than half of Earth's terrestrial surface is now plowed, pastured, fertilized, irrigated, drained, fumigated, bulldozed, compacted, eroded, reconstructed, manured, mined, logged, or converted to new uses.*" Such human activities have drastic and long-lasting effects on soils, thus these effects need to be better assessed and incorporated into PTFs and modeling.

2.3 Pedotransfer Functions (PTFs)

2.3.1 Five Categories of PTFs

Five general categories of PTFs may be identified for improvements in systematic estimation of dynamic soil properties from basic soils and landscape information (Fig. 2.2):

- *PTF Type I (fundamental PTFs)* shows physical or quasi-physical based relationships between desired soil properties (e.g., organic carbon content, soil bulk density, or K_{sat}) and their underlying fundamental controlling variables;
- *PTF Type II (continuous PTFs)* includes numerically continuous basic soil properties (dynamic or relatively static properties) for estimating desired soil parameters;
- *PTF Type III (class PTFs)* considers descriptive class-based soil features (such as soil classification, morphology, horizon, or map unit) to improve PTF estimations;
- *PTF Type IV (topo- or geo-rectified PTFs)* adds landscape features (e.g., topography or parent material) into PTFs, hence connecting the pedon and the landscape scales;
- *PTF Type V (climate- or land use-adjusted PTFs)* emphasizes the influence of climate and land use/management impacts on PTFs.

In the following, we use the above framework to illustrate the estimation of soil organic carbon (OC) content, dry bulk density (BD), and saturated hydraulic conductivity (K_{sat}) values using five types of PTFs based on different levels of information inputs.

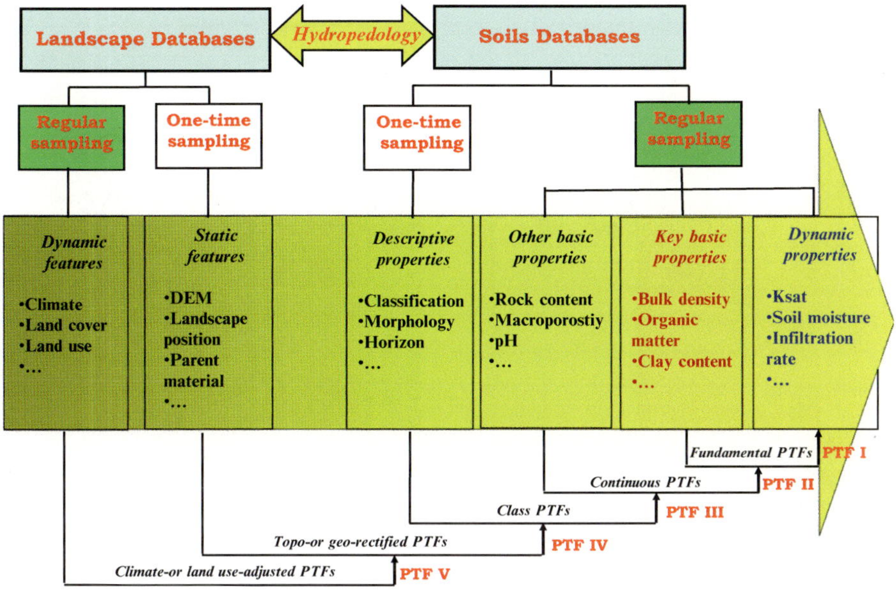

Fig. 2.2 Five general categories of PTFs (Type I to V) are identified for improvements in estimating dynamic soil properties from basic soils information and landscape features. All soil properties and landscape variables requiring regular sampling are dynamic in nature in the short-term or human timescales, while those indicated as requiring only one-time sampling are relatively stable or slowly-changing over long-term or geological timescales. Hydropedology functions as a bridge to link pedon-based soils databases and landscape-based environmental databases (Modified from Lin 2003)

2.3.2 PTF I: Fundamental PTFs

A major concern for PTFs is their general lack of fundamental mechanisms to relate desirable soil parameters to underlying controlling variables. Only a few quasi-physical PTFs exist for predicting soil hydraulic properties, such as those of Arya and Paris (1981), Haverkamp and Parlange (1986), and Arya et al. (1999), in which particle-size distribution is first translated into an equivalent pore-size distribution model and then related to water retention curve. However, even these PTFs may be questionable because of their unrealistic assumptions of capillarity bundle model. Here we will look at key variables that control soil OC, bulk density, and K_{sat}.

Soil OC or organic matter (OM) content is generally positively and linearly linked to clay and silt content in moderately well to well drained soils (Fig. 2.3a), which applies across soil textures and depths (Table 2.1). Variability (scatter of data points in Fig. 2.3a) is due to differences in (1) the type of clay minerals present (e.g., 2:1 silicates tend to stabilize more organic carbon), (2) site condition (e.g., cooler, high elevation locations being conducive to greater organic carbon accumulation), and (3) years since cultivation began (e.g., longer history of cultivation leading to lower organic carbon levels) (Brady and Weil 2004).

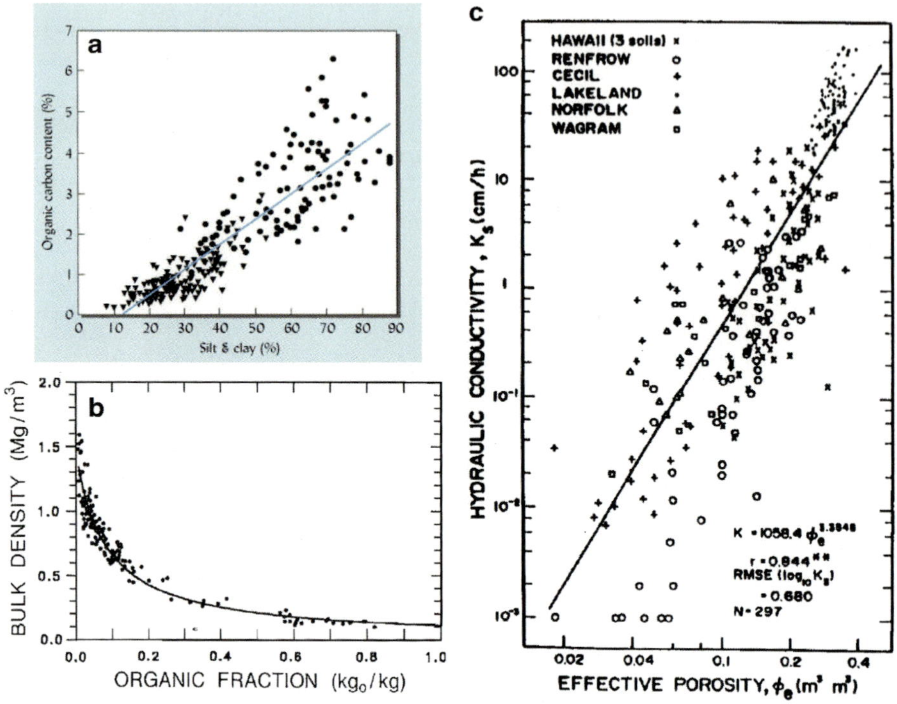

Fig. 2.3 Some fundamental relationships of soil OC, BD, and Ksat: (**a**) Soils high in silt and clay tend to contain high levels of organic carbon for surface soils in 279 tilled maize fields in subhumid regions of Malawi (*triangles*) and Honduras (*circles*) (After Brady and Weil 2004); (**b**) Soil bulk density as a function of organic fraction (n = 480) (After Federer et al. 1993); and (**c**) Saturated hydraulic conductivity as a function of effective porosity (After Ahuja et al. 1989)

Soil BD is controlled predominantly by OC and clay content, with OC always being negatively related to BD (Fig. 2.3b) and clay content being generally positively linked to BD (Table 2.2). Soil BD has also been commonly estimated using a combination of OM and mineral soils, with organic material density being 0.11–0.25 and mineral material density being 1.0–1.7 (see I-2 PTFs in Table 2.2).

Soil K_{sat} is clearly inversely related to BD and often clay content as well, in either linear or exponential relationships, whereas sand content and macroporosity have positive relationship with K_{sat} (Table 2.3). Both Cosby et al. (1984) and Ferrer-Julia et al. (2004) showed reasonable predictability of K_{sat} ($R^2 = 0.70$–0.87) using sand% and clay% based on 1,448 and 3,372 samples, respectively. Ahuja et al. (1984, 1989) showed that a modified Kozeny–Carman equation of the form $K_{sat} = C\phi_e^{\,n}$ is applicable to a wide range of soils from the southern region of the U.S., Hawaii, and Arizona (Fig. 2.3c), where C and n are empirical coefficients, and ϕ_e is effective porosity calculated as the saturated water content minus the water content at 33 kPa matric suction.

Table 2.1 Selected PTFs for estimating soil organic carbon (OC) content

PTF Type	PTF Equation	n	R^2	Horizon /Depth (m)	Soil types or orders	Region	Land use (topography)	Reference
I	OC=13.52+ 0.018×(clay%+silt%)	18	0.77	0.025	Various textures (sands, loams, clays)	Brazil	Grass, forest <2% slope	Zinnet al. (2005)
	OC =9.07+0.013 × (clay%+silt%)	18	0.53	0.075				
	OC=5.10+0.012×(clay%+silt%)	18	0.77	0.15				
	OC=4.08+0.011×(clay%+silt%)	18	0.84	0.25				
	OC=3.04+0.010×(clay%+silt%)	18	0.91	0.35				
	OC=2.68+0.007×(clay%+silt%)	18	0.80	0.55				
	OC=1.75+0.005×(clay%+silt%)	18	0.85	0.95				
II	OC=0.46+0.58×LOI-0.101×clay%-0.0085×depth	335	0.98	0-1.2	Alfisols, Spodosols	Belgium	Forest	De Vos et al.(2005a)
	OC=exp[(0.78-1.33×depth-0.28×color value +0.004× (clay%+silt%)+0.012×(clay%+silt%-60)×(depth-0.5)]	1579	0.46	0-2.6	7 soil orders	Australia	Forest, cultivated	Minasny et al. (2006)
VI	OM= 0.29 + 0.19 TWI – 0.07 (stream power index) – 0.002 (aspect)		0.48					Moore et al. (1993)
	OM = 3.174 + 1.775×TWI + 0.003×SL + 0.21×RPI	265	0.82	0-0.2	Anthrosols, Cambisols	China	Forest, cultivated	Guo et al. (2013)
V	OC = 4.09 -0.83×MAT + 0.022×MAT^2+ 0.127×MAP -0.00094 MAP^2+ 0.0006×MAPclay% + 0.0009×MAPsilt%	562	0.51	0-0.2	Central Plains soils excluding shallow, lithic, and aquic soils	U.S.	Rangeland	Burke et al. (1989)
	OC = 5.15 -0.75MAT +0.021MAT^2+ 0.058MAP -0.00046 MAP^2+ 0.00058MAPclay% + 0.00049MAPsilt%	383	0.54	0-0.2			Cultivated	
	OM = 23.3 +28.2(crop type in periodI)+6.9(crop type in periodV)	45	0.71	0-0.15	Inceptisols	Netherlands	Cropped, grass	Pulleman et al. (2000)

The one with a shaded color is the main one discussed in the text

Notes: *LOI* loss on ignition (g/kg), *MAP* mean annual precipitation (mm), *MAT* mean annual temperature (°C), *n* number of samples, *OM* organic matter (%), Period I: 63–31 year; Period V: 3–1 year; *RPI* relative position index, *SL* slope length (m), *TWI* topographic wetness index

Table 2.2 Selected PTFs for estimating soil dry bulk density (BD)

PTF Type	PTF equation	n	R^2	Horizon(Depth, m)	Soil types or orders	Region	Land use (topography)	Reference
I-1	$\log(BD \times 100) = 2.01 - 6.4 \times 10^{-4}(\log OM) - 0.22 \times (\log OM)^2$	103	0.96	A, B	Podzols	U.S.	Forest	Curtis and Post (1964)
	$BD = 1.62 - 0.06 \times OM$	30	0.70	Natural humicgley soils	Gleysols	U.S.	/	Saini (1966)
	$BD = 1.53 - 0.06 \times OM$	40	0.65	Imperfectly drained soils				
	$BD = 1.52 - 0.06 \times OM$	40	0.40	Well drained soils				
	$BD = 1.48 - 0.68 \times (\log OM)$	80	0.82	0-0.05	Sandy loam	Australia, UK	Pasture/ grassland	Jeffrey (1970)
	$BD = 1/(0.63 + 0.04 \times OM)$	136	0.71	-	-	U.S.	Forest	Drew (1973)
	$BD = 1.56 - 0.73 \log LOI$	212	0.40-0.96	Surface (0-0.3)	Gleysols, Histosols,	UK, Sweden, Australia, New Zealand	Woodland, grassland, meadow, heathland, moorland	Harrison and Bocock (1981)
	$BD = 1.73 - 0.77 \log LOI$	538	0.55-0.90	Subsurface (0.02-0.9)	Podzols,			
	$BD = 100 \times (0.026/OC)$	288	0.93	0-0.60	-	U.S.	Wetlands	Gosselink (1984)
	$\ln BD = -2.39 - 1.32 \times \ln OM - 0.17 \times (\ln OM)^2$	60	0.75	O, E, B, C	Spodosols	U.S.	Forest	Huntington et al. (1989)
	$BD = 1.83 * e^{-0.121 * OC^{0.5}}$	55	0.81	A, B, C	Podzols	U.S.	Forest	Alexander (1989)
	$\ln BD = -1.22 - 0.77 \times OM - 0.60 \times \ln OM - 0.06 \times (\ln OM)^2$	288	0.73	-	Loamy sand	Canada	Forest	Prevost and Marcel (2004)
	$\ln BD = -0.23 - 2.08 \times OM - 0.14 \times \ln OM$	126	0.85	-	Silt			
	$BD = 1.775 - 0.173 \times OM^{0.5}$	1082	0.57	A, E, B, C (0-1.2)	Entisols, Inceptisols, Spodosols, Alfisols	Belgium	Forest	De Voset al. (2005b)
	$BD = -1.98 + 4.10 \times OM - 1.23 \times \ln(OM) - 0.10 \times [\ln(OM)]^2$	125	0.82	0.01-0.18	Cambisols, Podzols	Canada	Forest	Perie and Quimet (2008)
	$\ln BD = 0.014 - 0.189 \times \ln OC - 0.066 \times (\ln OC)^2$	3513	0.67	0-0.30	Andosols	Japan	/	Nanko et al. (2014)
I-2	$BD = 100/[(OM/K1) + (100 - OM)/K2]$, $K1 = 0.25, K2 = 1.40$	/	/	0.05-0.08	Podzols	UK	Forest	Stewart et al. (1970)
	$BD = 100/[(OM/K1) + (100 - OM)/K2]$, $K1 = 0.21, K2 = 1.33$	77	/	/	/	/	/	Adams (1973)
	$BD = 100/[(OM/K1) + (100 - OM)/K2]$, $K1 = 0.22, K2 = 1.0\text{-}1.7$	2721	/	/	/	U.S.	/	Rawls (1983)
	$BD = (K1 \times K2)/[(K2 \times OM) + (1 - OM) \times K1]$, $K1 = 0.11, K2 = 1.45$	480	/	O, A, B	/	U.S.	/	Federer et al. (1993)
II	$BD = 2.92 - 0.23 \times OC^{0.5} + 4.27 \times WP - 36.90 \times WP^2 + 66.63 \times WP^3 - 1.44 \times (WP/clay\%)^{0.1}$	386	0.70	Upland soils	Alfisols; AridisolsEntisols, Inceptisols, Mollisols, Ultisols, Vertisols	U.S.	Cultivated, uncultivated	Alexander (1980)
	$BD = 4.92 - 0.25 \times OC^{0.5} - 3.72 \times (WP/clay\%)^{0.1} - 0.36 \times silt\% + 1.38 \times sand\%^2 - 1.46 \times sand\%^3$	335	0.63	Alluvial soils				
	$BD = 14.70\ (\text{field soil water content})^{-0.73}$	30	0.91	0-0.15	/	Canada	Grass, cereal, vegetable, uncultivated	Leirop (1981)
	$BD = 0.02 + 0.05 \times pH$	30	0.51					
	$BD = 0.15 + 0.0054 \times ash$	39	0.72					
	$BD = 1.74 - 0.22 \times OC^{0.5} + 0.005 \times clay\% - 0.02 \times WP$	19226	0.58	A,B, C	/	U.S.	/	Manrique and Jones (1991)
	$BD = 1.77 - 0.25 \times OC^{0.5} - 0.02 \times WP$	5018	0.59	A				
	$BD = 1.77 - 0.28 \times OC^{0.5} + 0.005 \times clay\% - 0.022 \times WP$	10143	0.61	B				
	$BD = 1.77 - 0.26 \times OC^{0.5} + 0.005 \times clay\% - 0.024 \times WP$	4065	0.40	C				

PTF equation	n	R^2	Horizon /Depth (m)	Soil types or orders	Region	Land use (topography)	Reference
BD = 1.52 - 0.0038×clay% - 0.05×OC - 0.045×pH-W + 0.001×sand%	323	0.56	A,B, C	Alfilsols, Ultisols, Oxisols	Brazil	Forest	Bernoux et al. (1998)
BD = 1.52 - 0.0028×clay% - 0.06×OC - 0.05×pH-W + 0.0028×sand%	121	0.57	A				
BD = 1.68 - 0.0048×clay% - 0.11×OC - 0.11×pH-W + 0.06×pH-K + 0.006×CEC	188	0.60	B, C				
BD = 2.40 - 0.32×ln(OM) - 0.03×clay% - 0.14×pH	/	0.62	/	/	Ghana	Forest	Salifu et al. (1999)
BD = 1.70 - 0.0031×silt% + 0.0026×clay% - 0.11×OC	140	0.83	A	/	Lithuania	/	Leonavičiutė (2000)
BD = 0.99 - 0.0059×ln(silt%) + 0.07×ln(clay%) + 0.09×ln(sand%) - 0.08×ln(OC)	56	0.47	E				
BD = 1.07 + 0.03×ln(silt%) + 0.04×ln(clay%) + 0.08×ln(sand%) - 0.06×ln(OC)	296	0.50	B				
BD = 1.07 + 0.01×ln(silt%) + 0.08×ln(clay%) + 0.09×ln(sand%) - 0.06×ln(OC)	455	0.46	BC+C				
BD = 1.59 - 0.084×OC - 0.0018×silt% - 0.002×clay-free silt% + 0.0001clay%×sand% - 0.0466×clay%.silt% + 0.0008×depth	937	0.56	/	/	U.S.	Forest; Prairie	Calhoun et al. (2001)
lnBD = 0.31 - 0.19×OC + 0.021×clay%- 0.000476×clay%^{2} - 0.00432×silt%	224	0.62	0.08-1.10	/	India	Forest, cultivated, bareland	Kaur et al. (2002)
BD = 1.69 - 0.20×OC$^{0.5}$ - 0.013×WP + 0.0079×clay% - 0.0007×silt%+ 0.00014×depth	47015	0.44	Variety of horizons	11 soil orders of Soil Taxonomy except Gelisols	U.S.	Variety of land uses	Heuscher et al.(2005)
BD = 1.42 - 0.15×OC + 0.44×sand% + 0.40×OC×clay% + 0.61×clay%×sand% - 0.52×OC×clay%^{2} - 0.32×sand%^{3}	337	0.68	Top soil (0-0.25)	Histosols	Sweden	Arable land	Katterer et al. (2006)
BD = 1.37-0.22×OC+2.34×clay% + 0.14×sand% + 0.61×OC×sand% - 4.74×clay%^{2} - 0.88×clay%×sand% + 0.0057×OC^3 - 0.11×OC^2×clay% - 0.06×OC^2×sand% + 0.62×OC×clay%^{2} - 0.38×OC×sand%^{2} + 1.93×clay%^{3}	1283	0.59	Subsoil (>0.25)				
BD = 1.56 - 0.0005×clay% - 0.01×OC + 0.0075× Sum of basic cations	1396	0.66	A(0-0.3), B+C(0.3-1)	Acrisols, Fluvisols,Luvisols	Brazil	Forest, grassland	Benites et al. (2007)
BD = 1.65 - 0.12×OC - 0.0042×clay% - 0.0036×sand% + 0.03×pH	348	0.55	A(0-0.30), B+C(0.30-1.00)	Cambisols, Fluvisols, Gleysols, Podzols, Regosols, Solonchaks, Vertisols	Tunisia	/	Brahim et al. (2012)
lnBD = 0.43 - 0.04×OM$^{0.5}$ - 0.0007×OM - 0.02×total nitrogen + 0.0001×clay%	1560	0.79	/	/	China	/	Han et al. (2012)
BD = 1.59 - 0 .47e$^{(0.06×OC)}$ - 0.08×ln(depth)	34	0.44	Volcanic material	14 soil groups of World Reference Base	Europe	/	Hollis et al. (2012)
BD = 0.81 + 0.82e$^{(-0.28×OC)}$ + 0.0014×sand% - 0.001×clay%	333	0.62	Cultivated topsoils				
BD = 1.13 - 0.11×lnOC + 0.055×ln(depth) + 0.0022×sand%	64	0.40	Compact subsoil				
BD = 0.70 + 0.75e$^{(-0.23×OC)}$ + 8.7×10^{-4}×sand% - 5.1×10^{-4}×clay%	925	0.63	All other mineral horizons				
BD = 1.49 − 0.33 × Ln(OC%)	67	0.68	All organic horizons				

Those with a shaded color are main ones discussed in the text

Notes: *Ash* determined by ashing soil samples in a muffle furnace at 375 °C for about 16 h, *CEC* cation exchange capacity (cmol/kg), *K1* density for organic materials, *K2* density for mineral materials, *LOI* loss on ignition (g/kg), *n* number of samples, *OC* organic carbon (g/kg), *OM* organic matter (%), *pH-W* pH measured in water (1:1), *pH-K* pH measured in KCl 1 M (1:1), *WP* wilting point (soil water tension at −15 bar)

Table 2.3 Selected PTFs for estimating soil saturated hydraulic conductivity (K_{sat})

PTF Type	PTF Equation	n	R^2	Horizon (Depth, m)	Soil types or orders	Region	Land use	Reference
I-1	$K_{sat} = 2.54 \times 10^{(-0.6 + 0.01\,sand\% - 6.40\times10^{-3}\,clay\%)}$	1448	0.87	A, B, C	Variety of textures	U.S.	Pasture, forest, cultivated	Cosby et al. (1984)
	$K_{sat} = 15.67 \times e^{(-0.20\,clay\%)}$	42	0.77	Ap, Bt (0-1.40+)	Ultisols	U.S.	Coastal Plains	Puckett et al. (1985)
	$K_{sat} = 30.38 \times e^{(-0.14\,clay\%)}$	577	0.45	/	Variety of textures	U.S.	/	Dane and Puckett (1994)
	$K_{sat} = 25.89 - 0.03\,clay\%^2$	37	0.44	Ap, E, Bt, BC (0-3)	Loam	U.S.	Interfluve along a hillslope	Shaw et al. (2000)
	$K_{sat} = 0.92\,e^{0.049\,sand\%}$	3372	0.70	Variety of horizons	Variety of textures (9 soil groups of WRB)	Spain	Variety of land uses	Julia et al. (2004)
	$K_{sat} = 173.40\,(silt\% + clay\%)^{-1.48}$	564	0.42	/	Variety of textures (7 soil orders)	India	/	Adihikary et al. (2008)
I-2	$K_{sat} = 42918 \times BD^{-21.26}$	12	0.86	A (0-20)	Variety of textures	U.S.	Variety of land uses	OCSCD et al. (2001)
	$K_{sat} = 0.36 - 0.16BD$	216	0.35	A, B, C (0.10-0.85)	Variety of textures	Australia	Forest	O'Connell et al. (2002)
I-3	$K_{sat} = 1058.4 \times \Phi e^{3.35}$	297	0.71	A, B, C (0-2)	Inceptisols, Ultisols	U.S.	Cropland	Ahuja et al. (1989)
	$K_{sat} = -4.68 + 4.37$ (number of macropores)	102	0.43	A, E, Bt (0-0.56)	Silt loam	U.S.	Prairie	Udawatta et al. (2008)
	$K_{sat} = -0.36 + 3743.88\,\Phi_{macro}$	120	0.58	0-0.50	Silt laom	U.S.	Pasture	Kumar et al. (2010)
II-1	$\ln K_{sat} = [19.52\Phi_t - 8.97 - 2.82\times10^{-2}\,clay\% + 1.81\times10^{-4}\,sand\%^2 - 9.41\times10^{-3}\,clay\%^2 - 8.40\Phi_t^2 + 7.77\times10^{-2}\,sand\%\times\Phi_t - 2.98\times10^{-3}\,sand\%^2\times\Phi_t^2 - 1.95\times10^{-2}\,clay\%^2\times\Phi_t^2 + 1.73\times10^{-5}\,sand\%^2\times clay\% + 2.73\times10^{-2}\,clay\%\times\Phi_t + 1.43\times10^{-3}\,clay\%^2\times\Phi_t - 3.50\times10^{-6}\,clay\%^2 * sand\%]$	230	/	/	Variety of textures	U.S.	/	Brakensiek et al. (1984)
	$\ln K_{sat} = 1.01 - 0.08\,sand\% + \frac{-3.99 + 3.67\times10^{-2}\,sand\% + 11.03\times10^{-4}\,clay\% + 8.75\times10^{-4}\,clay\%^2}{0.33 - 7.51\times10^{-3}\,sand\% + 1.76\times10^{-1}\log clay\%}$	230	0.95	/	Variety of textures	U.S.	/	Saxton et al. (1986)
	$K_{sat} = 0.04e^{(20.62 - 0.96\ln clay\% - 0.66\ln sand\% - 0.46\ln OM - 8.43BD)}$	182	0.20	Ap/AB/B/Bt/C	Variety of textures	Belgium	/	Vereecken et al. (1990)
	$\log K_{sat} = 9.56 - 0.81\log silt\% - 1.09\log clay\% - 4.64BD$	368	0.68	0.20-0.60	Sandy loam	U.S.	/	Jabro et al. (1992)
	$K_{sat} = 5.40 \times (-0.07\,sand\% - 0.17\,clay\%)$	42	/	/	Variety of textures	U.S.	/	Campell and shiozawa (1994)
	$K_{sat} = 0.07\,sand\% + 0.07\,silt\% + 0.04\,clay\% + 0.33OM - 4.69BD$	108	/	0.025-0.10	Variety of textures	U.S.	/	Salchow et al. (1996)
	$\ln K_{sat} = 7.76 + 0.04\,silt\% + 0.93$ (topsoil or not) $- 0.98\,BD - 4.80\times10^{-4}\,clay\%^2 - 3.22\times10^{-4}\,silt\%^2 + 1\times10^{-3}\,silt\%^{-1} - 0.07\,OM^{-1} - 0.64\ln silt\% - 0.01\,BD\times clay\% - 0.17\,BD\times OM + 0.30clay\% - 0.03silt\%)$	1136	0.19	/	/	Europe	/	Wosten et al. (1999)
	$\ln K_{sat} = 45.80 - 14.34BD + 1.48\times10^{-3}\,silt\%^2 - 27.5\,BD^{-1} - 0.89\ln \ln silt\% - 0.34\ln OM$	832	/	/	Sand	Europe	/	Wosten et al. (2001)
	$\ln K_{sat} = -42.60 + 8.71\,OM + 91.90\,BD - 20.79\,BD^2 - 0.21\,OM^2 - 0.02\,clay\%\times OM - 5.38\,BD\times OM$	832	/	/	Loam, clay	Europe	/	Wosten et al. (2001)
	$K_{sat} = -4.99 + 0.057\,sand\% - 0.013\,clay\% - 0.013\,OM$	3372	0.72	Variety of horizons	Variety of textures (9 soil groups of WRB)	Spain	Variety of land uses	Ferrer-Julia et al. (2004)
	$\lg K_{sat} = 3.50 - 2.81BD + 2.10\,sand\%$	112	0.79	0-3.00	/	U.S.	/	Balland et al. 2008
	$\ln K_{sat} = 13.26 - 1.91\ln sand\% - 0.97\ln silt\% - 0.06\,clay\% - 1.71\ln OM + 2.89\,OM - 8.03\ln BD$	36	0.66	0-2.00	Variety of textures	China	Bareland	Li et al. (2007)
	$K_{sat} = 2.38 - 1.20\,BD + 0.03\,OM - 0.01\,clay\%$	408	0.42	0-0.40	Variety of textures	Malaysia	Rice grown coastal plain	Aimum and Amin (2009)

	$\ln K_{sat} = 8.75 - 3.31\ BD + 0.89 \ln sand\%$	42	0.28	0-0.45	Variety of textures	U.S.	Forest/grassland/ pasture/cropland	Arrington et al. (2013)
	$\ln K_{sat} = 8.16 - 0.086\ clay\% - 0.03\ sand\% + 0.32 \ln dg + 3.01 \ln \Phi e + 1.91 Df$	428	0.55	/	Variety of textures	Australia	/	Minasny and McBratney. (2000)
	$\ln K_{sat} = -2.37 + 3.85\ BD + 0.09\ OM - 6.20 \ln BD - 0.34 \ln OM - 2.33 \ln clay\% - 0.41 \ln dg$	404	0.46	0-0.40	Variety of textures	Malaysia	Rice grown coastal plain	Aimum and Amin (2009)
	$\log K_{sat} = 2.77 + 0.01 + 5.00 \times 10^{-3}\ silt\% + 5.00 \times 10^{-3}\ clay\% - 975.71\ BD + 5 \times 10^{-3} OC - 0.26\ (\text{field capacity}) + 0.84\ (\text{permanent wilting point})$	102	0.54	0-0.20	Clay	India	Flat terrain	Patil et al. (2010)
II-2	$K_{sat} = 32.50 - 42.77\ sand\% - 95.81\ (\text{initial water content}) + 1160.26\ OC + 466.05\ \Phi_{macro} - 499.1\ \Phi_{root}$	96	0.88	A, E, B, C	Variety of textures (4 soil orders)	U.S.	Prairie, costal plants	Lin et al. (1999)
	$\ln K_{sat} = -8.10 + 1.18\ sand\% + 8.77\ BD + 37.89\ \Phi_{>30} - 4.28\ sand\%^2 - 3.22\ BD^2 - 20.78 \Phi_{>30}^2 + 5.52\ sand\% \times BD - 21.33\ sand\% \times \Phi_{>30} - 14.89 BD \times \Phi_{>30}$	195	0.51	0-1.50	Variety of textures	Turkey	Irrigated farmland	Merdun et al. (2006)
	$K_{sat} = 109.11 + 3.85\ (\text{number of macropores}) - 84.20\ BD$	102	0.45	A, E, Bt (0-0.56)	Silt loam	U.S.	Prairie	Udawatta et al. (2008)
	$K_{sat} = 10.90 - 3.54\ (\text{number of macropores}) + 6292.15 \Phi_{macro}$	120	0.60	0-0.50	Silt laom	U.S.	Pasture	Kumar et al. (2010)
	$K_{sat} = -21.90 + 56.02\ \Phi_{macro} - 2.87\ \Phi_{largest} - 49.74\ \Phi_t$	100	0.79	0-0.40	Silt laom	U.S.	Corn planted	Kim et al. (2010)
	$\ln K_{sat} = -2.19 + 3.28\ (\text{path number}) + 33.90 * \Phi_{macro(3D)}$	28	0.75	Ap, Bt (0-0.36)	Silt loam	U.S.	Cropland, pasture	Luo et al. (2010)
	$K_{sat} = 37.19 + 66.68 \Phi_{macro} - 22.22\ BD$	27	0.77	Ap, Bt (0-0.40)	Sodic soils	China	Cropland	Yu et al. (2014)
	$K_{sat} = -1.93 + 103.29\ \Phi_{macro} - 0.50 \ln \Phi_{irregular}$	27	0.70					
III	$K_{sat} = -22.43 - 38.26\ MI_t + 33.82\ MI_m + 21.07\ MI_{sg} + 47.54\ MI_{st} + 102.36\ MI_{pq} + 45.45\ MI_{ps} + 33.27\ MI_{rq}$	96	0.86	A, E, B, C	Variety of textures (4 soil orders)	U.S.	Prairie, costal plants	Lin et al. (1999)
	$K_{sat} = 0.16 - 0.11\ BD + 0.06\ \text{factor Hueclass} = 3 + 0.04\ \text{factor Hueclass} = 4 - 3 \times 10^{-3}\ (\text{rock fragment\%}) + 0.03\ \text{Value}$	216	0.35	A, B, C (0.10-0.85)	Several textures	Australia	Forest	O'Connell et al. (2002)
VI	$\ln K_{sat} = 18.10 - 2.64\ BD - 0.03\ silt\% - 0.05\ clay\% - 0.02\ pH + 0.03\ elevation$	100	/	0-0.10	Silt loam, loam, sandy clay loam	India	Hilly forest watershed	Santra and Das et al. (2008)
	$\log K_{sat} = 1.17 + 0.04\ silt\% + 0.69 \log sand\% + \frac{0.87}{sand}\% - 0.03 BD \times silt\% - 9.95 \times 10^{-6}\ OC \times elevation$	382	0.36	0-0.05	Clay loam	China	Loess plateau	Wang et al. (2012)

Those with a shaded color are main ones discussed in the text

Notes: Φ_t total porosity calculated from bulk density, Φ_e effective porosity, Φ_{macro} macroporosity, $\Phi_{macro(3D)}$ volumetric fraction of pores with equivalent cylindrical diameter >0.75 mm, Φ_{root} porosity at root-soil interface, $\Phi_{>30}$ porosity with diameter >30 μm, $\Phi_{largest}$ area of the largest pore based on scanned image, $\Phi_{irregular}$ area of pores with shape factor 2–5, *BD* bulk density, *Df* fractal dimension, *dg* geometric mean particle size, MI_t textural class index, MI_m initial moisture state index, MI_{sg} ped grade index, MI_{st} ped shape index, MI_{pq} macropore quantity index, *MIps* Macropore size index, *MIrq* root abundance index, *MIrs* root size index, *number of macropores* number of pores with diameter >1,000 μm, *n* number of samples, *OC* organic carbon (g/kg), *OM* organic matter (%), *path number* number of independent and continuous macropore path between two boundaries based on CT scanned three dimensional images, *topsoil or not* qualitative variable having the value of 1 (topsoil) or 0 (subsoil). Value: Munsell color value

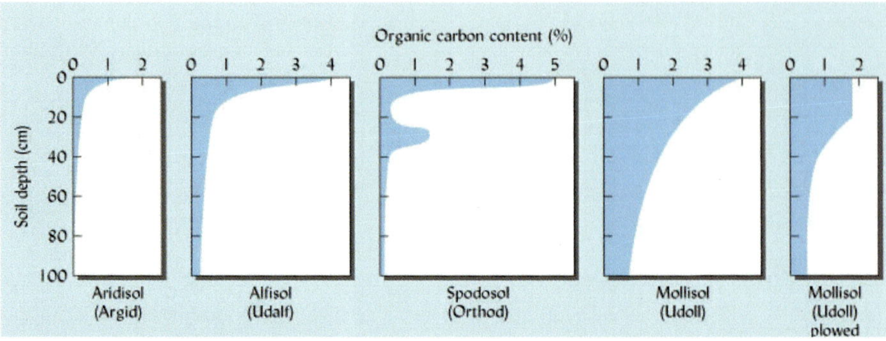

Fig. 2.4 Vertical distribution of organic carbon in well-drained soils of four soil orders: Aridisol in dry-region, Alfisol and Spodosol under forests, and Mollisol under grassland. The bulge of organic carbon in the Spodosol subsoil is due to illuvial humus in the spodic horizon. Note the impact of cultivation of C loss in Mollisol (After Brady and Weil 2004). Note that these C depth distribution patterns are indicative of water percolation features in each of these four soil orders

2.3.3 PTF II: Continuous PTFs

Most of existing PTFs are of this type, where a wide variety of basic soil properties (either dynamic or relatively static) are used to estimate desirable soil parameters through various statistical relationships.

Soil OC content has been commonly estimated using texture, depth, and even color (Table 2.1). Based on 1,579 samples from seven soil orders in Australia, Minasny et al. (2006) developed a negative exponential profile depth function to describe soil OC content at different depths: $\mathrm{OC} = \mathrm{C}_a \exp(-k\ z) + \mathrm{C}_b$, where C_a is the difference in OC content between the surface and the lowest depth, C_b is OC content at the bottom of the profile, z is the absolute value of depth from the soil surface, and k is the rate of OC decrease with depth. The three parameters (C_a, C_b, $k \geq 0$) vary with soil types, and thus could be link to soil classification as illustrated in Fig. 2.4.

In addition to OC and texture, soil BD estimation can be improved with additional inputs of soil water content, pH, depth, and even CEC, sum of basic cations, and total nitrogen (Table 2.2). Several Type II PTFs have been developed using huge databases (up to 47,015 samples) from across the U.S. to estimate soil BD in A, B, and C horizons based on OC, clay content, and wilting point (Manrique and Jones 1991), or with additional variables of silt content and soil depth (Heuscher et al. 2005) (Table 2.2).

Soil $\mathrm{K}_{\mathrm{sat}}$ has been estimated using a variety of combinations of BD, texture, OC, macropores, with or without additional properties (such as total porosity, effective porosity, porosity at root-soil interface, field capacity, wilting point, fractal dimension, geometric mean particle size, and others) (Table 2.3). In recent years, the addition of macropore-related features into $\mathrm{K}_{\mathrm{sat}}$ estimation has been increasingly recognized as a key control on $\mathrm{K}_{\mathrm{sat}}$. Interestingly, Ferrer-Julia et al. (2004) developed a one-variable (sand content) and three-variable (sand, clay, and OM contents) PTFs for estimating K_{sat} across a great variety of climatic and physiographic conditions (with a predominance of soils developed under semi-arid conditions) (Table 2.3). These PTFs did not include macropores, but still obtained R^2 of 0.70–0.72.

2.3.4 PTF III: Class PTFs

Descriptive class-based soil features are abound in soil survey databases, which have been underused in improving PTFs. Class variables, such as soil horizons (A, B, C, or specific diagnostic horizons), soil structural features (e.g., granular, blocky, prismatic peds), soil classification and related map units (e.g., soil series or great groups), and even clay mineralogy (e.g., kaolin vs. smectite), could help separate or group different soil types into a range of expected properties. This has important implications for data propagation in soil survey databases. For instance, soil series and related mapping units could be used to projector expected soil properties from one landscape to another with similar environmental conditions. Soil morphology is another information source for improving PTFs. This is because soil morphology is diagnostic of many soil properties, such as OC, water table, redox, soil moisture regime, and even BD and K_{sat}. However, while qualitative use of soil morphology has been widely applied for a long time (Brown 1954; Fox 1985), quantification of such data has been generally lacking. Lin et al. (1999) developed a first step in quantitative use of soil morphology (including descriptors of textural class, initial moisture state, pedality, macroporosity, and roots) and used that to develop PTFs. Interestingly, there are two relevant large programs in existence within the USDA NRCS: one is soil mechanics with lots of matrix K_{sat} determined (with consideration of rock fragment in some cases) and the other is official soil profile descriptions containing in situ soil profiles information (including soil structure and sometimes macropores-related features). It would be beneficial to construct PTFs to combine these two programs for estimating K_{sat} in diverse field soils.

Although no specific Type III PTFs have been developed for soil OC and BD, there are indirect ways of using class variables for better estimation of these soil properties. For example, with the knowledge of soil orders, we could expect contrasting soil OC depth distributions as illustrated in Fig. 2.4 for Aridisols, Alfisols, Spodosols, and Mollisols (both natural and plowed). Manrique and Jones (1991) showed that, for soils grouped according to their taxonomic classification, multiple-regression models based on soil properties could provide a relatively accurate alternative for predicting BD. With knowledge of soil horizons (A, E, B, or C; topsoil or subsoil; organic or mineral horizons; tilled layer, compact layer or volcanic layer etc.) or soil profiles (e.g., well drained or imperfectly drained soils; upland soils or alluvial soils), we can use different PTFs to estimate soil BD, as shown in Table 2.2. Calhoun et al. (2001) showed that using a combination of continuous variables (laboratory data) and nominal variables (site/state factor and morphological class descriptors) improved PTFs for estimating BD.

For K_{sat}, a couple of Type III PTFs have been developed (Table 2.3). One is by Lin et al. (1999) using morphometric indices (numericalized class variables) of textural class, initial moisture state, pedality, macroporosity, and roots. The other is by O'Connell and Ryan (2002) using soil color hue and value, together with BD and rock fragment content. These PTFs covered a wide range of textures and land uses, with the first one by Lin et al. (1999) for cropped and pastured soils, and the second one by O'Connell and Ryan (2002) for forested soils.

2.3.5 PTF VI: Topo- or Geo-rectified PTFs

Among the common soil-forming factors, topography and parent material are considered as sites factors that exert local controls on soil properties (such as OC and clay distribution along hillslopes). This is in contrast to climate and land use/land cover as flux factors considered in PTF Type V (see next section).

As illustrated by Moore et al. (1993), Rawls and Pachepsky (2002), and others, topography helps the understanding of causation and correlation of soil properties with landscape positions. Hence, "topo-rectifying" PTFs through taking into account topographic attributes would improve landscape-based PTFs. Table 2.1 shows a couple of PTFs for estimating OC solely based on topographic attributes, while Table 2.3 shows a couple of PTFs for estimating K_{sat} using a combination of elevation, soil texture, and BD. Harrison and Bocock (1981) showed that slope and elevation significantly influenced soil bulk density. Because of this, they recommended that for greater precision in estimating soil bulk density, locally derived PTFs would be preferred rather than general PTFs (such as those Type I PTFs listed in Table 2.2).

Franzmeier (1991) found that grouping soil properties used in PTF by lithomorphic classes significantly increased the reliability of predicting K_{sat}. Calhoun et al. (2001) showed the advantages of using parent material as a grouping criterion for separate PTFs. However, such geo-rectified PTFs are not commonly seem in the literature.

2.3.6 PTF V: Climate or Land Use Adjusted PTFs

Currently most PTFs has been generated under humid climatic conditions (Ferrer-Julia et al. 2004). Table 2.1 shows two PTFs (one for cultivated and the other for rangeland) for estimating soil OC using mean annual precipitation and temperature, along with soil texture. However, there are no climate-adjusted PTFs for BD or K_{sat} estimation currently in the literature.

Any soil may be changed significantly by land use and management practices, even though taxonomic soil classification may remain unchanged. For example, Droogers and Bouma (1997) studied a prime agricultural soil in the Netherlands and found that the organic matter content of a conventionally tilled variant had significantly decreased as compared with a variant subject to organic farming.

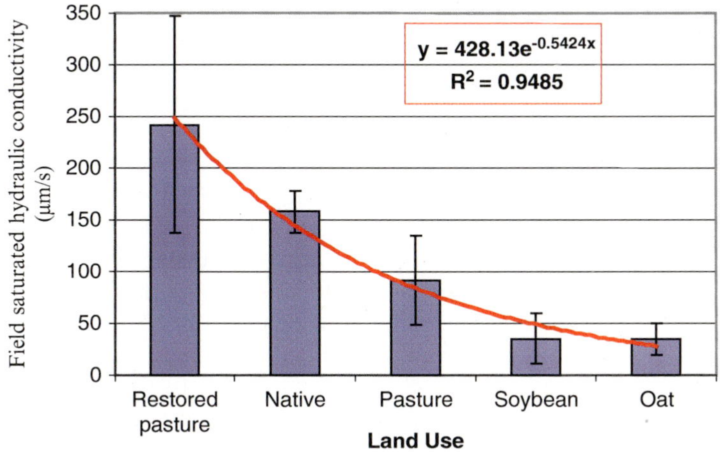

Fig. 2.5 Land use impacts on field saturated hydraulic conductivity in the surface soil of the Houston Black (a typical Vertisol)

Also, grassland had even higher organic matter contents even though soil classifications of these phenoforms were identical. Pulleman et al. (2000) showed that organic matter contents of these phenoforms could be predicted by regression analysis as a function of previous land use (Table 2.1). Distinguishing different phenoforms for a given soil series can refine PTFs under different human impacts, which will undoubtfully enhance soil modeling and prediction. Sonneveld et al. (2002) made such an analysis for a major sandy soil in the Netherlands.

While there is not a shortage of literature on land use impacts on soil properties, Type V PTFs that quantitatively relate soil properties to land use and management are yet to be better developed. Figure 2.5 illustrates the land use impacts on surface soil K_{sat} in a Vertisol, with an exponential decreasing trend from restored pasture and native vegetation to soybean and oat crop management systems. The differences among these land uses are reflected in cropping systems that tend to result in lower OM inputs, less structural stability, higher crusting/runoff potentials, lower macroporosity, and lower root density with increased degree of human perturbations.

2.4 Outlooks

2.4.1 Databases and SINFERS

One critical point made by Wosten et al. (2001) and Lilly and Lin (2004) is that major progress in PTFs is not to be expected from new statistical methods but rather from better databases. While empirical, regression, or functional approaches continue to be used in developing PTFs, new methods are increasingly being explored for developing and using PTFs, including artificial neural networks (e.g., Schaap and Bouten 1996; Minasny et al. 1999), regression trees (e.g., McKenzie and Jacquier 1997; Rawls and Pachepsky 2002), and the Group Method of Data Handling (e.g., Pachepsky and Rawls 1999). However, the success of any mathematical/statistical techniques will be heavily dependent on the quality, quantity, comparability, and transferability of the original data stored in the databases. Without the foundation of reliable and systematic databases, no matter how sophisticated the techniques used in deriving or using PTFs, the outputs would be futile and misleading (Lilly and Lin 2004). In the case of OC, BD, and K_{sat}, as reviewed in this chapter, any further attempts to improve their estimation by developing more complex regression equations from a number of soil properties are unwarranted (except in special circumstances), as it is simpler and much more reliable to use relevant fundamental PTFs (Type I PTFs) or to directly measure these soil properties if feasible.

On the other hand, many existing global and national databases (such as NCSS, UNSODA, WISE, HYPRES, ISRIC-WISE, and HWSD) should be better utilized to develop more robust PTFs, particularly in association with soil classifications, landscape features, soil profile descriptions, and land uses. In addition, developing a comprehensive database into an expert system that can be combined with the selection, use, and uncertainty quantification of PTFs would be beneficial. McBratney et al. (2002) proposed the concept of SINFERS, where a set of properly and logically conjoined PTFs serve as the knowledge rules for inference engines. Such a SINFERS takes known measurements with a given level of (un) certainty and infers desirable unknown data with minimal inaccuracy allowed in the system. The SINFERS has a source, an organizer, and a predictor; together they serve as a decision-support system for appropriate use of PTFs with uncertainty estimation. The sources are collections of soil databases and PTFs, which could additionally include geospatial data. The organizer arranges and categorizes the PTFs with respect to their required inputs and soil types from which they were generated. The inference engine is a collection of "if-then" type of logical rules for selecting appropriate PTFs with the minimum variance possible. Uncertainty of the prediction is assessed using Monte Carlo simulations, which can be quantified in terms of the model uncertainty and input data uncertainty (McBratney et al. 2002; McBratney and Minasny 2004).

Recognizing the importance of use-dependent and other dynamic soil properties, the USDA-NRCS has been considering the possible development of a dynamic soil properties database. Such a database, once developed, would significantly enhance the utility of soil survey databases and the development of dynamic PTFs. There are considerable benefits of formulating reliable PTFs for propagating data in global and national soil survey databases, especially if hydropedology framework is utilized to connect dynamic soil properties with soil survey databases. Some key issues for such a dynamic soil properties database should be carefully considered, including:

- *What to measure:* Essential soil variables, such as OC, BD, K_{sat}, and many others, must be included, along with the recoding of location, landscape features, map unit, soil classification, land use, as well as time and initial soil moisture condition;

- *Where and when to measure:* Taking into account soil-landscape spatial and temporal variability, appropriate spatial stratification and seasonal dynamics must be considered to determine site selection, replication number, and timing of measurements;
- *How deep to measure:* Appropriate depths or soil horizons should be identified for measurements to capture depth function down to parent materials. At the minimum, surface soil (A horizon) and subsoil (B horizon) should be included;
- *How to measure:* Common methodologies must be adopted for systematic and comparable data collections, documentations, and interpretations. Simple and yet reliable methods should be selected. Establishing a long-term monitoring network in benchmark soil-landscape units (catenae) would be complementary and highly valuable, including possible automatic data collections and synthesis. Such a network should be synergized with existing monitoring sites that provide climatic data (e.g., SCAN sites);
- *How to utilize:* Collected dynamic soil properties must be quality assured and quality controlled so that robust PTFs can be developed and applied. Data propagation and pattern recognition of soil survey databases can then be linked to these PTFs.

2.4.2 Hydropedoinformatics and Integrated Soil-Landscape Analysis

To integrate knowledge, scales, and databases of interactive pedologic and hydrologic processes and properties, and to streamline soil-landscape data capture, storage, visualization, analysis, and modeling, hydropedoinformatics has been suggested (Lin et al. 2005). The NCSS has provided over 100 years of soil inventory, measurement, and evaluation, and currently maintains several national databases (such as the SSURGO and the STATSGO, official soil series descriptions, soil characterization laboratory database, and soil climate monitoring network). These databases can be coordinated and utilized in both the development of PTFs and the construction of integrated hydropedoinformatic system, particularly if the above-discussed dynamic soil properties database can be established. Some opportunities of such a system are suggested below:

- Making connections between hydropedology and digital soil mapping is an exciting area, which can improve the connection between soil spatial mapping and process-based understanding. Precision soil mapping is of increasing demand for site-specific applications (such as precision agriculture, landscape hydrology, and urban development). Modern soil maps also need improvements for functional characterizations rather than focusing on soil taxonomy. Soil maps thus can no longer be static; rather, derivative and dynamic maps, tailored for a specific function or purpose, must be generated and updated on a regular basis.

- Classical block diagrams of soil-landscape relationships commonly seen in soil surveys could be used to indicate landscape hydrology, illustrating water flow direction and water table dynamics. These block diagrams could be further linked to watersheds to provide valuable conceptual frameworks of water movement over the landscape in different regions. Enhanced 3-D block diagrams with added information of water table dynamics, water flow paths, hydric soils, restrictive layers, and PTFs could significantly increase the values of soil survey products.
- Soil map units are better considered as landscape units rather than individual soil types (Wysocki et al. 2000). It is therefore desirable to go beyond the pure description of soil taxonomic classifications (as traditional soil surveys did); rather, it would serve diverse applications better by depicting soil functions, particularly those hydrologic functions critical to various local land uses. Hydropedologic Functional Unit (HFU) has been suggested as a means of cartographically representing important landscape-soil-hydrology functions, which is defined as a soil-landscape unit having similar pedologic and hydrologic functions (Lin et al. 2008). The goal of the HFU is to subdivide the landscape into similarly functioning hydropedologic units by grouping various geomorphic units that have similar storage, flux, pathway, and/or residence time of water. These units can be identified and delineated using traditional soil survey data in conjunction with various digital data, geophysical investigations, and in situ monitoring.
- An iterative loop of mapping, measurements (or monitoring), and modeling can provide an integrated and evolutionary approach to addressing soil-landscape complexity. Such an approach allows the development of adaptive strategy and the refinement of maps, measurements, and/or models as knowledge and database are accumulated. Measurements or monitoring generally collects point-based data, while modeling often attempts to cover large areas. A key to connect these two is the fabric of the subsurface over the landscape that should be mapped at appropriate spatial and temporal resolution so that meaningful extrapolation and upscaling of point-based data could be made. Mapping also provides a means of diagnosing and stratifying the landscape for determining optimal location and number of measurements. It is important to note that geostatistical functions should be derived from landscape stratified units and not indiscriminately across a broad landscape without prior partitioning of the causes of variability. Hydropedology, process-based understanding, and soil-landscape relationships can help improve the use of geostatistics and related PTFs.

2.5 Summary

This chapter introduces the idea of hydropedology as an intertwined branch of soil science and hydrology that provides a useful framework for connecting dynamics soil properties with soil survey databases. Soil survey databases and dynamic soil properties are briefly reviewed. Next, five categories of PTFs are illustrated for estimating soil organic carbon, dry bulk density, and K_{sat} based on (1) fundamental relationships, (2) essential soil variables, (3) class variables, (4) topo- or geo-rectifications, and (5) climate or land use adjustments. These is a need to continue the probe into the underlying mechanisms for PTFs. Quantitative soil morphometric indices could be further utilized to tap into vast soil profile descriptions to refine PTFs. Distinction between major management types within the same soil series (i.e., genoform vs. phenoform) could further enhance PTFs that involve soil series and land use as carriers of dynamic soil properties. Finally, some outlooks are discussed for further improvements in PTFs developments and applications, including some guidelines for developing a dynamic soil properties database alongside traditional soil survey databases. Hydropedoinformatics is also suggested for integrated soil-landscape analysis.

References

Adams WA (1973) The effect of organic matter on the bulk and true densities of some uncultivated podzolic soils. J Soil Sci 24:10–17

Adhikary PP, Chakraborty D, Kalra N, Sachdev CB, Patra AK, Kumar S, Tomar RK, Chandna P, Raghav D, Agrawal K, Sehgal M (2008) Pedotransfer functions for predicting the hydraulic properties of Indian soils. Aust J Soil Res 46:476–484

Ahuja LR, Naney JW, Green RE, Nielsen DR (1984) Macroporosity to characterize spatial variability of hydraulic conductivity and effects of land management. Soil Sci Soc Am J 48:699–702

Ahuja LR, Cassel DK, Bruce RR, Barnes BB (1989) Evaluation of spatial-distribution of hydraulic conductivity using effective porosity data. Soil Sci 148:404–411

Aimrun W, Amin MSM (2009) Pedo-transfer function for saturated hydraulic conductivity of lowland paddy soils. Paddy Water Environ 7:217–225

Alexander EB (1980) Bulk densities of California soils in relation to other soil properties. Soil Sci Soc Am J 44:689

Alexander EB (1989) Bulk density equations for southern Alaska soils. Can J Soil Sci 69:177–180

Arnold RW, Wilding LP (1991) The need to quantify spatial variability. In: Mausbach MJ, Wilding LP (eds) Spatial variabilities of soils and landforms, SSSA special publication #28. Soil Science Society of America, Inc., Madison, pp 1–8

Arrington KE, Ventura SJ, Norman JM (2013) Predicting saturated hydraulic conductivity for estimating maximum soil infiltration rates. Soil Sci Soc Am J 77:748–758

Arya LM, Paris JF (1981) A physico-empirical model to predict the soil moisture characteristics from particle-size distribution and bulk density data. Soil Sci Soc Am J 45:218–227

Arya LM, Leij FJ, van Genuchten MT, Shouse PJ (1999) Scaling parameter to predict the soil water characteristics from particle-size distribution data. Soil Sci Soc Am J 63:510–519

Balland V, Pollacco JAP, Arp PA (2008) Modeling soil hydraulic properties for a wide range of soil conditions. Ecol Model 219:300–316

Batjes NH (1996) Development of a world data set of soil water retention properties using pedotransfer rules. Geoderma 71:31–52

Batjes NH (2008) ISRIC-WISE harmonized global soil profile dataset (ver 3.1). Report 2008/02, ISRIC – World Soil Information, Wageningen (with dataset)

Benites VM, Machado PLOA, Fidalgo ECC, Coelho MR, Madari BE (2007) Pedotransfer functions for estimating soil bulk density from existing soil survey reports in Brazil. Geoderma 139:90–97

Bernoux M, Cerri C, Arrouays D, Jolivet C (1998) Bulk densities of Brazilian Amazon soils related to other soil properties. Soil Sci Soc Am J 62:743

Bouma J (1989) Using soil survey data for quantitative land evaluation. Adv Soil Sci 9:177–213

Bouma J, van Lanen HAJ (1987) Transfer functions and threshold values: from soil characteristics to land qualities. In: Beek KJ et al (ed) Proc. ISSS/SSSA workshop on quantified land evaluation. Washington, DC. 27 Apr–2 May 1986. Int. Inst. for Aerospace Surv. and Earth Sci. Publ. no. 6. ITC Publ. Enschede, The Netherlands. pp 106–110

Brady NC, Weil RR (2004) Elements of the nature and properties of soils, 2nd edn. Pearson-Prentice Hall, Upper Saddle River

Brahim N, Bernoux M, Gallali T (2012) Pedotransfer functions to estimate soil bulk density for Northern Africa: Tunisia case. J Arid Environ 81:77–83

Brakensiek DL, Rawls WJ, Stephenson GR (1984) Modifying SCS hydrologic groups and curve numbers for rangeland soils. ASAE paper no. PNR-84203, St. Joseph

Brown G (1954) Soil morphology and mineralogy: a qualitative study of some gleyed soils from North-west England. J Soil Sci 5:145–155

Buol SW, Southard RJ, Graham RC, McDaniel PA (2001) Soil genesis and classification, 5th edn. Iowa State University Press, Ames

Burke IC, Yonker CM, Parton WJ, Cole CV, Schimel DS, Flach K (1989) Texture, climate, and cultivation effects on soil organic matter content in U.S. grassland soils. Soil Sci Soc Am J 53:800–805

Burrough P, McDonnell R (1998) Principles of geographic information systems. Oxford University Press, Oxford

Calhoun FG, Calhoun FG, Smeck NE, Slater BL, Bigham JM (2001) Predicting bulk density of Ohio soils from morphology, genetic principles, and laboratory characterization data. Soil Sci Soc Am J 65:811

Campbell GS, Shiozawa S (1994) Prediction of hydraulic properties of soils using particle size distribution and bulk density data. In: Van Genuchten MT et al (eds) Proceedings of the international workshop on indirect methods for estimating the hydraulic properties of unsaturated soils. University of California, Riverside, pp 317–328

Cosby BJ, Hornberger GM, Clapp RB, Ginn TR (1984) A statistical exploration of the relationships of soil-moisture characteristics to the physical-properties of soils. Water Resour Res 20:682–690

Curtis RO, Post BW (1964) Estimating bulk density from organic-matter content in some Vermont forest soils1. Soil Sci Soc Am J 28:285–286

Dane JH, Puckett W (1994) Field soil hydraulic properties based on physical and mineralogical information. In: Van Genuchten MT et al (eds) Proceedings of the international workshop on indirect methods for estimating the hydraulic properties of unsaturated soils. University of California, Riverside, pp 317–328

Daniels RB (1988) Pedology, a field or laboratory science? Soil Sci Soc Am J 52:1518–1519

De Vos B, Van Meirvenne M, Quataert P, Deckers J (2005a) Predictive quality of pedotransfer functions for estimating bulk density of forest soils. Soil Sci Soc Am J 69:500

De Vos B, Vandecasteele B, Deckers J, Muys B (2005b) Capability of loss-on-ignition as a predictor of total organic carbon in non-calcareous forest soils. Commun Soil Sci Plant Anal 36:2899–2921

Drew LA (1973) Bulk density estimation based on organic matter content of some Minnesota soils. Sci. Jour. Ser. Paper No. 8333 of the University of Minnesota. Agr. Expt. Sta

Droogers P, Bouma J (1997) Soil survey input in exploratory modeling of sustainable soil management practices. Soil Sci Soc Am J 61:1704–1710

FAO/IIASA/ISRIC/ISS-CAS/JRC (2012) Harmonized world soil database (version 1.2). FAO/IIASA, Rome/Laxenburg

Federer CA, Turcotte DE, Smith CT (1993) The organic fraction–bulk density relationship and the expression of nutrient content in forest soils. Can J Forest Res 23:1026–1032

Ferrer-Julia M, Estrela-Monreal T, Sanchez del Corral-Jimenez A, Garcia-Melendez E (2004) Constructing a saturated hydraulic conductivity map of Spain using pedotransfer functions and spatial prediction. Geoderma 123:257–277

Fox CA (1985) A morphometric system for describing the micromorphology of organic soils and organic layers: further quantitative and qualitative characterization. Can J Soil Sci 65:695–706
Franzmeier DP (1991) Estimation of hydraulic conductivity from effective porosity data for some Indiana soils. Soil Sci Soc Am J 55:1801–1803
Goryachkin SV (2005) Studies of the soil cover patterns in modern soil science: approaches and tendencies. Eurasian Soil Sci 38:1301–1308
Gosselink JG, Gosselink JG, Hatton R, Hopkinson CS (1984) Relationship of organic carbon and mineral content to bulk density in Louisiana marsh soils. Soil Sci 137:177–180
Grossman RB, Harms DS, Seybold CA, Herrick JE (2001) Coupling use-dependent and use-invariant data for soil quality evaluation in the United States. J Soil Water Conserv 56:63–68
Guo P-T, Wu W, Sheng Q-K, Li M-F, Liu H-B, Wang Z-Y (2013) Prediction of soil organic matter using artificial neural network and topographic indicators in hilly areas. Nutr Cycl Agroecosyst 95:333–344
Han G-Z, Zhang G-L, Gong Z-T, Wang G-F (2012) Pedotransfer functions for estimating soil bulk density in China. Soil Sci 177:158–164
Harrison AF, Bocock KL (1981) Estimation of soil bulk-density from loss-on-ignition values. J Appl Ecol 18:919–927
Haverkamp R, Parlange JY (1986) Predicting the water-retention curve from particle size distribution: 1. Sandy soils without organic matter. Soil Sci 142:325–339
Heuscher SA, Brandt CC, Jardine PM (2005) Using soil physical and chemical properties to estimate bulk density. Soil Sci Soc Am J 69:51–56
Heuvelink GBM, Webster R (2001) Modelling soil variation: past, present, and future. Geoderma 100:269–301
Hole FD (1976) Soils of Wisconsin, Wisconsin Geology National History Survey Bulletin 87, Soils Series. 62. University of Wisconsin Press, Madison
Hollis JM, Hannam J, Bellamy PH (2012) Empirically-derived pedotransfer functions for predicting bulk density in European soils. Eur J Soil Sci 63:96–109
Huntington TG, Huntington TG, Johnson CE, Johnson AH, Siccama TG (1989) Carbon, organic matter, and bulk density relationships in a forested spodosol. Soil Sci 148:380–386
Jabro JD (1992) Estimation of saturated hydraulic conductivity of soils from particle-size distribution and bulk-density data. Trans Asae 35:557–560
Jalabert SSM, Martin MP, Renaud JP, Boulonne L, Jolivet C, Montanarella L, Arrouays D (2010) Estimating forest soil bulk density using boosted regression modelling. Soil Use Manage 26:516–528
Jeffrey DW (1970) A note on the use of ignition loss as a means for the approximate estimation of soil bulk density. J Ecol 58:297–299
Jenny H (1941) Factors of soil formation – a system of quantitative pedology. McGraw-Hill, New York
Katterer T, Andren O, Jansson P (2006) Pedotransfer functions for estimating plant available water and bulk density in Swedish agricultural soils. Acta Agric Scand Sect B – Plant Soil Sci 56:263–276
Kaur R, Kumar S, Gurung HP (2002) A pedo-transfer function (PTF) for estimating soil bulk density from basic soil data and its comparison with existing PTFs. Soil Res 40:847–858
Kim H, Anderson SH, Motavalli PP, Gantzer CJ (2010) Compaction effects on soil macropore geometry and related parameters for an arable field. Geoderma 160:244–251
Kumar S, Anderson SH, Udawatta RP (2010) Agroforestry and grass buffer influences on macropores measured by computed tomography under grazed pasture systems. Soil Sci Soc Am J 74:203–212
Lagacherie P, McBratney AB, Voltz M (eds) (2007) Digital soil mapping. An introductory perspective. Developments in soil science, vol 31. Elsevier, Amsterdam
Leij F, Alves WJ, van Genuchten MTh, Williams JR (1996) The UNSODA unsaturated soil hydraulic database. User's manual version 1.0. EPA/600/R-96/095. National Risk Management Laboratory, Office of Research and Development, Cincinnati, OH

Leonavičiutė N (2000) Predicting soil bulk and particle densities by pedotransfer functions from existing soil data in Lithuania. Geografijos metrasˇtis 33:317–330
Li Y, Chen D, White RE, Zhu A, Zhang J (2007) Estimating soil hydraulic properties of Fengqiu County soils in the North China plain using pedo-transfer functions. Geoderma 138:261–271
Lilly A (1997) A description of the HYPRES database (Hydraulic properties of European soils). pp 161–184. In: Bruand A, Duval O, Wösten JHM, Lilly A (eds) The use of pedotransfer functions in soil hydrology research. Proc. second workshop of the project using existing soil data to derive hydraulic parameters for simulation modelling in environmental studies and in land use planning. Orleans, 10–12 Oct 1996. pp 161–184
Lilly A, Lin HS (2004) Using soil morphological attributes and soil structure in pedotransfer functions. In: Pachepsky Y, Rawls W (eds) Development of pedotransfer functions in soil hydrology. Elsevier, Amsterdam, pp 115–142
Lin HS (2003) Hydropedology: bridging disciplines, scales, and data. Vadose Zone J 2:1–11
Lin HS (2011) Three principles of soil change and pedogenesis in time and space. Soil Sci Soc Am J 75:2049–2070
Lin HS (ed) (2012) Hydropedology: synergistic integration of soil science and hydrology. Academic/Elsevier, Amsterdam
Lin HS, McInnes KJ, Wilding LP, Hallmark CT (1999) Effects of soil morphology on hydraulic properties: II. Hydraulic pedotransfer functions. Soil Sci Soc Am J 63:955–961
Lin HS, Bouma J, Wilding L, Richardson J, Kutilek M, Nielsen D (2005) Advances in hydropedology. Adv Agron 85:1–89
Lin HS, Bouma J, Pachepsky Y, Western A, Thompson J, van Genuchten MT, Vogel H, Lilly A (2006) Hydropedology: synergistic integration of pedology and hydrology. Water Resour Res 42:W05301. doi:10.1029/2005WR004085
Lin HS, Brook E, McDaniel P, Boll J (2008) Hydropedology and surface/subsurface runoff processes. In: Anderson MG (Editor-in-Chief) Encyclopedia of hydrologic sciences. Wiley. doi:10.1002/0470848944.hsa306
Luo LF, Lin HS, Li SC (2010) Quantification of 3-D soil macropore networks in different soil types and land uses using computed tomography. J Hydrol 393:53–64
Manrique LA, Jones CA (1991) Bulk density of soils in relation to soil physical and chemical properties. Soil Sci Soc Am J 55:476–481
Martin MP, Lo Seen D, Boulonne L, Jolivet C, Nair KM, Bourgeon G, Arrouays D (2009) Optimizing pedotransfer functions for estimating soil bulk density using boosted regression trees. Soil Sci Soc Am J 73:485–493
McBratney AB, Minasny B (2004) Soil inference systems. In: Pachepsky Y, Rawls W (eds) Development of pedotransfer functions in soil hydrology. Elsevier, Amsterdam, pp 323–346
McBratney AB, Minasny B, Cattle SR, Vervoort RW (2002) From pedotransfer functions to soil inference systems. Geoderma 109:41–73
McBratney AB, Santos MLM, Minasny B (2003) On digital soil mapping. Geoderma 117:3–52
McKenzie NJ, Jacquier DW (1997) Improving the field estimation of saturated hydraulic conductivity in soil survey. Aust J Soil Res 35:803–825
McKenzie NJ, Ryan PJ (1999) Spatial prediction of soil properties using environmental correlation. Geoderma 89:67–94
Merdun H, Cinar O, Meral R, Apan M (2006) Comparison of artificial neural network and regression pedotransfer functions for prediction of soil water retention and saturated hydraulic conductivity. Soil Tillage Res 90:108–116
Minasny B, McBratney AB (2000) Evaluation and development of hydraulic conductivity pedotransfer functions for Australian soil. Aust J Soil Res 38:905–926
Minasny B, McBratney AB, Bristow KL (1999) Comparison of different approaches to the development of pedotransfer functions for water-retention curves. Geoderma 93:225–253
Minasny B, McBratney AB, Mendonça-Santos ML, Odeh IOA, Guyon B (2006) Prediction and digital mapping of soil carbon storage in the lower Namoi valley. Soil Res 44:233–244
Moore ID, Gessler PE, Nielsen GA, Peterson GA (1993) Soil attribute prediction using terrain analysis. Soil Sci Soc Am J 57:443–452

Mualem Y (1986) Hydraulic conductivity of unsaturated soils: predictions and formulas. In: Klute A (ed) Methods of soil analysis. Part 1, 2nd edn. Agronomy Society of America, Madison, pp 799–823

Nanko K, Ugawa S, Hashimoto S, Imaya A, Kobayashi M, Sakai H, Ishizuka S, Miura S, Tanaka N, Takahashi M, Kaneko S (2014) A pedotransfer function for estimating bulk density of forest soil in Japan affected by volcanic ash. Geoderma 213:36–45

Nordt LC, Jacob JS, Wilding LP (1991) Quantifying map unit composition for quality control in soil survey. In: Mausbach MJ, Wilding LP (eds) Spatial variabilities of soils and landforms, SSSA special publication #28. Soil Science Society of America, Inc., Madison, pp 183–197

O'Connell DA, Ryan PJ (2002) Prediction of three key hydraulic properties in a soil survey of a small forested catchment. Aust J Soil Res 40:191–206

Ocean County Soil Conservation District (OCSCD), Schnabel Engineering Associates, Inc., USDA Natural Resources Conservation Service (2001) Impact of soil disturbance during construction on bulk density and infiltration in ocean county, New Jersey. Ocean County Soil Conservation District, Ocean County. http://www.soildistrict.org/

Pachepsky Y, Rawls WJ (1999) Accuracy and reliability of pedotransfer functions as affected by grouping soils. Soil Sci Soc Am J 63:1748–1757

Pachepsky Y, Rawls W (eds) (2004) Development of pedotransfer functions in soil hydrology. Elsevier, Amsterdam

Patil NG, Rajput GS, Nema RK, Singh RB (2010) Predicting hydraulic properties of seasonally impounded soils. J Agric Sci 148:159–170

Périé C, Ouimet R (2008) Organic carbon, organic matter and bulk density relationships in boreal forest soils. Can J Soil Sci 88:315–325

Prevost M, Marcel P (2004) Predicting soil properties from organic matter content following mechanical site preparation of forest soils. Soil Sci Soc Am J 68:943

Puckett WE, Dane JH, Hajek BF (1985) Physical and mineralogical data to determine soil hydraulic-properties. Soil Sci Soc Am J 49:831–836

Pulleman MM, Bouma J, van Essen EA, Meijles EW (2000) Soil organic matter content as a function of different land use history. Soil Sci Soc Am J 64:689–693

Rawls WJ (1983) Estimating soil bulk density from particle size analysis and organic matter content. Soil Sci 135:123–125

Rawls WJ, Pachepsky YA (2002) Soil consistence and structure as predictors of water retention. Soil Sci Soc Am J 66:1115–1126

Richter DD, Mobley ML (2009) Monitoring earth's critical zone. Science 326:1067–1068

Rode AA (1947) The Soil Forming Process and Soil Evolution. Israel Program for Scientific Translations, Jerusalem (Translated into English by J.S. Joffe, 1961.)

Saini GR (1966) Organic matter as a measure of bulk density of soil. Nature (Lond) 210:1295–1296

Salchow E, Lal R, Fausey NR, Ward A (1996) Pedotransfer functions for variable alluvial soils in southern Ohio. Geoderma 73:165–181

Salifu KFM, Meyer WL, Murchison HG (1999) Estimating soil bulk density from organic matter content, pH, silt and clay. J Trop For 15:112–120

Santra P, Das BS (2008) Pedotransfer functions for soil hydraulic properties developed from a hilly watershed of Eastern India. Geoderma 146:439–448

Saxton KE, Rawls WJ, Romberger JS, Papendick RI (1986) Estimating generalized soil-water characteristics from texture. Soil Sci Soc Am J 50:1031–1036

Schaap MG, Bouten W (1996) Modelling water retention curves of sandy soils using neural networks. Water Resour Res 32:3033–3040

Schaap MG, Leij FJ, van Genuchten MT (2001) Rosetta: a computer program for estimating soil hydraulic parameters with hierarchical pedotransfer functions. J Hydrol 251:163–176

Schaetzl R, Anderson S (2005) Soil – genesis and geomorphology. Cambridge University Press, Cambridge

Sequeira CH, Wills SA, Seybold SA, West LT (2014) Predicting soil bulk density for incomplete databases. Geoderma 213:64–73

Shaw JN, West LT, Radcliffe DE, Bosch DD (2000) Preferential flow and pedotransfer functions for transport properties in sandy Kandiudults. Soil Sci Soc Am J 64:670–678

Soil Survey Staff (1993) Soil survey manual. U.S. Dept. Agri. Handbook No. 18. U.S. Government Printing Office, Washington, DC

Sonneveld MPW, Bouma J, Veldkamp A (2002) Refining soil survey information for a Dutch soil series using land use history. Soil Use Manage 18:157–163

Stewart VI, Adams WA, Abdulla HH (1970) Quantitative pedological studies on soils derived from Silurian mudstones. J Soil Sci 21:248–255

Targulian VO (2005) Elementary pedogenic processes. Eurasian Soil Sci 38:1255–1264

Targulian VO, Goryachkin SV (2004) Soil memory: types of record, carriers, hierarchy and diversity. Revista mexicana de ciencias geológicas 21:1–8

Tietje O, Hennings V (1996) Accuracy of the saturated hydraulic conductivity prediction by pedotransfer functions compared to the variability within FAO textural classes. Geoderma 69:71–84

Tugel AJ, Herrick JE, Brown JR, Mausbach MJ, Puckett W, Hipple K (2005) Soil change, soil survey, and natural resources decision making: a blueprint for action. Soil Sci Soc Am J 69:738–747

Udawatta RR, Anderson SH, Gantzer CJ, Garrett HE (2008) Influence of prairie restoration on CT-measured soil pore characteristics. J Environ Qual 37:219–228

USDA-Natural Resources Conservation Services (NRCS) (1994) State Soil Geographic (STATSGO) data base: data use information. National Coil Survey Center, Lincoln

USDA-Natural Resources Conservation Services (NRCS) (1995) Soil Survey Geographic (SSURGO) data base: data use information. National Coil Survey Center, Lincoln

van Genuchten MTh, Feike L, Wu L (eds) (1999a) Proceedings of the international workshop on characterization and measurement of the hydraulic properties of unsaturated porous media, Riverside

van Genuchten MT, Schaap MG, Mohanty BP, Simunek J, Leij FJ (1999b) Modeling flow and transport processes at the local scale. In: Feyen J, Wiyo K (eds) Modelling of transport process in soils at various scales in time and space. Wageningen Pers, Wageningen, pp 23–45

Vereecken H, Maes J, Feyen J (1990) Estimating unsaturated hydraulic conductivity from easily measured soil properties. Soil Sci 149:1–12

Verhagen J, Bouma J (1998) Defining threshold values for residual N-levels. Geoderma 85:199–213

Vogel HJ, Roth K (2003) Moving through scales of flow and transport in soil. J Hydrol 272:95–106

Wang Y, Shao M a, Liu Z (2012) Pedotransfer functions for predicting soil hydraulic properties of the Chinese Loess Plateau. Soil Sci 177:424–432

Wosten JHM, Lilly A, Nemes A, Le Bas C (1999) Development and use of a database of hydraulic properties of European soils. Geoderma 90:169–185

Wosten JHM, Veerman GJ, de Groot WJM, Stolte J (2001) Water retention and hydraulic conductivity characteristics of top and subsoils of the Netherlands: the staring-series (In Dutch.) ALTERRA Report Nr.153. ALTERRA, Wageningen

Wysocki D, Schoeneberger P, LaGarry H (2000) Geomorphology of soil landscapes. In: Sumner ME (ed.-in-chief) Handbook of soil science. CRC Press, Boca Raton, pp E-5–E-39

Yu HL, Yang PL, Lin HS, Ren SM, He X (2014) Effects of sodic soil reclamation using flue gas desulphurization gypsum on soil pore characteristics, bulk density, and saturated hydraulic conductivity. Soil Sci Soc Am J (In press). doi:10.2136/sssaj2013.08.0352

Zhu AX, Hudson B, Burt J, Lubich K, Simonson D (2001) Soil mapping using GIS, expert knowledge, and fuzzy logic. Soil Sci Soc Am J 65:1463–1472

Zinn YL, Lal R, Resck DVS (2005) Texture and organic carbon relations described by a profile pedotransfer function for Brazilian Cerrado soils. Geoderma 127:168–173

Part II
Data Analysis in Soil Physics and Pedotransfers Functions

Chapter 3
State-Space Analysis in Soil Physics

Ole Wendroth, Yang Yang, and Luís Carlos Timm

Abstract Over the past three decades, state-space models have been used in soil physics mostly to describe spatial processes of transport- or biomass-related state variables. The objective of this contribution on behalf of the second Brazilian Soil Physics Meeting is to provide an introduction into the opportunities of state-space models and to explain their conceptual differences and advantages compared to current widely used analytical approaches that do not account for space/time covariance behavior, measurement or model uncertainty. An overview on the diversity of state-space model applications is provided. The opportunities of state-space models for designing and analyzing experiments with and without treatments especially emphasizing non-randomized experiments are addressed and illustrated. Expanding state-space models for scale-transfer issues and describing spatio-temporal processes simultaneously are pointed out.

Keywords Spatial process • Kalman filter • Scale-dependent treatment design • Spectral analysis

3.1 Introduction

Observations in soils and plants that are growing in soils vary in space and time. If sampled close enough in space, observations show some degree of continuity known as spatially structured behavior. In that case, observations manifest a spatial process at the given sampling scale (Blöschl and Sivapalan 1995). In order to better

O. Wendroth (✉) • Y. Yang
Department of Plant and Soil Sciences, University of Kentucky, Lexington, KY 40546-0091, USA
e-mail: owendroth@uky.edu

L.C. Timm
Rural Engineering Department, Faculty of Agronomy, Federal University of Pelotas, CP 354, 96001-970 Capão do Leão, RS, Brazil

W.G. Teixeira et al. (eds.), *Application of Soil Physics in Environmental Analyses: Measuring, Modelling and Data Integration*, Progress in Soil Science, DOI 10.1007/978-3-319-06013-2_3,

manage soil, crop and the environment, the relationships between different soil and crop spatial and temporal processes need to be diagnosed. In many cases this diagnosis is not trivial and often times complicated because besides those few variables that can be observed over space or time, many others affect the spatial and temporal behavior of processes. It is impossible to experimentally capture all processes affecting each other and to become able to precisely describe the main process of interest based on a collection of observations of a number of state variables.

The purpose of this chapter is to introduce the background for the relevance of state-space analysis by an example, to give an overview on the historical development and a review on the application of state-space analysis in soil physics, and to illustrate some opportunities for applying state-space analysis to reveal important process information from field experiments.

In many cases, agricultural scientists try to derive "well-defined" or unique response functions between input and output, for example, crop yield and a given input or treatment from the relatively small vicinity of an experimental field, while replicate plots with the same treatments are laid out across the field. In a similar fashion, soil scientists observe the behavior of a given state variable within or across mapping units in order to document how a variable differs, e.g., with spatial changes in soil texture. If an obvious response that would be manifested in a correlative relationship and small standard error bars between replicates cannot be obtained, experimental results are often dropped. When studying processes in the larger vicinity of a farmer's field, it is often times found that indeed no uniform response exists or the relationship does not turn out as clear as expected based on our knowledge. This behavior is caused by underlying processes that are not included in the investigation but they effectively modify the response manner between the observed processes across the field.

In Fig. 3.1a, an example of two spatial field processes A_i and B_i is given where subscript i refers to the location along a transect where a series of observations was taken. Both series proceed over a domain of 600 m. In the zone between 10 and 80 m, A_i has a generally decreasing manner while B_i rather fluctuates at the same level. The series are obviously not related in this zone. From 90 m to 220 m, the progresses of both variables are characterized by an overall increase and appear very similar. Whereas over the first 220 m A_i is generally larger than B_i, from 230 to 490 m, their behavior relative to each other is shifted, manifested in all B_i values being larger than A_i in this zone (Fig. 3.1a) indicated also in the plot of their differences relative to each other (Fig. 3.1b). In the final 100 m, both series decrease in general while A_i is again larger than B_i. Summarizing, over specific segments of the domain, we observe that both series are spatially associated. Besides these associations, they change their relative magnitude to each other. This change in magnitude is emphasized by illustrating their differences in Fig. 3.1b. Over wide parts of the domain, the differences between both variables behave uniform. As a consequence of the shifts of their magnitude relative to each other, the scatter diagram shown in Fig. 3.1c does not reveal any relationship between the two series, manifested by an r^2 of 0.004. Despite the local shifts and the low value obtained for the coefficient of determination, the conclusion is drawn that a spatial relationship between both variables exists, however, this relationship is not uniform.

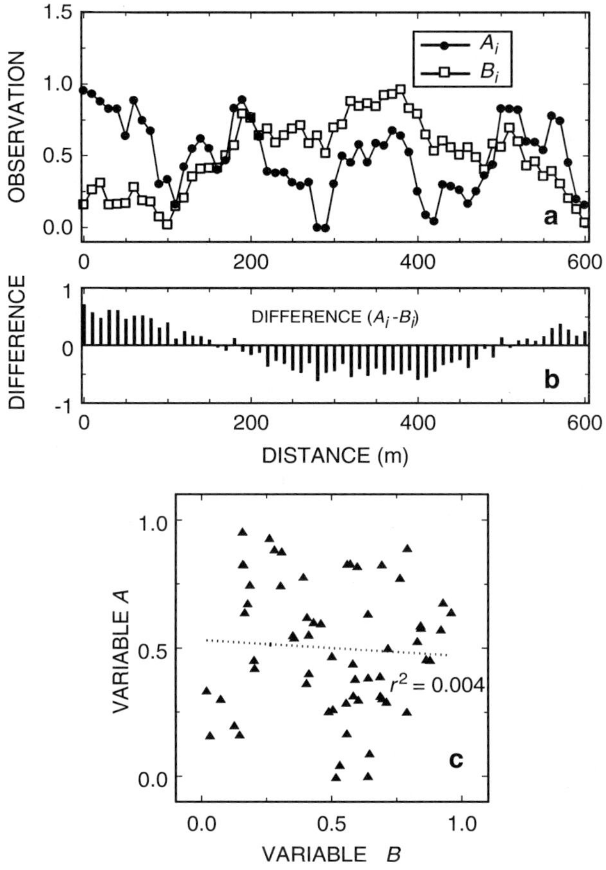

Fig. 3.1 Spatial processes of two series A_i and B_i along a 600-m transect (**a**), their differences between each other (**b**), and their scatter diagram, regression and coefficient of determination (**c**)

The result of the correlation analysis (Fig. 3.1c) suggests no relationship between A_i and B_i. Our intuition though tells us that there exist local relationships but the process behavior of both variables over the entire domain seems to be affected by further spatial processes. Therefore, in the following step the two series are analyzed for their basic spatial behavior.

In Fig. 3.2a, the autocorrelation function is presented according to Shumway (1988)

$$r(h) = \frac{\mathrm{cov}[A_i(x), A_i(x+h)]}{\sqrt{\mathrm{var}[A_i(x)]}\sqrt{\mathrm{var}[A_i(x+h)]}} \tag{3.1}$$

the numerator being the autocovariance of a series for a given lag distance, and the denominator being the product of their variances, resulting in the sample variance. The autocorrelation function indicates the spatial correlation of each variable with itself, observed over increasing lag or separation distance. While the autocorrelation function for series A_i drops in a steep fashion to zero and then systematically

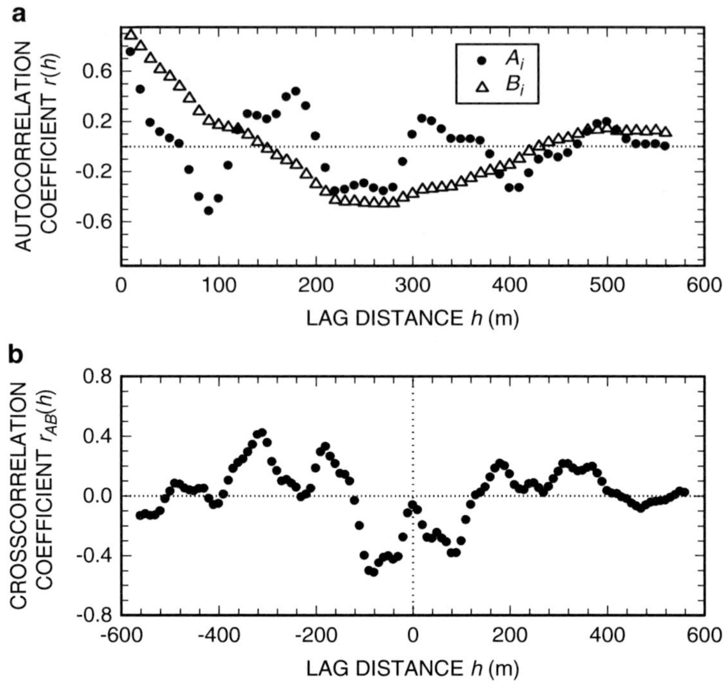

Fig. 3.2 Autocorrelation functions of series A_i and B_i (**a**), and their crosscorrelation function (**b**)

fluctuates between positive and negative correlation coefficients, the decay of the autocorrelation for series B_i is longer and reflects the overall smooth increase and decrease in the original data of series B_i (Fig. 3.1a). The peaks occurring in the autocorrelation function of series A_i at lengths of h = 180 m, around 300 m and 500 m reflect cyclic fluctuations in the original series.

The cross correlation function defined as (Shumway 1988)

$$r_c(h) = \frac{\mathrm{cov}[A_i(x_i), B_i(x_i + h)]}{\sqrt{\mathrm{var}[A_i(x_i)]}\sqrt{\mathrm{var}[B_i(x_i + h)]}} \tag{3.2}$$

is formally very similar to the autocorrelation function, except that the relationship of two processes is considered for different lag distances, and it extends over both positive and negative lag distances. The cross-correlation coefficient at lag distance $h = 0$ equals the Pearson correlation coefficient. Besides the minor relationship between A_i and B_i observed at the same location, i.e., $r_c(h = 0)$, the crosscorrelation function (Shumway 1988) illustrates the existence of both positive and negative correlations observed at different lag distances (Fig. 3.2b). Apparently processes occur across the domain at different scales.

In order to further investigate their scale behavior, their power spectra $S(f)$ – known to integrate the autocorrelation function to the frequency domain – effectively

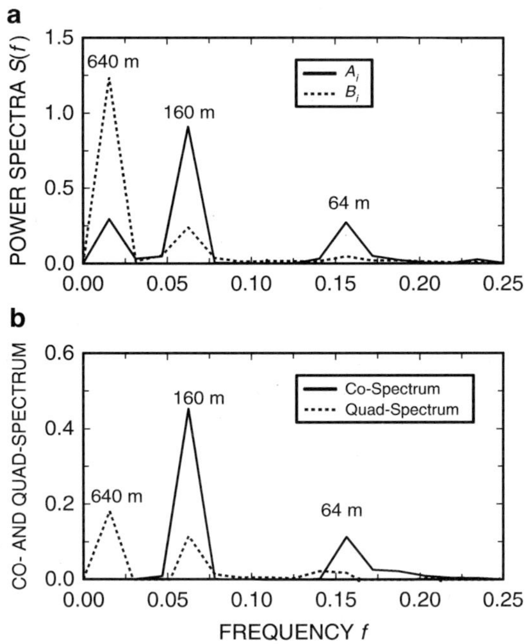

Fig. 3.3 Power spectra of series A_i and B_i (**a**), and their co- and quad-spectra (**b**)

depict variation components occurring at different scales (Kachanoski and De Jong 1988; Nielsen and Wendroth 2003; Wendroth et al. 2011b). Spectra were computed according to Shumway (1988)

$$S(f) = 2\int_0^\infty r(h)\cos(2\pi f h)dh \tag{3.3}$$

with f denoting the frequency being the inverse of the period or wavelength λ. Results are presented in Fig. 3.3a. The main periodic variation component of series B_i is indicated in the peak at a frequency $f = 0.0156$ manifesting the large-scale trend component with a cycle length of 640 m. The wavelength of this variation component is longer than the length of the sampled domain because the sampled part did not cover one complete cycle. This signal reflects the overall increase and decrease of the original series and reveals the shape of the autocorrelation function (Fig. 3.2a). Two smaller peaks occur at $f = 0.0625$ and 0.1563 caused by cyclic variations at shorter wavelengths of 160 and 64 m, respectively. The power spectrum for series A_i can be characterized by dominating cyclic component at 160 m, and less pronounced variations at a larger scale (640 m) and a smaller scale (64 m). The co- and quad-spectrum ($Co(f)$ and $Q(f)$, respectively) computed according to (Shumway 1988).

$$Co(f) = 2\int_0^\infty r_c(h)\cos(2\pi f h)dh \tag{3.4}$$

and

$$Q(f) = 2\int_0^\infty r_c'(h)\sin(2\pi f h)dh, \tag{3.5}$$

respectively, reflect common scales of variation between two series, and their collective consideration informs about their synchronous behavior of any existing phase shifts. In Eqs. 3.4 and 3.5, $r_c(h)$ and $r_c'(h)$ reflect the respective ways of averaging the left and right hand sides of the crosscorrelation function. For further details see Nielsen and Wendroth (2003). Cross-spectra are shown in Fig. 3.3b identifying 160 m as the most common scale of variation between the two series. Notice the positive correlation coefficients at lag distances around h = +160 m and h = −160 m in the crosscorrelation function (Fig. 3.2b).

With this initial spatial analysis, a step is undertaken that not only exhibits the structured, obviously non-random correlation behavior of both series. Moreover, the relationships between the two variables are of a complex nature that cannot be revealed by ordinary Pearson correlation. Having gained an insight into their characteristic variation components prevailing at one common and two different spatial scales, we learn that a relationship between two processes may not necessarily be revealed by applying correlation analysis but by answering the question: Moving from one point to the next, to what extent is the change of one process related to the change of the other, and do the changes in both processes from location i to $i+1$ in fact coincide? Focusing on spatial associations across neighboring locations is an appropriate approach to diagnose the relationship between variables rather than applying mere correlation analysis and ignoring the spatial information contained in the data. As illustrated in the behavior of both processes A_i and B_i shifting their levels relative to each other across different zones is an often existing but unrecognized phenomenon in field data. The existence of statistical methods has also been documented through Figs. 3.2 and 3.3 which are eligible and effective in revealing the presence of spatial relationships between observed series.

The way of considering a spatial relationship inherent in one or between two variables based on their point-to-point process is called *autoregressive*. Autoregressive models have been applied in many different disciplines and areas for analyzing and describing data series in the temporal domain. For observations collected as time series it appears very logic that a system status observed at a time is closely related to the status observed at a previous time step. For time series, autoregressive model descriptions are chosen because they allow handling sudden jumps in the observed series that cannot be explained easily, but without substantially changing the relationship between processes relative to their previous states. Instead, based on the past, they consistently describe the ongoing process after such a jump or irregularity.

Another important consideration in the description of processes is the fact that observations of series are only a limited reflection of the real process. The observed process includes some quantity of observational noise. The quality of the detection of the process depends on the fluctuation of discrete observations. Filtering noise adhered to observations to reveal the process has been a challenge and a motivation to develop mathematical algorithms.

3.2 Development of State-Space Analysis and the Kalman Filter

More than 50 years ago, Kalman (1960) and Kalman and Bucy (1961) introduced a new theory for a filtering and prediction technique of serial observations based on the concept of state and state-transition, i.e., first-order differential equations. The new concept was called state-transition method and was able to capture stationary and non-stationary processes. While the process of a series was described by estimating its state, the error of the estimation was quantified simultaneously. Moreover, the observation of the system state was affected by a random error and did not have to be assumed completely true during the estimation of coefficients in the state-equation. Hence, this technique was first applied in systems where the change of state was described while both the observation and the model error were separated from the signal, i.e., the observed series was filtered.

The simple concept of analyzing and describing a system's state from one point in time or in space to the next has long ago been applied in hydrologic sciences, economic disciplines etc., and it is extremely relevant for soil science, agronomy and related disciplines when plant and soil processes are analyzed in spatial or temporal domains. Rather than trying to estimate or predict the state of a system at a given point in time or space based on a set of input information for the same point – a goal that is impossible to be reached successfully unless there is a huge amount of input information available – the relevant question is now: Given the history of the development of the system's status, i.e., two or more processes that affect each other, what is the change of one variable from one point to the next with the given change of another underlying variable? The approach based on this type of question results in satisfactory predictions of the process despite limited input. To some portion, unsampled information is already integrated by basing the prediction on the previous system's state. In other words, at the local scale, two series may proceed while their changes are closely related with each other. At a larger scale their relationship changes due to some other factor interfering and altering the level of one or both variables and their way of coinciding with each other. In this case, a classical correlation analysis indicates a weak relationship, while state-space analysis captures the variables' association despite other factors altering their relationship.

As the example of series A_i and B_i indicated above, their correlation coefficient was very small (Fig. 3.1c). In the following, both series were analyzed in a state-space approach to reveal their spatial relationship. Shumway and Stoffer (1982) implemented the Kalman Filter in an autoregressive state-space model. Statistical parameters were estimated in the expectation maximization (EM) algorithm.

One of the key elements of a state-space model is the state-equation

$$Z_i = \mathbf{\Phi} Z_{i-1} + \omega_i \tag{3.6}$$

that can have the form of an autoregressive model in which the state vector Z_i at location i is related to its previous state at i-1 through the transition matrix $\mathbf{\Phi}$. Only to a limited extent can the process be described appropriately by this autoregressive model type and by the variables included in the vector because the model's reflection of the reality is limited, and the number of variables included in the state vector remain incomplete. The difference between the true state of the vector and the estimated value is the model error ω_i with covariance matrix Q. This error is assumed to be zero mean and uncorrelated. Owing to the fact that besides the model error, the system's description is affected also by an observational error, the other element necessary in a state-space system is the observation equation:

$$Y_i = M_i Z_i + \upsilon_i \tag{3.7}$$

Through this equation, the true state of the system Z_i is embedded in the observation through a measurement or design matrix M_i. The rationale for this equation is the fact that any observation we take is associated with an observational or measurement error υ_i. Similar to the model error, the measurement error is a zero mean uncorrelated process with variance R. In the EM algorithm, the autoregressive coefficients in the transition matrix, the initial state X_0 and its variance P_0, the model and measurement variances Q and R are optimized simultaneously. This optimization occurs by minimizing the differences between predictions and observations. The steps taken so far in the analysis are the ones commonly used in parameter estimation in ordinary autoregressive models. The advantage of the Kalman filter implemented in the autoregressive model relies on the fact that before the next step is predicted, the current state is statistically updated by weighting the current prediction against the current observation. This weighting coefficient is called Kalman gain K_g (Gelb 1974; Katul et al. 1993) and its magnitude depends on the measurement and the model variance. If relative to the model variance, the measurement variance is small, more weight will be on observations. If on the other hand, observations are noisy while the model is relatively reliable, more weight is put on the prediction. For detailed descriptions of the filter equations, see Shumway (1988), Katul et al. (1993) and Nielsen and Wendroth (2003). Notice, the approach to consider different error components in spatial or temporal process description bears the opportunity that in case of considerable measurement error, a relatively reliable model can help to identify and capture the process. And on the other hand, for many cases in which the model description remains incomplete, a sound measurement scheme can compensate for conceptual shortcomings in the underlying model theory.

After completing a forward recursion step, the result is a set of predictions, based on the autoregressive relationship and the updating step in the Kalman filter at each location where an observation exists. Before the next iteration, the predicted status

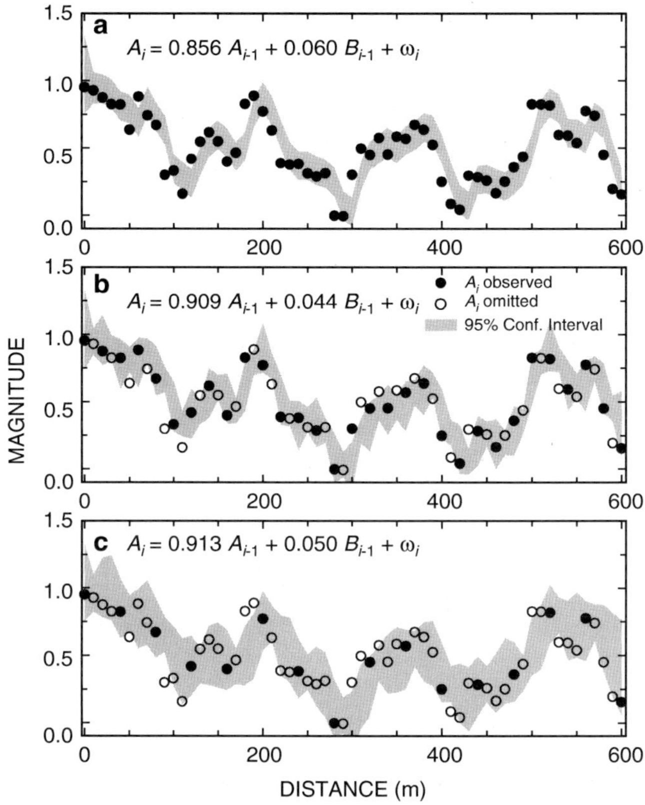

Fig. 3.4 Autoregressive state-space models for series A_i based on A_i and B_i. In (**a**), all observations of A_i and B_i were considered, in (**b**), all observations of B_i and every other observation of A_i, and in (**c**) all observations of B_i and every fourth observation of A_i

is processed in another filter, called the smoothing gain in backward recursions while the objective function, i.e., the log-likelihood, is computed already after completing the forward recursion and before the backward smoothing recursions.

This algorithm was applied to series A_i and B_i. Prior to the state-space analysis, both series A_i and B_i were normalized using (Eq. 10.21 in Nielsen and Wendroth 2003).

$$ysc_i = \frac{y_i - \left(\mu_y - 2\sigma_y\right)}{4\sigma_y} \tag{3.8}$$

where the scaled value ysc_i is computed based on the observed value y_i, the mean of the series μ_y, and its standard deviation σ_y (Nielsen and Wendroth 2003). In the three estimation scenarios presented in Fig. 3.4, the observation density of series A_i was modified. In Fig. 3.4a all observations shown in Fig. 3.1a were included. The process of observed series A_i is well captured by the 95 % confidence range of the state-space model estimation. Leaving out every other observation in series A_i the state-space model still captures the process of the entire series well, while the confidence interval of estimation increases (Fig. 3.4b). A further increase in the confidence interval occurs when only every fourth observation is considered by the model, i.e., the prediction is updated only for one out of four observations (Fig. 3.4c). Since the observations of both series fluctuate between 0 and 1 with a mean of 0.5 and a standard deviation of 0.25, the coefficients in the state equation reflect approximately the percentage contribution of both variables to the estimate. The contribution of process A_i itself is approximately 93 %, the one for B_i around 7 % (Fig. 3.4a). The contribution does not change substantially among the three different scenarios.

3.3 State-Space Applications in Soil Physics

Morkoc et al. (1985) were the first who applied autoregressive state-space models for describing the spatial process of soil water content and soil temperature in a field soil that had been irrigated with "low-salt" and "high-salt" water in two neighboring sprinkler lines. In the center of this field, the irrigated water with different salt content that was applied from the two sprinkler lines overlapped each other. Along the transect perpendicular to the irrigation treatments, both series showed inverse spatial behavior, i.e., in zones where soil water content gradually increased from one location to the next, soil surface temperature decreased, and vice versa. While the negative correlation between soil water content and soil temperature was obvious although not very tight, the process of both series could be precisely estimated in state-space analysis. Morkoc et al. (1985) studied different spatial sampling scenarios and showed that the process of soil surface temperature could be well captured based on temperature and water. The process of soil water was also precisely described based on observations of soil water content and surface temperature at previous locations. This study revealed opportunities for sampling decisions in presence of measurement noise, spatial variability and model uncertainty. While in this study, a large-scale process caused by the different irrigation water qualities and the overlapping zone was underlying the spatial behavior of soil water content and temperature, autoregressive state-space models were applied in a number of studies in which no obvious large-scale process existed.

Nielsen and Alemi (1989) were the first who described the spatial process of crop yield as affected by underlying processes using an autoregressive state-space approach. In their example, besides other factors nematode infestation strongly affected cotton yield. The authors conceded that factors other than nematode infestation had a strong impact on cotton yield. However, already focusing on the process of nematode infestation allowed a smooth estimation of the spatial cotton yield distribution.

Soil water is known to vary in time and space. Or and Hanks (1992) developed a space-time state-space method in order to describe soil water dynamics in space and time. In their approach, soil moisture status and its associated errors were predicted in time for a number of locations in a field. For times, when spatial measurements existed, these measurements were kriged. The resulting spatial estimates and their associated errors were treated as measurements and measurement errors which were then used to update the soil water status estimated in the temporal domain. Although this approach has not been widely applied, it is a very promising and effective way to model space-time water fields, and for designing measurement schemes. The increasing availability of remote sensing scenes, makes the combination of spatial and temporal domains in predictions very attractive, not only for predicting soil water but also other relevant processes such as biomass production.

The analysis of spatial variation of crop development and yield has been a challenge for a very long time. Its deterministic description is a task that would be extremely hard to accomplish because of process interactions and spatially varying relationships of crop growth and the underlying processes. Empirical descriptions based on multiple regression approaches are successful only to a very limited extent as numerous studies have shown (e.g., Bruce et al. 1990). A very complex issue in such approaches is the spatial representativity of measurements, manifesting their usefulness as a covariate. In summary, to capture all biotic and abiotic processes in the life of a plant to explain why in one location of a field the crop grows faster and produces higher yields than in another location is an endeavor that has very limited or no success. Instead, integrating process behavior in relatively simply observable, indicative state variables and observing how one process behaves in the spatial or the temporal domain and how it is related to other processes is a very promising way, even in the presence of unsampled processes that change the response behavior between variables as illustrated in the example above (Fig. 3.1).

In order to quantify symbiotic nitrogen fixation of legume crops, it is a common approach to grow a reference crop with a similar vegetation period. Alternatively, the more expensive but more precise isotope dilution method can be applied. In 1984, a field experiment in a variable field soil was conducted in Seibersdorf, Austria, for the purpose of quantifying nitrogen fixation from the atmosphere (Reichardt et al. 1987; Kirda et al. 1988). Due to the pronounced spatial variability in this field soil, classical statistical approaches revealed huge variation interfering with the usual nitrogen fixation quantification and prohibiting a trustworthy mean. When the authors identified variables indicative of soil variation and reflecting the yield variation, and implemented them in the same state-space approach as used by Morkoc et al. (1985), the spatial variability of both the test and the reference crop was described (Wendroth et al. 1992). In order to achieve this result, for the quantification of the supply of plant-available nitrogen for the reference crop, and of the activity of symbiotically fixing bacteria the volumetric stone content was considered. The stone content varied considerably in space and reached a magnitude of up to 50 %. In zones where the stone content was high, the volume of soil was limited by stones, and the inorganic nitrogen was more rapidly consumed by the reference crop ryegrass than in zones with no or few stones. Therefore, the spatial pattern of ryegrass yield corresponded negatively to the abundance of stones. In contrast, the N-fixing alfalfa crop yielded even high in zones with large volumes of stones. In these zones, inorganic nitrogen might have been taken up quickly but symbiotically fixed N secured N supply. For sure, the stone content did by far not capture all processes important for ryegrass and alfalfa growth. However, it integrated several major yield-limiting processes into one indicative state variable that was efficient for the description of spatial yield variability in both crops (Wendroth et al. 1992).

Katul et al. (1993) combined a simple physically-based soil water transport equation with a non-linear Kalman filtering approach to inversely estimate empirical coefficients of the hydraulic conductivity – water content relationship ($K(\theta)$). In this estimation procedure, the state of the soil water storage in a particular soil layer of an internal drainage experiment and the associated state variance were described in short time steps. Each time, an observation became available, the system's state and its variance were updated with the Kalman gain. The two main sources of uncertainty, i.e., the measurement variance and the model variance were estimated simultaneously. The latter reflects the limitations of the simple transport equation, and the two-parameter form of Gardner's (1958) hydraulic conductivity function. This estimation scheme was later adapted for multiple soil layers (Wendroth et al. 1993). In addition, dividing the time series in two domains related to the drainage processes through structural macropores, and through finer matrix pores increased the precision in the estimation of soil water content time series. In these two contributions, the opportunity of compensating limitations in the model through more frequent measurements became obvious. On the other hand, limitations in measurements, i.e., with respect to their frequency and quality, can be compensated to some degree by a simple state equation and the associated variance propagation.

Assouline (1993) implemented a Kalman filter approach for simultaneously solving the water, heat and salt balance equations for the estimation of evaporation as one of the most important water balance components of Lake Kinneret. Each of the elements of the water balance equations was associated to an error. These uncertainties were utilized for estimating the system's state of water balance components on a monthly basis. Assouline (1993) pointed out that the use of the Kalman filter improved the estimation of individual balance components and their temporal behavior over the study period of 5 years substantially.

Spatial relationships between a very important soil structural property, i.e., aggregate stability, soil organic matter content, clay content and soil water content were derived by Timm et al. (2004) for a 84-m-long transect. They compared their results to those obtained with multiple linear regression analysis and found higher coefficients of determination between measurements and estimations obtained with the state-space analysis than ordinary regression. Timm et al. (2004) pointed out that the difference in estimation quality was caused by employing neighboring observations and their relationships in the analysis instead of assuming observations were distributed randomly. The other reason for the higher estimation quality of the state-space technique is the consideration of observation and model noise and their inclusion in the updating step in the Kalman filter.

3.4 Recent Applications of State-Space Analysis

Remember, in the original work of Morkoc et al. (1985), spatial processes of soil water content and soil temperature were described for a domain that received treatments of water with different levels of salt concentration. State-space analysis can generally be applied in studies of spatial relationships across a domain that is treated uniformly or has received no treatment at all. In this case the inherent variation of known and unknown processes and their relationships is quantifiable with state-space analysis. State-space analysis can also be applied in systems that have received one or several treatments at the same or at different spatial scales (Wendroth et al. 2011b). In this case, treatments can be considered as spatial boundary conditions changing across the spatial domain. Schwen et al. (2013) investigated a system in which the influence of different land use and rainfall characteristics on bromide leaching and soil moisture state variables was studied. As will be discussed later, a specialty of the experimental layout was the cyclic arrangement of treatments across the experimental field. With their analysis, Schwen et al. (2013) showed that the change of soil water pressure head before and after the irrigation experiment strongly depended on land use and the irrigation water applied at different amounts in different parts of the field. Moreover, for different measurement resolutions, they described the vertical center of Br^- mass used as a leaching depth indicator as a function of center of mass, land use, irrigation amount and intensity at the previous location. Besides this state-space description based on experimental boundary conditions, Schwen et al. (2013) also conducted a scenario analysis of relevant soil variables affecting the leaching behavior of bromide. Change in soil water status, bulk density and clay content – three indicative variables for complex processes – were efficient in the description of spatial bromide leaching behavior. The best spatial description of bromide leaching resulted from the combination of irrigation amount, bulk density and soil clay content. This study reveals that underlying soil spatial variability – here manifested in changing soil clay content – does not have to be considered as a disturbing phenomenon in the identification of treatment effects but is part of the spatial leaching process description.

Based on another field study on solute leaching, Yang et al. (2013) conducted a detailed analysis of the spatial processes of bromide concentration across an area with different land use in which particular zones received the same amount of precipitation however at different intensities. Bromide concentration results were scaled according to Eq. 3.8, and displayed for ten 10-cm layers along a 48-m transect (Fig. 3.5). Besides identifying the impact of different surface boundary conditions on bromide distribution in the profile and across the domain using state-space analysis, Yang and Wendroth (2014) compared the relationships between the spatial bromide concentration distributions in different layers. Parts of the results of their state-space analysis of scaled bromide concentration processes are displayed in Table 3.1. In the left side, the bromide concentration at a particular soil depth is described based on the layer directly above. The average squared deviation (*SD*)

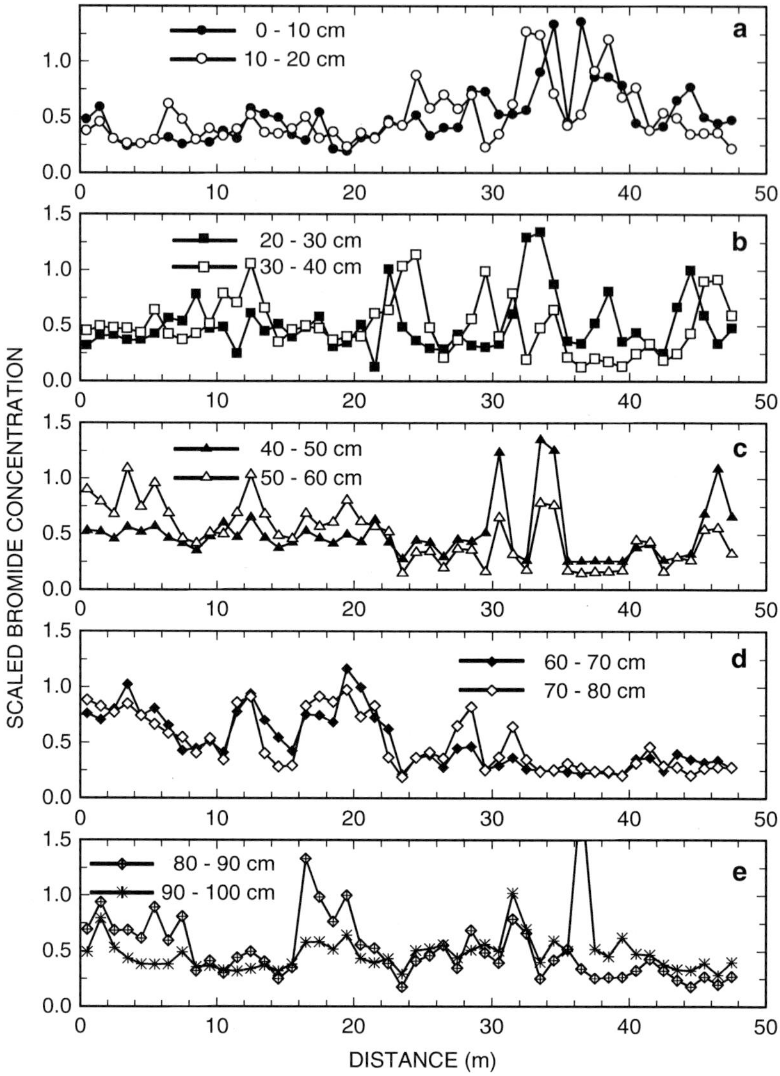

Fig. 3.5 Scaled bromide concentration measured along a 48-m- transect for ten 10-cm depth increments

between observed and estimated scaled bromide concentration and the corrected Akaike information criterion (*AICc*) were applied as quality criteria of the estimation (Shumway and Stoffer 2000). During the systematic analysis of spatial bromide processes, it became apparent that the bromide concentration in the layer above was not the best covariate for describing the spatial bromide process in a particular layer. Yang et al. (2013) found that the best covariate supporting the estimation of bromide distribution in the upper 50 cm of the profile was the bromide

Table 3.1 Autoregressive state-space equations for describing the spatial process of bromide concentration at different layers

Depth (cm)	Autoregressive state-space equation (based on layer above)	*SD*	*AICc*	Autoregressive state-space equation (based on flow domain)	*SD*	*AICc*
				Upper flow domain		
10–20				$Br15_{\mathrm{i}} = 0.542 Br15_{i-1} + 0.421 Br5_{i-1} + \omega_i$	0.0009	−5.915
20–30	$Br25_{\mathrm{i}} = 0.610 Br25_{i-1} + 0.319 Br15_{i-1} + \omega_i$	0.0051	−4.142	$Br25_{\mathrm{i}} = 0.497 Br25_{i-1} + 0.447 Br5_{i-1} + \omega_i$	0.0009	−5.847
30–40	$Br35_{\mathrm{i}} = 0.711 Br35_{i-1} + 0.276 Br25_{i-1} + \omega_i$	0.0058	−4.009	$Br35_{\mathrm{i}} = 0.830 Br35_{i-1} + 0.132 Br5_{i-1} + \omega_i$	0.0042	−4.341
40–50	$Br45_{\mathrm{i}} = 1.137 Br45_{i-1} - 0.143 Br35_{i-1} + \omega_i$	0.0506	−1.847	$Br45_{\mathrm{i}} = 0.760 Br45_{i-1} + 0.206 Br5_{i-1} + \omega_i$	0.0205	−2.751
				Lower flow domain		
60–70				$Br65_{\mathrm{i}} = 0.903 Br65_{i-1} + 0.042 Br55_{i-1} + \omega_i$	0.0003	−6.983
70–80	$Br75_{\mathrm{i}} = 0.579 Br75_{i-1} + 0.342 Br65_{i-1} + \omega_i$	0.0008	−5.999	$Br75_{\mathrm{i}} = 0.177 Br75_{i-1} + 0.786 Br55_{i-1} + \omega_i$	0.00005	−8.711
80–90	$Br85_{\mathrm{i}} = 0.572 Br85_{i-1} + 0.354 Br75_{i-1} + \omega_i$	0.0044	−4.285	$Br85_{\mathrm{i}} = 0.375 Br85_{i-1} + 0.591 Br55_{i-1} + \omega_i$	0.0017	−5.261
90–100	$Br95_{\mathrm{i}} = 0.996 Br95_{i-1} - 0.007 Br85_{i-1} + \omega_i$	0.0450	−1.967	$Br95_{\mathrm{i}} = 0.840 Br95_{i-1} + 0.146 Br55_{i-1} + \omega_i$	0.0345	−2.229

In the left column, Br concentration is described based on the previous observation of Br concentration in the same and in the previous layer. In the right hand side, Br concentration is based on the previous observation of Br concentration in the same layer and in the top layer of the flow domain

SD Average of squared differences between measured and estimated concentration values. *AICc* refers to the corrected Akaike Information criterion. The number behind *Br* refers to the center of the depth interval. Hence, *Br*25 refers to Bromide concentration in the layer 20–30 cm

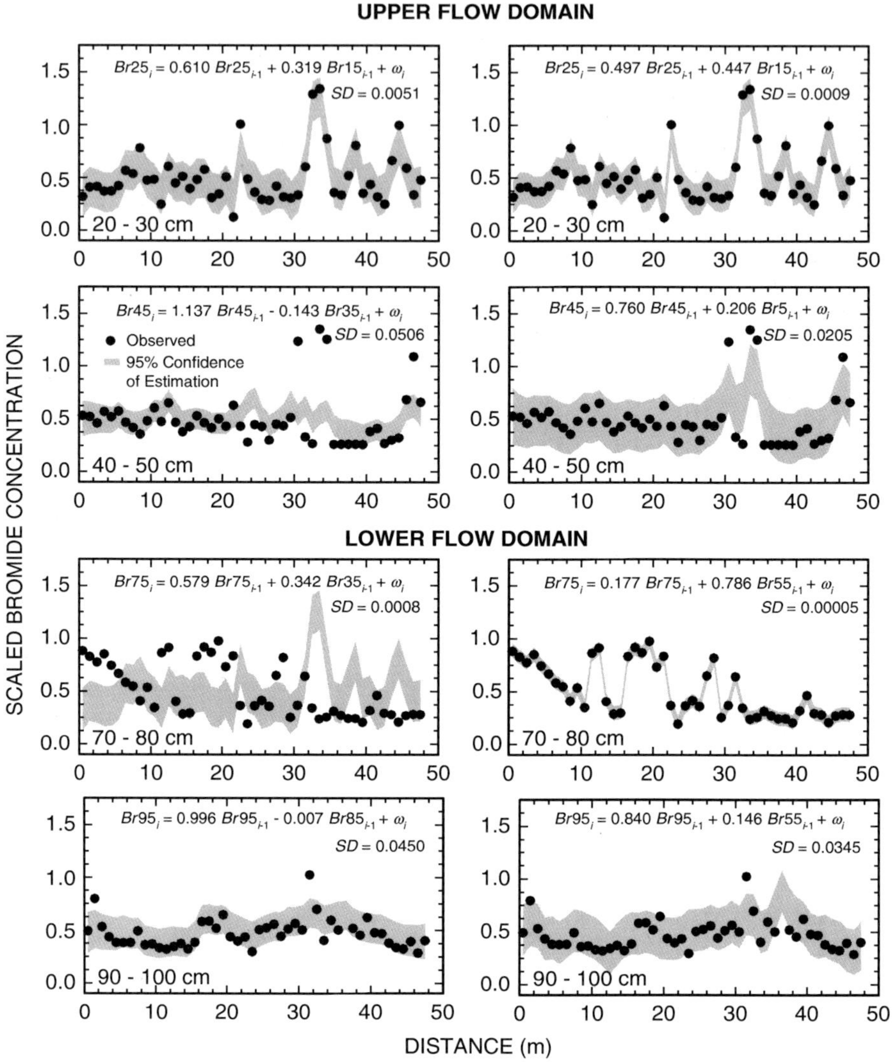

Fig. 3.6 Autoregressive state-space model results for selected layers. On the *left side*, the concentration in the layer directly above was the co-variate, on the *right side*, layers in the top of the flow domain were the covariates. The *number behind Br* indicates the center depth of the respective compartment, i.e., *Br*25 refers to Br concentration in the 20–30 cm depth

concentration process in the top 0–10 cm layer (Table 3.1 and Fig. 3.6). Moreover, for describing the spatial bromide process in the deeper layers, not the surface layer but the 50–60 cm layer was the most effective covariate in the state vector (Table 3.1 and Fig. 3.6). In the left hand side of Fig. 3.6, scaled bromide concentration distribution as estimated using state-space analysis is displayed for the 20–30, 40–50, 70–80, and the 90–100 cm depth, while these and the respective layers directly above were included in the state vector. In contrast, state-space results shown for the same layers in the right hand side of Fig. 3.6 are based on the top layers in the respective flow domains, i.e., 0–10 cm and 50–60 cm depth, respectively. For the first time, state-space analysis revealed that the flow processes in this field soil were governed by two different flow domains, i.e., a domain in the upper 50 cm, and a domain below 50 cm depth. Apparently, with water and solutes passing the 40–50-cm depth in this field, the flow dynamics changed. A possible reason for this transition could be the fact, that this soil had been ploughed several years before this experiment, and this particular depth still exhibits some compaction phenomena. On the other hand, a soil textural change observed at the same depth might be another cause for flow and transport exhibiting different patterns in the upper and in the lower domain.

3.5 Additive State-Space Model – Opportunities for Treatment Experiments in Variable Field Soils

In the majority of applications, autoregressive state-space models have been utilized for describing first-order autoregressive processes. In a few cases, spatial series were described as second-order processes. Shumway (1988) provided modifications in the coding of the EM algorithm, and Nielsen and Wendroth (2003) presented the methodology and its application. Wendroth et al. (2011a) described the spatial distribution of bromide concentration in the soil as a second-order process. Shumway (1988) illustrated further examples of models that emphasized any particular lag in the description of a spatial series. Such models showed to be efficient if any lag other than the first lag is meaningful for estimating the series. A special case of autoregressive state-space models can be applied for the analysis of series that exhibit cyclic variations, which are in addition affected by underlying trends. Shumway (1988) illustrated such a specific state-space model, i.e., an additive state-space model, with the analysis of a time series of quarterly income data. This series consisted of annual quarterly cycles, underlain by a long term trend. In order to analyse such a series, the two main components, i.e., the trend and the cyclic components are separated. Depending on the priorities, once the variation of the series has been decomposed into its components, these components and their causes can be further addressed in ongoing analysis.

In a spatial additive state space model, the observed process y_s is defined as (Shumway 1988)

$$y_s = x_{s1} + x_{s2} + \nu_s \tag{3.9}$$

where x_{s1} and x_{s2} are two spatial components, and ν_s is the zero mean uncorrelated observation error. The first component x_{s1} is a trend described as a first order univariate autoregressive process (Shumway 1988)

$$x_{s1} = \phi x_{s-1,1} + w_{s1} \tag{3.10}$$

and the second component is the cyclic process

$$x_{s2} = -x_{s-1,2} - x_{s-2,2} - \ldots - x_{s-(k-1),2} + w_{s2} \tag{3.11}$$

where k is the length of the cycle. In this equation x_{s2} is composed of k-1 elements of the cycle because at k, a new cycle begins. The errors w_{s1} and w_{s2} are zero-mean uncorrelated errors of both state equations. Shumway (1988) provided details of the coding of the additive autoregressive version of the EM algorithm.

In a field-scale tracer experiment, Wendroth et al. (2011b) applied different rainfall characteristics as treatments in a non-random, cyclic fashion, similar to the experiments by Bazza et al. (1988) and Shillito et al. (2009). This design allowed studying the impact of rainfall amount, intensity and application time delay at different scales in a variable field soil.

As a result, large and small-scale components of bromide tracer leaching depth were successfully depicted. The short periods of variation in leaching depth were assigned to the application time delay, described as process x_{s2}. The large-scale component of the process x_{s1} was described subsequently in a multivariate autoregressive state-space model, in which precipitation amount and intensity, varying at large scales were identified as the main underlying processes.

A very common agronomic topic of interest is the response of crop yield to the amount of nitrogen fertilizer applied. To quantify the yield response function to the nitrogen fertilizer rate, experiments are usually conducted with treatments and their replicates being spatially randomized. In the presence of substantial soil variation, an effect of N rate on yield may not be identified. Here, we provide an example where nitrogen treatments are not randomly laid out but in a periodic pattern. Six rates between 0 and 168 kg/ha were applied in plots of 30 m width and 10 m length in a farmer's field in Kentucky. The sinusoidal pattern of nitrogen application rate is displayed with the associated yield of winter wheat in Fig. 3.7. Yield data fluctuate in a periodic shape similar to that of the nitrogen application rate. A tendency exists that at distances >200 m, the wheat yield in general decreases compared to the zone at distances <200 m. In the following analysis, the periodic yield variations should be separated from the trend component of another process that is unknown.

In Fig. 3.8, the wheat grain yield, the cyclic component caused by the fertilizer application pattern x_{s2}, and the large-scale component x_{s1} are shown. While the cyclic series x_{s2} proceeds at a similar level across the field with estimations fluctuating in the same fashion as the fertilizer application pattern, the large-scale component x_{s1} reflects the decrease of the overall process at distances >200 m. The fertilizer response can easily be grasped from this illustration. Now, the reason for the large scale process x_{s1} is of particular interest. Among several variables sampled in this field, the following three series display a similar pattern compared to that of x_{s1}: Elevation (Fig. 3.9a) and clay content at 0–30 and 60–90 cm depth (Fig. 3.9b), while the clay in 60–90 cm shows a closer relation to x_{s1} than in 0–30 cm. The elevation and the clay content at the lower depth are employed in a multivariate autoregressive state-space model to describe the trend component in

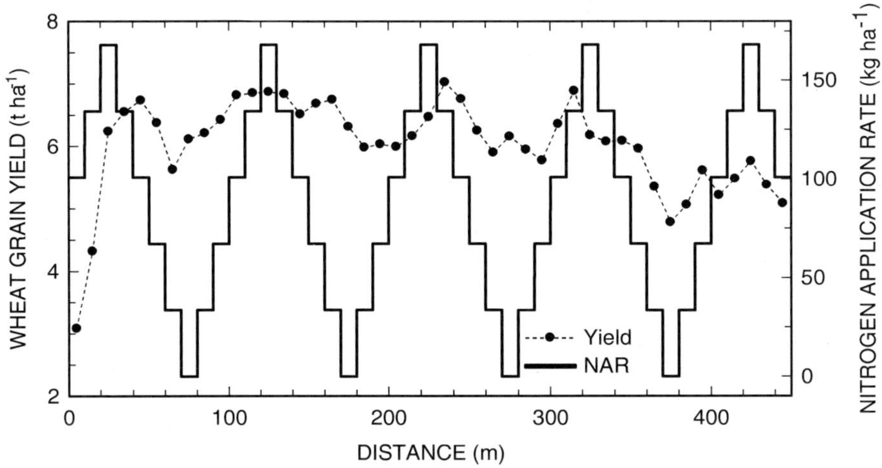

Fig. 3.7 Nitrogen application rate (NAR) and wheat grain yield along a 450-m transect in a farmer's field in western Kentucky

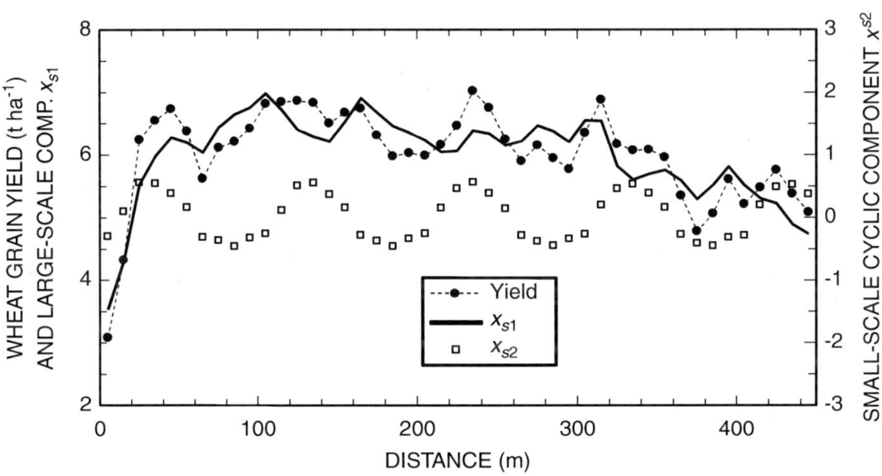

Fig. 3.8 Winter wheat grain yield, large-scale and small-scale yield components. These components were quantified in additive state-space models along a transect in a farmer's field in western Kentucky

the yield response to nitrogen fertilizer application rate. The process of x_{s1} could be satisfactorily described (Fig. 3.10a), and in a case where only one of four known values x_{s1} were given to the model, the clay in 60–90 cm depth supported the estimation of x_{s1} along the catena (Fig. 3.10b).

One of the conclusions of experiments with cyclic treatment arrangements is that treatment experiments can be conducted in heterogeneous landscapes. Underlying influences that cause the response behavior to be non-uniform can be depicted in a

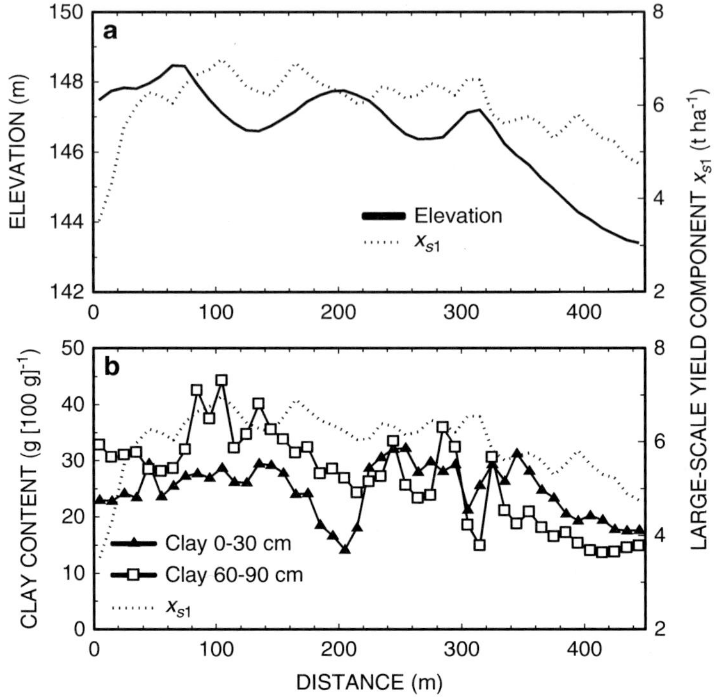

Fig. 3.9 Large-scale yield component and elevation (**a**) and clay content in 0–30 and 60–90 cm depth (**b**) along a transect in a farmer's field in western Kentucky

Fig. 3.10 Autoregressive state-space models of the scaled large-scale component of wheat yield based on all observations of the large-scale component, elevation and the clay content in 60–90 cm (Clay90) (**a**), and based on every fourth large-scale component observation and all observations of Clay90 (**b**)

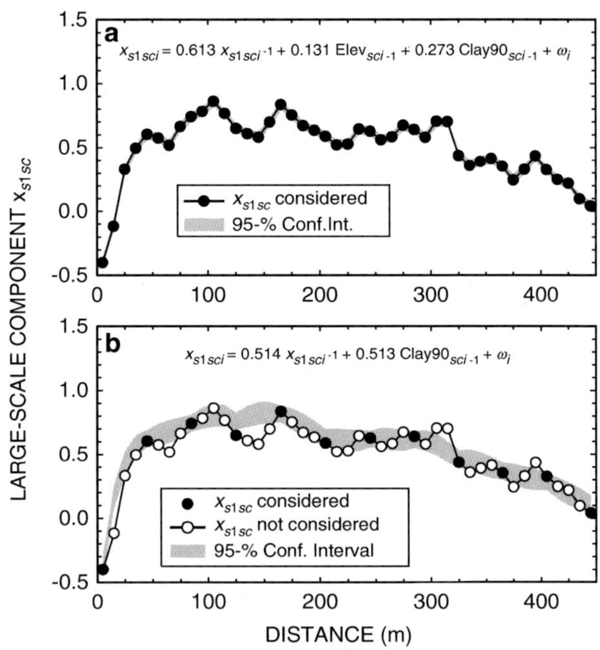

combination of additive and autoregressive state-space models. In the majority of cases, a homogeneous soil to conduct a randomized experiment simply does not exist. Moreover, the approaches outlined and explained here based on examples amplify the opportunity of conducting experiments in landscapes and farmers' fields. The studies above and those conducted by Shillito et al. (2009), Wendroth et al. (2011b), Schwen et al. (2013), and Yang et al. (2013) present opportunities to establish field experiments in variable landscapes and to apply adequate statistical methods that support the diagnosis of such systems despite the presence of underlying soil trends.

3.6 Outlook

Basing the diagnosis of spatial and temporal series on differential equations implies huge opportunities for the design and the analysis of studies with and without treatments. The opportunity to simultaneously estimate model parameters and model and observation uncertainties allows to combine simple models and observations for an effective prediction and updating scheme in the description of soil water processes, biomass development, soil carbon dynamics, various metabolic reactions and their dependence on water and temperature. Accounting for observation and estimation uncertainties allows stochastic integration of specific processes in indicative state variables. The initial example on the analysis of processes A_i and B_i suggests that decomposing variation scales in frequency-based approaches followed by autoregressive state-space models based on neighborhood relationships between state variables is a very promising way to a better understanding of field processes and their variability. The opportunities implied in state-space models as up- and downscaling scenarios of spatial and temporal processes have to be explored. State-space models with their updating and smoothing steps have a great potential to analyze space-time fields of soil and crop state variables, and help to identify the dominating processes in both time and space domain. Pedotransfer functions – developed to provide regional information – do not include yet spatial or temporal covariance information. Implementing coregionalization techniques such as state-space models in the multivariate estimation of regional transport parameter distribution would elevate the relevance and precision of pedotransfer functions.

References

Assouline S (1993) Estimation of lake hydrologic budget terms using simultaneous solution of water, heat, and salt balances and a Kalman filtering approach: application to Lake Kinneret. Water Resour Res 29:3041–3048

Bazza M, Shumway RH, Nielsen DR (1988) Two-dimensional spectral analyses of soil surface temperature. Hilgardia 56:1–28

Blöschl G, Sivapalan M (1995) Scale issues in hydrological modelling: a review. Hydrol Process 9:251–290

Bruce RR, Snyder WM, White AW, Thomas AW, Langdale GW (1990) Soil variables and interactions affecting prediction of crop yield. Soil Sci Soc Am J 54:494–501

Gardner WR (1958) Some steady-state solutions of the unsaturated moisture flow equation with application to evaporation from a water table. Soil Sci 85:228–232

Gelb A (1974) Applied optimal estimation. Massachusetts Institute of Technology Press, Cambridge, MA, 374 pp

Kachanoski RG, De Jong E (1988) Scale dependence and the temporal persistence of spatial patterns of soil water storage. Water Resour Res 24:85–91

Kalman RE (1960) A new approach to linear filtering and prediction problems. Trans ASME J Basic Eng 8:35–45

Kalman RE, Bucy RS (1961) New results in linear filtering and prediction theory. Trans ASME J Basic Eng 83:95–108

Katul GG, Wendroth O, Parlange MB, Puente CE, Nielsen DR (1993) Estimation of in situ hydraulic conductivity function from nonlinear filtering theory. Water Resour Res 29:1063–1070

Kirda C, Hardarson G, Zapata F, Reichardt K (1988) Spatial variability of root zone soil water status and of fertilizer N uptake by forage crops. Soil Technol 1:223–234

Morkoc F, Biggar JW, Nielsen DR, Rolston DE (1985) Analysis of soil water content and temperature using state-space approach. Soil Sci Soc Am J 49:798–803

Nielsen DR, Alemi MH (1989) Statistical opportunities for analyzing spatial and temporal heterogeneity of field soils. Plant Soil 115:285–296

Nielsen DR, Wendroth O (2003) Spatial and temporal statistics – sampling field soils and their vegetation. Catena, Reiskirchen, 416 p

Or D, Hanks RJ (1992) Spatial and temporal soil water estimation considering soil variability and evapotranspiration uncertainty. Water Resour Res 28:803–814

Reichardt K, Hardarson G, Zapata F, Kirda C, Danso SKA (1987) Site variability effect on field measurement of symbiotic nitrogen fixation using the 15N isotope dilution method. Soil Biol Biochem 19:405–409

Schwen A, Yang Y, Wendroth O (2013) State-space models describe the spatial variability of bromide leaching controlled by land use, irrigation, and pedologic characteristics. Vadose Zone J. doi:10.2136/vzj2012.0196

Shillito RM, Timlin DJ, Fleisher D, Reddy VR, Quebedeaux B (2009) Yield response of potato to spatially patterned nitrogen application. Agric Ecosyst Environ 129:107–116

Shumway RH (1988) Applied statistical time series analysis. Prentice Hall, Englewood Cliffs, 379 pp

Shumway RH, Stoffer DS (1982) An approach to time series smoothing and forecasting using the EM algorithm. J Time Ser Anal 3:253–264

Shumway RH, Stoffer DS (2000) Time series analysis and its applications. Springer, New York, 549 pp

Timm LC, Reichardt K, Oliveira JCM, Cassaro FAM, Tominaga TT, Bacchi OOS, Dourado-Neto D, Nielsen DR (2004) State-space approach to evaluate the relation between soil physical and chemical properties. Rev Bras Cienc Solo 28:49–58

Wendroth O, Al-Omran AM, Kirda C, Reichardt K, Nielsen DR (1992) State-space approach to spatial variability of crop yield. Soil Sci Soc Am J 56:801–807

Wendroth O, Katul GG, Parlange MB, Puente CE, Nielsen DR (1993) A nonlinear filtering approach for determining hydraulic conductivity functions in field soils. Soil Sci 156:293–301

Wendroth O, Koszinski S, Vasquez V (2011a) Soil spatial variability. In: Huang PM, Li YC, Sumner ME (eds) Handbook of soil science, 2nd edn. CRC Press, Boca Raton, pp 10-1–10-25

Wendroth O, Vasquez V, Matocha CJ (2011b) Field experimental approach to bromide leaching as affected by scale-specific rainfall characteristics. Water Resour Res 47:W00L03. doi:10.1029/2011WR010650

Yang Y, Wendroth O (2014) State-space approach to field-scale bromide leaching. Geoderma 217–218:161–172. doi.org/10.1016/j.geoderma.2013.11.013

Yang Y, Wendroth O, Walton RJ (2013) Field-scale bromide leaching as affected by land use and rain characteristics. Soil Sci Soc Am J. doi:10.2136/sssaj2013.01.0018

Chapter 4
Spatial Relationships Between Soil Water Content and Hydraulic Conductivity in a Highly Structured Clay Soils

Sidney Rosa Vieira, Célia Regina Grego, George Clarke Topp, and Willian Daniel Reynolds

Abstract The water, element essential for life, is present as a large reservoir in the soil. Soil water content varies in space as a consequence of the variability of other related properties and its study is necessary to know how this variation occurs in space and time. The objective of this study was to assess the spatial and temporal variability of soil water content and its relationships with hydraulic conductivity. Soil water content was measured with a TDR equipment and the hydraulic conductivity was measured with a constant head well permeameter at 15 and 50 cm depth, in a nearly flat 1.2 ha field at the Central Experimental Farm of the Agriculture Canada, Ottawa. The soil is classified as a Rideau soil series, is primarily clay to silty clay texture. Sampling was made on a 10 m spacing square grid with 164 sampling points. Soil water content was measured on 33 dates during the frost free months in 1987, 1988 and 1989. The rate of change in soil water content was calculated for the periods which did not have any rain. The spatial variability was analyzed examining the descriptive statistical parameters, the parameters of the models fitted to individual semivariograms as a function of time and the maps obtained by kriging interpolation. It was concluded that soil structure played a very important role on the spatial distribution of hydraulic conductivity and water content. Temporal stability of water in the soil is quite different during the year probably because of the hydraulic conductivity role. Parabolic trend removal

S.R. Vieira (✉)
Agronomy Institute (IAC), Campinas, São Paulo, Brazil
e-mail: sidney@iac.sp.gov.br

C.R. Grego
Embrapa Satellite Monitoring, Campinas, São Paulo, Brazil

G.C. Topp
Eastern Cereal and Oilssed Research Centre, Agriculture and Agri-Food Canada, Ottawa, ON, Canada

W.D. Reynolds
Greenhouse and Processing Crops Research Centre, Agriculture and Agri-Food Canada, Harrow, ON, Canada

W.G. Teixeira et al. (eds.), *Application of Soil Physics in Environmental Analyses: Measuring, Modelling and Data Integration*, Progress in Soil Science, DOI 10.1007/978-3-319-06013-2_4,

worked well all through the year probably because of the topography. The behavior of spatial variability the rate of change in water of soil did not repeat in different times of the year studied.

Keywords TDR • Temporal stability • Semivariogram • Geostatistics

4.1 Introduction

The water, element essential for life, is present as a large reservoir in the soil. Their distribution in space is variable due to fluctuations of other related properties, so knowledge of this variation in space and time becomes important in studies of soil physical properties. Soil water changes over landscape have been identified in several studies. Vauchaud et al. (1985) found temporal stability of frequency distribution of water in the soil, Kachanoski and De Jong (1988) studied the temporal persistence of spatial patterns of soil water storage, Vieira et al (1997) reported on the scaling of spatial variability over time and Gonçalves et al (1999) observed a temporal stability of soil water content stored in an irrigated area. During the last few decades geoestatistical methods have been intensively used for descriptions of spatial variability in depth (Nielsen et al. 1983; Vieira et al. 1983; Vieira 2000; Grego and Vieira 2005; Vieira et al 2008). Most of the authors reported that geoestatistics is useful for evaluate the spatial dependence, use this spatial dependence to interpolate values for the places where they have not been measured, and to provide information to construct maps and design sampling strategies as a function of the spatial variability. The difference of statistics and geostatistics is that classical statistics require normality and spatial independence of the dates while geostatistics requires correlation and spatial dependence.

Variability in time, and in particular, with repeating patterns is being a challenge to soil research. Adequate information about soil properties showing spatial distribution stable in time could contribute to significantly reduce the number of measurements (Vauchaud et al. 1985; Kachanoski and De Jong 1988). Soil water content analyzed in consecutive sampling dates have been reported of having temporal stability (Vauchaud et al. 1985; Gonçalves et al. 1999). Vauchaud et al. (1985) addressed the occurrence of temporal stability of the spatial distribution of soil water content and concluded that the places where higher water content occur in one moment may remain that way at other moments. Vieira et al. (1991) expanded this concept and considered the scaling of semivariograms to simultaneously examine the spatial variability in consecutive sampling dates. Scaling semivariograms of several variables measured over the same field provides a simple but powerful integration method (Vieira et al. 1988; Vieira et al. 1991; Vieira et al. 1997) in the sense that the more the semivariograms scale the more similar the variability of the corresponding variables are.

Therefore, if semivariograms scale, it indicates that not only the mean values and dispersion coefficients occur at the same locations but all variability repeats in time, although the absolute values may be different. Besides, the analysis of the parameters of the models fitted to the semivariograms as a function of the time of successive samplings may be of help in assessing the temporal stability of the spatial variability. Thus, in order to analyze temporal and spatial variability of water content it is important calculate and compare scaled semivariograms for different dates, compare semivariogram model parameters and construct and compare contour maps for different dates and variables.

The objective of this chapter was:

- To describe the spatial variability as well as the temporal stability of the spatial distribution for soil water content.
- To use geostatistics to evaluate the spatial variability of soil water content and of hydraulic conductivity.
- To analyze the temporal stability of the spatial dependence of soil water content and its relations with the soil hydraulic conductivity.

4.2 Methodology

4.2.1 Study Area and Soil Sampling

The study was developed in a Rideau clay loam soil (Gleyed Melanic Brunisol) with clay loam to clay surface texture, in the Central Experimental Farm of Agriculture Canada, Ottawa, with grasses vegetation maintained low. One hundred and sixty four TDR (time domain reflectometry) rods for measuring volumetric water content were installed. Therefore, soil water content measurements could be obtained on the same points as many times as wanted since the TDR rods remained in-place at the same points in the field. The TDR rods were installed at 45° angle and kept there for 3 years. All the soil water content contents measured were done with the TDR method according to Topp and Davis (1985).

All water content measurements over the field were collected within a two-hour time period. The triangular field, kept vegetated with natural pasture grass, measured 110 m in the x-direction (base) and 220 m in the y-direction (height), was divided into a 10×10 m grid with 164 points. The grid and topographic map of the study area are shown in Fig. 4.1.

The study period extended from early September of 1987 until early May 1989 during the ground surface frost-free months. The dates of samplings in 1987 were September 11, 14, 17, 21, 24, and October 29 with two samplings on the same day with different TDR instruments. There were 19 samplings dates in 1988: May 06, 13 and 30, June 06, 13, 21 and 27, July 04, 15, 22 and 28, August 11, 18 and 25, September 02, 08, 15, 22 and 29, and October 06. The dates of samplings in 1989 were April 14, 18, 21, 25 and 28 and May 01. Figure 4.2 shows the precipitation in mm, during the study period.

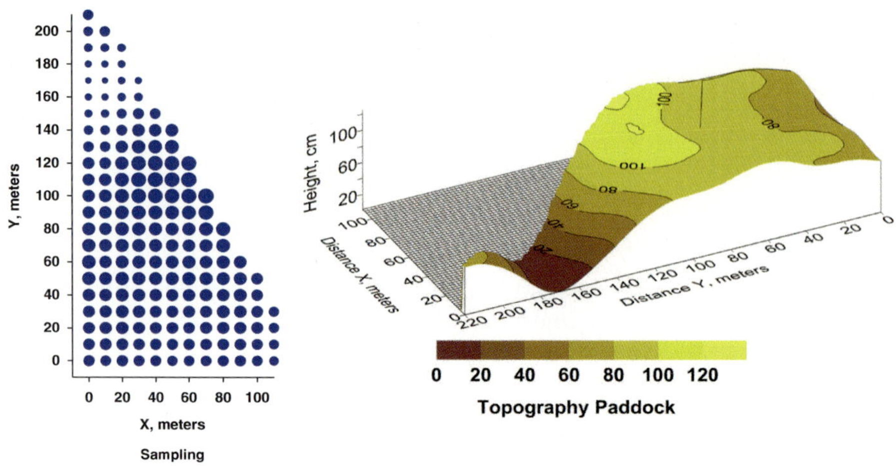

Fig. 4.1 Sampling scheme and topography of the area

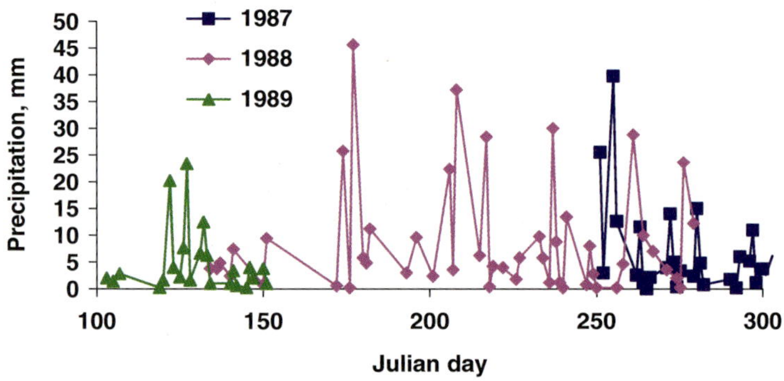

Fig. 4.2 Rainfall as a function of time during the years 1987 until 1989

The sampling dates were converted to Julian days in order to make the parameters plot a function of time during the year.

Figure 4.2 shows a distribution of rainfall during the entire research period for the three years. Except for a few peaks of very high rainfall most of them are somewhere around 5 and 10 mm. Therefore, except for some days right after the peaks of rainfall, it is expected that the major contribution for soil water content is the snow melt.

The field saturated hydraulic conductivity, K_{fs}, was measured in each one of the locations shown in Fig. 4.1 using the constant head well permeameter (Reynolds and Elrick 1986) and calculated using the one head approach as recommended by Reynolds et al. (1992). At the time the hydraulic conductivity were measured, TDR readings were made in each one of the 164 sampling points in order to have the initial soil water content.

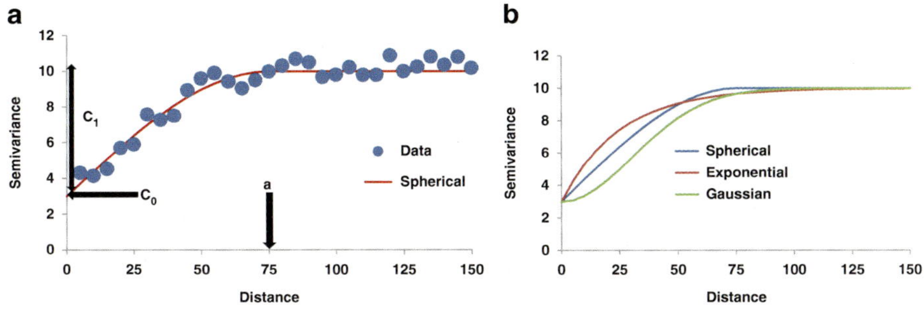

Fig. 4.3 (**a**) Characteristics of a typical semivariogram with C_0 (nugget effect), C_1 (structural variance) and *a* (range); (**b**) Most commonly used models: spherical, exponential and Gaussian

4.2.2 Spatial Variability Theory

In order to be able to evaluate the spatial variability with geostatistics the basic requirements are: Cartesian coordinates, intrinsic hypothesis, semivariogram showing a well-defined sill and similarity between neighbors. The equations that represent hypothesis for geostatistics are:

$$E\{Z(x_i)\} = m \tag{4.1}$$

Where a random function $Z(x_i)$ is stationary of order two if the expected value, $E\{Z(x)\}$ exists and does not depend on the position x.

In addition to the condition in Eq. 4.1, the Eq. 4.2 represent a random function $Z(x_i)$ when is a intrinsic hypothesis. The increment $[Z(x_i) - Z(x_i + h)]$ has a finite variance, and does not depend on xi for all vectors h.

$$VAR[Z(x_i) - Z(x_i + h)] = E[Z(x_i) - Z(x_i + h)]^{'2} \tag{4.2}$$

The function γ (h) is the semivariogram (Matheron 1963) estimated by

$$\gamma^*(h) = \frac{1}{2\,N(h)} \sum_{i=1}^{N(h)} [Z(x_i) - Z(x_i + h)]^2 \tag{4.3}$$

where N(h) is the number of pairs of measured values $Z(x_i)$, $Z(x_i + h)$, separated by a vector h (distance).

The main characteristics of a semivariogram are shown in Fig. 4.3a.

Experimental semivariograms need to be fitted to some mathematical model which must meet the criteria of conditional positive definiteness MacBratney and Webster (1986). Amongst all the variety of models which satisfy that condition, the fitting parameters that describe them are: the nugget effect C_0, the sill $(C_0 + C_1)$, (C_1 is the structured variance coefficient to be defined later), and the range of spatial dependence *a*.

The models most frequently used (McBratney and Webster 1986) are the spherical, the exponential and the gaussian models. A graph of these models is shown in Fig. 4.3b for illustration purposes.

When data shows a trend (semivariogram without a sill), according to Vieira et al (2010b), an alternative is fit a trend surface using minimization of the sum of squares of the deviations and subtract this surface from the original data generating a residual variable.

For instance, the equation for a parabolic trend surface is:

$$\mathrm{Zest}(x, y) = \mathrm{A0} + \mathrm{A1X} + \mathrm{A2Y} + \mathrm{A3X2} + \mathrm{A4XY} + \mathrm{A5Y2} \tag{4.4}$$

Where Zest(x, y) is a estimate value for parabolic trend surface to measured values (A0,. . ., A5Y2)

Thus, the residuals, Zres(x, y), can be calculated

$$\mathrm{Zres}(x, y) = \mathrm{Z}(x, y) - \mathrm{Zest}(x, y) \tag{4.5}$$

With the objective of comparing the variability of different samplings, Vieira et al. (1997) proposed a scaling technique for the semivariogram expressed by:

$$\gamma\mathrm{sch} = \gamma\mathrm{Lh}\alpha\mathrm{L}, \mathrm{L} = 1, 2, \ldots, \mathrm{m} \tag{4.6}$$

Where α i is the scaling factor for semivariance (γ) and m is the number of measured variables.

The scale factor α, is a constant that can take the value of the calculated variance, the sill (the highest value of the semivariance) or the square of the mean values. The scaling concept proposed by Vieira et al. (1997) may be helpful in the analysis of the temporal stability of the spatial variability of soil water content corresponding to different sampling dates for the same location. Within each data set (year of sampling) the scaled semivariograms for all sampling dates can be plotted together in order to make comparisons and verify if and when the spatial variability loses temporal stability. When data of several dates coalesce into a unified semivariogram structure, it is possible to take advantage of variables having the same spatial structure and, hence, reduce the number of semivariograms needed to analyze and draw interpretations regarding their spatial variability.

Two facts are very important about the above theoretical concepts: (i) The semivariogram γ (h) is assumed to be isotropic, i.e., either there is no significant anisotropy or there is a transformation to remove the anisotropy before scaling is applied; (ii) γ (h) can take the value of the calculated variance Var(z) no matter whether it represents the true variance or not, since the scaling factor is simply a number chosen to make the semivariograms coalesce into a single curve. The reason why scaled semivariograms may provide an adequate way to analyze temporal evolution of the spatial variability is that it will be noticeable when the spatial variability pattern changes, and thus the possible cause can be examined.

Therefore, scaling is used in this paper only for the comparison between semivariograms for different samplings on the same field.

The examination of the spatial variability for soil water content as the time during the year progresses may reveal information about some water content threshold value at which the soil hydraulic conductivity begins to cause changes in the spatial variability patterns.

4.2.3 Water Content Change Over Successive Sampling Dates

The rate of change of water content over time, $\Delta\ (\theta_i, \theta_j)$ can be expressed as the difference in water content (% vol.) measured at times i and j divided by the time interval (day).

$$\Delta\,\theta i, \theta J = (\theta i - \theta j)(ti - tj) \tag{4.7}$$

The rate of change in water content over time, $\Delta\ (\theta_i, \theta_j)$ was calculated for the period when there was no rain. Therefore, all the change in water content was caused strictly by water loss to the atmosphere. When the soil has enough water to satisfy the atmospheric demand without restrictions, assuming that the energy available for the evaporation process is not variable in space at the scale of the field under study (110x220 m), then the variability of the evaporation rate is likely to reflect the soil surface conditions. Besides, because the soil cover constituted primarily of pasture grasses and it was maintained at low height, then not much spatial variability due to vegetation evaporation is expected for this field.

4.3 Result and Discussion

The descriptive statistical moments for hydraulic conductivity at 15 and 50 cm depths and soil water content measured in 164 points are shown in Table 4.1. The hydraulic conductivity for the two depths showed very high coefficients of skewness and

Table 4.1 Descriptive statistical moments for hydraulic conductivity and soil water content measured at 15 and 50 cm depth

Variable	Num	Mean	Variance	CV	Min	Max	Skew	Kurt
Hydraulic Conductivity – 15 cm (logs)	164	−677	291	−2,520	−1,135	−285	009	−042
Soil water content – 15 cm	164	2,372	3,520	2,502	1,070	3,640	−012	−065
Hydraulic Conductivity – 50 cm (logs)	164	−826	270	−1,991	−1,340	−495	−079	016
Soil water content – 50 cm	164	3,951	5,174	1,821	1,690	5,360	−033	−038

Table 4.2 Semivariogram model parameters for hydraulic conductivity and soil water content measured at 15 and 50 cm depth

Variable	Model	C_0	C_1	a	r^2	RMSE
Hydraulic Conductivity 15 cm (logs)	Exponential	228	087	7,211	045	0.005
Soil water content 15 cm	Spherical	000	2,908	7,000	080	0.139
Hydraulic Conductivity 50 cm (logs)	Spherical	052	203	7,000	084	0.008
Soil water content 50 cm	Gaussian	1,670	2,997	7,000	094	0.080

kurtosis. For that reason it was decided to work on the logs rather than the original values. This transformation, although it is always better not to need it, it does make the coefficients of skewness and kurtosis to approach 0 (zero). In this condition the data has a frequency distribution close to the normal and it makes the analysis easier according to Vieira (2000).

It can be seen in Table 4.1 that mean hydraulic conductivity is much higher at 15 cm than at 50 cm depth which is caused by the strong blocky structure at depths near the surface and massive structure at deeper depths. On the other hand the moisture content is much higher at 50 cm than at 15 cm depth. The coefficients of variation are within the normal range expected for physical soil data of this type. Similar results have been reported by Vieira et al. (1988). One aspect of Table 4.1 which is initially difficult to explain is the large range in water content (approximately 25 % for 15 cm depth and 36 % for 50 cm depth). It should be considered that after the snow melt redistribution of water not very much change should be expected for moisture content at 50 cm depth this without a crop growing cultivated in this area.

The hydraulic conductivity and soil water content at 15 and 50 cm depth showed enough spatial dependence to influence on the interpolation of values at any spacing finer than that they were measured (10 m) without bias and with minimum variance, as the parameters for the models fitted to the semivariogram (Table 4.2) indicate. The range of spatial dependence was 70.0 m or more for all of them. The spatial dependence for the hydraulic conductivity for the 15 cm depth was somewhat weak due to the randomness of the macro pores at the surface layer which causes a lot variable in the water flow.

The Fig. 4.4a, b show the scaled semivariograms for hydraulic conductivity and soil water content, respectively, at 15 and 50 cm depth. The Rideau soil at this site has a very strong blocky structure at the first 25 cm. On the other hand, at 50 cm depth this soil has a very massive structure or no structure at all. Therefore, at 15 cm depth the water moves much faster in between the structure blocks while the soil water content is mostly retained inside the blocks. That is the main reason why the scaled semivariograms for 15 cm depth are so different because they represent different soil physical processes. On the other hand, at 50 cm depth the water moves faster where there is more water and for that reason, the spatial variability of soil water content is so similar to that of the hydraulic conductivity. According to Vieira et al. (1997) the soil physical process is the main reason why scaled semivariograms are so alike.

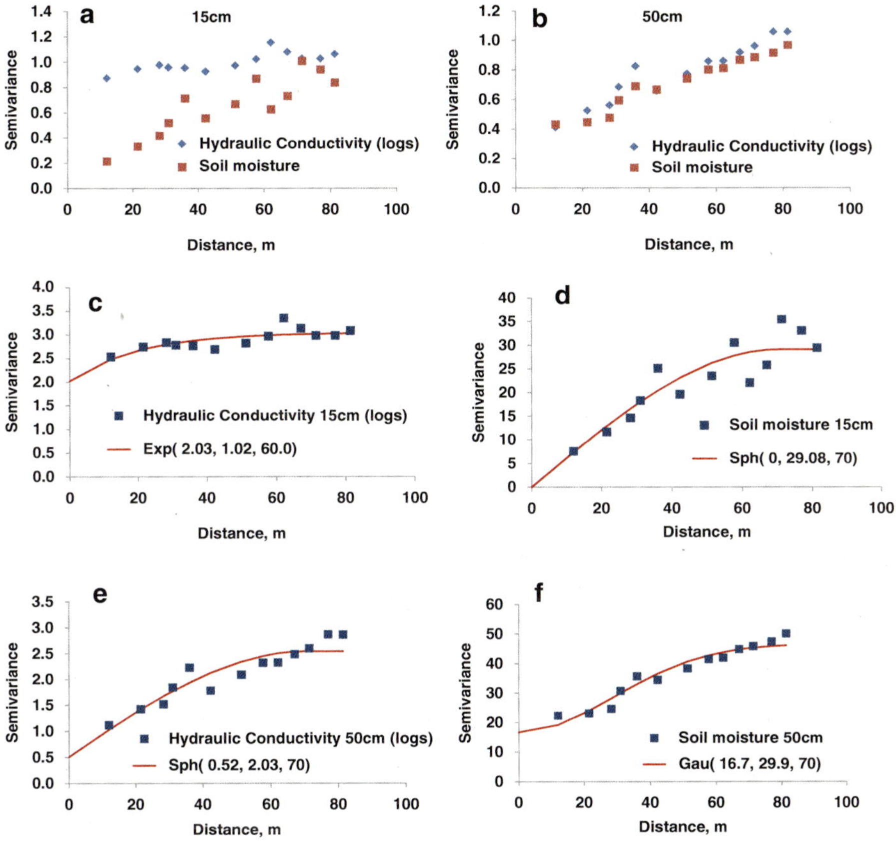

Fig. 4.4 Semivariograms for hydraulic conductivity and water content: (**a**) scaled semivariogram for hydraulic conductivity and water content at 15 cm depth; (**b**) scaled semivariogram for hydraulic conductivity and water content at 50 cm depth; (**c**) semivariogram for hydraulic conductivity (logs) at 15 cm depth; (**d**) semivariogram for water content at 15 cm depth; (**e**) semivariogram for hydraulic conductivity (logs) at 50 cm depth; (**f**) semivariogram for water content at 50 cm depth

Figure 4.4c–f. show that the models chosen to fit the semivariograms worked very well for all of them as the models follow the experimental semivariograms with strong spatial dependence as it is also indicated by the r^2 values shown in Table 4.2.

The spatial dependence expressed by the semivariograms parameters shown in Table 4.2 are validated by the jack knifing procedure whose parameters are shown in Table 4.3. According to Vieira et al. (2010a) the main parameters in Table 4.3 are the mean error and the variance of the errors, which should be, respectively, 0 (zero) and 1 (one). Thus it can be concluded that all 4 semivariograms are validated as their parameters approach the ideal values.

The maps for hydraulic conductivity and soil water content measured at the time the hydraulic conductivities were measured are shown in Fig. 4.5. The region in the

Table 4.3 Jack knifing results for the semivariogram model parameters for hydraulic conductivity and soil water content measured at 15 and 50 cm depth

Variable	Neighbors	A	b	r^2	Mean error	Variance of errors	IJK	RMSE	Dist
Hydraulic Conductivity 15 cm (logs)	16	−6.17	0.09	0.25	0.00	0.97	−1.68	1.65	61
Soil water content 15 cm	16	4.19	0.82	0.91	0.00	1.08	−0.18	2.41	61
Hydraulic Conductivity 50 cm (logs)	12	−3.91	0.53	0.73	−0.01	1.24	−0.51	1.12	51
Soil water content 50 cm	12	17.93	0.55	0.76	0.00	1.12	−0.57	4.67	51

field with low water content values for 15 cm depth falls exactly in the region of higher topographic heights (Fig. 4.1b). It is amazing to see that this same region had also low soil water content for the 50 cm depth. The maps for hydraulic conductivity and water content for 15 cm depth illustrate the reason why their semivariograms are so different when they are scaled which simply is that they represent a very distinct variability. On the other hand, the maps for hydraulic conductivity and soil water content for 50 cm depth have variability almost exactly the reverse, with low values of conductivity where the soil water content is high. That was exactly the idea reported by Vieira et al. (1997) that if the semivariograms scale well it is an indication that the two variables either vary almost equally or almost opposite to each other.

The scaled semivariograms for rate of change in soil water content shown in Fig. 4.6 indicate that the variability for water loss is not stable in time. Similar results were found by Grego et al. (2006) showing that the scaled semivariograms for soil water content may be used to indicate time stability of the spatial variability. Five out of twenty nine semivariograms showed pure nugget effect. All the nuggets happened after a large previous rain. Twenty three models were spherical and one was exponential.

The in Fig. 4.7 show that the rate of change of water content does not repeat in time. It should be noticed that these rate of change in water content are for the three years period which means that there might be climate variability involved in these numbers. The intervals at the beginning of the year (April, May) probably reflect the amount of snow cover. Beginning of May 1988 and beginning of April 1989 seem to agree with each other probably of snow melt. The two rates for 1988 (05/30-06/06 and 06/06-13/06) do not seem to have anything in common and rather are completely the opposite. Beginning of September 1987 seem to agree with end of April 1989 but are not related to topography and/or surface texture but rather with the hydraulic conductivity at 50 cm depth. Buttafuoco et al. (2005) showed a high temporal correlation between the soil water contents measured at different times, declining as the interval between the observations increases. The autumn rain events on dry soil produced an erratic distribution pattern of water in the soil and the kriged maps of soil water revealed the dynamics of soil water redistribution owing to evapotranspiration or rainfall.

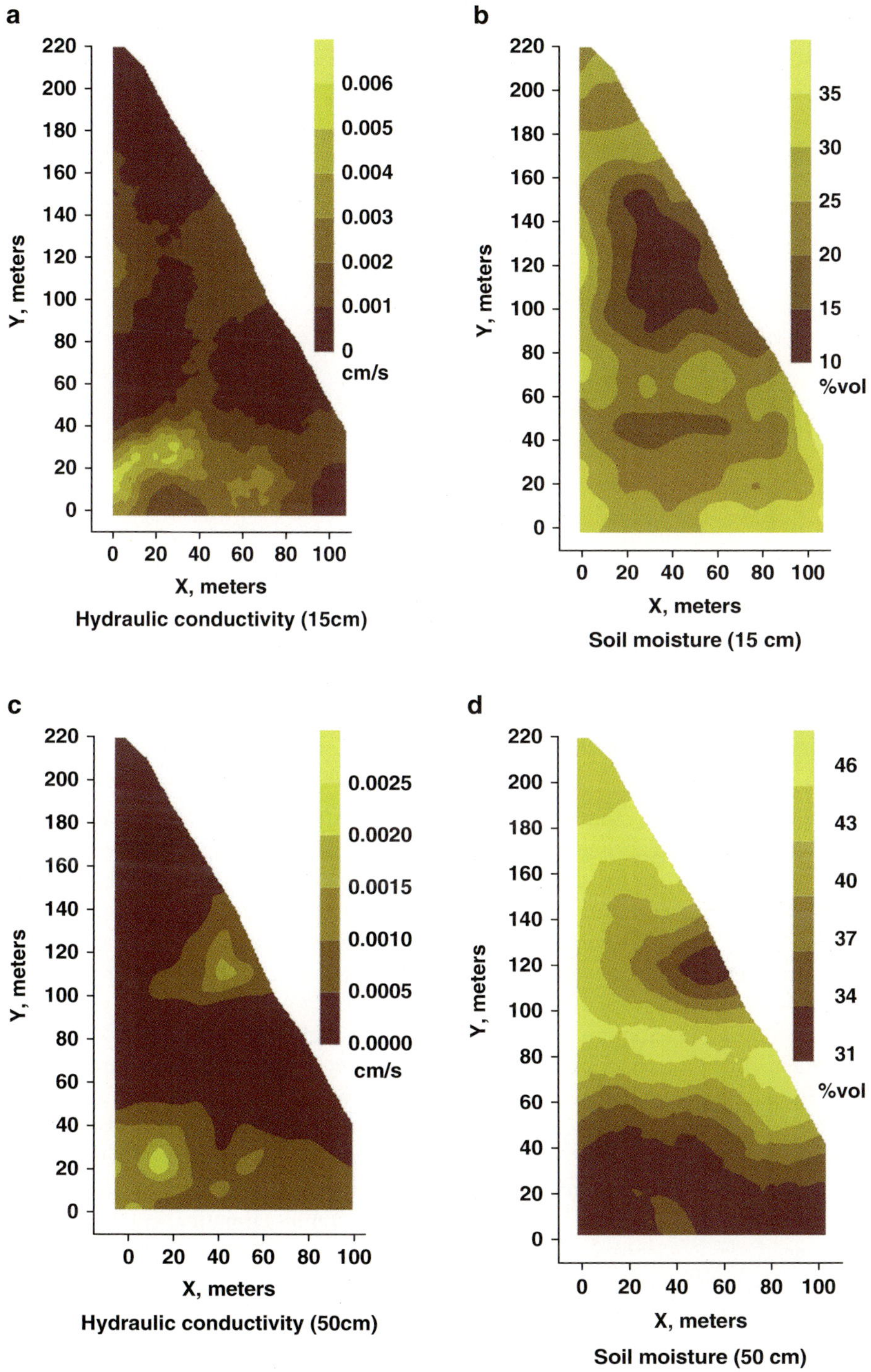

Fig. 4.5 Maps for hydraulic conductivity (**a** and **b**) and soil water content (**c** and **d**) at 15 and 50 cm depth

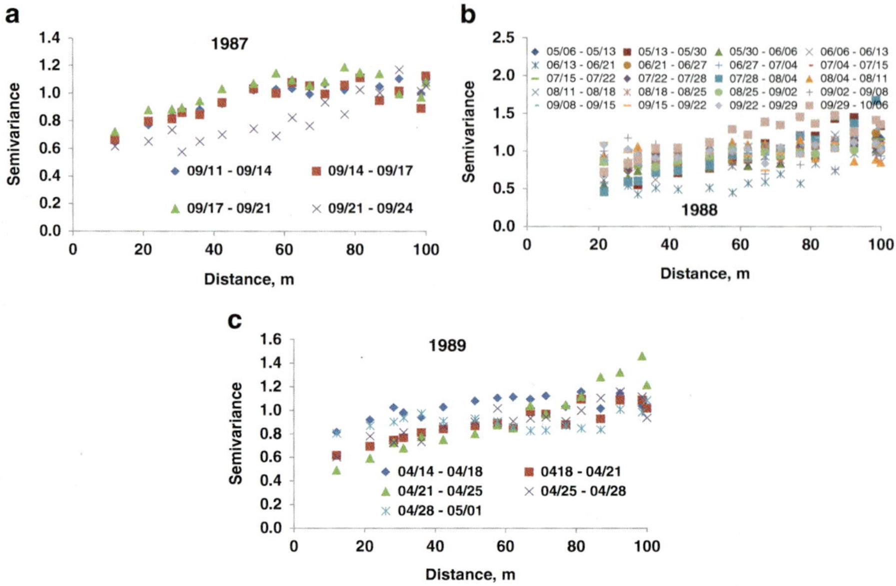

Fig. 4.6 Scaled semivariograms for rate of change in soil water content over the 3 years: 1987 (**a**); 1988 (**b**); 1989 (**c**)

4.3.1 Scaling of Semivariograms

As the semivariograms for the original soil water content data measured during the three years of the experiment showed a very strong trend (Fig. 4.4) violating the intrinsic hypothesis of geostatistics (Journel and Huijbregts 1978), parabolic trend surface equations were fitted to the data and subtracted from them. The residuals generated by the difference between originals and trend surface produced semivariograms that showed a very well defined sill, and for this reason the residuals were used in the remaining analysis.

Once the semivariogram has a defined sill, a model must be fit to its experimental values in order to provide a continuous function over the whole distance range.

Residuals of parabolic trend for water content at the beginning of the autumn in 1987 scale very well in time. It seems that the scaling works better for one week interval. Between 14 and 17 of September there was no rain but Sep14 and Sep17 still scale well. For the entire frost free year of 1988 the scaling does not seem to work too well probably because it involves very different climatical conditions over the year.

In this sense Ávila et al. (2011) used the temporal stability concept to compare the spatial structure of the fitted semivariogram models and was possible to detect a strong spatial dependence of soil moisture and spatial pattern of the scaled

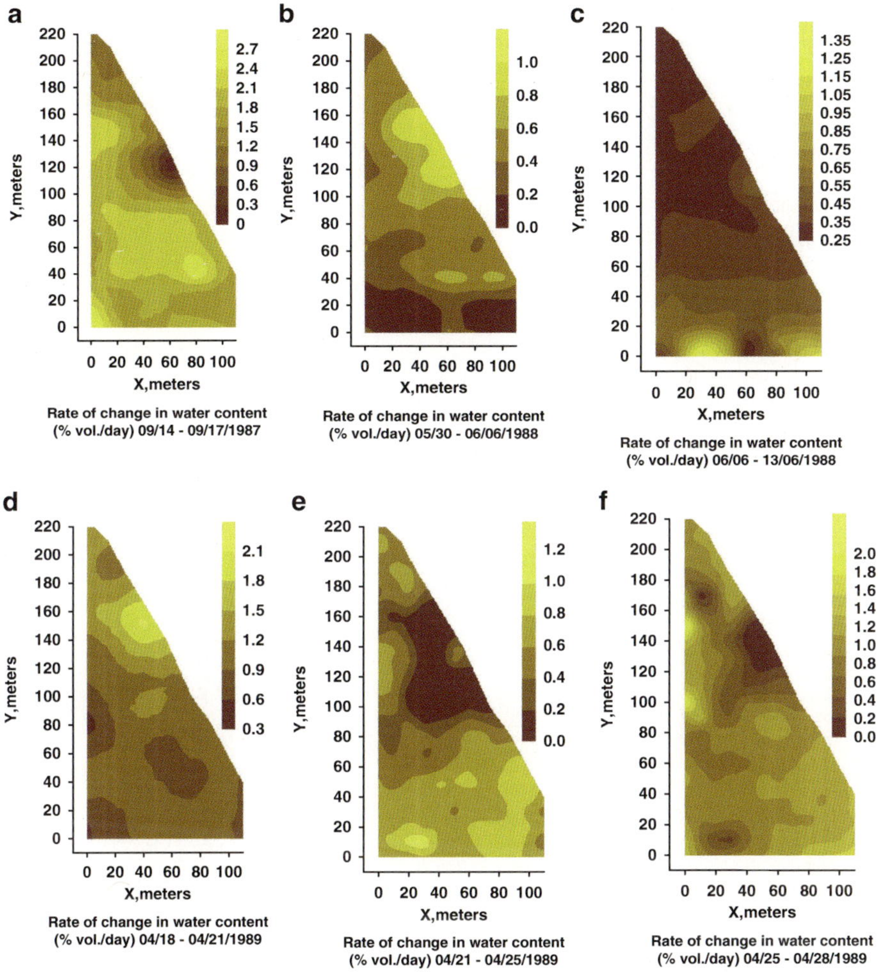

Fig. 4.7 Maps for rate of change in soil water content for the 6 periods without rain in 1987 (**a**), 1988 (**b**, **c**) and 1989 (**d**, **e**, **f**)

semivariograms was similar for summer and fall, and different for winter and spring. In the present work it appears that the spatial behavior for water content is different at the beginning and at the end of the year. Even for a short period of time right after snow melt (March or April) the spatial behavior seem to change from one week to the next. It seem like two weeks at this time of the year is the limit for the stability of spatial variability. It should be noticed also that the semivariograms were already close to each other before the scaling (Fig. 4.8). The parabolic trend removing process produced semivariograms with very well defined sills for all three years.

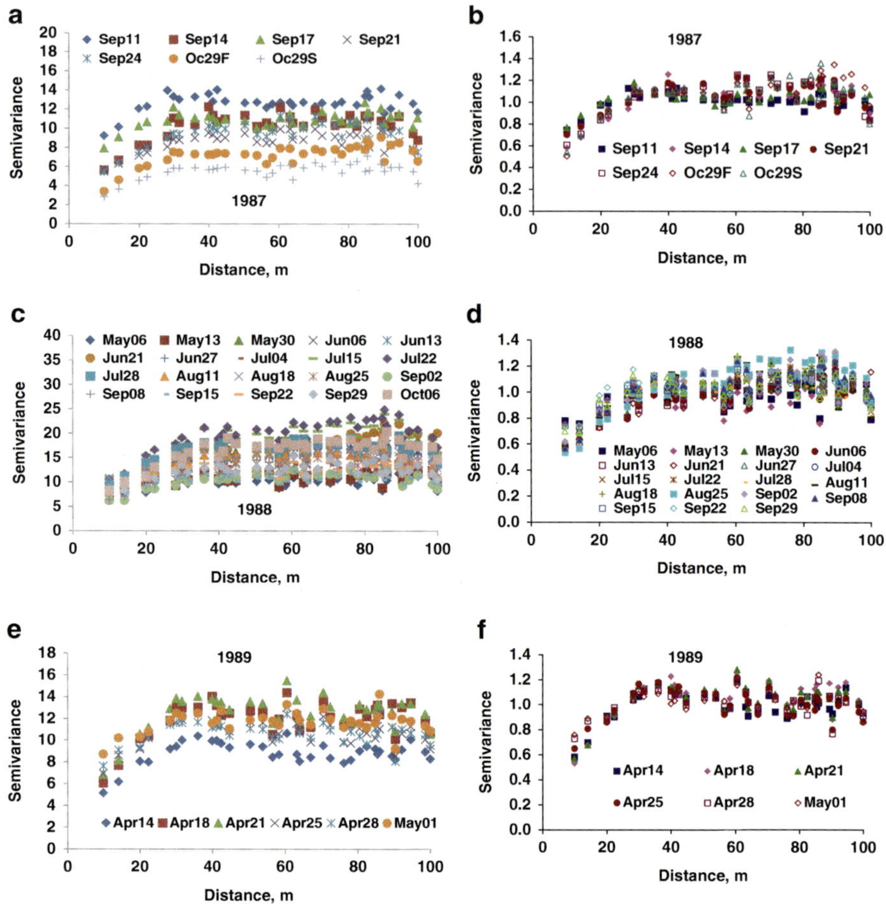

Fig. 4.8 Non scaled and scaled semivariograms for soil water content for the 3 years: 1987 (**a**, **b**); 1988 (**c**, **d**); 1989 (**e**, **f**)

4.3.2 Soil 'memory'

The length of time that soil water content expresses correlation with soil water content at an initial time is what may be called as soil 'memory'. The reason for this is in the question 'how long does the soil "remember" the initial soil water content? In order to illustrate this idea with these three year data, the correlation between water content at some initial time and subsequent soil water content was calculated and plotted in the graphs in Fig. 4.9. For comparison purposes, these calculations were done for both the original values and the residuals of a parabolic trend. The initial times were September/11, May/06 and April/14, respectively for

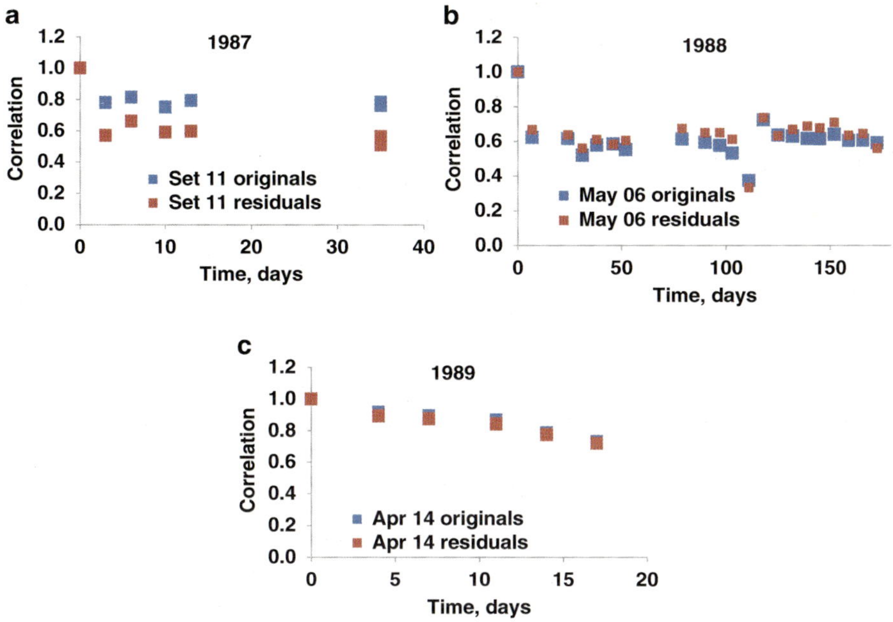

Fig. 4.9 Persistence of correlation between soil water content for the initial sampling date and soil water content in subsequent dates in 1987 (**a**), 1988 (**b**) and 1989 (**c**)

1987, 1988 and 1989. The correlation between any soil water content and the initial one was never lower than 0.6, even for 1988 when the length of time is larger than 150 days. That means that the time stability of the water content is quite high. For 1989, when the data covers only the beginning of the frost free year (April) and the soil is very wet due to snow melt, the correlation remains at approximately 0.8.

4.4 Conclusions

The following conclusions can be extracted in accordance with the text:

- Soil structure played a very important role on the spatial distribution of hydraulic conductivity and water content.
- Temporal stability is quite different during the year probably because of the hydraulic conductivity role.
- Parabolic trend removal worked well all through the year probably because of the topography.
- Rate of change in water content with time did not repeat variability in different times of the year.
- Correlation of water content remains for a long time (more than 100 days).
- Unique field topography and surface texture may have been the reason for the results.

References

Ávila LF, Mello CR, Mello JM, Silva AM (2011) Padrão espaço-temporal da umidade volumétrica do solo em uma bacia hidrográfica com predominância de Latossolos. Rev Bras Ciênc Solo Viçosa 35:1801–1810

Buttafuoco G, Castrignano A, Busoni E, Dimase AC (2005) Studying the spatial structure evolution of soil water content using multivariate geostatistics. J Hydrol 311:202–218

Gonçalves ACA, Folegatti MV, Silva AP (1999) Estabilidade temporal da distribuição espacial da umidade do solo em área irrigada por pivô central. Rev Bras Ciênc Solo 23:55–164

Grego CR, Vieira SR (2005) Variabilidade espacial de propriedades físicas do solo em uma parcela experimental Revista Brasileira de Ciência do Solo (Impresso). Viçosa 29:169–177

Grego CR, Vieira SR, Antonio AM, Della Rosa SC (2006) Geostatistical analysis for soil water content under no tillage cropping system. Sci Agric 63:341–350

Journel AG, Huijbregts CJ (1978) Mining geostatistics. Academic, San Diego, 600 p

Kachanoski RG, De Jong E (1988) Scale dependence and the temporal persistence of spatial patterns of soil water storage. Water Resour Res 24:85–91

Matheron G (1963) Principles of geostatistics. Econ Geol (Lancaster) 58:1246–1266

Mcbratney AB, Webster R (1986) Choosing functions for semivariograms of soil properties and fitting them to sampling estimates. J Soil Sci 37:617–639

Nielsen DR, Tillotson PM, Vieira SR (1983) Analysing field – Measured soil water properties agricultural. Water Manage 6:63–109

Reynolds WD, Elrick DE (1986) A method for simultaneous in situ measurements in the vadose one of field saturated hydraulic conductivity, sorptivity, and the conductivity-pressure head relationship. Ground Water Monit Rev 6:84–95

Reynolds WD, Vieira SR, Topp GC (1992) An assessment of the single-head analysis for the constant head well permeameter. Can J Soil Sci (Ottawa) 72:489–501

Topp GC, Davis JL (1985) Measurement of soil water content using time domain reflectometry (TDR): a field evaluation. Soil Sci Soc Am J 49:574–582

Vauchaud G, De Silane P, Balabanis A, Vauclin M (1985) Temporal stability of spatially measured soil water probability density function. Soil Sci Soc Am J 49:822–827

Vieira SR (2000) Uso de geoestatística em estudos de variabilidade espacial de propriedades do solo. In: Novais RF (ed) Tópicos em Ciência do Solo. Sociedade Brasileira de Ciência do Solo, Viçosa, pp 3–87

Vieira SR, Hatfield JL, Nielsen DR, Biggar JW (1983) Geoestatistical theory and application to variability of some agronomical properties. Hilgardia 51(3):1–75

Vieira SR, Reynolds WD, Topp GC (1988) Spatial variability of hydraulic properties in a highly structured clay soil. In: Validation of flow and transport models for the unsaturated zone Ruidoso. Conference Proceedings Ruidoso, Novo Mexico: Department of Agronomy and Horticulture, New Mexico State University, pp 471–483

Vieira SR, Lombardi Neto F, Burrows IT (1991) Mapeamento da chuva máxima provável para o Estado de São Paulo. Rev Bras Ciênc Solo 15:93–98

Vieira SR, Nielsen DR, Biggar JW, Tillotson PM (1997) The scaling of semivariograms and the kriging estimation. Rev Bras Ciênc Solo 21:525–533

Vieira SR, Grego CR, Toop GC (2008) Analyzing spatial and temporal variability of soil water content. Bragantia 67:463–469

Vieira SR, Carvalho JRP, Ceddia MB, Paz González A (2010a) Detrending non stationary data for geostatistical applications. Bragantia 69(Suplemento):1–8

Vieira SR, Carvalho JRP, Paz González A (2010b) Jack knifing for semivariogram validation. Bragantia 69(Suplemento):97–105

Chapter 5
State-Space Approach to Understand Soil-Plant-Atmosphere Relationships

Luís Carlos Timm, Klaus Reichardt, Cláudia Liane Rodrigues de Lima, Leandro Sanzi Aquino, Letiane Helwig Penning, and Durval Dourado-Neto

Abstract This chapter presents two different state-space approaches to evaluate the relation between soil and plant properties using examples of sugarcane, coffee and forage. These state-space approaches take into account sampling positions and allow a better interpretation of the data in relation to the field. Concepts of autocorrelation and crosscorrelation functions are first introduced, followed by theoretical aspects of both state-space approaches. More emphasis is given to the last one based on the Bayesian formulation, which gives more attention to the evolution of the estimated observations. It is concluded that the use of these dynamic regression models improve data analyses, being therefore recommended for several studies involving time and space data series, related to the performance of a given soil-plant-atmosphere system.

L.C. Timm (✉)
Rural Engineering Department, Faculty of Agronomy, Federal University of Pelotas, CP 354, 96001-970 Capão do Leão, RS, Brazil
e-mail: lctimm@ufpel.edu.br

K. Reichardt
Soil Physics Laboratory, CENA/USP, Piracicaba, SP, Brazil

C.L.R. de Lima
Soil Science Department, Faculty of Agronomy, Federal University of Pelotas, Capão do Leão, RS, Brazil

L.S. Aquino
Agronomy Post-Graduate Program, Faculty of Agronomy, Federal University of Pelotas, Capão do Leão, Brazil

L.H. Penning
Soil and Water Management and Conservation Post-Graduate Program, Faculty of Agronomy, Federal University of Pelotas, Capão do Leão, RS, Brazil

D. Dourado-Neto
Crop Production Department, ESALQ/USP, Piracicaba, SP, Brazil

W.G. Teixeira et al. (eds.), *Application of Soil Physics in Environmental Analyses: Measuring, Modelling and Data Integration*, Progress in Soil Science, DOI 10.1007/978-3-319-06013-2_5,

Keywords Time series • Regression models • Soil variability • Spatial variability • Temporal variability • Dynamic models • State-space models

5.1 Introduction

There is a great class of physical, chemical and biological phenomena that, when observed and numerically quantified, can result in sequences of data distributed along time and/or space. Time data sequences are called time series. Examples are: (a) monthly average values of air temperature, (b) yearly average values of rainfall, (c) yearly sugarcane yield and (d) yearly soil organic carbon content. Similarly, space data sequences are called space series. Examples are: (a) soil temperature values collected across a landscape at the same time, (b) soil water content values collected across a corn field on the same day, (c) sugarcane yield values measured across a field during a single harvest and (d) soil pH values collected across a pasture at a fixed time.

Up to the 1960ies, research in agronomy relied mostly on classical statistics (analysis of variance, mean, coefficient of variation, regression analysis, etc), that presupposes the independence of observations among themselves and within the analysis ignores the sampling locations in the field. Commonly, agronomic experiments are carried out ignoring the fact that observations might be spatially or temporally dependent. More recently it has been emphasized that adjacent observations of a given variable are not necessarily independent, and that the variability has to be taken into consideration in their statistical analysis. Nielsen and Alemi (1989) comment that observations within and among treatments might not, in fact, be independent among themselves, rendering the experimental design inadequate.

Soil spatial variability occurs at different scales and is related to variations of the parent material, climate, relief, organisms and time, i.e., related to the processes of soil formation and/or effects of management practices adopted for each agricultural use (McGraw 1994). Statistical tools like autocorrelograms, crosscorrelograms, semivariograms, spectral analysis, kriging, cokriging, autoregressive models, ARIMA models, state-space models, etc, are now frequently used to study the spatial variability of soil attributes, and can potentially lead to management practices that allow a better understanding of the interactive processes within the soil-plant-atmosphere system (Vieira et al. 1983; Wendroth et al. 1997, 2001, 2003; Dourado-Neto et al. 1999; Timm et al. 2003a, b, 2006, 2011; Parfitt et al. 2009; Jia et al. 2011; Oliveira et al. 2011; Liu et al. 2012).

The concern about the spatial variability of soil properties is expressed in several reports related to agronomy. Until recently, the most detailed studies of spatial variability of soil properties indicated limitations of the classical methods of Fisher's statistics. In general, normality and independence of observations are not tested, even knowing that the independence must be assured a priori, i.e., before sampling. All variability is assumed to be residual, being due to uncontrolled factors. Recently, applied statistical tools take into account the structure of the spatial dependence of the observations. This approach has lead to an improvement in the understanding of physical, chemical and biological processes that control the soil-plant-atmosphere system and, therefore, to the adoption of better management practices with less environmental impact. Having this in mind, the basic principles of concepts and tools most commonly used to analyze and characterize the spatial variability of agronomic data sets are here presented.

A discrete time or temporal series can be considered as a set of n observations Y described by

$$Y(t_i) \qquad i = 1, 2, 3, \ldots\ldots, n \tag{5.1}$$

collected at equidistant times t

$$t_i - t_{i-1} = \alpha \tag{5.2}$$

and manifest a serial dependence among themselves. Series collected continuously during the total experimental time interval ($n\alpha$) have to be transformed into discrete series through a "sampling" procedure at equidistant time intervals. The interval α (also called lag h) between observations is in general chosen by the scientist, however, in several situations is defined by the available data set. According to Tukey (1980), the basic objectives for analyzing a time series are: (a) modeling the process under consideration, (b) obtaining conclusions in statistical terms and (c) evaluation of the ability of the model in terms of forecast.

The models used to describe temporal series are stochastic processes controlled by probabilistic laws. The choice of these models depends on several factors such as the behavior of the phenomenon or the "a priori" knowledge we have about its nature, and the objective of the analysis. From a practical point of view, the choice also depends on the existence of good estimation methods and on the availability of adequate softwares.

A temporal (spatial) series can be analyzed in two ways: (1) in the time (space) domain and (2) in the frequency domain. In both cases it is desired to construct models for the series based on known concepts. For the time (space) domain models, the analysis should identify the stationary components (aleatory or purely random variables) and the not stationary components which define the mean function of the process under consideration. In the time (space) domain the models are parametric with a finite number of parameters. Among the parametric models are the autoregressive models (AR), moving average models (MA), autoregressive moving average models (ARMA), autoregressive integrated moving average models (ARIMA) and the state-space models. For the frequency domain, the models are non-parametric, and the procedures involve the decomposition of the series into frequency components, which present a spectrum being as a fundamental characteristic. Among these models used to analyze periodic phenomena of the data, spectral and cross spectral analyses have several applications in the soil-plant-atmosphere system.

When interested in the analysis of a series in the time (or space) domain, one of the most frequent assumptions is that the series is stationary, which means that the series propagates in an aleatory or purely random way along time (or space) with their statistical properties (mean and variance) being constant, reflecting some sort of a stable equilibrium. Many series, however, manifest some sort of non-stationarity. Hence, whenever a statistical procedure relies on the assumption of stationarity, it is usually necessary to transform the original data in order to satisfy the stationarity assumption.

With the simple definition of time series given above, "Time Series Analysis" is a well-defined area within statistics, not excluding the classical analyses, and questions answered by one cannot necessarily be answered by the other.

5.2 Autocorrelation Function ACF

After sampling a variable Y, mean and variance are calculated to provide quantitative information of the sampled population and to plan subsequent statistical analysis. In many cases, observations are not independent of each other, and by calculating an autocorrelation coefficient, the level of auto-dependence can be evaluated. For stationary processes (those in which the static properties are independent of space or time), the covariance between observations is a function of the number of lags h between their sampling points. Time series are collected along time at intervals of α (Eq. 5.2) in minutes, hours, months, etc, and space series along transects (or grids) at spacings h ($x_i - x_{i-1} = h$), in cm, m, km, etc. The covariance C(h) between such variables given by Salas et al. (1988) is

$$C(h) = \frac{1}{n-h}\sum_{i=1}^{n-h}\left[Y(x_{i+h}) - \overline{Y}\right]\left[Y(x_i) - \overline{Y}\right] \tag{5.3}$$

where $\overline{Y}$ is the arithmetic mean of the Y data set and n is the number of observations Y.

If C(h) is normalized dividing it by the variance s^2 of the sample, we obtain coefficients r(h) of the autocorrelation function

$$r(h) = \frac{C(h)}{s^2} \tag{5.4}$$

which manifest values between +1 and −1 for $i = 1$ to $i = n-h$. It is important to note that for the calculation of r(h), the observations Y have to be collected at regularly spaced h intervals. The values of r(h) for $h = 0$, which represents the correlation between $Y(x_i)$ and $Y(x_i)$ is obviously equal to 1. For the first neighbor pairs $Y(x_i)$ and $Y(x_{i+1})$ for a distance of one lag ($h = 1$), a value of r(1) can be obtained using Eqs. (5.3) and (5.4). The same procedure is used for second neighbor pairs [$Y(x_i)$ and $Y(x_{i+2})$], and further neighbors ($h = 3, 4, \ldots$) obtaining a r(h) value for each h. Plotting r as a function of h we obtain the autocorrelogram of the variable Y.

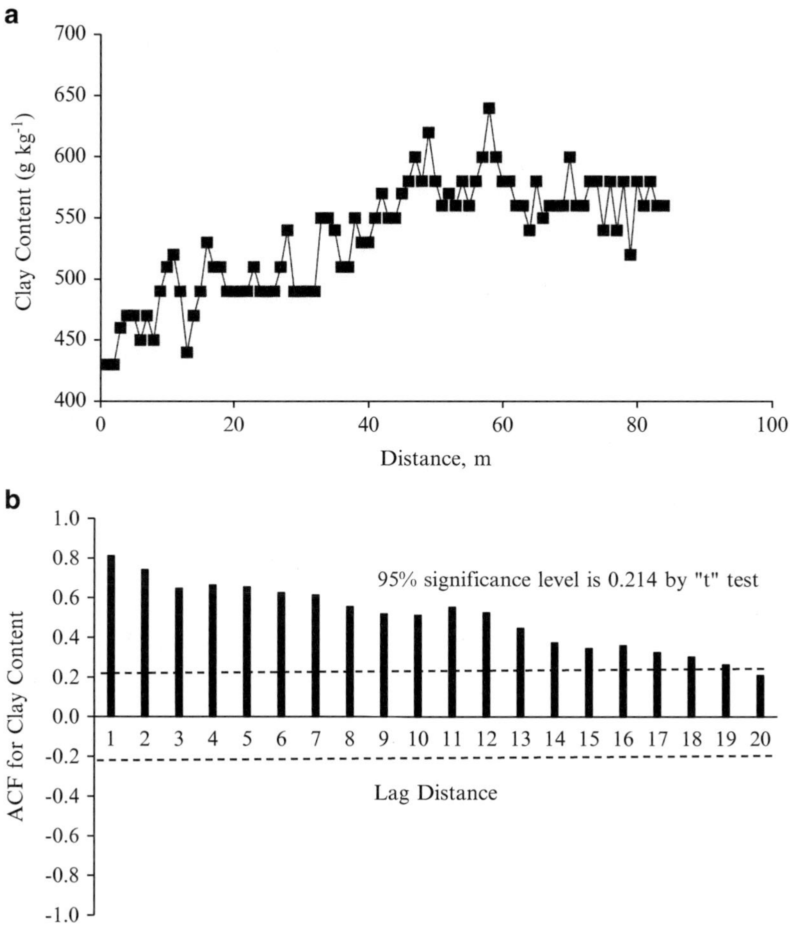

Fig. 5.1 (**a**) Clay content data distribution along the 84 m spatial transect; (**b**) Autocorrelation function (ACF) for soil Clay Content indicating that neighbor observations of clay contents are spatially dependent up to 19 lags (19 m in this example) using t test at 95 % of significance level

The next step is the calculation of the fiducial intervals of r, to recognize if they are significant or not, and in this way define the length interval in which the spatial dependence of the variable is significant. One way to measure the autocorrelation confidence interval CI is using the accumulated probability function (e.g., ±1.96 for a 95 % probability level) for the normalized distribution function (Davis 1986), and the number of observations n. Therefore,

$$\mathrm{CI} = \pm \frac{\mathrm{p}}{\sqrt{\mathrm{n}}} \tag{5.5}$$

Figure 5.1 shows as an example the (A) spatial distribution of clay contents along a 84 m spatial transect, and (B) autocorrelogram for clay content data indicating that neighbor observations of clay contents are spatially dependent up to 19 lags (19 m in this example) using the t test at the 95 % significance level.

5.3 Crosscorrelation Function CCF

Having two sets of variables $Y(x_i)$ and $W(x_i)$ observed at the same locations x_i (or same times t_i), their spatial crosscorrelation structure can be analyzed calculating coefficients of crosscorrelation. Although each variable has its own autocorrelogram, an analysis of their crosscorrelation indicates to which distance (or time interval) one is related to the other. The coefficient r_c of the crosscorrelation function will be also a function of h, and describes the degree of linear association between both variables (Nielsen and Wendroth 2003).

The coefficients of the crosscorrelation function $r_c(h)$, between the variables Y and W, separated by distances αh, or by a lag number h, are calculated with

$$rc(h) = \frac{cov_{YW}(h)}{s_Y \times s_W} \tag{5.6}$$

where

$$cov_{YW}(h) = \frac{1}{n-h}\sum_{i=1}^{n-h}\left[Y(x_i) - \overline{Y}\right]\left[W(x_{i+h}) - \overline{W}\right] \tag{5.7}$$

and s_Y^2 is the variance of Y

$$s_Y^2 = \frac{1}{n-1}\sum_{i=1}^{n}\left[Y(x_i) - \overline{Y}\right]^2 \tag{5.8}$$

and s_W^2 is the variance of W

$$s_W^2 = \frac{1}{n-1}\sum_{i=1}^{n}\left[W(x_i) - \overline{W}\right]^2 \tag{5.9}$$

A plot of r_c as a function of h represents the crosscorrelogram. For $h = 0$ (observations taken at the same position x_i), the value $r_c(0)$ given by Eq. (5.6) is the linear regression coefficient obtained through classical statistics. For the first neighbor pairs $[Y(x_i), W(x_{i+1})]$ collected at a distance α in one direction ($h = 1$), we obtain the coefficient r_c (1), and for the other direction ($h = -1$) the coefficient $r_c(-1)$. This is because in the case of two variables, each of them has different neighbors for each direction, i.e., we have two different pairs – (Y_i, W_{i+1}) and (Y_i, W_{i-1}). The same procedure is used for more distant neighbors, obtaining values

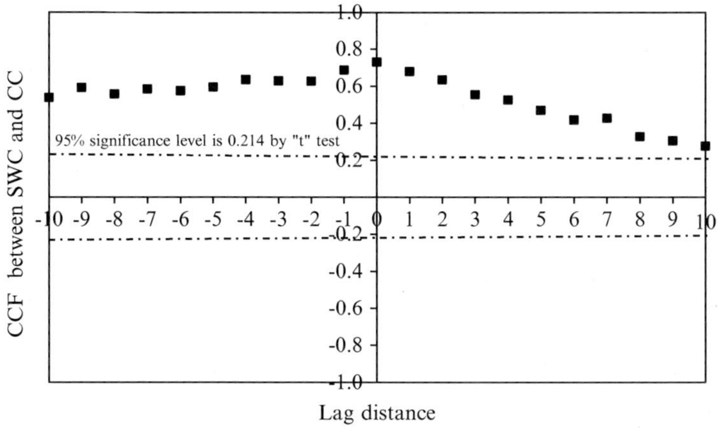

Fig. 5.2 Crosscorrelation function (CCF) between soil water content and clay content values along a 84 m spatial transect. *SWC* soil water content data set, *CC* clay content data set

of $r_c(h)$ and $r_c(-h)$. A crosscorrelogram indicates how far two different observations are spatially related (Wendroth et al. 1997).

According to Nielsen and Wendroth (2003), it is more difficult to estimate the significance of $r_c(h)$ as compared to $r(h)$. Significance tests like the t test are usually based on the assumption that the observed values of $Y(x_i)$ and $W(x_i)$ are normally distributed and independent among themselves. Taking this into consideration, the significance level of r_c is, in general, given by

$$t = \sqrt{\frac{(n - h) - 2}{1 - r_c^2}} \tag{5.10}$$

The level of significance of the test is obtained by comparing the value of t in Eq. (5.10) with critical values of t for (n-2) degrees of freedom. The crosscorrelation function is, in general, not symmetric, i.e., $r_c(h) \neq r_c(-h)$. However, when there is a known physical relation between Y and W, the crosscorrelation function will tend to symmetry (Nielsen and Wendroth 2003). Based on this argument, Timm et al. (2004) analyzed the spatial crosscorrelation structure between soil water content and clay content data sets collected along a 84 m spatial transect, estimating the level of significance of the crosscorrelation coefficients $[r_c(h)]$ by Eq. 5.5. Figure 5.2 shows that there is a strong spatial correlation between soil water content and clay content values at least 10 lags in both directions, using the t test at the 95 % significance level.

5.4 The State-Space Approach

The state-space model of a stochastic process involving j data sets $Y_j(t_i)$, all collected at the same time, is based on the property of Markovian systems that establishes the independence in the future of the process under consideration in relation to its past, once given the present state. In these systems, the state of the process under consideration condenses all information of the past needed to forecast the future. According to this definition, the State-Space Approach comes from the Time Series Analysis, however, it has been applied to study the spatial variability of data sets collected along spatial transects. From our knowledge, Morkoc et al. (1985) and Shumway (1988) were the first researchers to apply this approach in agronomy aiming to evaluate spatial data series.

The state-space approach is a combination of two systems of equations. The first is the observation equation

$$\mathbf{Y_j}(\mathbf{x_i}) = \mathbf{M_{jj}}(\mathbf{x_i})\, \mathbf{Z_j}(\mathbf{x_i}) + \mathbf{v}_{\mathbf{Y_j}}(\mathbf{x_i}) \tag{5.11}$$

where the observation vector $\mathbf{Y_j(x_i)}$ of the process under consideration is generated as a function of the state vector $\mathbf{Z_j(x_i)}$. The second is the state equation

$$\mathbf{Z_j}(\mathbf{x_i}) = \boldsymbol{\phi}_{\mathbf{jj}}\, \mathbf{Z_j}(\mathbf{x_{i-1}}) + \mathbf{u}_{\mathbf{Z_j}}(\mathbf{x_i}) \tag{5.12}$$

where the non observed state vector $\mathbf{Z_j(x_i)}$ is dynamically evolved. The observation vector $\mathbf{Y_j(x_i)}$ at position i is related to the state vector $\mathbf{Z_j(x_i)}$ at the same position i through the observation matrix $\mathbf{M_{jj}(x_i)}$ and by the observation error $\mathbf{v_{Yj}(x_i)}$ (Eq. 5.11). On the other hand, the state vector $\mathbf{Z_j(x_i)}$ at position i is related to the same vector at position i-1 through the state coefficient matrix $\boldsymbol{\phi}_{\mathbf{jj}}\mathbf{(x_i)}$ (transition matrix) and an error associated to the state $\mathbf{u_{Zj}(x_i)}$ with the structure of a first order autoregressive model. It is assumed that $v_j(x_i)$ and $u_j(x_i)$ are normally distributed and independent as well as being non correlated among themselves for all lags.

The matrix $\mathbf{M_{jj}}$ in Eq. (5.11) comes from the set of j linear observation equations

$$\begin{array}{l} Y_1(x_i) = m_{11}Z_1(x_i) + m_{12}Z_2(x_i) + \ldots\ldots\ldots + m_{1j}Z_j(x_i) + v_{Y_1}(x_i) \\ Y_2(x_i) = m_{21}Z_1(x_i) + m_{22}Z_2(x_i) + \ldots\ldots\ldots + m_{2j}Z_j(x_i) + v_{Y_2}(x_i) \\ \quad\vdots \qquad\qquad \vdots \qquad\qquad\quad \vdots \qquad\qquad \vdots \qquad\qquad \vdots \qquad\qquad \vdots \\ Y_j(x_i) = m_{j1}Z_1(x_i) + m_{j2}Z_2(x_i) + \ldots\ldots\ldots + m_{jj}Z_j(x_i) + v_{Y_j}(x_i) \end{array}$$

which can be written in the matrix form

$$\begin{bmatrix} Y_1(x_i) \\ Y_2(x_i) \\ \vdots \\ Y_j(x_i) \end{bmatrix} = \begin{bmatrix} m_{11} & m_{12}\ldots\ldots\ldots m_{1j} \\ m_{21} & m_{22}\ldots\ldots\ldots m_{2j} \\ \vdots & \vdots \qquad\quad \vdots \\ m_{j1} & m_{j2}\ldots\ldots\ldots m_{jj} \end{bmatrix} \times \begin{bmatrix} Z_1(x_i) \\ Z_2(x_i) \\ \vdots \\ Z_j(x_i) \end{bmatrix} + \begin{bmatrix} v_{Y_1}(x_i) \\ v_{Y_2}(x_i) \\ \vdots \\ v_{Y_j}(x_i) \end{bmatrix}$$

The matrix $\boldsymbol{\phi}_{\mathbf{jj}}$ in Eq. (5.12) comes from the following set of state equations

$$\begin{aligned}
Z_1(x_i) &= \phi_{11}Z_1(x_{i-1}) + \phi_{12}Z_2(x_{i-1}) + \dots\dots\dots + \phi_{1j}Z_j(x_{i-1}) + u_{Z_1}(x_i)\\
Z_2(x_i) &= \phi_{21}Z_1(x_{i-1}) + \phi_{22}Z_2(x_{i-1}) + \dots\dots\dots + \phi_{2j}Z_j(x_{i-1}) + u_{Z_2}(x_i)\\
&\vdots\\
Z_j(x_i) &= \phi_{j1}Z_1(x_{i-1}) + \phi_{j2}Z_2(x_{i-1}) + \dots\dots\dots + \phi_{jj}Z_j(x_{i-1}) + u_{Z_j}(x_i)
\end{aligned}$$

or in the matrix form:

$$\begin{bmatrix} Z_1(x_i) \\ Z_2(x_i) \\ \vdots \\ Z_j(x_i) \end{bmatrix} = \begin{bmatrix} \phi_{11} & \phi_{12}\cdots\cdots\phi_{1j} \\ \phi_{21} & \phi_{22}\cdots\cdots\phi_{2j} \\ \vdots & \vdots \quad\quad \vdots \\ \phi_{j1} & \phi_{j2}\cdots\cdots\phi_{jj} \end{bmatrix} \times \begin{bmatrix} Z_1(x_{i-1}) \\ Z_2(x_{i-1}) \\ \vdots \\ Z_j(x_{i-1}) \end{bmatrix} + \begin{bmatrix} u_{Z_1}(x_i) \\ u_{Z_2}(x_i) \\ \vdots \\ u_{Z_j}(x_i) \end{bmatrix}$$

The above equations contain distinct perturbations or noises, one associated with observations $v_{Yj}(x_i)$ and the other with state $u_{Zj}(x_i)$. According to Gelb (1974), the development of methods to process noise-contaminated observations can be credited to the work carried out by Gauss and Legendre (around 1800) who, independently, developed the method of the minimum squares for linear models. Plackett (1950) developed a recursive solution for the minimum square method into linear models. Kalman (1960) using a state-space formulation, developed a very good recursive filter for estimations in stochastic, dynamic linear systems, being well known today as the Kalman Filter (KF). According to Gelb (1974), a good estimator is a computational algorithm that processes observations in order to find a minimum estimate (following some sort of optimization criterion) of the state error of a system, using: (a) the knowledge of the dynamics of the observations and of the system, (b) assuming statistical inferences for the noises associated to observations and to states and (c) knowledge of the initial condition of the information. In summary, given a dynamic system of equations that describes the behavior of the vectors of state and of observations, the statistic models that characterize the observational and state errors, and an initial condition of the information, the KF performs the sequential actualization of the state vector at space i-1 (or time t−1) to space i (or time t). It can therefore be said, that the KF is essentially a recursive solution that permits a sequential processing of the observations, within the original method of the minimum squares of Gauss. It should be noted however, that another algorithm has to be used [for example, the algorithm of maximum likelihood (EM) thoroughly discussed in Shumway (1988), Shumway and Stoffer (2000, 2011)] so that, together with the KF, the problem of noise-contaminated observations can be solved (Gelb 1974).

Depending on the objectives of a study involving the state-space methodology, one can have three different types of estimates: (a) when the time (or space) at which an estimate is wished coincides with the last observed value, the problem is said to be one of filtering; (b) when the time (or space) of interest is inside the set of observations, i.e., the complete set of data is used to estimate the point of interest, the problem is said to be one of smoothing; and (c) when the time (or space) of interest is after the last observation, the problem is said to be one of forecasting.

From this it can be seen that any linear or non-linear model (Katul et al. 1993; Wendroth et al. 1993) can be represented in the state-space formulation, i.e., by a system of two equations: one for the observation vectors and another for the evolution of the state vectors.

The linear system of dynamic equations (5.11 and 5.12) through space has been presented here in a generalized form of the state-space approach. Now we shall present two different ways of using this approach, the first presented by Shumway (1988) and Shumway and Stoffer (2000, 2011) which has been used by several researchers in agronomy, giving emphasis to the equation of the evolution of the state of the system (Eq. 5.12); and the second, introduced by West and Harrison (1989, 1997) which is still not so frequently used in agronomy, giving emphasis to the observation equation (5.11). Applications of these two different state-space approaches can be found in Morkoc et al. (1985), Hui et al. (1998), Dourado-Neto et al. (1999), Timm et al. (2003a, b, 2004, 2006, 2011), Wendroth et al. (2003, 2006), Joschko et al. (2009), Jia et al. (2011), Oliveira et al. (2011), Liu et al. (2012), among others.

5.4.1 Shumway's State-Space Approach

This approach, presented by Shumway (1988) and Shumway and Stoffer (2000, 2011), gives more attention to the equation of the evolution of the state of the system, where the matrix of the transition coefficients ϕ in Eq. (5.12) is a matrix of dimension jxj that indicates the spatial measure of the linear association among the j variables of interest. These coefficients are optimized through a recursive procedure, using an algorithm of the KF type (Gelb 1974; Shumway and Stoffer 1982) in which the method of maximum likelihood is used together with the mean maximization algorithm of Dempster et al. (1977). In this case, Eqs. (5.11 and 5.12) are solved assuming initial values for the mean and the variance of each variable in the covariance matrix **R** of the noise of the observations, for the covariance matrix **Q** of the noise associated with the state vector, for the matrix ϕ of the transition coefficients, and for the observation matrix **M**. Because Shumway (1988) considers the matrix **M** as being an identity matrix for all positions i (or for all times t) Eq. (5.11) becomes

$$\mathbf{Y_j}(\mathbf{x_i}) = \mathbf{Z_j}(\mathbf{x_i}) + \mathbf{v}_{\mathbf{Y_j}}(\mathbf{x_i}) \tag{5.11a}$$

During the development of the software ASTSA (Applied Statistical Time Series Analysis) which is used for the analysis of time (space) series, the matrix **M** is an identity matrix during all steps of variable estimation through space or time (Shumway 1988). This shows the greater emphasis of this approach in being referenced to the equation of state evolution (Eq. 5.12), and not to the observation equation (Eq. 5.11). More specific details can be found in Shumway (1988), Shumway and Stoffer (2000), Nielsen and Wendroth (2003) and Shumway and Stoffer (2011).

An application of the Shumway's state-space approach will be presented here using space and time data series. Space data series are from a sugarcane field experiment and time data series are from a coffee field experiment both carried out in Piracicaba, SP, Brazil.

Example 1: Application of the Shumway's State-Space Approach Using Space Data Series

The state-space methodology is illustrated to evaluate the sugarcane soil-plant system, using six variables measured along a spatial transect of 84 sampling points. Sugarcane is a semi-perennial crop, once planted being harvested for canes to extract sugar after every year, up to 5–7 years. After the first harvest, called planted cane, the following crops grown from sprouted rhizomes are called ratoons. The crop was planted on a 15 × 100 m area, with a spacing of 1.4 m between rows, using a randomized block design with four treatments and four replicates each. All treatments with borders were disposed along crop lines, so that a spatial transect of 84 sampling points (Fig. 5.3) could be obtained to measure several variables of interest.

The crop variable used here is the number of cane stalks (NCS) per meter of row (Fig. 5.4), which represents the yield and is related to physical and chemical soil properties, such as availability of phosphorus (P, Fig. 5.5), calcium (Ca, Fig. 5.6), magnesium (Mg, Fig. 5.7), clay content (CC, Fig. 5.1a) and aggregate stability (AS, Fig. 5.8). We present a qualitative discussion of Shumway's (1988) state-space formulation, to identify the variables that most contribute to this spatial process, in order to obtain information for a better management of the available soil resources and improve sugarcane yield. Autocorrelation function ACF (Eq. 5.4) for each series indicate spatial correlation structures for various numbers of lags. As an example we show in Fig. 5.9 the cases of NCS (Fig. 5.9a) and P (Fig. 5.9b). The ACF is a tool that reflects the local variation between neighbor observations separated by a varying number of lags, being used to check the spatial dependence of a given data series.

As the interest of this study was the spatial correlation structure between two variables and, having this in mind the crosscorrelation function CCF (Eq. 5.6) was calculated and the crosscorrelogram plotted for number of cane stalks and all other variables, P, Ca, Mg, CC, and AS (Fig. 5.10 is one example). These crosscorrelograms justify the application of the Shumway's (1988) state-space formulation in different scenarios. Table 5.1 presents thirty one different combinations of these variables yielded state equations and coefficients that were used to evaluate the behavior of the model with respect to the estimates of the observed values, looking for points that fall out of the fiducial limits. Three of them are shown as examples in Figs. 5.11, 5.12 and 5.13. The normalization through Eq. (5.13) has also the advantage of eliminating differences of the order of magnitude of the values of

Fig. 5.3 Schematic experimental design showing the 15 cane lines, each 100 m long, indicating the *central line* used to measure physical and chemical soil properties. *B* border, *T* treatments, *R* replicate

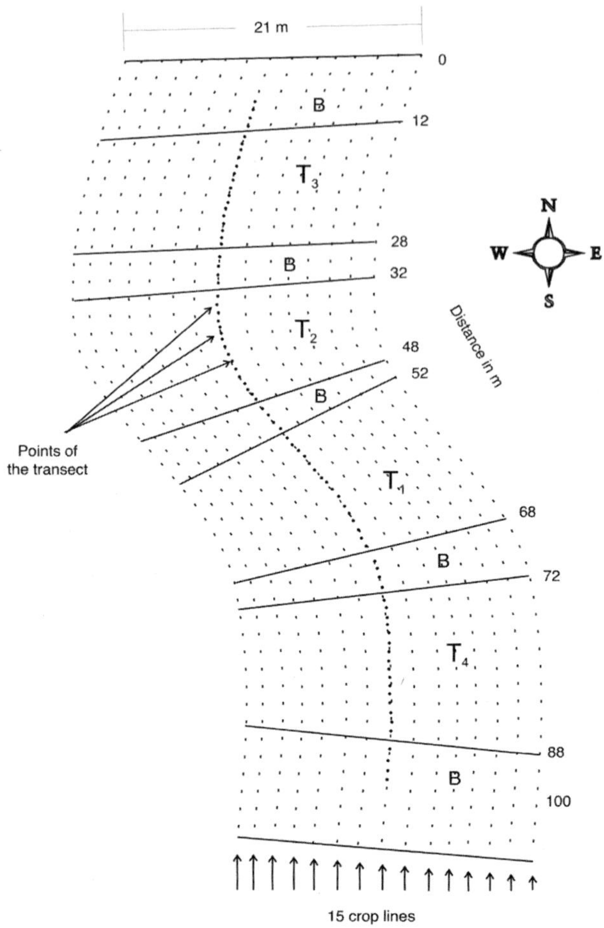

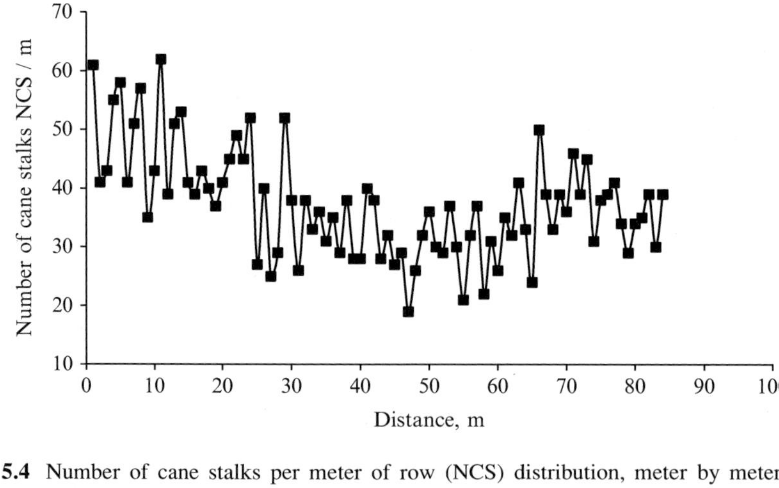

Fig. 5.4 Number of cane stalks per meter of row (NCS) distribution, meter by meter, along 84 point transect

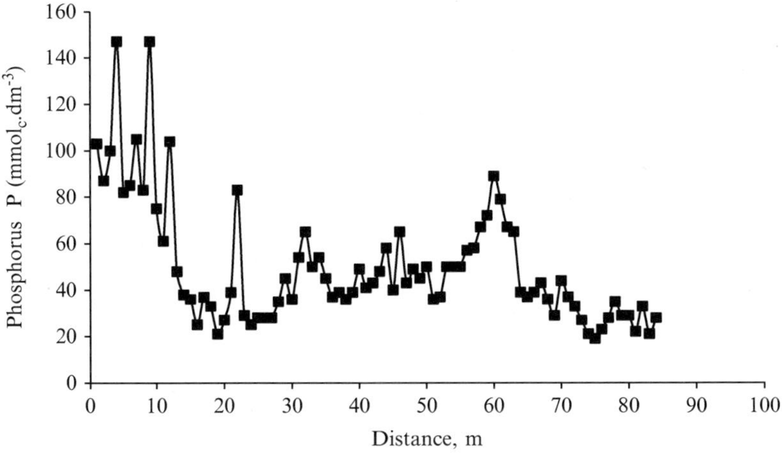

Fig. 5.5 Available soil phosphorus (P) distribution, meter by meter, along 84 point transect

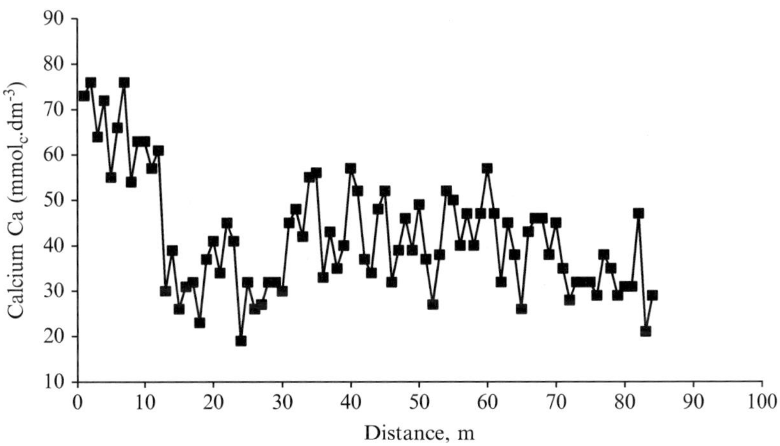

Fig. 5.6 Available soil calcium (Ca) distribution, meter by meter, along 84 point transect

each variable. As mentioned before, the values of the coefficients of the matrix $\boldsymbol{\phi}$ (Eq. 5.12) identify the contribution of each variable and give a measure of the ability of the model in describing the studied process. They are optimized via algorithms of KF type, comparing predicted values with observed values, at each actualization step leading to a Kalman gain, making it possible to separate observations into signal and noise, with the guarantee that the noise is a random process of zero mean. In this way, the state-space model integrates the influence of local effects when a new observation is available, providing a greater flexibility in relation to a global adjustment of the coefficients (multiple static regression analysis). This allows for a better and more precise degree of explanation of the variables under study.

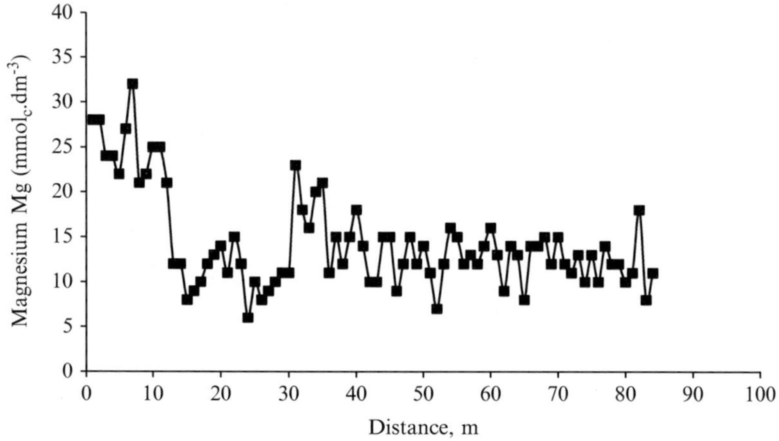

Fig. 5.7 Available soil magnesium (Mg) distribution, meter by meter, along 84 point transect

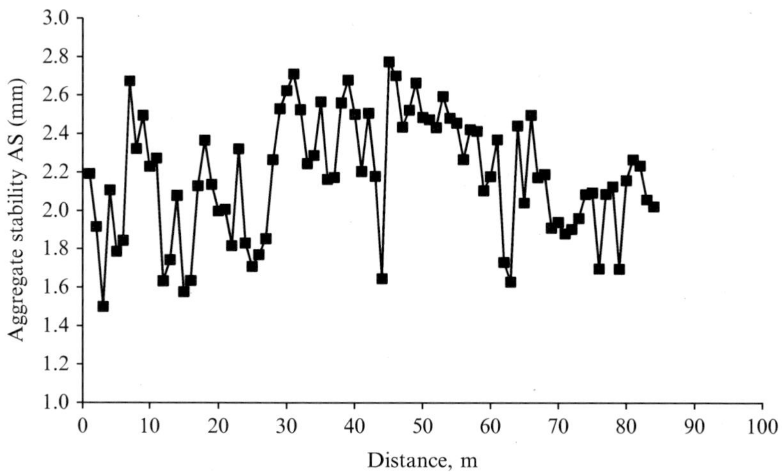

Fig. 5.8 Aggregate stability distribution, meter by meter, along the 84 point transect

From Eq. 5.12, it can be seen that the transition coefficients $\boldsymbol{\phi}$ relate the variable Z_j at position i to its value at position i-1. Hui et al. (1998) show that when data is normalized before the application of the state-space methodology, the magnitude of the coefficients $\boldsymbol{\phi}$ become directly proportional to the contribution of each variable in the estimation of $Z_j(x_i)$. They suggest the following normalization of $Z_j(x_i)$ to obtain $z_j(x_i)$:

$$z_j(x_i) = \frac{Z_j(x_i)\text{-}\left[\overline{Z_j}\text{-}2s\right]}{4s} \tag{5.13}$$

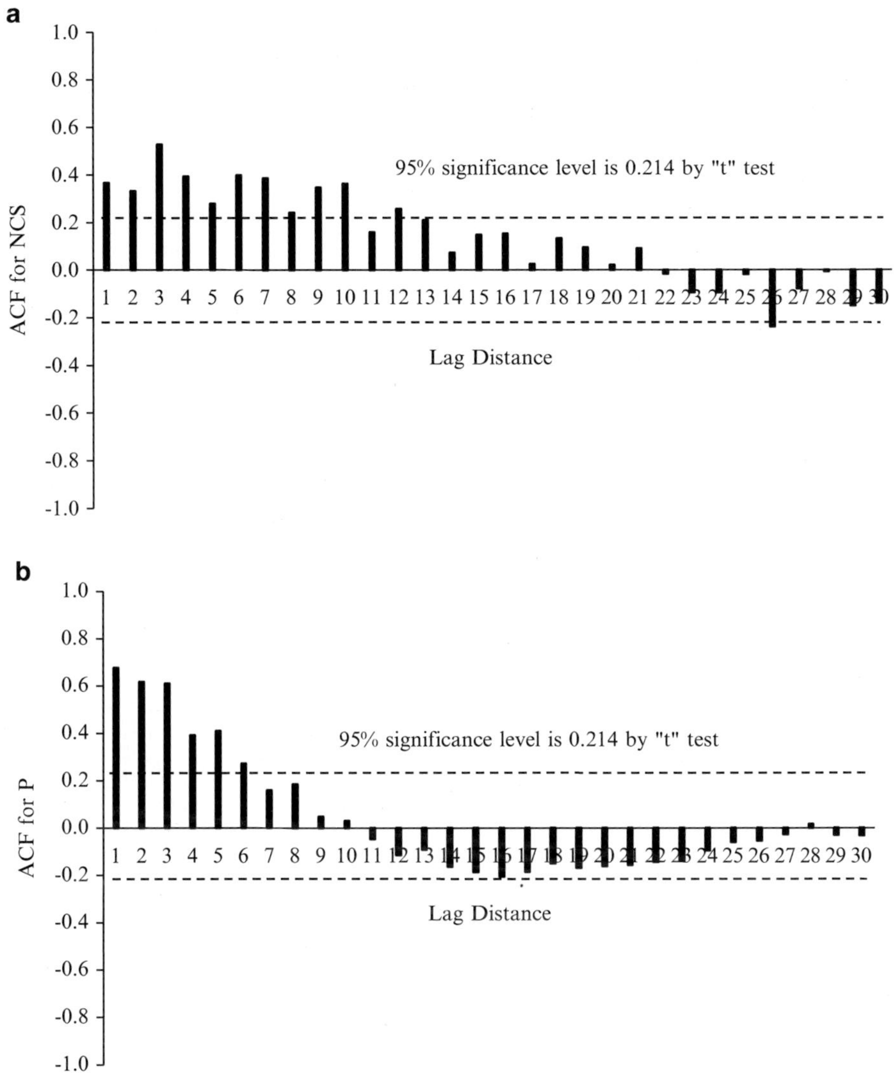

Fig. 5.9 (**a**) Calculated autocorrelation function (ACF) for number of cane stalks per meter of row (NCS); and (**b**) Calculated autocorrelation function (ACF) for phosphorus content (P)

$\overline{Z_j}$ and s are the mean value and the standard deviation, respectively, of the original data $Z_j(x_i)$.

This above example is more completely reported in Timm et al. (2003a) and Reichardt and Timm (2012).

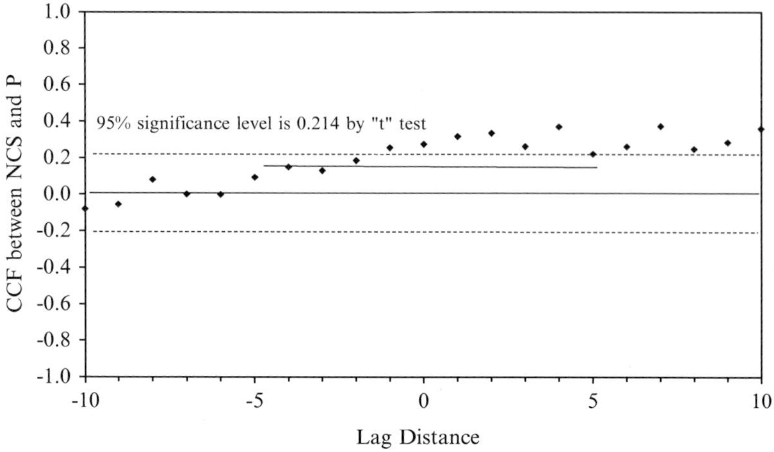

Fig. 5.10 Calculated crosscorrelogram function (CCF) between number of cane stalks per meter of row (NCS) and soil phosphorus (P) data

Example 2: Application of the Shumway's State-Space Approach Using Time Data Series

This example illustrates the application of the Shumway's state-space approach (called here as state-time approach) using time data series collected in a coffee crop grown in Piracicaba, SP, Brazil (22° 42′ 30″ S; 47° 38′ 00″ W, 580 m asl). The coffee (*Coffea arabica* L.), was of the cultivar "Catuaí Vermelho" (IAC-144) and is a perennial crop, 3 to 5 years old during the experimental period, which is the beginning of the yearly coffee production cycles. The spacing between plants was 0.75 m and between rows 1.5 m. Rows were kept bare chemically and manually, as commonly done in coffee plantations. The soil is a Rhodic Kandiudalf (Soil Survey Division Staff 1993), locally called "Nitossolo Vermelho Eutroférrico" (EMBRAPA 2006); and the climate is of the Cwa type (Köppen 1931), with dry winter.

Soil water contents θ (i) were measured along a horizontal domain x_i (m) at 15 locations (i = 1, 2,. . .,15), and at five depths z_k (m), 0.2, 0.4, 0.6, 0.8, and 1.0 m from surface (k = 1, 2,. . .,5), every 14 days, at times t_j (j = 1, 2, 3,. . .,52) covering a two year period starting on September 01, 2003. Soil water content measurements obtained with a neutron probe (model CPN 503 DR) were not taken at regular spacings along a leveled contour line of the horizontal domain corresponding to a coffee row, but following the distribution of five fertilizer plots arranged within a 0.2 ha coffee field. Details of the fertilizer trial can be found elsewhere (Fenilli et al. 2007). Measurements of θ were made using aluminum neutron probe access tubes installed below crop canopies.

Slow neutron counting data were transformed into soil water contents (θ) using calibration curves established as suggested by Reichardt et al. (1997), taken as valid

Table 5.1 State-space equations of number of stalk canes using soil phosphorus (P), soil calcium (Ca), soil magnesium (Mg), soil clay content (CC), and soil aggregate stability (AS) data and values of R^2 from linear regression between estimated and measured values of NCS

Equation	R^2
$NCS_i = 0.857NCS_{i-1} - 0.106P_{i-1} + 0.026Ca_{i-1} + 0.267Mg_{i-1} + 0.163CC_{i-1} - 0.221AS_{i-1} + u_{NCSi}$	0.502
$NCS_i = 0.876NCS_{i-1} + 0.098P_{i-1} - 0.144Ca_{i-1} + 0.133Mg_{i-1} + 0.0286CC_{i-1} + u_{NCSi}$	0.532
$NCS_i = 0.971NCS_{i-1} - 0.005P_{i-1} + 0.148Ca_{i-1} - 0.108Mg_{i-1} - 0.016AS_{i-1} + u_{NCSi}$	0.502
$NCS_i = 0.898NCS_{i-1} + 0.352P_{i-1} - 0.279Ca_{i-1} + 0.040CC_{i-1} - 0.028AS_{i-1} + u_{NCSi}$	0.521
$NCS_i = 0.817NCS_{i-1} - 0.094P_{i-1} + 0.314Mg_{i-1} + 0.165CC_{i-1} - 0.218AS_{i-1} + u_{NCSi}$	0.579
$NCS_i = 0.785NCS_{i-1} - 0.197Ca_{i-1} + 0.420Mg_{i-1} + 0.159CC_{i-1} - 0.182AS_{i-1} + u_{NCSi}$	0.521
$NCS_i = 0.951NCS_{i-1} + 0.068P_{i-1} + 0.048Ca_{i-1} - 0.079Mg_{i-1} + u_{NCSi}$	0.483
$NCS_i = 0.920NCS_{i-1} + 0.050P_{i-1} + 0.002Ca_{i-1} + 0.018CC_{i-1} + u_{NCSi}$	0.488
$NCS_i = 0.902NCS_{i-1} + 0.250P_{i-1} - 0.192Ca_{i-1} + 0.027AS_{i-1} + u_{NCSi}$	0.483
$NCS_i = 0.926NCS_{i-1} + 0.069P_{i-1} - 0.025Mg_{i-1} + 0.019CC_{i-1} + u_{NCSi}$	0.508
$NCS_i = 0.968NCS_{i-1} + 0.123P_{i-1} - 0.084Mg_{i-1} - 0.018AS_{i-1} + u_{NCSi}$	0.457
$NCS_i = 0.925NCS_{i-1} + 0.095P_{i-1} + 0.082CC_{i-1} - 0.115AS_{i-1} + u_{NCSi}$	0.503
$NCS_i = 0.879NCS_{i-1} - 0.012Ca_{i-1} + 0.097Mg_{i-1} + 0.027CC_{i-1} + u_{NCSi}$	0.503
$NCS_i = 0.937NCS_{i-1} + 0.160Ca_{i-1} - 0.094Mg_{i-1} - 0.018AS_{i-1} + u_{NCSi}$	0.489
$NCS_i = 0.959NCS_{i-1} + 0.056Ca_{i-1} + 0.062CC_{i-1} - 0.088AS_{i-1} + u_{NCSi}$	0.512
$NCS_i = 0.918NCS_{i-1} + 0.103Mg_{i-1} + 0.099CC_{i-1} - 0.134AS_{i-1} + u_{NCSi}$	0.494
$NCS_i = 0.942NCS_{i-1} - 0.039P_{i-1} + 0.087Ca_{i-1} + u_{NCSi}$	0.472
$NCS_i = 0.946NCS_{i-1} + 0.127P_{i-1} - 0.085Mg_{i-1} + u_{NCSi}$	0.489
$NCS_i = 0.912NCS_{i-1} + 0.058P_{i-1} + 0.019CC_{i-1} + u_{NCSi}$	0.526
$NCS_i = 0.923NCS_{i-1} + 0.062P_{i-1} + 0.004AS_{i-1} + u_{NCSi}$	0.477
$NCS_i = 0.938NCS_{i-1} + 0.148Ca_{i-1} - 0.097Mg_{i-1} + u_{NCSi}$	0.491
$NCS_i = 0.923NCS_{i-1} + 0.050Ca_{i-1} + 0.018CC_{i-1} + u_{NCSi}$	0.524
$NCS_i = 0.929NCS_{i-1} + 0.065Ca_{i-1} - 0.003AS_{i-1} + u_{NCSi}$	0.446
$NCS_i = 0.904NCS_{i-1} + 0.063Mg_{i-1} + 0.023CC_{i-1} + u_{NCSi}$	0.530
$NCS_i = 0.914NCS_{i-1} + 0.060Mg_{i-1} + 0.015AS_{i-1} + u_{NCSi}$	0.492
$NCS_i = 0.972NCS_{i-1} + 0.035CC_{i-1} - 0.018AS_{i-1} + u_{NCSi}$	0.478
$NCS_i = 0.915NCS_{i-1} + 0.073P_{i-1} + u_{NCSi}$	0.510
$NCS_i = 0.924NCS_{i-1} + 0.065Ca_{i-1} + u_{NCSi}$	0.501
$NCS_i = 0.920NCS_{i-1} + 0.068Mg_{i-1} + u_{NCSi}$	0.501
$NCS_i = 0.958NCS_{i-1} + 0.030CC_{i-1} + u_{NCSi}$	0.528
$NCS_i = 0.962NCS_{i-1} + 0.026AS_{i-1} + u_{NCSi}$	0.528

All observations have been scaled using Eq. 5.13

$$(NCS)_i = 0.8571*(NCS)_{i-1} - 0.1055*(P)_{i-1} + 0.0262*(Ca)_{i-1} + 0.2665*(Mg)_{i-1} + 0.1625*(CC)_{i-1} - 0.2214*(AS)_{i-1} + u_{(NCS)i}$$

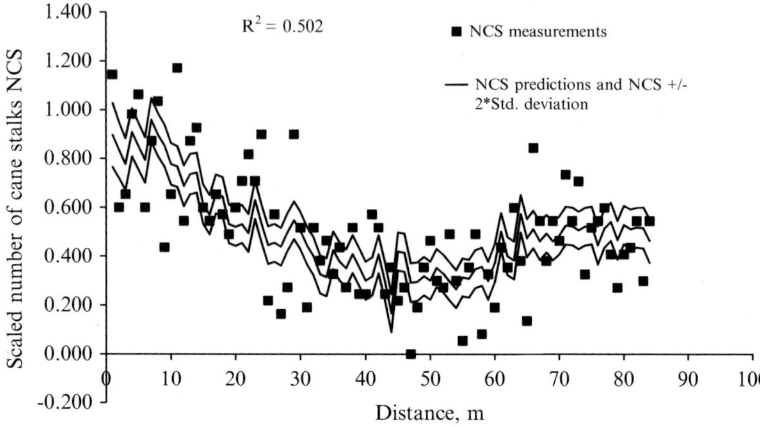

Fig. 5.11 State-space analysis of scaled number of cane stalks data as a function of scaled number of cane stalks, scaled phosphorus content, scaled calcium content, scaled magnesium content, scaled clay content, and scaled aggregate stability, all at location i-1

$$(NCS)_i = 0.9183*(NCS)_{i-1} + 0.1033*(Mg)_{i-1} + 0.0992*(CC)_{i-1} - 0.1338*(AS)_{i-1} + u_{(NCS)i}$$

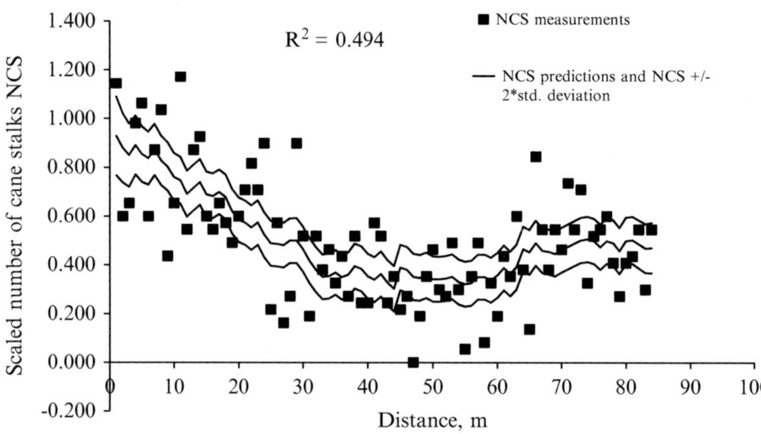

Fig. 5.12 State-space analysis of scaled number of cane stalks data as a function of scaled number of cane stalks, scaled magnesium content, scaled clay content, and scaled aggregate stability, all at location i-1

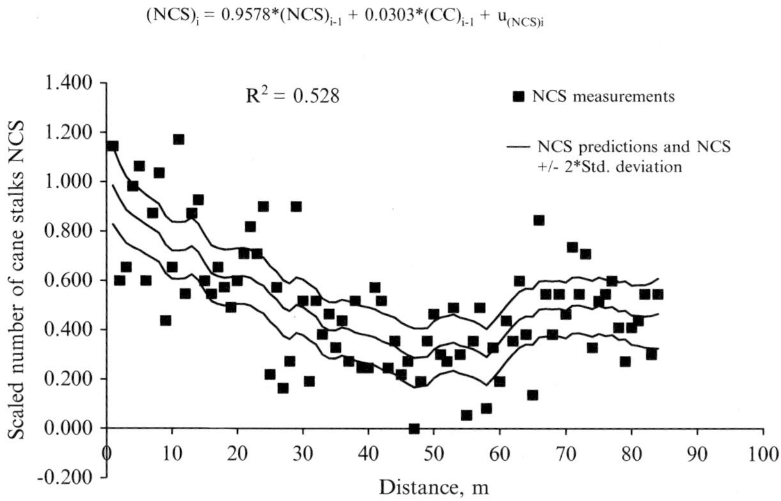

Fig. 5.13 State-space analysis of scaled number of cane stalks data as a function of scaled number of cane stalks and scaled clay content, all at location i-1

over all depths. Soil water storages at times j and positions i, $S_j(i)$ (mm) for the 0 – 1.0 m soil layer were calculated from $\theta_{t,x}(k)$ data by the trapezoidal rule:

$$S_j(i) = \left[1.5\theta_{i,j}(1) + \theta_{i,j}(2) + \theta_{i,j}(3) + \theta_{i,j}(4) + 0.5\theta_{i,j}(5)\right]\frac{1000}{5} \tag{5.14}$$

with $\Delta z = 0.2$ m. Soil water contents $\theta_{i,j}(1)$ measured at the depth 0.2 m (k = 1) were considered to cover a layer of $1.5\Delta z = 0.3$ m which includes soil surface. This first measurement made at the depth of 0.2 m was evaluated to be deep enough not to lose slow neutrons to the atmosphere. $\theta_{i,j}(5)$ measured at 1.0 m (k = 5) covered $0.5\Delta z = 0.1$ m since the lower level of the control volume for water balances was set at 1.0 m, and the total depth L was taken as 1,000 mm to obtain data in mm. The coffee root system was assumed not to reach depths below z = 1.0 m, which was confirmed by Silva et al. (2009).

The time variability structure of the mean soil water storage $\overline{S_K}$ data was studied using the state-time approach (Shumway 1988; Shumway and Stoffer 2000; Shumway and Stoffer 2011) which provides opportunities for a suitable identification of temporal relations between soil-atmosphere-plant variables taking into account their temporal association. The state-time analysis characterizes the state of a system (set of p unobservable variables) at time t to its state at a time t-k, k = 1, 2, 3, . . ., 52, in our study. For k = 1, the state-space approach is described as follows [note here that the state equation (Eq. 5.12) is presented substituting x by t and the error associated to the state u is substituted by w]:

$$Z_t = \phi Z_{t-1} + w_{Z_t} \tag{5.15}$$

Z_t and Z_{t-1} being the state vector (a set of p unobservable variables) at time t and t−1, respectively; ϕ is the p x p matrix of state coefficients, which is the matrix of the regression coefficients; and w_{Z_t} noises of the system for t = 1, 2, 3,. . ., k. If these Z variables were observable, this would be the usual structure of a vector autoregressive model, in which the coefficients of the matrix ϕ could be estimated by multiple regression techniques, taking Z_t and Z_{t-1} as the dependent and independent variables, respectively. In the case of the state-time model, however, the true state of the variables is considered "embedded" in an observation equation:

$$Y_t = MZ_t + \nu_{Y_t} \tag{5.16}$$

the observation vector Y_t being related to the state vector Z_t by an observation matrix M (usually an identity matrix, p x p, as mentioned before) and an observation noise vector ν_{Y_t}. The noises w_{Z_t} and ν_{Y_t} are assumed to be independent of each other. The state coefficients of the matrix ϕ and noise variances of Eq. 5.15 are estimated through a recursive procedure given by Shumway and Stoffer (1982). The Z_t data are scaled with respect to their mean and standard deviation (s), according to Eq. 5.13. As cited before, this transformation allows state coefficients of the matrix ϕ have magnitudes directly proportional to their contribution to each state variable used in the analysis. The software Applied Statistical Time Series Analysis (ASTSA) (Shumway 1988) was used for applying the state-space approach.

Concomitantly to $S_k(i)$ measurements, Silva et al. (2006) evaluated time series of evapotranspiration $ET_k(i)$, rainfall $P_k(i)$, supplementary sprinkler irrigation $I_k(i)$, surface runoff $RO_k(i)$, and soil water drainage fluxes $Q_k(i)$ below the 1.0 m depth, to establish complete water balances, which were used in the state-time and multiple regression analyses. Irrigation was applied only during the dry winter, in just a few events when the available water capacity reached about 25 % of its maximum. For the analysis, the irrigation (I) was added to the rainfall (P). For a few 14 day intervals with no rainfall (P = 0) during the rainy season, a negligible value of P = 0.1 mm was assumed for this variable, avoiding the zeros to permit the state-time analysis. It is important to mention that classical multiple regression is based on mean values of each variable throughout the time being investigated and that the magnitudes of each variable at a given time compared to their respective values at a previous or future time are neglected.

The state–time analysis allows a better insight of the relations among the climate variables that determine soil water storage S. So, in order to better understand the temporal relations between S, P, and ET, a discussion is made comparing the state-time analysis to the classical multiple regression using the same state variables. Figure 5.14a, b show the multiple regression and state-time equations and the value of their coefficients of determination (r^2) from linear regressions between estimated and measured values of scaled (Eq. 5.13) soil water storage. Classical multiple regression is based on mean values of each variable throughout the time being investigated, in which the magnitudes of each variable at a given time compared to their respective values at a previous or future time are neglected, so that no more than 35.8 % of the variance of the biweekly-measured soil water storage data was

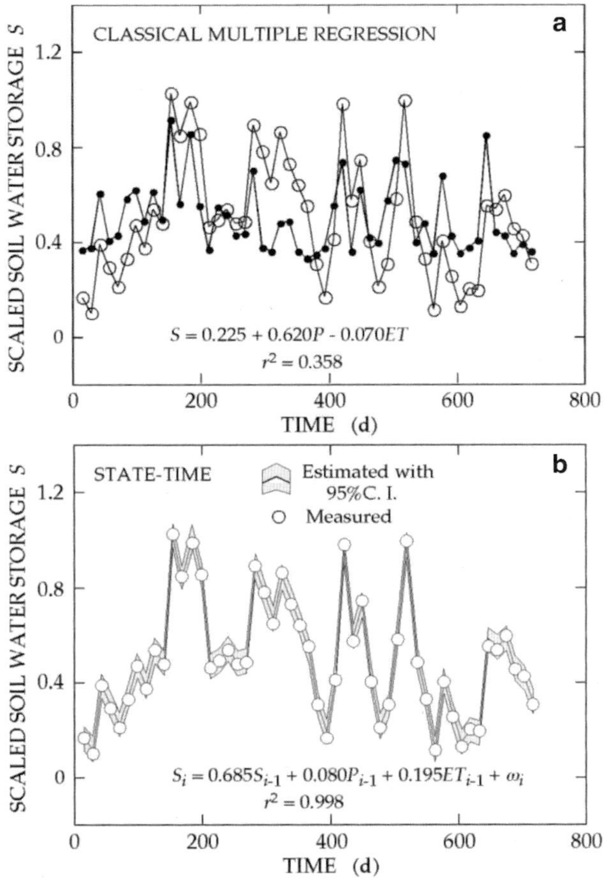

Fig. 5.14 Estimates of soil water storage measured biweekly for 714 days using (**a**) classical multiple regression and (**b**) state-time analysis

explained from the measurements of precipitation and evapotranspiration (Fig. 5.14a). Estimated values by regression are much less variable than those measured, and consistently underestimate the larger and overestimate the smaller measured values.

When the temporal associations among soil water storage, precipitation, and evapotranspiration data were considered, 99.8 % of the variance of the soil water storage was explained from the use of the state-time analysis (Fig. 5.14b). We note that nearly 70 % of the previous value S_{i-1} contributes to that of S_i while preceding values P_{i-1} and ET_{i-1} contribute only 8 % and 20 %, respectively.

The major experimental consideration influencing the utility of state-time analyses is the time interval between successive measurements that allows the possibility of state variables to be temporally associated. In other words, measurements taken during very short time intervals will tend to be autocorrelated or cross correlated with each other. However, with increasing time, the state variables change their magnitudes as environmental conditions change. We know that a water balance for a given soil profile is the result of five processes that occur as a

Table 5.2 Autocorrelation and cross correlation coefficients for state variables soil water storage S, precipitation P, and evapotranspiration ET

	Autocorrelation coefficient r(h)≪			Crosscorrelation coefficient r_c(h)≪		
lag h≪≪	S	P	ET	S vs P	ET vs S	ET vs P
0	1	1	1	0.595	0.153	0.359
1	0.551	0.163	0.558	0.370	0.005	0.507
2	0.257	0.119	0.444	0.203	−0.053	0.316
3	0.038	0.024	0.344	0.033	0.036	0.375
4	−0.005	0.082	0.185	−0.050	0.159	0.072
5	0.025	0.081	0.099	−0.027	0.126	−0.028

≪The 95 % significance level of r and r_c is 0.2745. ≪≪A lag of h = 1 is equal to 14 days

function of time – precipitation plus irrigation, surface runoff, evapotranspiration, storage of water in the soil profile and the drainage of water from the soil profile. Each of these processes quantified by Silva et al. (2006), who provide data for this study indicated that surface runoff was negligible over the two year period and that the drainage of water from the soil profile has yielded accurate measurements of water storage S_i in the profile. Hence, neglecting surface runoff, the use of only three state variables (S, P and ET) in the state-time analysis accounts for the physical processes responsible for a quantitative estimate of S. The temporal autocorrelation and cross correlations functions (Eqs. 5.4 and 5.6, respectively) given in Table 5.2 indicate that ET, S and P have autocorrelation lengths of 3, about 2 and less than 1 lag, respectively. In other words, values of ET are related to each other during more than 3 consecutive sampling dates (42 days), those of *S* during no more than 2 consecutive sampling dates (28 days) and those of P are essentially not related to each other between consecutive sampling dates (14 days). All three values of lag are reasonable, including that for precipitation. Indeed, the general nature of rainfall is more seasonal and does not consistently repeat its relative magnitude with a 2-week periodicity through a 2-year period.

Examining the cross correlation coefficients in Table 5.2, it can be seen that ET is related to P for more than 3 consecutive sampling dates (42 days) and that S is related to P for at least 2 consecutive sampling dates (28 days). The fact that ET and S showed essentially not to be related to each other between sampling dates is not obvious since in many occasions the actual value of ET was much below the potential value. However, during the 2-year period, regardless of the daily and biweekly fluctuations of local weather conditions, every effort was made to irrigate the field in a timely manner to provide adequate amounts of water stored in the root zone.

There are several methods available to examine the reliability of state-time analyses (see for instance, Shumway and Stoffer 2011). Here, we choose (on the basis of the information in Table 5.2) as an example to observe the impact of omitting increasing numbers of observations from the calculations of the state variable being estimated. Figure 5.15 shows, as an example, the results when soil water storage is estimated with all measurements of P and ET, but with increasing numbers of its biweekly measurements omitted from the state-time analysis.

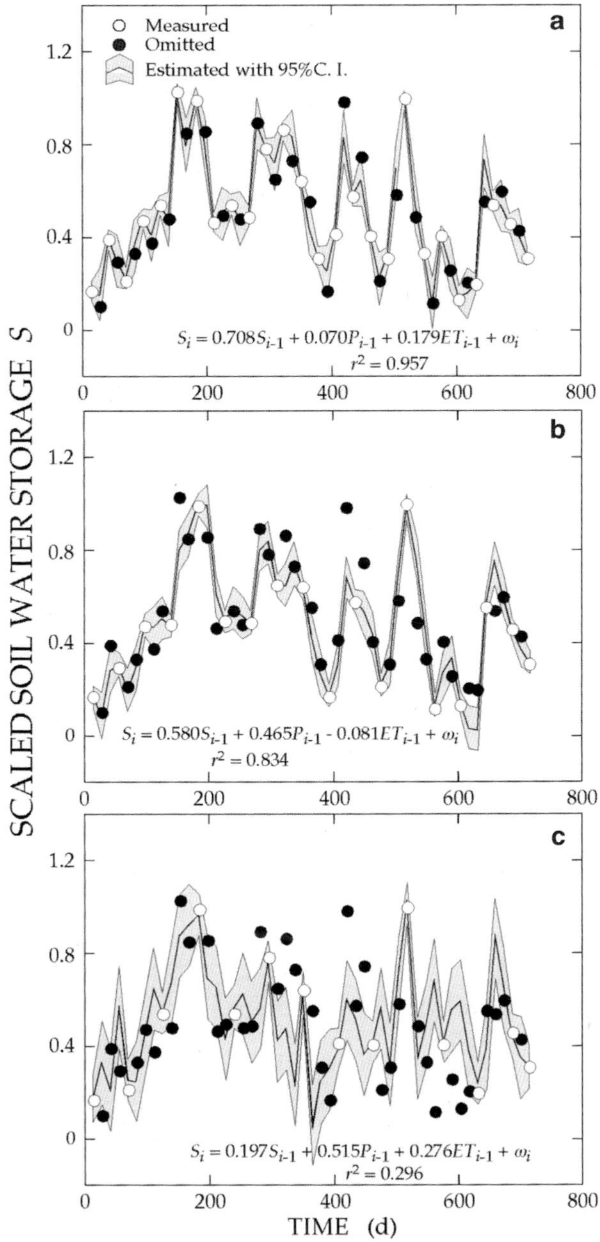

Fig. 5.15 Soil water storage measured biweekly for 714 days estimated from measurements of precipitation and evapotranspiration with (**a**) one-half, (**b**) two-thirds and (**c**) three-fourths of the soil water storage observations omitted from the state-time analysis

Figure 5.15a illustrates the results when one-half of the observations of soil water storage were not considered in the calculations. Comparing Figs. 5.14b and 5.15a, it can be seen that the coefficient of determination r^2 decreased slightly from 0.998 to 0.957 and that the width of the confidence intervals increased. At each time step when a measured value of S is omitted from the calculation, its forward prediction cannot be compared to its observation, and hence, an update based on its temporal association is precluded and causes a larger confidence interval. State-time estimates in Fig. 5.15b made while ignoring two out of every three observations of soil water storage are not as good as those illustrated in Fig. 5.15a. Nevertheless, a linear regression between state-time estimated and measured values of S yielded a coefficient of determination $r^2 = 0.834$. However, notice that about five values omitted in the calculations fall outside of the confidence interval as a result of the state-time analysis judging they did not belong to the distribution of S values used in the calculation. State-time estimates in Fig. 5.15c made while ignoring three out of every four observations of soil water storage are definitely not reliable. A linear regression between estimated and measured values of S yielded a coefficient of determination r^2 of only 0.296 and about 16 values omitted in the calculations fall outside of the confidence interval. There are two primary reasons why the state-time estimates illustrated in Fig. 5.15c do not agree with reality. First, during a time period of 56 days (4 lags and nearly equal to 2 months), values of soil water storage are no longer temporally related to each other during the 2-year experiment (Table 5.2) – a requirement of state-time analyses. Second, the amounts of water that eventually drained from the 1-m soil profile from large rainfalls robustly accounted for in the state variable P occurring within time spans of 56 days could not be ignored. Note in Figs. 5.14b and 5.15a–c, as the relative number of ignored observations of soil water storage increased, the magnitude of the transition coefficient of S_{i-1} decreases with estimates of S_i depending progressively on the values of P_{i-1}. In other words, with fewer and fewer temporal observations of S_i available, reliable estimates of S_i depend more and more on the temporal association between soil water storage and precipitation. This dependence is entirely reasonable inasmuch as changes in soil water storage are generally related directly to amounts of precipitation infiltrating the soil surface during relatively short time periods. Notice, however, that no such consistent trend was manifested during these short time periods by the transition coefficients of ET_{i-1}. This fluctuation is reasonable inasmuch as changing local weather conditions can easily cause major shifts in evapotranspiration that do not impose major changes in average soil water storage.

This example given above is more completely reported in Timm et al. (2011) and more details about the experimental field area can be found in Silva et al. (2006).

Timm et al. (2003b) mentioned that the use of Kalman filter based methods (state-space), in principle, sound appropriate mainly for filtering the measurement noise (Eq. 5.11) and the model uncertainty of the state equation (Eq. 5.12). By standard state-space approach (Shumway 1988; Shumway and Stoffer 2000, 2011) we understand that the state-variables are just a filtered (noiseless) version of the observed variables and that these noise-free quantities follow a (vector) first order autoregressive model with constant parameters. This is the special way by which the state-space approach has been employed in the soil science literature (Timm et al. 2004; Joschko et al. 2009; Jia et al. 2011; Oliveira et al. 2011; Liu et al. 2012). Although largely used in practice for soil research and related areas, we argue that this basic approach presents limitations or drawbacks, the main ones being the following:

1st) Since the parameters of the auto-regressive process are constant, they do not express directly the spatial heterogeneity (soil variability) present in the transect data. The main model parameters are global or space-independent (just average values) and not local or space-dependent as it should be if we want to express the relationship of the variables at each point in space, that is, to express the soil heterogeneity in an explicit way;
2nd) The model is restricted to an auto-regressive process, introducing unnecessary constraints into the relationship among the variables, limiting therefore the model's building process. For instance, it does not permit to express the relationship among different variables at the same point in space, but only at lagged points, which also restricts the model's building process;
3rd) Other limitations are related to the Kalman Filter implementation via the EM (Expectation-Maximization) algorithm (Dempster et al. 1977; Shumway 1988; Shumway and Stoffer 2000, 2011) which do not provide standard errors of parameter estimates in its basic formulation, unless some extra computations are considered, as for instance, using the SEM algorithm, that is, the EM algorithm with estimation of standard errors (Meng and Rubin 1991; Shumway and Stoffer 2000, 2011).

In fact, the third mentioned limitation is more related to the filter computational implementation than to the method itself, and therefore, can be avoided in part if we choose carefully the software to be used or if we make some extra calculations in order to include the Hessian matrix, which is related to the standard errors. For instance, the ASTSA software in its first version (Shumway 1988) does not present standard errors of auto-regressive parameter estimates, but these estimates could be implemented with some extra computations (see, for example, Meng and Rubin 1991). However, the first and the second mentioned limitations are more important, since they are related to modeling flexibility and parameter interpretation.

In order to overcome the mentioned points about the standard procedures in use, an alternative state-space approach was presented by Timm et al. (2003b), based on dynamic (space-varying coefficients) regression models, known as dynamic linear models (DLM) (West and Harrison 1989, 1997). The proposed models should incorporate the spatial variability (soil heterogeneity) present in the data set through the evolution of the regression parameter along the spatial transect, according to a Markovian process (random walk). The consideration of a dynamic model of a regression type as an alternative to the standard state-space approach (Shumway 1988; Shumway and Stoffer 2000, 2011) has the merit of overcoming the two main difficulties just mentioned. The choice of predictors (regressors), in Timm's et al. (2003b) approach, is more flexible (not restricted to the auto-regressive form) and the regression coefficients are space-dependent. Although theoretically simple, in practice this modeling process requires some data transformations, as it will be illustrated in the next sections.

5.4.2 West and Harrison's (1989, 1997) State-Space Approach

The Bayesian formulation presented by West and Harrison (1989, 1997) and originally published by Harrison and Stevens (1976) has not frequently been used in agronomy (Reichardt and Timm 2012). In this case a general parametric formulation is used by which the observations are linearly related to parameters (Eq. 5.11) that have a dynamic evolution according to a random walk (Eq. 5.12), with the possibility of the incorporation of uncertainties associated to the model itself and to the parameters of the model. The probabilities of the model and its parameters are continuously updated in time/space using the Bayes theorem (Cantarelis 1980). The acceptance and use of this approach in agronomy has not been as quick as expected, particularly by those without a deep knowledge in statistics, due to the difficulties in establishing values (or their law of variation) for the parameters "$v_j(x_i)$ (Eq. 5.11)" and "$u_j(x_i)$ (Eq. 5.12)". To make this approach more accessible, Ameen and Harrison (1984) used discount factors to calculate the covariance matrix of the noise parameters $u_j(x_i)$. Discount factors relate to the relevance of the observations during the evolution of time/space – with the most recent information usually being more relevant in the modeling process. The smaller the discount factor, the less importance is given to previous information. Hence, the use of these factors assures that the stochastic influence on the evolution of the parameters (Eq. 5.12) is not directly made explicit through the noise $u_j(x_i)$. The stochastic influence is derived by the combination of a relation that establishes only the deterministic evolution of $Z_j(x_i)$ and the random process guaranteed by the discount matrix.

In this state-space approach, the state equation describes the evolution of the regression coefficients β through a random walk vector

$$\beta_i = \beta_{i-1} + u_i \tag{5.17}$$

where u_i are normally distributed with zero mean and constant variance W [$u_i \sim N(0,W)$] and are non-correlated among themselves (white noise) (Timm et al. 2003b). The regression coefficients vector β is related to the observable response variable Y through the observation equation

$$Y_i = F_i\beta_i + v_i \tag{5.18}$$

where F_i is a known matrix containing the regressors, which reduces to a vector for unidimensional responses, and v_i are non-correlated errors with zero mean, constant variance and normal distribution.

The dynamic regression model (Eqs. 5.17 and 5.18) is a local and not a global model, because it contains variable β coefficients having the subscript i. These coefficients vary along space/time according to a Markovian evolution (first order autoregressive process, not having to be stationary), being therefore called "state variables of the system" (West and Harrison 1989, 1997; Pole et al. 1994). Hence, we have a basic regression equation (observation equation, Eq. 5.18), and a second equation (evolution equation, Eq. 5.17) which characterizes the form of the variation of these state parameters along space. Parameters are estimated in an optimal way through algorithms of the Kalman Filter type (extensions of the basic Kalman Filter). The equations of estimation are sequential, comprising the observational equations of actualization (via Bayes theorem and observation equation) and the spatial/time actualization equations (consequence of the evolution equation).

In this approach more emphasis is given to Eq. 5.11. In the dynamic regression, the β coefficients are considered as a state vector (Eq. 5.12) following a random walk process. The transition matrix ϕ is unity and the observation matrix $\mathbf{M}$ that relates the state vector to the observation vectors, being formed by the regression coefficients.

Example 1: Application of the West and Harrison's State-Space Approach Using Space Data Series

An application of the West and Harrison's state-space approach will be presented here using space data series. Space data series are from an oat crop (*Avena strigosa*) field experiment carried out in Jaguariúna, SP, Brazil. As mentioned before, the objective of this study was to propose another methodology using a dynamic regression model with coefficients changing along the space as an alternative to the standard state-space model.

Soil samples used in this study were collected in Jaguariuna (22° 41′ S and 47° W), SP, Brazil, on a Typic Haplustox in May 1999. Samples were taken along a line (transect) located in the middle of two adjacent contour lines. The transect samples, totaling 97, were collected in the plow layer (0–0.20 m) at points spaced 2 m apart. The transect soil had been limed, received phosphate (broadcasted and incorporated) and was planted to an oat crop, three months before soil sampling. Samples were air dried, granted to pass a 2 mm sieve and analyzed for soil organic carbon (SOC) by the Walkey-Black method (Walkley and Black 1934) and for total soil nitrogen (TSN) by the Kjeldhal method (Bremner 1960).

The proposed method to build and implement a model for soil quality data analysis (along a spatial transect) was based on the following main assumptions:

(i) Along the transect with inherent soil variation, local characteristics are better represented by a local model (and not a global one) with space-varying coefficients expressing the heterogeneity of the site;
(ii) Different variables measured by different scales or units, in order to be analyzed or modeled in conjunction and in an efficient way, could be conveniently transformed into a dimension-free scale, using Eq. 5.13. However, a method that does not involve data transformation is preferable since it allows an easier interpretation;
(iii) Some soil quality variables (as for instance, total soil nitrogen) are time-consuming and expensive to be measured, but can be well correlated to other variables easier to be measured (as for example, the soil organic carbon).

The proposed method for soil quality data analysis through space-varying regression models was based on the following procedure:

1st stage: Data transformation

The data should be transformed using Eq. 5.13.

2nd stage: Model Building and Fitting

Step 2.1 – Define which variable was the basic regressor (in this case, the "easy to measure" variable was soil organic carbon in its filtered version) and which one was the response or dependent variable (in this case, the "expensive to measure" variable was total soil nitrogen in its original version). The lagged version of the dependent variable can eventually be used as a second regressor;

Step 2.2 – Once the model variables have been previously defined and prepared (transformation, etc.), then fit the space-varying coefficient regression model as a special dynamic model with regression components, using any implementation of the DLM, as for instance, the BATS system (**B**ayesian **A**nalysis of **T**ime **S**eries; West and Harrison 1997).

The alternative modeling approach to the soil data was the space-varying coefficient regression (state-space) model in which the state-vector was formed by the dynamic regression coefficients β_i , as follows,

$$\text{observation equation}: Y_i = F_i\beta_i + v_i, \qquad v_i \sim N(0; V)$$

$$\text{state equation}: \qquad \beta_i = \beta_{i-1} + u_i, \qquad u_i \sim N(0; W),$$

where $Y_i = TSN_i$ is the response variable, $F_i = (1, TSN_{i-1}, SOC_i)$ are the considered regressors, v_i and u_i are defined as before (with the exception that v_i is now a scalar), and the regression coefficient state-vector follows a random-walk type of evolution. This non-standard state-space model, known as dynamic regression model (which is a special case of the so called dynamic linear model) was implemented as a DLM, using the BATS system (Pole et al. 1994; West and Harrison 1997).

The available data consisted of two spatial series with 97 observations each: the total soil nitrogen series and the soil organic carbon series (Fig. 5.16). The series do not present any exceptional or extreme points such as outliers or likewise (Fig. 5.16a, b). For both series along the transect points there is a detectable soil heterogeneity previously assumed (assumption **i**), since the process level and the process variability are not totally stable through the space. Therefore, the assumption (**i**) that the process presents local characteristics changing in space seems to be reasonable.

The data information was initially explored through an analysis of its correlation structure. The correlation structure between the two series (Crosscorrelation Function CCF – Eq. 5.6) is shown at Fig. 5.17c, in which the high value (near 0.80) for CCF at lag 0 suggests that the carbon series can be a good predictor for the nitrogen series. This is in accordance with our assumption (**iii**) about using a variable easy to

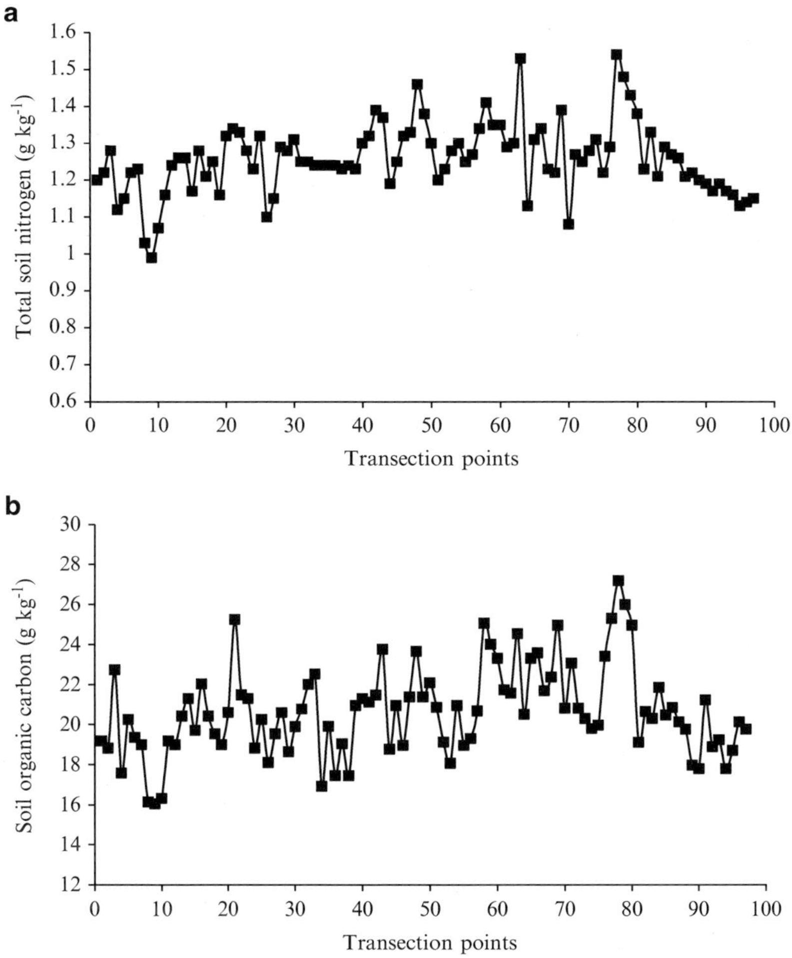

Fig. 5.16 Spatial data series: total soil nitrogen TSN (**a**) and soil organic carbon SOC (**b**) along the 97 point transect

measure such as the soil organic carbon to predict a more expensive one such as the soil total nitrogen. Also, the significant values for the Autocorrelation Function (ACF – Eq. 5.4) at the first three lags for both series (Fig. 5.17a, b) show the presence of spatial correlation and an autoregressive structure for observations distant until 6 m in this study. This data can be analyzed through a state-space model, since it is based on a vector AR structure (Fig. 5.17a–c).

The dynamic (space-varying) regression model for the nitrogen series is here implemented in two different versions: in the first one (called Model I) the regressors are the lagged nitrogen and lagged carbon, and in the second one (called Model II) the regressors are carbon and lagged nitrogen. Since soil carbon is more correlated to soil nitrogen than lagged carbon (as shown in Fig. 5.17c), we expect that model II will fit better to the data than model I.

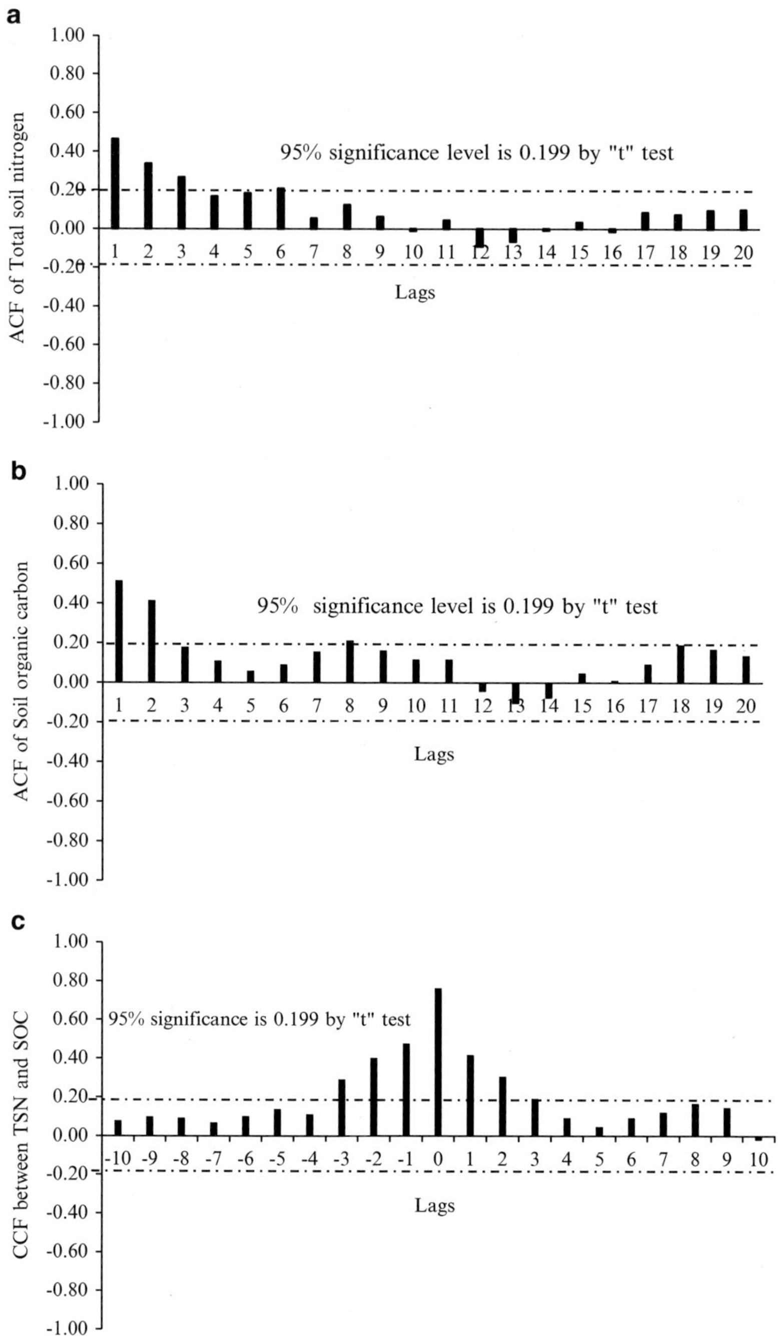

Fig. 5.17 Estimated spatial correlation structure: ACF for total soil nitrogen TSN (Fig. 17a), ACF for soil organic carbon SOC (Fig. 17b) and CCF between total soil nitrogen TSN versus soil organic carbon SOC (Fig. 17c)

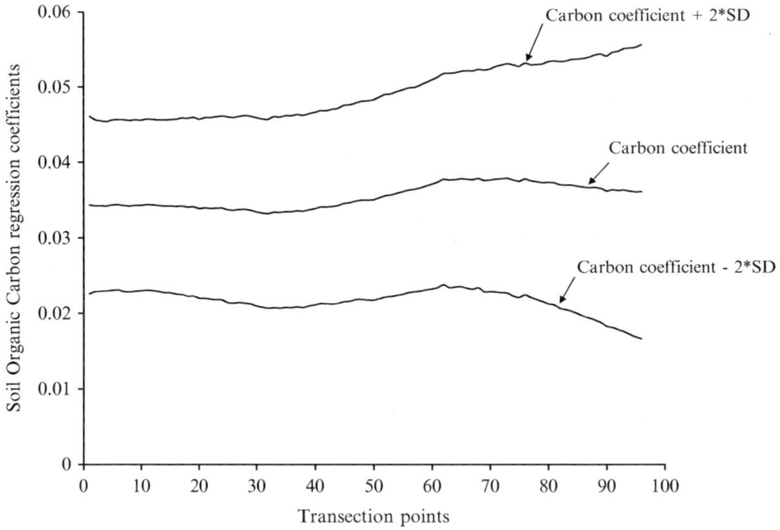

Fig. 5.18 Space-varying parameter estimates (soil organic carbon coefficient)

The space-varying regression model has been fitted to the nitrogen-carbon data, and some parameter estimation (point and interval) and goodness of fit results are obtained. For the Model I, R^2 measure is 0.943 and for the proposed Model II is 0.997, i.e., the proposed method (Model II) has better fitting performance when compared to Model I. There is an advantage of qualitative nature for the proposed method since the data present local characteristics (level and variability) that change along the transect points, it is not surprising that the soil nitrogen-carbon model parameter estimates change in space (Fig. 5.18). In this figure, the interval parameter estimates are also presented for each point, giving more complete information on the soil carbon-nitrogen relationship.

The model II for the nitrogen series with its two standard deviation confidence intervals is presented in Fig. 5.19, together with the original data, showing clearly the good fitting performance, with all the data points inside the confidence intervals.

The fact that the alternative state-space approach provides very good fitting performance (R^2 coefficient greater than 0.99) is not surprising since it has a local characteristic, having the state variable adapted to the data at each point via Kalman Filter. On the other hand, it is very different as compared to the standard state-space approach presented by Shumway (1988) and Shumway and Stoffer (2000, 2011) regarding to the extraction of qualitative information from the data, where the two main differences are related to parameter definition and interpretation. As explained here, the key parameters for the standard state-space are the autoregressive coefficients (with the mentioned shortcomings) and the key quantities for the alternative state-space approach are the space-varying soil carbon-nitrogen relationship expressing the soil spatial heterogeneity along the transect.

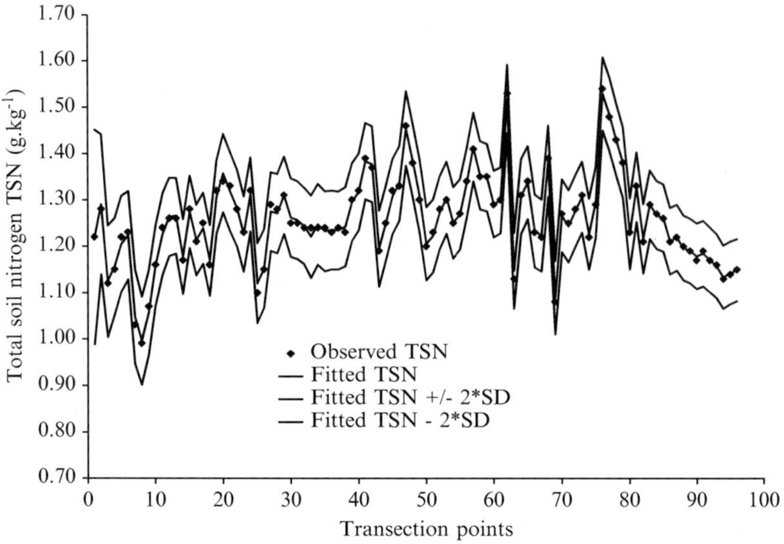

Fig. 5.19 Observed total soil nitrogen series and its model fitting with 2*standard deviation (SD) confidence interval from Model II

This above example is more completely reported in Timm et al. (2003b). An earlier example of the application of the West and Harrison's state-space approach can be found in Timm et al. (2000).

5.4.3 Application of Both State-Space Approaches to Forecast Space Data Series

As mentioned by Tukey (1980), one of the basic objectives for analyzing a time series is to evaluate the ability of the model in terms of forecast. Based on this and on the fact that there were no published manuscripts in the literature to evaluate the potential of both state-space approaches to forecast spatial series in the agronomic area, Timm et al. (2006) evaluated the potential of both state-space approaches to forecast total soil nitrogen (TSN) from soil organic carbon (SOC) collected along a spatial transect and to compare their performances between themselves and to compare both state-space approaches with standard uni and multivariate regression models, used as reference. Artificial neural networks (both feed-forward and recurrent neural networks) were also used in this comparison. The prediction performance of all models was evaluated in terms of the distance between the observed and predicted values of total soil nitrogen TSN and the statistical measures considered were the Mean Square Error (MSE) and the Mean Absolute Percentage Error (MAPE). More details about the theoretical aspects of each class of model can be found in Timm et al. (2006).

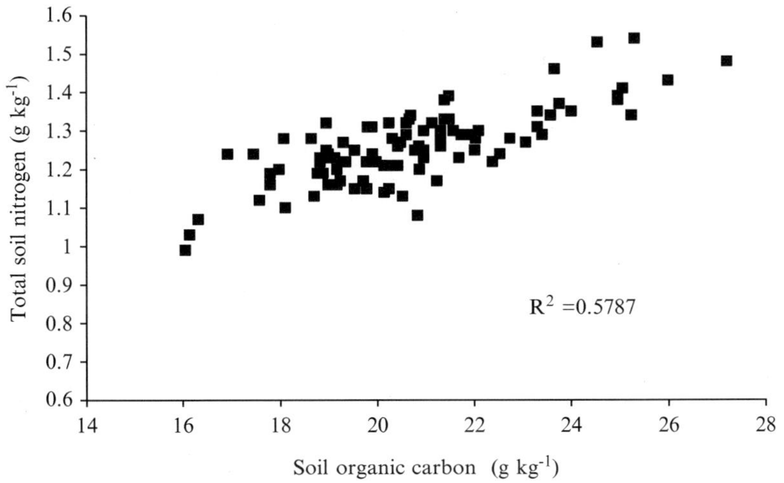

Fig. 5.20 Dispersion diagram between the total soil nitrogen TSN and soil organic carbon SOC data set showing a strong linear dependence between the variables

The TSN and SOC data sets used in Timm et al. (2006) come from Timm et al. (2003b) and consisted of the same two spatial series with 97 observations of each variable (Fig. 5.16). The spatial structure of each variable was evaluated using the autocorrelation function (Fig. 5.17a, b) and the spatial dependency between the TSN and SOC data sets was calculated by the crosscorrelation function (Fig. 5.17c). Figure 5.20 shows a linear relation (R^2 coefficient of 0.5787) between the TSN and SOC collected at the same spatial point of the transect. Additionally, the analysis of Fig. 5.20 looks promising for the use of neural network models, suggested by a possible non-linear relation between the TSN and SOC. From this figure it can be seen that the slope in the extremes of the dispersion diagram is higher than the slope in the middle and, therefore, the relation can be expressed as a linear segment, to represent the global non-linear relation. Therefore, the analysis of Figs. 5.16, 5.17, and 5.20 suggest that the total soil nitrogen at each point could be reasonably predicted by the soil organic carbon at the same spatial point and by the total soil nitrogen at the nearest neighbor.

Timm et al. (2006) evaluated the prediction performance of all adjusted models in two versions: (a) for the first, the last 10 transect points of TSN were omitted in order to make their prediction (Table 5.3), and (b) for the second, the first 10 points of TSN were omitted with the same objective (Table 5.4). As already mentioned, the statistical measures considered for comparisons between models were the MSE and the MAPE.

Table 5.3 shows that among the models without latent variables, i.e., among the true regression models, the original VAR model gives the worst results (independent of the statistical measure considered, MSE = 0.00713 and MAPE = 0.0639). In this case, unlike other models it uses the lagged SOC as a regressor variable and not the SOC value at the same point, which has a stronger linear relation with TSN as

Table 5.3 Predictive performance (10 last transect points) of standard regression, of state-space and of neural network models, for total soil nitrogen TSN

Prediction models			Statistical measures	
			MSE	MAPE
Without latent variable	Scalar regression	Standard linear	0.00388	0.04301
		AR (1) error	0.00389	0.04279
	Vector auto-regression	Standard VAR	0.00713	0.06390
		Corrected VAR	0.00350	0.03905
	No-parametric regression	GAM/splines	0.00435	0.04359
		GAM/lowess	0.00361	0.04084
With latent variable	Artificial neural networks	Feedforward	0.00313	0.03727
		Recurrent	0.00279	0.03599
	State-space models	Standard	0.00096	0.02302
		Dynamic	0.00288	0.03960

AR (1) error = Regression model with 1st order auto-regressive error; Standard VAR = Standard Vector Auto-Regression model; Corrected VAR = Corrected Vector Auto-Regression model; GAM = Generalized Additive Model; Standard = Standard state-space model (Shumway 1988; Shumway and Stoffer 2000, 2011); Dynamic = Dynamic state-space model (West and Harrison 1989, 1997); MSE = Mean Square Error; and MAPE = Mean Absolute Percentage Error

Table 5.4 Predictive performance (10 first transect points) of standard regression, of state-space and of neural network models, for total soil nitrogen TSN

Prediction models			Statistical measures	
			MSE	MAPE
Without latent variable	Scalar regression	Standard linear	0.00483	0.04665
		AR (1) error	0.00475	0.04601
	Vector auto-regression	Standard VAR	0.00713	0.06390
		Corrected VAR	0.00423	0.04358
	No-parametric regression	GAM/splines	0.00793	0.05583
With latent variable	Artificial neural networks	Feedforward	0.00344	0.03898
		Recurrent	0.00213	0.02827
	State-space models	Standard	0.00314	0.04192
		Dynamic	0.00407	0.04589

AR (1) error = Regression model with 1st order auto-regressive error; Standard VAR = Standard Vector Auto-Regression model; Corrected VAR = Corrected Vector Auto-Regression model; GAM = Generalized Additive Model; Standard = Standard state-space model (Shumway 1988; Shumway and Stoffer 2000, 2011); Dynamic = Dynamic state-space model (West and Harrison 1989, 1997); MSE = Mean Square Error; and MAPE = Mean Absolute Percentage Error

shown in the Figs. 5.17c and 5.20. The corrected VAR shows the best results among the regression models, for which the minimum values of MSE (=0.00350) and MAPE (=0.0395) were found. This model has the SOC as a more appropriate predictor measured at the same point in space. This is consistent because the model, although being a global one, is presented as a bi-dimensional system. It is composed of two equations which treat the dynamics of the relation between TSN and SOC in the soil in a more adequate way, i.e., there is no a hierarchy between variables. They are both treated in the same way, considered as random variables. The standard linear regression model (scalar model) is, also, a global model. It, however, is presented as a unidimensional system with a hierarchical treatment between TSN and SOC variables (only the variable TSN is considered a random variable). Therefore, both statistical performance measures (MSE = 0.00388 and MAPE = 0.04031) gave higher values as compared to the corrected VAR model. The GAM model (in its best performance version, i.e., the lowess) presented a MSE of 0.00361, slightly higher than the corrected VAR (MSE of 0.00350), however, similar values as compared in terms of the MAPE measure (=0.04084). Such performance of this model is due to the fact that it incorporates non-linear (Fig. 5.20) and local characteristics, i.e., it does not ignore the spatial correlation structure between the TSN and SOC variables along the transect taking into consideration the information the variables carry in function of their neighborhood.

Among the models with latent variables, the standard state-space model (linear and local characteristics), independent of the statistical measure considered, presented the best prediction performance for the 10 last transect points (MSE = 0.00096 and MAPE = 0.02302). It was followed by the non-linear recurrent neural network model (MSE = 0.00279 and MAPE = 0.03599). Such performance can be due to the fact that the state-space model expresses the local linear behavior of the 10 last TSN values of the transect (Fig. 5.16a). The recurrent neural network takes into account the local characteristic of the spatial dependence structure between TSN and SOC data (Fig. 5.17c) by data feedback expressing in this way the point-to-point spatial variability of TSN. On the other hand, the feedforward neural network is a non-linear and global model which does not express this point to point fluctuation (Fig. 5.16a). In the same way, the static linear regression models, intensively used in agronomy, are global models whose regression coefficients are mean values which do not change along space, therefore not expressing the point to point fluctuations of the variable under study. This can lead to misunderstandings that might induce inadequate procedures of soil management (Nielsen and Alemi 1989). Beyond this, the response of the variable is not unique along the experimental transect, frequently yielding low coefficients of determination when compared to the dynamic models, as shown by Timm et al. (2003a, 2004).

In Table 5.4 we considered only one version of the GAM models (GAM/splines) because the other version (GAM/lowess, implemented by the SAS software) has a restriction with respect to the regression value for prediction (it must be inside the interval of used data), which is not satisfied for these particular data sets. Analyzing Table 5.4, the best predictive performance of the corrected VAR model (MSE = 0.00423 and MAPE = 0.04358) can also be seen for the 10 first TSN values of the spatial transect in relation to the other regression models without latent variables. The best performance, however, to predict TSN among the latent variable models (and among all models) was given by the recurrent neural network (MSE = 0.00213 and MAPE = 0.02827).

Tables 5.3 and 5.4 also indicate that the use of dynamic linear models (state-space models) which take into account the local spatial dependence structure, as well as the feedforward (no-linear and global characteristics) and recurrent (no-linear and local characteristics) neural networks give the best TSN predictions of the 10 last and 10 first TSN values of the spatial transect, i.e., the statistical measured values of MSE and MAPE were lower when compared to the regression models (without latent variables) considered as standard methods for comparison studies. Both state-space and neural network models have in their essence the philosophy of the use of state variables which are not observed directly during the different processes which occur simultaneously in the complex atmosphere-plant-soil system. They however belong to the used algorithms for practical implementation of these models.

5.5 Final Considerations

The use of statistical tools like the state-space approach is a tentative to describe the spatial or time association among different variables measured along space or time and can, therefore, provide a better understanding of the complex relation between variables related to the Soil-Plant-Atmosphere system. Since the response of one variable is not unique in an area, in several cases when applying statistical multiple regressions, low coefficients of determination are found. Furthermore, this analysis gives us a global adjustment of the coefficients that do not represent point to point variations and that can lead to interpretations that induce to inadequate management procedures.

Because dynamic regression models represented in the state-space form are relatively recent tools, they have not yet been frequently used to quantify soil-plant-atmosphere relations. Having been introduced in the sixties of the last century, their greater implementation occurred only since the late eighties (West and Harrison 1989, 1997). Being models of local adjustment, it is possible to precisely estimate regression coefficients for each sampled location. This possibility tends to alleviate the problem of spatial/time variability encountered in precision agriculture. Until today, the more commonly used static models are models of global behavior with their regression coefficients being average values not varying along space/time. They provide an opportunity to describe the spatial or time association among different variables measured across a field or at different times and can, therefore, provide a better understanding of the complex relations between soil, plant, and atmosphere variables. We conclude that dynamic regression models of state-space form are more adequate for the study of soil-plant-atmosphere relationships mainly because they account for the local spatial and temporal character of agronomic processes.

The representation of models in terms of state-space in agronomy is still relatively recent and, from what is known today, this is a very promising tool to study the relations among several variables linked to the dynamic system comprised by the soil, the plant, and the atmosphere.

Acknowledgments We wish to express thanks to the Abdus Salam International Centre for Theoretical Physics (ICTP), Trieste, Italy, to the Brazilian Research Council (CNPq) and to the Brazilian Federal Agency for Improvement of Graduate Education (CAPES) for the scholarships and funding.

References

Ameen JRM, Harrison PJ (1984) Discount weighted estimation. J Forecast 3:285–296

Bremner JM (1960) Determination of nitrogen in soil by the Kjeldahl method. J Agric Sci 55:11–33

Cantarelis NS (1980) An investigation into the properties of Bayesian forecasting models. Thesis, Warwick University

Davis JC (1986) Statistics and data analysis in geology, 2nd edn. Wiley, New York

Dempster AP, Laird NM, Rubin DB (1977) Maximum likelihood from incomplete data via the EM algorithm. J R Stat Soc Ser B 39:1–38

Dourado-Neto D, Timm LC, Oliveira JCM et al (1999) State-space approach for the analysis of soil water content and temperature in a sugarcane crop. Sci Agr 56:1215–1221

EMBRAPA – Empresa Brasileira de Pesquisa Agropecuária, Centro Nacional de Pesquisa de Solo (2006) Sistema brasileiro de classificação de solos. Embrapa Solos, Rio de Janeiro

Fenilli TAB, Reichardt K, Dourado-Neto D et al (2007) Growth, development and fertilizer N recovery for a coffee crop growing in Brazil. Sci Agric 64:541–547

Gelb A (1974) Applied optimal estimation. Massachusetts Institute of Technology, Cambridge

Harrison PJ, Stevens CF (1976) Bayesian forecasting (with discussion). J R Stat Soc Ser B 38(3):205–267

Hui S, Wendroth O, Parlange MB et al (1998) Soil variability – infiltration relationships of agroecosystems. J Balkan Ecol 1:21–40

Jia X, Shao M, Wei X et al (2011) Estimating total net primary productivity of managed grasslands by a state-space modeling approach in a small catchment on the Loess Plateau, China. Geoderma 160:281–291

Joschko M, Gebbers R, Barkusky D et al (2009) Location-dependency of earthworm response to reduce tillage on sandy soil. Soil Tillage Res 102:55–66

Kalman RE (1960) A new approach to linear filtering and prediction theory. Trans ASME J Basic Eng 8:35–45

Katul GG, Wendroth O, Parlange MB et al (1993) Estimation of in situ hydraulic conductivity function from nonlinear filtering theory. Water Resour Res 29:1063–1070

Köppen W (1931) Grundriss der Klimakunde. Walter D E Guyter & Co., Berlin

Liu ZP, Shao MA, Wang YQ (2012) Estimating soil organic carbon across a large-scale region: a state-space modeling approach. Soil Sci 177:607–618

McGraw T (1994) Soil test level variability in Southern Minnesota. Better Crops, Potash & Phosphate Inst 78(4):24–25

Meng XL, Rubin DB (1991) Using EM to obtain asymptotic variance-covariance matrices: the SEM algorithm. J Am Stat Assoc 86:899–909

Morkoc F, Biggar JW, Nielsen DR et al (1985) Analysis of soil water content and temperature using state-space approach. Soil Sci Soc Am J 49:798–803

Nielsen DR, Alemi MH (1989) Statistical opportunities for analyzing spatial and temporal heterogeneity of field soils. Plant Soil 115:285–296

Nielsen DR, Wendroth O (2003) Spatial and temporal statistics: sampling field soils and their vegetation. Catena Verlag, Reiskirchen

Oliveira MP, Tavares MHF, Opazo MAU et al (2011) Selecting statistical models to study the relationship between soybean productivity and soil physical attributes. Braz J Soil Sci 35:97–104

Parfitt JMB, Timm LC, Pauletto EA et al (2009) Spatial variability of the chemical, physical and biological properties in lowland cultivated with irrigated rice. Braz J Soil Sci 33:819–830

Plackett RL (1950) Some theorems in least squares. Biometrika 37:149–157

Pole A, West M, Harrison J (1994) Applied Bayesian forecasting and time series analysis. Chapman & Hall, London

Reichardt K, Timm LC (2012) Solo, Planta e Atmosfera: conceitos, processos e aplicações, 2nd edn. Manole, São Paulo

Reichardt K, Portezan-Filho O, Bacchi OOS et al (1997) Neutron probe calibration correction by temporal stability parameters of soil water content probability distribution. Sci Agric 54:17–21 (Special issue)

Salas JD, Delleur JW, Yevjevich V et al (1988) Applied modeling of hydrologic time series. Water Resources Publications, Littleton

Shumway RH (1988) Applied statistical time series analyses. Prentice Hall, New York

Shumway RH, Stoffer DS (1982) An approach to time series smoothing and forecasting using the EM algorithm. J Time Ser Anal 3:253–264

Shumway RH, Stoffer DS (2000) Time series analysis and its applications. Springer, New York

Shumway RH, Stoffer DS (2011) Time series analysis and its applications: with R examples, 3rd edn. Springer, New York

Silva AL, Roveratti R, Reichardt K et al (2006) Variability of water balance components in a coffee crop grown in Brazil. Sci Agric 63:153–165

Silva AL, Bruno IP, Reichardt K et al (2009) Soil water extraction by roots and Kc for the coffee crop. R Bras Eng Agric Ambient 13:257–261

Soil Survey Division Staff (1993) Soil survey manual. US Government Printing Office, Washington, DC

Timm LC, Fante Junior L, Barbosa EP et al (2000) A study of the interaction soil – plant using state-space approach. Sci Agric 57:751–760

Timm LC, Reichardt K, Oliveira JCM et al (2003a) Sugarcane production evaluated by the state-space approach. J Hydrol 272:226–237

Timm LC, Barbosa EP, Souza MD et al (2003b) State-space analysis of soil data: an approach based on space-varying regression models. Sci Agric 60:371–376

Timm LC, Reichardt K, Oliveira JCM et al (2004) State-space approach to evaluate the relation between soil physical and chemical properties. Braz J Soil Sci 28:49–58

Timm LC, Gomes DT, Barbosa EP et al (2006) Neural network and state-space models for studying relationships among soil properties. Sci Agric 63:386–395

Timm LC, Dourado-Neto D, Bacchi OOS et al (2011) Temporal variability of soil water storage evaluated for a coffee field. Soil Res 49:77–86

Tukey JW (1980) Can we predict where time series should go next? In: Brillinger DR, Tiao GC (eds) Directions in time series. Inst. Math. Statist, Hayward, pp 1–31

Vieira SR, Hatfield TL, Nielsen DR et al (1983) Geostatistical theory and application to variability of some agronomical properties. Hilgardia 51:1–75

Walkley A, Black IA (1934) An examination of the Degtjareff method for determining soil organic matter and a proposed modifications of the chromic acid titration method. Soil Sci 37:29–38

Wendroth O, Katul GG, Parlange MB et al (1993) A nonlinear filtering approach for determining hydraulic conductivity functions. Soil Sci 156:293–301

Wendroth O, Reynolds WD, Vieira SR et al (1997) Statistical approaches to the analysis of soil quality data. In: Gregorich EG, Carter MR (eds) Soil quality for crop production and ecosystem health. Elsevier, Amsterdam, pp 247–276

Wendroth O, Jürschik P, Kersebaum KC et al (2001) Identifying, understanding, and describing spatial processes in agricultural landscapes – Four case studies. Soil Tillage Res 58:113–127

Wendroth O, Reuter HI, Kersebaum KC (2003) Predicting yield of barley across a landscape: a state-space modeling approach. J Hydrol 272:250–263
Wendroth O, Koszinski S, Pena-Yewtukhiv E (2006) Spatial association between soil hydraulic properties, soil texture and geoelectric resistivity. Vadose Zone J 5:341–355
West M, Harrison J (1989) Bayesian forecasting and dynamic models. Springer, London
West M, Harrison J (1997) Bayesian forecasting and dynamic models, 2nd edn. Springer, London

Chapter 6
Pedotransfer Functions for Brazilian Soils

Alexandre Hugo Cezar Barros and Quirijn de Jong van Lier

Abstract The growing need for PTFs frequently leads to the use of inadequate PTFs and databases, especially when dealing with data from different regions or climates, such as in tropical and temperate climate zones. Most PTFs have been developed for soils from temperate climates, which tend to be very different from tropical soils. In fact, PTFs are not suitable for extrapolation: they should not be applied to soils that are outside the range or region where the PTFs ware developed. In this chapter the current status of PTF development in Brazil is reviewed, and suggestions for research aiming future improvements are made.

Keywords Development PTF • Review • Soil data • Brazil

6.1 Introduction

The term "Pedotransfer Function" (PTF) was introduced in soil science in the 1980s to substitute older expressions (Bouma and Van Lanen 1986; Bouma 1989), like pedofunction (Lamp and Kneib 1981) or transfer function. The aim of PTFs is to transform available information into required data, and PTFs consist of mathematical expressions to estimate soil properties from other soil properties that are more regularly measured (McBratney et al. 2002).

The first attempts to develop PTFs date from the early twentieth century and were done by Briggs and McLane (1907), refined by Briggs and Shantz (1912),

A.H.C. Barros (✉)
Brazilian Agricultural Research Corporation (EMBRAPA) – National Centre of Soil Research, Rua Antonio Falcão, 402, Boa Viagem, Recife, PE C.P. 51020-240, Brazil
e-mail: alexandre.barros@embrapa.br

Q. de Jong van Lier
Center for Nuclear Energy in Agriculture, University of São Paulo (CENA/USP), São Paulo, Brazil
e-mail: qdjvlier@usp.br

W.G. Teixeira et al. (eds.), *Application of Soil Physics in Environmental Analyses: Measuring, Modelling and Data Integration*, Progress in Soil Science,
DOI 10.1007/978-3-319-06013-2_6,

who proposed a correlation between soil water content and texture. After the introduction of the concepts of field capacity and permanent wilting point (Veihmeyer and Hendrickson 1927) it became more and more common to correlate particle size distribution, porosity, density and organic matter content to water content and plant water availability. In this sense, the most complete study was performed in the early 1980s by Williams et al. (1983) who classified soil water retention based on texture class.

The first reported PTF from Brazil was developed by Arruda et al. (1987) to determine the available water from texture information. During the 1990s PTFs became more accepted in Brazil, and the first PTFs for soils from the semiarid northeast of Brazil were developed (Da Silva et al. 1990). During this period, soil hydraulic parameters were analyzed using statistical techniques to generate PTFs to estimate soil water availability (Masutti 1997; Coelho et al. 1998; Tomasella and Hodnett 1998). Bernoux et al. (1998) developed the first baseline for predicting bulk density from soil properties for soils across the Amazon basin.

During the last decades the use of PTFs increased remarkably (Wösten et al. 2001; Pachepsky and Rawls 2004; Shein and Arkhangel'skaya 2006; Weynants et al. 2009; Vereecken et al. 2010; McBratney et al. 2011). In Brazil, Mendonça-Santos and Santos (2003) and Minasny et al. (2003) reviewed concepts, use and development of PTFs. These publications focused mainly on development methods, accuracy and uncertainty of PTFs, as well as qualitative and quantitative evaluations of PTF applications. The number of papers published on the subject PTF increased strongly between 2000 and 2010 (McBratney et al. 2011), due to the scarcity of data to be used in several applications involving soil hydrology. At present, PTFs are widely used in soil science, hydrology, meteorology and agronomy.

The growing need for PTFs frequently leads to the use of inadequate PTFs and databases, especially when dealing with data from different regions or climates, like in the case of tropical and temperate climate zones. Most PTFs have been developed for temperate climates (Wösten et al. 2001; Tomasella and Hodnett 2004), and soils from these regions tend to be very different than tropical soils (Tomasella and Hodnett 1998; Tomasella et al. 2000; Minasny and Hartemink 2011). Since 1970 there are reports on PTFs for tropical soils, but they are completely outnumbered by reports on PTFs for temperate climates (Minasny and Hartemink 2011).

In oxisols, the ultimate weathering state of a tropical soil, 1:1 clay minerals and iron and aluminum oxides predominate. These minerals form structural units that are very different from those observed in soils with 2:1 clay minerals, most common in temperate regions. Additionally, silt contents can be much higher in temperate climate soils, whereas clay contents in kaolinitic tropical soils range normally from 60 % to 90 %. Soils with more than 60 % of clay are considered as low permeability heavy clays and are regarded as "non-agricultural soils" in temperate climates (Carsel and Parrish 1988). When compared to typical temperate clayey soils, kaolinitic tropical soils showed: high permeability (Ksat usually 10–1,000 mm h^{-1}), low available water capacity (70 mm m^{-1}), and almost 80 % of the plant available water between −10 and −1,000 kPa (Demattê 1988; Correa 1984; Tomasella and Hodnett 1996). On the other hand, Sobieraj et al. (2001) stated that the hydraulic behavior of tropical soils cannot be explained by soil mineralogy, but the poor performance of temperate PTF in predicting saturated hydraulic conductivity results from the lack of ability to reproduce the effects of macroporosity in tropical soils.

The field capacity concept is based on dynamic soil properties as well as characteristics of the entire soil profile. Nevertheless, field capacity is often understood statically as the water content at some predefined matric potential, where values of −6 kPa, −10 kPa and −33 kPa are most frequently used. Reichardt (1988) and Souza and Reichardt (1996) discuss the validity of these values for Brazilian soils. Anyway, direct field observations from drainage experiments are very rare, and for PTF development in Brazilian soils the static matric potential approach is always used.

It is not recommended to apply a PTF under conditions that are different from those for which it was developed. To illustrate this, Dalbianco (2009) showed that most of the published equations to estimate hydraulic conductivity underestimated values when applied to a watershed in Rio Grande do Sul state. He tested the equations proposed by Cosby et al. (1984), Brakensiek et al. (1984), Puckett et al. (1985), Saxton et al. (1986), Dane and Puckett (1992), Campbell and Shiozawa (1994) and Saxton and Rawls (2006), which were all developed in the USA.

Even with the worldwide increase in attention for PTFs, in Brazil its use is quite restricted and PTFs are most commonly used to estimate crop water availability using field capacity and permanent wilting point (PWP) (Arruda et al. 1987; Masutti 1997; Coelho et al. 1998; Tomasella and Hodnett 1998; Lopes-Assad et al. 2001; Rossato 2001; Oliveira et al. 2002; Peraza 2003; Ottoni 2005; Mello et al. 2005; Gevaerd and Freitas 2006; Urach 2007; Reichert et al. 2009; Andrade and Stone 2010; Costa 2012; Dos Santos et al. 2013). Reports about testing PTFs and performing a functional evaluation are uncommon, as are correlations to other applications of soil science like soil fertility and soil management. One of the reasons for the slow advance of studies involving PTF in Brazil may reside in the lack of organization and availability of databases containing information on soil hydraulic properties.

In this chapter the current status of PTF development in Brazil will be reviewed, emphasizing research developed, mostly published in local journals, theses and reports. This review is intended to demonstrate the development of PTFs for Brazilian soils and suggests future improvements.

6.2 Soil Data and Information

To develop a PTF, a database is needed containing a large number of data from several soil types, well organized and operational, able to provide information at a high level of reliability (Wösten et al. 2001). The developing of databases with harmonized support scales and experimental protocols could significantly improve the quality of PTF predictions (Vereecken et al. 2010). However, it is very common that data and information about soils, needed for PTF development, are included in not easily accessible technical reports from soil surveys. These reports give only general information about field morphology, soil texture and structure, and about some physical and chemical soil properties. Each of the reporting institutions or organizations uses its own protocols, making it difficult to standardize rules and methods, especially for older surveys.

Soil maps of the Brazilian territory are available at the scale of 1:1.000.000, as a result of the Radambrasil project and soil map of Brazil, at a scale of 1:5.000.000. Besides these maps, a large number of exploratory surveys have been performed at scales between 1:50.000 and 1:750.000. However, in spite of the technological level, the information about soils in Brazil is scarce, mainly due to the absence of a database at the required scale (Lopes-Assad 2006). The scale of soil surveys is a key factor to determine the precision of PTF estimations, and maps at scales of 1:10.000 and 1:25.000 allow better results when compared to surveys at a scale of 1:100.000 (Leenhardt et al. 1994).

In general, PTFs in Brazil are developed using a relatively small number of data, or employing international databases as support to validate PTFs. International databases most frequently used are the UNsaturared SOil Hydraulic DAtabase – UNSODA (Leij et al. 1996; Nemes et al. 2001), HYdraulic PRoperties of European Soils – HYPRES (Wösten et al. 1999), United States Department of Agriculture – USDA – Natural Resource Conservation Service (ncsslabdatamart.sc.egov.usda.gov), ISRIC- Soil information system – World Inventory of Soil Emission Potentials – WISE (Batjes 2008). The location of the ISRIC-WISE soil profiles of Brazil is illustrated in Fig. 6.1. The data can be downloaded at [www.isric.org/data/isric-wise-global-soil-profile-data-ver-31].

A program to organize available data on soil properties in Brazil originated from the Brazilian Agricultural Research Corporation (EMBRAPA), that developed SigSolos (Chagas et al. 2004). At present, the data from SigSolos are being incorporated in a new internet system by EMBRAPA (www.bdsolos.cnptia.embrapa.br/consulta_publica.html). This new system is designed using the new Brazilian soil taxonomy (SiBCS) as well as data format criteria (Oliveira et al. 2009), containing at present data from more than 7,000 soil profiles and almost 30,000 horizons (Fig. 6.2).

In the context of the project "assessment of soil organic carbon stocks and changes at national scale that relates to carbon sequestration", sponsored by the Global Environmental Facility (GEF), a digital soil properties database of the Amazon part from Radambrasil was developed, providing information on soil physical and chemical attributes for that region. The data set was digitized from 23 bulletins (volumes) produced by the project Radambrasil. The database is for freely for non-commercial scientific and educational use (Bernoux and Cerri 2003). Batjes et al. (2007) prepared a sound soil data set for the Brazilian Amazon to be used in modeling. This was essential to allow development of a uniform protocol for handling the inputs for the Global Environment Soil Organic Carbon Modeling.

A database of soil physical and hydraulic properties is being composed by the Department of Hydrology of the Geological Survey of Brazil (CPRM). The study involves four main stages, from data collection to the elaboration of the database model and PTFs development. As of September 2013, 8,983 water retention and 1,253 saturated hydraulic conductivity datasets have been collected (According to Ottoni, M.V, CPRM, written communication, September, 2013).

Another database for Brazilian soils was developed from soil survey information, mainly from Radambrasil (Cooper et al. 2005). Due to the lack of a solid and uniform soil classification system, this database only allows qualitative evaluations. In order to increase the applicability of this large database, Benedetti et al. (2008) proceeded to reorganize the classification of 5,479 soil profiles (10,950 horizons).

The absence of a national soil database caused many programs to organize soil data individually and based on classical information like texture, grain size distribution (sand, silt and clay content) and organic carbon, as well as bulk density. Tomasella et al. (2000) and Rossato et al. (2004) organized soil data from approximately 900 profiles, mainly from exploratory soil surveys at scale 1:1.000.000 from the Radambrasil project. This project resulted in the largest database of soil hydraulic and physical properties ready to be used (Fig. 6.1), although most soils are from the south and southeast regions of Brazil. Other initiatives to organize soil data were made for Brazil's Northeastern region (Barros et al. 2013; Oliveira et al. 2002), Southern region (Peraza 2003; Urach 2007; Reichert et al. 2009; Costa 2012); Southeastern region (Rossato et al. 2004; Vaz et al. 2005); Central-West Brazil (Andrade and Stone 2010) and Northern Brazil (Mello 2007).

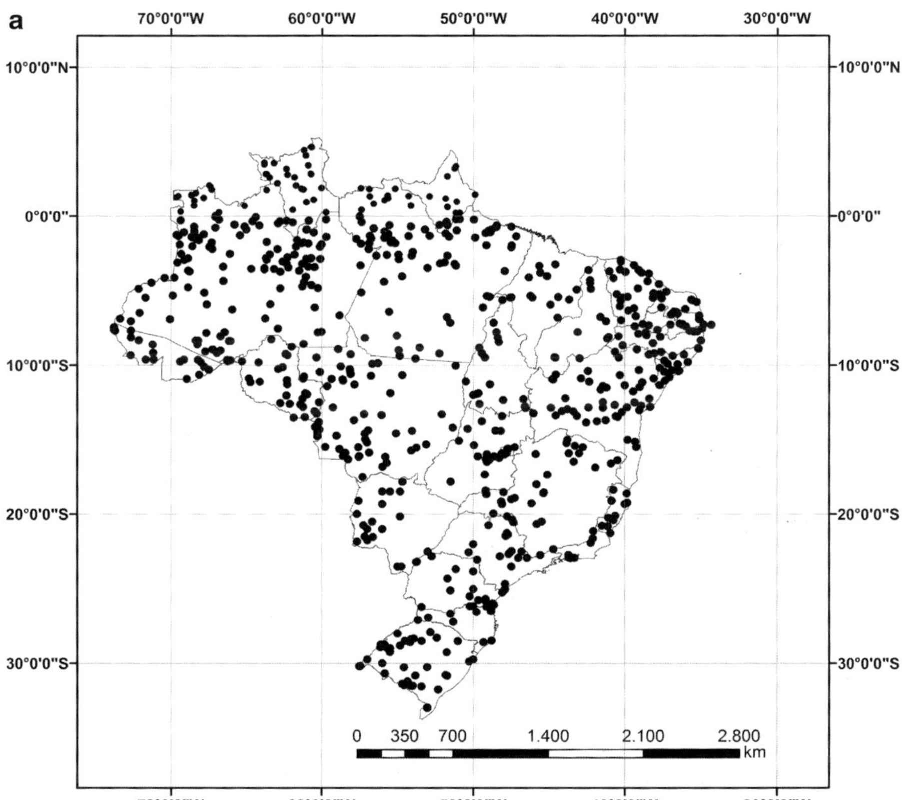

Fig. 6.1 Soil map of Brazil with the location of the soil profiles that compose the database of (**a**) ISRIC-WISE and (**b**) Tomasella et al. (2000) and Rossatto et al. (2004)

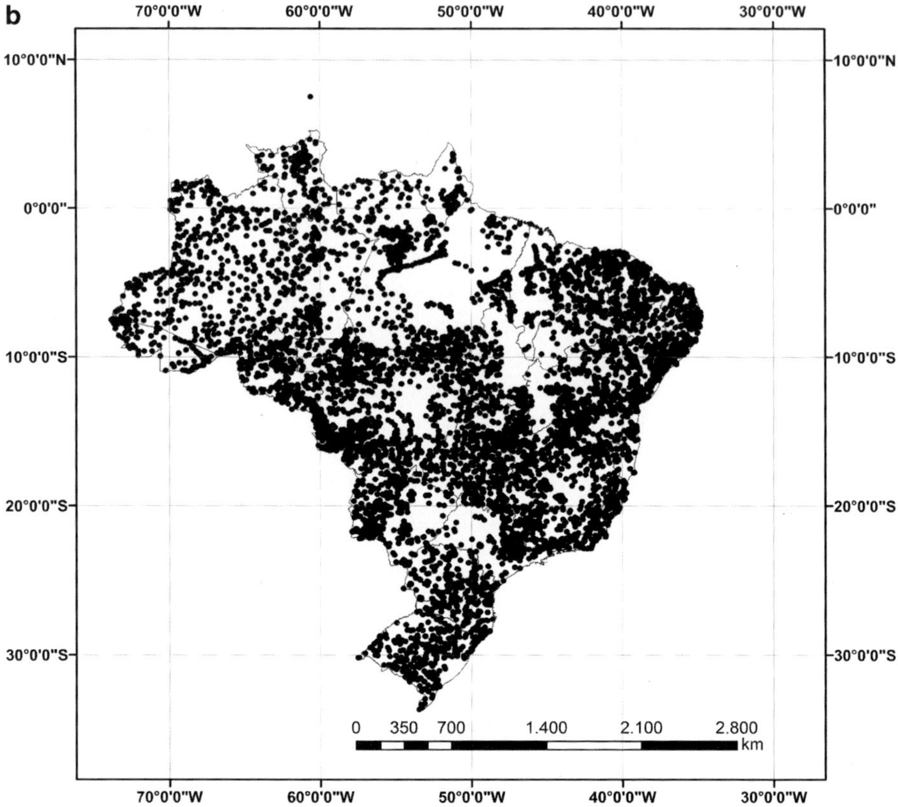

Fig. 6.1 (continued)

In Brazil, as in the rest of the world, most studies to evaluate PTF performance do not consider the quality and differences of methods used in PTF development. There is no data standard due to methodological problems. A common problem faced using soil databases from tropical countries is that water content at −10 or −33 kPa is measured on disturbed samples (Bell and Van Keulen 1996), due to the fact that samples collected during surveys aimed at mapping and classification exclusively, whereas bulk density and soil clods are not commonly collected (Minasny and Hartemink 2011).

In fact, more information is needed on the impact of the measurement methods once few studies have addressed the issue of assessing the techniques (Vereecken et al. 2010). The size of databases and the used methods are the most important factors that affect PTF precision and reliability (Pachepsky et al. 1999b; Wösten et al. 2001).

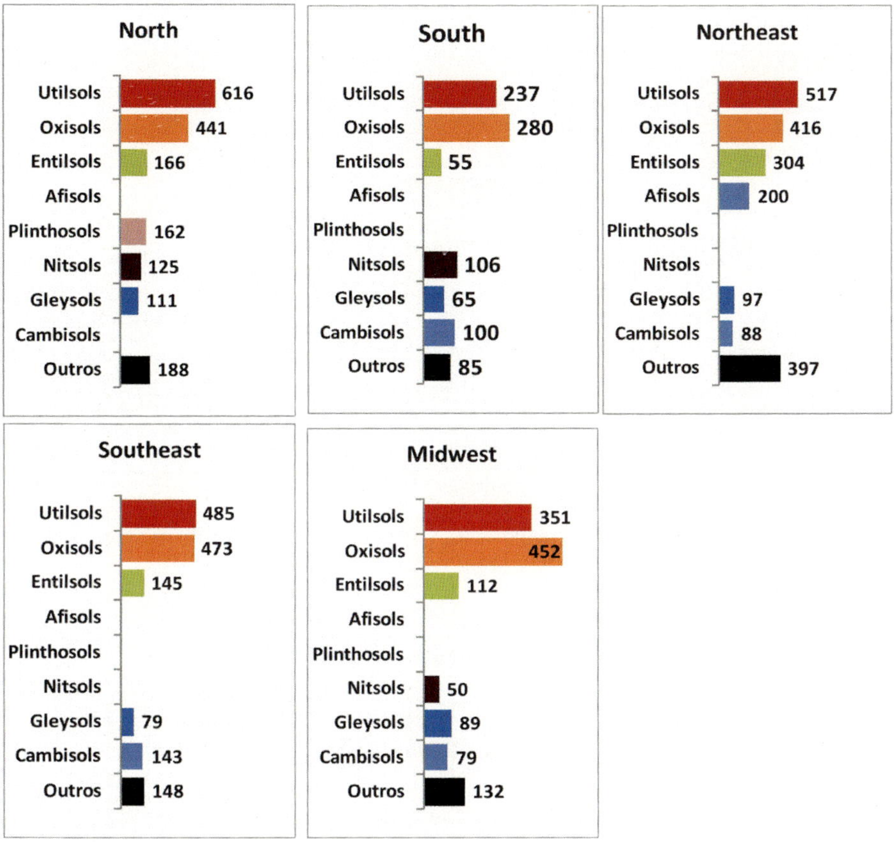

Fig. 6.2 Information about Brazilian soils organized per region and soil class by Embrapa (Oliveira et al. 2009)

6.3 Types of PTFs

PTFs can be developed to estimate chemical, physical and biological soil attributes, although usually they are developed to estimate hydraulic parameters – soil water retention and hydraulic conductivity. In general, PTFs are organized according to their type and application. Regarding application, two types exist: static PTFs, that are used to predict, among others properties, soil quality indicators, and dynamic PTFs, used to predict parameters to be used as input in simulation models (Fig. 6.3) (Minasny et al. 2003). With respect to PTF types, they can be classified in two groups: class PTFs, that estimate soil properties according to a soil class, like the prediction of soil water availability from texture data; and continuous PTFs, that predict soil properties as a continuous function of measured properties (Wösten et al. 1995).

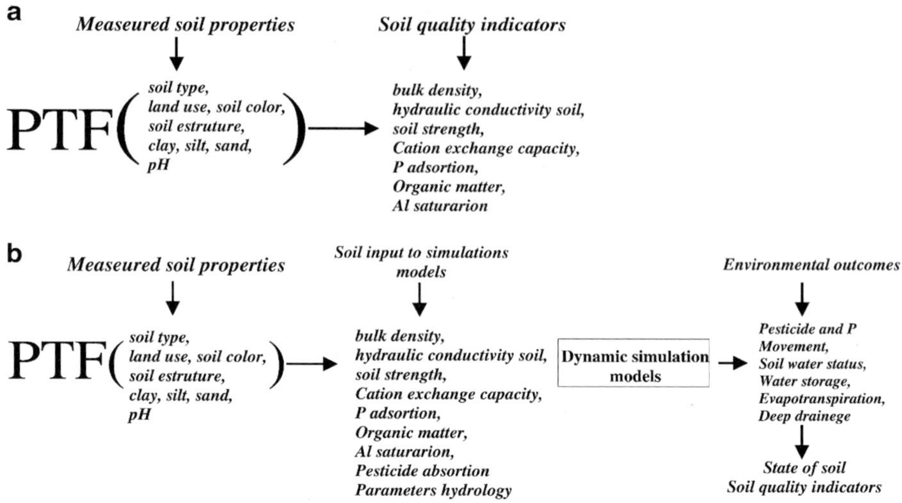

Fig. 6.3 (**a**) Static and (**b**) dynamic approaches using pedotransfer functions for predicting soil quality (Adapted from Minasny 2000)

PTFs can be empirical or physical-empirical. Empirical PTFs establish relations between predicting and predicted variables by statistical regression; Physical-empirical PTFs derive soil properties based on some physical mechanism or principle. For example, on predicting the soil water retention curve, Arya and Paris (1981) used particle size distribution, transforming it by converting solid mass fractions in water content and pore size distribution in matric potential using the capillarity Equation.

6.4 Methods for PTF Development

Several approaches can be employed for PTF development. Usually, data and PTFs are developed for a grouping of soils, organized by soil genesis, layering, texture, structure, bulk density, parent material and so on. Hodnett and Tomasella (2002) argued that it will never be possible to develop a reliable "universal" PTF because the worldwide range of soil properties is huge. In fact, grouping emerged as a strategy to stratify the resulting variability, thus enabling the development of better PTFs to predict hydraulic properties from easily available basic soil properties (Bruand 2004b).

Grouping has been done by horizons (Hall et al. 1977; Lamp and Kneib 1981); genetic horizons (Pachepsky et al. 1992); hydraulic horizons (Wösten et al. 1986); numerical soil classification (Williams et al. 1983); FAO soil class (Batjes 1996), texture class (Tietje and Hennings 1996) and parent material and morphology

(Franzmeier 1991). Tomasella and Hodnett (1997) discussed the possibility of extrapolating PTFs for hydraulic conductivity among groups of soils. Van den Berg et al. (1997) proposed to group soils at a worldwide level to develop PTFs. Bruand (2004b) presented a comprehensive outlook of the grouping strategies and concludes that texture, soil bulk density, and the type of parent material appear to be the most efficient criteria to improve accuracy of PTF.

Wösten et al. (1999) made a separation between topsoil and subsoil as primary grouping; then, texture was used as predictor. Gaiser et al. (2000) showed the importance of a mineralogical stratification for PTF development. Oliveira et al. (2002) grouped soils according to their clay activity to develop PTFs for the prediction of water content at specific matric potentials. Gonçalves et al. (1997) established a grouping according to the main texture classes from the FAO triangle and showed an increased accuracy for the prediction of soil water retention and hydraulic conductivity. Pachepsky et al. (1999b) grouped soils according to taxonomic unit, soil moisture regime, soil temperature regime, and soil textural class, improving both PTF accuracy and reliability to estimate soil water.

To develop a PTF, two basic principles should be considered: efficiency and uncertainty. The principle of efficiency consists of not estimating something that can easily be measured, whereas the principle of uncertainty states that a PTF should not be used unless its uncertainty level can be evaluated. Thus, if several PTFs are available for a certain problem, the one with the smallest variance should be used (Minasny et al. 2003).

Several methods are available for PTF development, being the most common the linear and multiple regression techniques. A linear relation between one or more dependent variables Y and k independent variables is assumed, X_j (j = 1,2,3,. . .k). The independent variables are also called explanatory variables or predictors, because, they are used to explain the variation of, and predict, the dependent variable Y. The linear regression model describes the relation between the k independent variables X_j and the dependent variable Y:

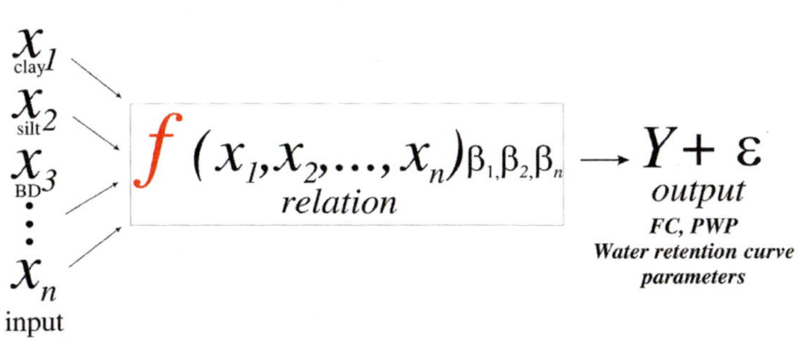

β_n , $n = 0,1,2,\ldots n$ are the regression coefficients, x_n , $n = 0,1,2,\ldots n$ are explanatory variables (e.g. contents of silt, clay, soil organic matter, etc.) and ε is the random error.

During the multiple regression procedure, properties with a small weight can be excluded, reducing the number of independent variables and avoiding the problem of multi-collinearity. Special routines like backward, forward, stepwise, stagewise and principle component analysis (PCA) are used when the number of predicting variables is very large and to verify the linear dependence between variables.

Three steps are recommended to develop a PTF: exploratory analysis, model structuring and model validation. The exploratory analysis aims to identify the existence of relations between variables. The second step consists of the development of the regression model and the third step is the validation and test of the model.

The selection of predictors should be done carefully: databases contain normally several possible predictors that are interrelated. To reduce this kind of problem, special techniques to select predictors can be used, like data mining and group method of data handling (GMDH), artificial neural networks (ANN), extended nonlinear regression (ENR), regression trees, fuzzy logic, and cross validation procedures (Minasny and McBratney 2002). Several of these techniques are sophisticated regressions that do not assume any known model. These techniques have been used to develop PTFs, and include the nearest neighbor methods (Nemes et al. 2006; Jagtap et al. 2004), support vector machines (Lamorski et al. 2008; Twarakavi et al. 2009), and multi-scale Bayesian neural networks (Jana et al. 2007). The use of these methods is discussed in Pachepsky and Schaap (2004) and in Pachepsky and Rawls (2004).

Each of these regression methods has its pros and cons. The advantage of using regression trees is related to the transparency of results, whereas neural networks are cap able of assuming any relation between parameters without considering, a priori, a supposition or model. According to Vereecken et al. (2010), in general, methods based on artificial neural networks show a better performance because they do not incorporate a predefined model. On the other hand, models generated by neural networks are black boxes and their algorithms cannot be analyzed too well. Several of these techniques have been explored in Brazil. Netto (2007), for example, developed a PTF using ANN to estimate water retention of soils from the Canchim watershed, county of São Carlos, Brazil. The ANN was efficient for developing the PTF. Tomasella et al. (2003) developed a PTF using group method of data handling (GMDH). Cagliari et al. (2011) employed ANN to estimate remaining phosphorus from pH values, exchangeable Al^{3+} content and sum of bases.

The described methods to establish empirical relationships between soil attributes do not consider spatial autocorrelation of soil properties. Nebel et al. (2010) used geostatistics to evaluate the applicability and possibility of the transfer of PTFs in order to estimate the gravimetric soil water content at fixed values of matric. The eight tested PTFs overestimated water contents at −33 kPa and underestimated at −1,500 kPa. Regarding the spatial structure, none of the PTFs altered the spatial pattern of data, but all of them resulted in average values and standard deviations that were different from the observed values.

Regression techniques (linear or non-linear) are the most common methods for PTF estimation of soil water retention. The following methods can be distinguished: (i) regression at fixed matric potentials, (ii) physical model; and (iii) retention curve parameter regression.

The regression technique at fixed matric potentials generates PTFs that allow estimating the water content at a fixed matric potential. Although this is a very

Table 6.1 Models of hydraulic properties

Model	Function		Parameters
Brooks and Corey (1964) *water retention curve*	$\theta(h) = \begin{cases} \theta_s \\ \theta_r + (\theta_s - \theta_s)\left(\frac{h_b}{h}\right)^{\lambda} \end{cases}$	*if* $h \geq h_b$ *if* $h < h_b$	$\theta_r = residual\ water\ content$ $\theta_s = saturation\ water\ content$ $h_b = air - entry\ value$ $\lambda = empirical\ parameter$
Hydraulic conductivity	$K(S_e) = \begin{cases} K_s \\ K_s + S_e^{\frac{2}{\lambda}+3} \end{cases}$	*if* $h < h_b$ *if* $h < h_b$	$S_e = \frac{\theta - \theta_r}{\theta_s - \theta_r}$
Campbel (1974) *water retention curve*	$\theta(h) = \begin{cases} \theta_s \\ \left(\frac{h_b}{h}\right)^{\frac{1}{b}} \end{cases}$	*if* $h \geq h_b$ *if* $h \geq h_b$	$b = fractal\ coefficient$
Hydraulic conductivity	$K(\theta) = \begin{cases} K_s \\ K_s + \left(\frac{\theta}{\theta_s}\right)^{2b+3} \end{cases}$	*if* $h \geq h_b$ *if* $h \geq h_b$	$S_e = \frac{\theta - \theta_r}{\theta_s - \theta_r}$
Van Genuchten (1980) *water retention curve*	$\theta(h) = \theta_r + \frac{\theta_r - \theta_r}{(1 + \alpha\|h\|^n)^{1-{}^1/_n}}$		$\alpha = empirical\ parameter$ $n = curve\ shape\ parameter$ $m = 1 - 1/n$ $\iota = empirical\ parameter$
Hydraulic conductivity	$K(S_e) = K_s S_e^{\iota}[1 - (1 - S_e^{1/m})^m]^2$		

common approach, it has the disadvantage not to result in an estimate for the entire water retention curve. It has the advantage of being very simple with respect to the regression techniques (Gonçalves 1994). An entire curve can be obtained by this technique if water contents at several matric potentials are estimated. Related techniques are shown by Gregson et al. (1987) and by Ahuja et al. (1985).

The physical model method includes schemes proposed by Arya and Paris (1981), Haverkamp and Parlange (1986), and Tyler and Wheatcraft (1989). Vaz et al. (2005) validated the Arya and Paris (1981) water retention model for Brazilian soils. The model consists of three parts: (i) the pore size distribution is estimated from the particle size distribution; (ii) the water content is estimated from the particle size distribution using the mass conservation equation; and (iii) matric potentials are estimated using the capillarity equation and pore size distribution.

Other physical-empirical models include those that predict the water retention curve from fractal models of soil structure (Anderson and McBratney 1995; Bird et al. 2000; Fazeli et al. 2010). The method by Campbell (1974), modified by Hutson and Cass (1987), has been used by some researchers to estimate the soil water retention curve (Tormena and Da Silva 2002). At present, most emphasis is given to the development of PTFs to estimate parameters of water retention equations by Brooks and Corey (1964), Campbell (1974) and Van Genuchten (1980) (Table 6.1).

Vereecken et al. (1992) infer that estimates of parameters of water retention equations, i.e. van Genuchten equations, enables comparisons between soils easier. On the other hand, recent studies showed that the prediction of single matric potentials yields better results. Comparing the two different methods for obtaining analytical expressions of the moisture retention curve, Tomasella et al. (2003) suggested that moisture contents at different matric potentials may be controlled by different soil properties and then, that using PTF for a specific matric potential allows a better combination of these properties. Pedotransfer functions providing point estimates of the water content or hydraulic conductivity tend to be more accurate than estimates based on parametric PTFs (Pachepsky et al. 1996).

The major part of PTFs developed for Brazil use the single matric potential method, mainly to estimate the plant available water between fixed boundaries at matric potentials of -10 or -33 kPa and $-1{,}500$ kPa. During the last years an increase in PTFs to predict parameters for equations by Van Genuchten (1980) and by Brooks and Corey (1964) is observed.

6.5 Criteria for PTF Evaluation

Wösten et al. (2001) defined accuracy as the comparison between predicted and measured values that were used to develop the PTF and reliability as the evaluation of PTFs using measured values that are different from those which were used to develop the PTF. There is, in fact, no common rule to compare PTFs, as the quality criterion may vary depending on the objective of application of the PTF. One of the principles of validating models states that complete testing is not possible (Balci 1997). Some authors use the same term for different criteria or denote the same criteria for different terms (Shein and Arkhangel'Skaya 2006). Schaap (2004) discussed the three most common criteria for PTF evaluation: optimization, precision and uncertainty.

Optimization is related to PTF calibration, and generally the least-squares method is used. Precision criteria are used to validate the model against an independent data set, and uncertainty criteria are used to characterize the probability distribution of predicted values. PTFs should be submitted to statistical validation (error, goodness of fit of the model, identification of model misspecification, examination of confidence intervals of regression coefficients and outliers detection), to test their performance in prediction, retained variables and regression coefficients; and to a functional validation (optimization), defined as a statistical test of PTFs in the context of a specific application, for example, water balance, solute transport in soils, initialization of water status in meteorology models, etc. (Vereecken et al. 1989). Wösten et al. (1986) present a similar concept and call it the functional criterion.

Some examples of functional evaluation of PTFs are presented in Barros (2010), who estimated the parameters (α, n, θr e θs) of the Van Genuchten (1980) soil water retention model and used these parameters to simulate maize and sorghum yields using the SWAP model (Kroes et al. 2008). Da Silva (2010) evaluated a PTF to estimate soil hydraulic conductivity and compared the obtained values with measured ones, concluding that the predictive performance of the PTF was weak. A similar result was obtained by Montenegro and Ragab (2012) who applied the PTF by Rawls et al. (1982) to estimate hydraulic conductivity of soils in northeast Brazil.

In order to improve the performance of PTFs, the following items should be emphasized: (i) the mathematical expression(s); (ii) the influence of basic input soil attributes; (iii) the functional evaluation; and (iv) alternative methods to obtain and fit the equations (Pachepsky et al. 1999a; Vereecken et al. 2010).

6.6 Statistical Tests to Evaluate PTFs

Imam et al. (1999) established two main categories of quantitative fitting, based on the difference between observed and estimated values (that is, residues). The first one includes the mean error (ME), the mean absolute error (MAE), sum of squared error (SSE), and root mean squared error (RMSE). The other one is based on the covariance of values, calculated by the correlation coefficient (R), determination coefficient (R^2) and the index of agreement (d) (Willmott et al. 1985). In general, for a model to be considered capable of producing good results it should be capable of producing small residues, show a high correlation between estimated and measured values and show absence of systematic errors (Belocchi et al. 2002).

Several studies in Brazil use the RMSE to compare PTFs (Table 6.2). The RMSE includes both systematic and random components. Then mean deviation (MD) and root mean square deviation (RMSD) of Tietje and Tapkenhinrichs (1993) is used to evaluate the goodness of fit for the water retention curve (Table 6.2). Wösten et al. (2001) recommend that when measured values are those used to develop the equation, the accuracy of the equation be evaluated.

The main statistics used for PTFs development are listed in Table 6.2. y_i represents the observed value, $\hat{y}_i$ the estimated value, and $\bar{y}_i$ is the average of observed values. N is the total number of observations. The values a and b are the interaction boundaries ($a = 0.1$ and b is set according to the smallest (most negative) potentials measured). A PTF can be considered precise if the variability of the values it estimates is not significantly different data from field descriptions and soil morphology data, and if the average of errors is not different from zero (Wösten et al. 2001). In Table 6.3 values of statistical evaluations for several PTFs developed in Brazil are presented.

Table 6.2 Main statistical parameters for PTF evaluation

Statistic	Symbol	Formulation
Difference-based statistics		
Sum of squares	***SSE***	$\frac{1}{N}\sum_{i=1}^{N}(y_i-\hat{y}_i)^2$
Root mean square error	***RMSE***	$\sqrt{\frac{1}{N}\sum_{i=1}^{N}(y_i-\hat{y}_i)^2}$
Mean error	***ME***	$\frac{1}{N}\sum_{i=1}^{N}(y_i-\hat{y}_i)^2$
Absolute mean error	***AME***	$\frac{1}{N}\sum_{i=1}^{N}\lvert y_i-\hat{y}_i\rvert$
The mean deviation	***MD***	$\frac{1}{N}\sum_{i=1}^{N}\frac{1}{b-a}\int_a^b(y_i-\hat{y}_i)d(\log_{10}h)$
Root mean squared deviation	***RMSD***	$\frac{1}{N}\sum_{i=1}^{N}\sqrt{\frac{1}{b-a}\int_a^b(y_i-\hat{y}_i)d(\log_{10}h)}$
Index of agreement (Wilmott)	***D***	$1-\frac{\sum_{i=1}^{n}(y_i-\hat{y}_i)^2}{\left[\sum_{i=1}^{n}(\lvert\hat{y}_i-\overline{\hat{y}_i}\rvert)+(\lvert y_i-\overline{y}_i\rvert)\right]^2}$
Correlation-based statistics		
Coefficient of determination	***R²*, *R***	$1-\frac{\sum_{i=1}^{N}(y_i-\hat{y}_i)^2}{\sum_{i=1}^{N}(y_i-\overline{y}_i)^2}$
Coefficient of correlation		$\frac{\sum_{i=1}^{N}\left[(\hat{y}_i-\overline{\hat{y}_i})^2(y_i-\overline{\hat{y}_i})^2\right]}{\sqrt{\sum_{i=1}^{N}\left[(\hat{y}_i-\overline{\hat{y}_i})^2(\hat{y}_i-\overline{\hat{y}_i})^2\right]}}$

6.7 Most Frequently Used Predictors

When developing PTFs, several data sources can be used as estimators, like soil laboratory data, data from field descriptions and soil morphology data (Minasny and Hartemink 2011). Many studies showed that water retention depends on soil texture, structure and composition (Rawls et al. 1991; Wösten et al. 2001; Pachepsky and Rawls 2004). Figure 6.4 shows how soil properties can compose PTFs to be used in several applications.

The most important features that strongly interfere with soil properties and are most frequently used in PTFs are soil texture (clay, silt and sand contents), bulk density and organic matter content.

Table 6.3 Typical examples of the water retention PTF accuracy in Brazilian PTFs

Source	Matric potential	RMSE	R^2
Arruda et al. (1987)	$w_{at\ -33}$ kPa		0.91
	$w_{at\ -1,500}$ kPa		0.95
Da Silva et al. (1990)	$w_{at\ -33}$ kPa		0.99
	$w_{at\ -1,500}$ kPa		0.99
Masutti (1997)	$w_{at\ -33}$ kPa	0.040	
	$w_{at\ -1,500}$ kPa	0.020	
Tomasella and Hodnett (1998)	$\theta_{at\ -33}$ kPa	0.040	
	$\theta_{at\ -1,500}$ kPa	0.060	
Coelho et al. (1998)	$\theta_{at\ -33}$ kPa		0.57 mean
	$\theta_{at\ -1,500}$ kPa		0.50 mean
Lopes Assad et al. (2001)	$w_{at\ -10}$ kPa		0.77
	$w_{at\ -1,500}$ kPa		0.72
Gaiser et al. (2000)	$\theta_{at\ -33}$ kPa		0.91
	$\theta_{at\ -1,500}$ kPa		0.92
Giarola et al. 2002	$\theta_{at\ -10}$ kPa		0.79
	$\theta_{at\ -1,500}$ kPa		0.81
Oliveira et al. (2002)	$w_{at\ -33}$ kPa		0.96
	$w_{at\ -1,500}$ kPa		0.95
Peraza (2003)	$w_{at\ -5}$ kPa		0.66
	$w_{at\ -10}$ kPa		0.67
	$w_{at\ -33}$ kPa		0.69
	$w_{at\ -100}$ kPa		0.72
	$w_{at\ -500}$ kPa		0.76
	$w_{at\ -1,500}$ kPa		0.76
Tomasella et al. (2003)	$\theta_{at\ porosity}$	0.040	
	$\theta_{at\ -6}$ kPa	0.035	
	$\theta_{at\ -10}$ kPa	0.031	
	$\theta_{at\ -33}$ kPa	0.021	
	$\theta_{at\ -100}$ kPa	0.020	
	$\theta_{at\ -1,500}$ kPa	0.025	
Mello et al. (2005)	$\theta_{at\ -6}$ kPa		0.80
	$\theta_{at\ -10}$ kPa		0.82
	$\theta_{at\ -33}$ kPa		0.89
	$\theta_{at\ -1,500}$ kPa		0.91
Alves et al. (2009)	$\theta_{at\ -33}$ kPa		0.92
	$\theta_{at\ -1,500}$ kPa		0.89
Reichert et al. (2009)	$\theta_{at\ -10}$ kPa		0.67
	$\theta_{at\ -33}$ kPa		0.62
	$\theta_{at\ -1,500}$ kPa		0.62
Urach (2007)	$w_{at\ -6}$ kPa		0.62
	$w_{at\ -10}$ kPa		0.65
	$w_{at\ -33}$ kPa		0.68
	$w_{at\ -100}$ kPa		0.72
	$w_{at\ -500}$ kPa		0.76
	$w_{at\ -1,500}$ kPa		0.77

(continued)

Table 6.3 (continued)

Source	Matric potential	RMSE	R^2
Andrade and Stone (2010)	$\theta_{at\ -6}$ kPa		0.76 mean
Nascimento et al. (2010)	$\theta_{at\ -33}$ kPa		0.67 mean
	$\theta_{at\ -1,500}$ kPa		0.72 mean
Michelon et al. (2010)	$w_{at\ 0}$ kPa	0.027	0.77
	$w_{at\ -1}$ kPa	0.016	0.87
	$w_{at\ -33}$ kPa	0.011	0.93
	$w_{at\ -100}$ kPa	0.017	0.90
	$w_{at\ -500}$ kPa	−0.031	0.71
	$w_{at\ -1,500}$ kPa	0.023	0.81
Rodrigues et al. (2011)	$w_{at\ -33}$ kPa		0.58
	$w_{at\ -1,500}$ kPa		0.59
Barros et al. (2013)	$\theta_{at\ -10}$ kPa	0.031	
	$\theta_{at\ -33}$ kPa	0.027	
	$\theta_{at\ -1,500}$ kPa	0.031	
Dos Santos et al. (2013)	$w_{at\ -33}$ kPa (0–20 cm)		0.64 mean
	$w_{at\ -1,500}$ kPa (0–20 cm)		0.59 mean
	$w_{at\ -33}$ kPa (40–70 cm)		0.66 mean
	$w_{at\ -1,500}$ kPa (40–70 cm)		0.65 mean

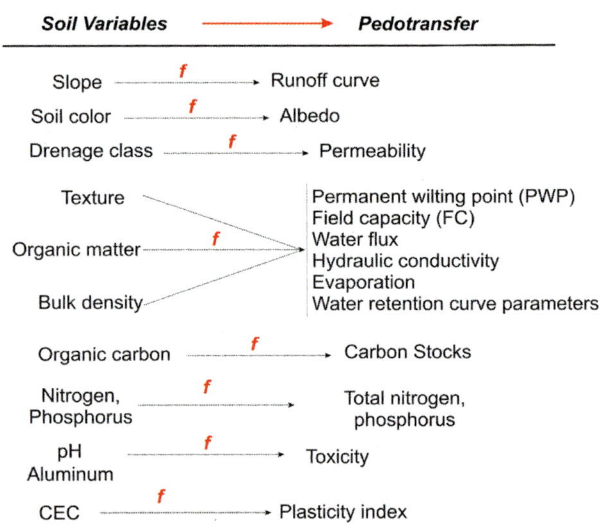

Fig. 6.4 Example of several soil variables used in PTFs

6.7.1 *Texture and Particle-Size Distribution*

Soil properties are largely determined by the size and shape distribution of its solid particles, determining the pore geometry. Water or air movement through a soil is determined by properties of the fluid as well as of the soil. Hydraulic properties of tropical soils tend to be very different than those of temperate climate soils due to

Table 6.4 Grain size classification systems as proposed by USDA/FAO, ISSS and SBCS

Fraction	USDA/FAO	ISSS	Brazil (SBCS)
	Diameter (mm)		
Very coarse sand	2–1	–	–
Coarse sand	1–0.5	2–0.2	2–0.2
Medium sand	0.5–0.25	–	–
Fine sand	0.25–0.10	0.2–0.02	0.2–0.05
Very fine sand	0.10–0.05	–	–
Sand	0.05–2	0.02–2	0.05–2
Silt	0.002–0.05	0.02–0.002	0.05–0.002
Clay	<0.002	<0.002	<0.002

the very low silt content of tropical soils (Tomasella et al. 2003). The type of clay mineral is another factor causing differences in water retention and flow between tropical and temperate soils (Minasny and Hartemink 2011).

Soil texture classes are defined based on size limits for sand, slit and clay particles. Two main grain size classification systems are internationally used, the system proposed by the International Society of Soil Science (ISSS) and the USDA/FAO system (Table 6.4). Most internationally published PTF adhere to the USDA/FAO system (Nemes and Rawls 2004). Comparisons between different representations of the particle-size distribution in PTF are important, but rare. More than 80 % of soil water retention can be explained by particle size distribution alone (Reichardt and Timm 2004). Most PTFs developed in Brazil use soil texture as predictor (Table 6.5).

Tomasella and Hodnett (1998) estimated soil water retention for tropical soils of Brazil using soil texture, and found an average accuracy of 0.062 $m^3\ m^{-3}$ along the water retention curve. Hodnett and Tomasella (2002) showed that averaging Van Genuchten parameters across textural classes led to significant differences between tropical and temperate soils for most textures. Bernoux et al. (1998) showed that bulk density can be predicted at a good level of accuracy by clay content. Another common use of the soil texture class information is when it is used as a grouping factor prior to developing PTF.

6.7.2 *Bulk Density*

Soil bulk density is considered an important variable for the predict of soil water retention and hydraulic conductivity; it is also used together with soil organic matter in PTFs that predict penetration resistance. Bulk density is an indicator of compaction and porosity and is needed as input parameter for water, sediment, and nutrient transport models. Rawls (1983) showed that, together with the amount of organic carbon, the particle-size distribution could be used for predicting the density of natural undisturbed soils. Nevertheless, in soil surveys bulk density is rarely determined, making improvement of PTF performance difficult.

Table 6.5 Predicted properties and selected predictors used in soils from Brazil

Predicted properties	Soil type	Region	Predictors (input variables)	References
Available water capacity $\theta_{at\ -33\ and\ -1,500\ kPa}$	Various	São Paulo	Silt, clay	Arruda et al. (1987)
$\theta_{at\ -33\ and\ -1,500\ kPa}$	Various	Semi arid Brazilian tropics	Texture, bulk density, organic matter	Da Silva et al. (1990)
Water retention curve van Genuchten parameters $\theta_{at\ -33\ and\ -1,500\ kPa}$	Oxisols Ferralsols	Part northeast Brazil	Clay, silt, sand, organic matter, CEC, Al_{dith} and Fe_{dith}	van den Berg et al. (1997)
Saturated hydraulic Conductivity	Red-yellow oxisols from the Amazon	Brazilian Amazon	Effective porosity θ_{at} −10 kPa	Tomasella and Hodnett (1997)
Unsaturated hydraulic conductivity	Various Oxisols	Brazilian Amazon	Water retention curve	Tomasella and Hodnett (1997)
Available water capacity $\theta_{at\ -33\ and\ -1,500\ kPa}$	Various	Forest zone, Pernambuco	Sand, clay and bulk density	Masutti (1997)
$\theta_{at\ -33\ and\ -1,500\ kPa}$	Oxisols, Ulyisols, Entisols (Gleysols)	Minas Gerais and Piaui	Clay, silt, sand and bulk density	Coelho et al. (1998)
Bulk density	Oxisols, Alfisols and Utilsols	Brazilian Amazon	Organic carbon, Sand, clay, silt, chemical characteristics, depth, sum of bases, CEC, Free Fe, pH, bulk density	Bernoux et al. (1998)
Bulk Density Water retention curve Brooks Corey parameters	Various	Brazilian Amazon	Sand, clay, silt, bulk density and porosity, Organic C	Tomasella and Hodnett (1998)
Available water capacity $\theta_{at\ -33\ and\ -1,500\ kPa}$	Crushed samples, low activity clay (LAC) and non LAC soils	Northeast (Ceará and Piauí)	Organic carbon, clay and silt	Gaiser et al. (2000)

Water retention curve van Genuchten parameters	Various	Brazil	Coarse sand, fine sand, silt, clay, organic C, moisture equivalent and bulk density	Tomasella et al. (2000)
Available water capacity $\theta_{at-33\ kPa}$	Ultisols	Rio de Janeiro	Clay, sand, OM, Microporosity	Fabian and Ottoni Filho (2000)
Available water capacity $\theta_{at-33\ and\ -1,500\ kPa}$	Oxisols	Southeast and Bahia	Total sand	Lopes-Assad et al. (2001)
Water retention curve (van Genuchten parameters)	Various	Brazil	Sand, silt, clay, organic carbon, pH, CEC and bulk density	Hodnett and Tomasella (2002)
Water retention curve Hutson and Cass parameters van Genuchten parameters	Oxisols	São Paulo	Bulk density, tillage and soil landscape position	Tormena and Da Silva (2002)
Available water capacity $\theta_{at-33\ and\ -1,500\ kPa}$	Oxisols	Brazilian Savannah (Cerrado)	Clay	Balbino et al. (2002)
Available water capacity $\theta_{at-33\ and\ -1,500\ kPa}$	Utilsols, Nitosols, Oxisols, Entisol and Alfisols	Pernambuco	Sand, silt, clay and bulk density	Oliveira et al. (2002)
Available water capacity $\theta_{at-33\ and\ -1,500\ kPa}$	Various	Pernambuco, Northeast	Sand, clay, silt and bulk density	Oliveira et al. (2002)
Available water capacity $\theta_{at-33\ and\ -1,500\ kPa}$ and Bulk density	Various	Rio Grande do Sul and Santa Catarina	Sand, clay, silt and Fe_2O_3	Giarola et al. (2002)
Available water capacity $\theta_{at-5,-10,-33,-100,-500\ and\ -1,500\ kPa}$	Various	Rio Grande do Sul	Clay, silt, sand and organic matter	Peraza (2003)
Water retention curve (Porosity, $\theta_{at-6,-33,-100,-1,500\ kPa}$)	Various	Brazil	Clay, silt, coarse sand, fine sand, bulk density and moisture equivalent	Tomasella et al. (2003)
$\theta_{at-33\ kPa}$	Various	Rio de Janeiro	clay, microporosity	Ottoni (2005)
Water retention curve van Genuchten parameters	Inceptisols (Cambisols)	Minas Gerais	Clay, sand, silt, organic matter, bulk density and depth	Mello et al. (2005)

(continued)

Table 6.5 (continued)

Predicted properties	Soil type	Region	Predictors (input variables)	References
Water retention curve Ayra and Paris model	Various	São Paulo e Rio Grande do Sul	Particle size distribution	Vaz et al. (2005)
Organic C	Oxisols	Brazilian Cerrado	Depth, clay and silt	Zinn et al. (2005)
Residual P	Oxisols, Ultisols	Brazil	pH in NaF	Alves and Lavorenti (2006)
Bulk density	Various	Rio de Janeiro	Organic C, clay content and sum of cations	Benites et al. (2007)
Water retention curve	Oxisols, Alfisols, Entisols, Ultisols	São Paulo	Clay, sand, silt, BD	Netto (2007)
Available water capacity $\theta_{at\,-6,-10,-33,-100,-500\text{ and }-1{,}500\text{ kPa}}$		Rio Grande do Sul	Sand, silt, clay and Bulk density	Urach (2007)
Water retention curve Penetration resistance	Ultisols	Parana	CO and bulk density	Fidalski and Tormena (2007)
Bulk Density	Various	Rondonia and Mato Grosso	Clay, silt, organic C, pH	Mello (2007)
Reference bulk density (proctor density, −800 kPa)	Various	Rio Grande do Sul	OM, BD, Clay and silt	Reichert et al. (2008)
Penetration resistance	Oxisols	São Paulo	Organic C, clay content and bulk density	Da Silva et al. (2008)
Water infiltration, saturated hydraulic conductivity and Van Genuchten parameters	Ultisols, Entisols, Oxisols and Vertisols	Rio Grande do Sul	Bulk density, texture and porosity	Fiorin (2008)
Resistence to the soil penetration	Entisols, Oxisols, Ultisols	São Paulo	Clay, water retention curve	Almeida (2008)
$\theta_{at\,-33\text{ and }-1{,}500\text{ kPa}}$	Various	Semiarid, Pernambuco	Silt, clay, sand	Alves et al. (2009)
Available water capacity $\theta_{at\,-33\text{ and }-1{,}500\text{ kPa}}$	Various	Rio Grande do Sul	Clay, silt, sand, BD and organic matter	Reichert et al. (2009)
$\theta_{at\ sat\,-1,-33,-100,-500\text{ and }-1{,}500\text{ kPa}}$	Various	Rio Grande do Sul	Bulk density, macroporosity, micro-porosity, fine sand fine and clay content	Michelon et al. (2010)

$\theta_{at-6\ kPa}$	Various	Brazilian Cerrado, Nort and Northeast	Silt, clay, sand and organic matter	Andrade and Stone (2010)
Available water capacity (AWC) $\theta_{at-33\ and\ -1,500\ kPa}$	Oxisols and Utilsols	Rio de Janeiro, Alagoas, Bahia, Pernambuco, Sergipe and Maranhão	Sand, silt, clay, CO and bulk density	Nascimento et al. (2010)
Maximum bulk density	Oxisols, Ultisols, Nitosols	Rio Grande do Sul	Clay and organic matter	Marcolin and Klein (2011)
Residual phosphorus	mainly Oxisols and Ultisols	Sao Paulo	*pH NaF, Sum Base (Ca, K, Mg) e Al3 +*	Cagliari et al. (2011)
Available water capacity $\theta_{at-33\ and\ -1,500\ kPa}$	Oxisols	Brazilian Cerrado	Clay, Silt, OM, BD	Rodrigues et al. (2011)
Soil carbon stocks	Entisols – aqu (Gleisols) (Plinthosols)	Distrito Federal	Organic carbon, bulk density, silt, clay, sand and depth	França (2011)
Organic C	Oxisols	Minas Gerais	Depth, Fe_2O_3, Clay and silt	Zinn et al. (2012)
Available water capacity $\theta_{at-33\ and\ -500\ kPa}$	Various	Rio Grande do Sul	Soil organic matter, coarse sand, fine sand and clay	Dos Santos et al. (2013)
Water retention curve van Genuchten parameters	Entisols, Oxisols, Ultisols, Alfisols and Inceptisols (Cambisols)	Northeast Brazil	Clay, silt, sand, organic matter and bulk density	Barros et al.(2013)

However, De Vos et al. (2005) tested 12 PTFs that predict bulk density and concluded that published PTFs showed large differences in performance and should be used with care, especially when applied in environments other than the ones in which they were calibrated.

Bulk density correlates negatively to Al_2O_3, clay, Fe_2O_3 and organic matter and positively to coarse and fine sand (Giarola et al. 2002). Similar results for correlations between bulk density and organic matter and texture were found also by Alexander (1980), Adams (1973), Rawls (1983) and Bernoux et al. (1998).

In tropical soils, high levels of iron and aluminum oxides are associated to the development of a typical granular structure, determining the bulk density values (Schwertmann and Taylor 1977).

In Brazil, Bernoux et al. (1998) and Tomasella and Hodnett (1998) presented the first PTFs for bulk density for Amazon soils, using texture, organic carbon and pH as predictor. For soils of the southern state Rio Grande do Sul, Marcolin and Klein (2011) used a PTF to estimate the maximum soil density using clay and organic matter content as predictors. From this maximum density they estimated the relative density, an indicator for compaction.

Da Silva et al. (2008) developed PTFs for water retention curves and penetration resistance using bulk density and showed a good performance related to monitoring the soil physical quality. Urach (2007) used bulk density as a predictor for water contents at matric potentials of −33 kPa e −1,500 kPa for soils from Rio Grande do Sul state.

6.7.3 Organic Matter

Soil organic matter is considered to be the main factor maintaining quality and productivity of agricultural soils, due to its influence on chemical and physical soil properties. The organic matter is the strongest predictor for bulk density. Adams (1973) showed that the amount of organic carbon had a significant effect on the bulk density of soils

An estimation of retention and availability of water in soils from Rio Grande do Sul state was performed by Reichert et al. (2009), who observed strong influence of clay and silt contents on field capacity and permanent wilting point, whereas the organic matter content showed an smaller effect. Vereecken et al. (2010) argued that soil organic matter reduced substantially the performance of PTF predicting the hydraulic conductivity compared with cases where only soil texture and bulk density were used. According to Costa (2012), who analyzed soils from the Santa Catarina state, field capacity and permanent wilting point for these soils depend strongly on grain size distribution and on organic matter content, with soils with a finer texture and/or higher organic matter content retaining more water.

Dalbianco (2009) observed that including soil properties that represent its structural state, like density and organic carbon, improves the predictive capacity of PTFs for saturated hydraulic conductivity. Giarola et al. (2002) found influence of organic matter in a PTF to predict field capacity and permanent wilting point.

6.7.4 Soil Morphological Attributes and Soil Structure

Visible morphological characteristics are always found in soil survey reports. They include color, type of soil horizon, texture, structure, consistence, mottling, concretions, position and steepness, presence of signs of erosion, etc. These descriptions are of qualitative nature, but the volume of information is very large and it can be used in PTF development. Lilly and Lin (2004) state that soil structure is a function of interaction between climate and texture, therefore PTFs that incorporate soil structure to predict hydraulic properties are more easily extrapolated between regions. Minasny et al. (2003) conclude that saturated hydraulic conductivity can be predicted by a PTF based on soil field morphology.

Although the importance of soil morphology as a predictor in PTFs was shown by many researchers (McKeague et al. 1982; McKenzie and MacLeod 1989; McKenzie and Jacquier 1997), not too much attention has been given to this fact. The main difficulty in including soil morphology in PTFs is the qualitative and subjective nature of its description. According to Vereecken et al. (2010), soil characteristics that refer to structure and morphology can play an important role in PTF performance.

Gaiser et al. (2000) used clay mineral composition to predict water retention at −33 and −1,500 kPa in 663 horizons from Brazil and Niger. The PTFs developed for soils containing predominantly low-activity clay differed considerably from those established for non low activity clays. Hodnett and Tomasella (2002) compared the Van Genuchten soil retention parameters for temperate and tropical soils. They showed differences between montmorillonitic and kaolinitic soils and suggested to include predictors related to mineralogy in PTFs (Bruand 2004a).

A research by Dalbianco (2009) showed that PTFs that use soil structure as predictor have a better predictive capacity for soil hydraulic conductivity, and that only texture information cannot explain all variations in hydraulic conductivity.

6.7.5 Other Variables

Soil surveys usually contain information at very small scales and soil attribute data like pH, CEC, bulk density, structure, pore size distribution and geometry, position in the landscape can be missing, limiting PTF development. Analyzing reports on PTF development it is seen that the number of predictors is generally small (Table 6.5). Tomasella and Hodnett (2004) evaluated several PTFs developed from a database with data from 771 horizons of tropical soils and showed that the PTFs did not perform well in estimating soil water retention. This is most probably due to the simplicity of PTFs, in other words, the limited number of predictors included.

Including more detailed information like terrain attributes and vegetation (Leij et al. 2004), soil structure (Pachepsky and Rawls 2004), water content at selected matric potentials, morphological properties or taxonomic information could improve the accuracy of PTFs (Medeiros 2012).

6.8 PTF Software in Brazil

Few initiatives exist on the development of software to use PTFs in Brazil. Prevedello (1999) developed the software SPLINTEX, allowing determining the Van Genuchten (1980) water retention parameters from PTF, including the possibility of estimation of field capacity and permanent wilting point. The validation of the Arya and Paris (1981) model performed by Vaz et al. (2004) for representative soils from São Paulo state was the basis for the development of the software Qualisolo, aiming to estimate soil water retention curves based on texture and bulk density data (Naime et al. 2006).

6.9 Conclusions

Based on the presented review, some conclusions can be drawn regarding the development of PTFs in Brazil:

1. PTF development requires large and reliable databases. This implies that well structured and easily accessible information systems of measured soil hydraulic characteristics need to be created. Possible predictors need to be characterized with more detail in databases.
2. The increase in efforts regarding PTF development and evaluation allow a better understanding of soil water dynamics in tropical soils. However, studies that compare different measurement methods for hydraulic properties and their suitability to develop PTF are still missing.
3. Estimates of hydraulic conductivity by PTF are relatively scarce. PTFs for model parameter estimation are predominant, mainly addressing the Van Genuchten (1980) soil water retention equation, as well as PTFs to estimate specifically field capacity and permanent wilting point.
4. The application of the predicted hydraulic properties depends on the required scale and resolution. For site-specific application, the general PTFs cannot be used to accurately simulate soil-water dynamics. In this case, local calibration is required.
5. Functional evaluation of PTFs is rarely performed. Future works in Brazil should test the application of hydraulic pedotransfer functions in predicting soil water balance, crop production, and spatial variability of soil hydraulic properties in the field. Virtually all crop growth simulation models used in Brazil predict the soil water balance based on approximations developed for temperate climates. There is not much effort to derive PTFs can be include in simulations model for tropical soils.
6. Field capacity cannot be estimated by a laboratory determination of water content at a fixed matric potential. This should be remembered when using PTFs for field capacity prediction. In reality, these indirect methods are attempts to eliminate laborious *in situ* experiments
7. PTFs are not suitable for extrapolation: they should not be applied to soils that are outside the range or region where the PTF was developed.

Acknowledgments The authors would like to thank Alexandre Ferreira do Nascimento and José Coelho de Araújo Filho for their comments and corrections that helped to improve the manuscript.

References

Adams WA (1973) The effect of organic matter on the bulk and true densities of some uncultivated podzolic soils. J Soil Sci 24:10–17

Ahuja LR, Naney JW, Williams RD (1985) Estimating soil water characteristics from simpler properties and limited data. Soil Sci Soc Am J 49:1100–1105

Alexander EB (1980) Bulk densities of California soils in relation to other soil properties. Soil Sci Soc Am J 44:689–692

Almeida CX (2008) Funções de pedotransferência para a curva de resistência do solo à penetração. Dissertação (mestrado), Universidade Estadual Paulista, Jaboticabal, UNESP, 48 p

Alves ME, Lavorenti A (2006) Remaining phosphorus estimate through multiple regression analysis. Pedosphere Soil Sci Soc China 16(5):566–571

Alves GVA, Masutti MM, Oliveira LN de (2009) Caracterização da água disponível a partir de parâmetros físico-hídricos em solos de referência do sertão pernambucano. In: IV Congresso de Pesquisa e Inovação da Rede Nacional de Educação e Tecnologia -CONNEPI, Belém

Anderson AN, McBratney AB (1995) Soil aggregates as mass fractals. Aust J Soil Res 33:757–772

Andrade RS, Stone FL (2010) Estimativa da umidade na capacidade de campo em solos sob Cerrado. Rev Bras Eng Agríc Ambient 15(2):111–116, Campina Grade, PB

Arruda FB, Zullo J, Oliveira JB (1987) Parâmetros de solo para calculo de água disponível com base na textura do solo. Rev Bras Sci Solo 11:11–15

Arya LM, Paris JF (1981) A physicoempirical model to predict soil moisture characteristics from particle-size distribution and bulk density data. Soil Sci Soc Am J 45:1023–1030

Balbino LC, Bruand A, Brossard MG, Hajnos M, Guimarães MF (2002) Changes in porosity and microaggregation in clayey Ferralsols of the Brazilian Cerrado on clearing for pasture. Eur J Soil Sci 53:219–230

Balci O (1997) Principles of simulation model validation, verification, and testing. Trans Soc Comput Simul 14:3–12

Barros AHC (2010) Desenvolvimento de funções de pedotransferencia e sua utilização em modelo agro-hidrologico. ESALQ/USP, Tese de Doutorado, Piracicaba, 148 p

Barros AHC, Jong D, van Lier Q, Maia AHM, Scarpare VF (2013) Pedotransfer functions to estimate water retention parameters of soils in northeastern Brazil. Rev Bras Ciênc Solo 37:379–391

Batjes NH (1996) Development of a world data set of soil water retention properties using pedotransfer rules. Geoderma 71:31–52

Batjes NH (2008) Mapping soil carbon stocks of Central Africa using SOTER. Geoderma 146:58–65

Batjes NH, Al-Adamat R, Bhattacharyya T, Bernoux M, Cerri CEP, Gicheru P, Kamoni P, Milne E, Pal DK, Rawajfih Z (2007) Preparation of consistent soil data sets for modelling purposes: secondary SOTER data for four case study areas. Agric Ecosyst Environ 122:26–34

Bell MA, Van Keulen H (1996) Effect of soil disturbance on pedotransfer function development for field capacity. Soil Technol 8:321–329

Belocchi G, Acutis M, Fila G, Donatelli M (2002) An indicator of solar radiation model performance based on a fuzzy expert system. Agron J 94:1222–1233

Benedetti MM, Sparovek G, Cooper M, Curi N, Carvalho Filho A (2008) Representatividade e potencial de utilização de um banco de dados de solos do Brasil. Rev Bras Cienc Solo 32:2591–2600

Benites VM, Machado PLOA, Fidalgo ECC, Coelho MR, Madari BE (2007) Pedotransfer functions for estimating soil bulk density from existing soil survey reports in Brazil. Geoderma 139:90–97

Bernoux M, Cerri CEP (2003) Digital soil properties database of the Amazon part from the RADAMBRASIL project (Version 1, 25 September 2005, tp://lba.cptec.inpe.br/lba_archives/CD/CD-208/Cerri/). In: Centro de Energia Nuclear Na Agricultura (CENA), Universidade de Sao Paulo

Bernoux M, Arrouays D, Cerri C, Volkoff B, Jolivet C (1998) Bulk densities of Brazilian Amazon soils related to other soil properties. Soil Sci Soc Am J 62:743–749

Bird NRA, Perrier E, Rieu M (2000) The water retention function for a model of soil structure with pore and solid fractal distributions. Eur J Soil Sci 51:55–63

Bouma J (1989) Using soil survey data for quantitative land evaluation, 1989. Adv Soil Sci 9:177–213

Bouma J, Van Lanen HAJ (1986) Transfer functions and threshold values: from soil characteristics to land qualities. In: Proceedings of the international workshop on quantified land evaluation procedures, 27/04–2/05/1986, Washington, DC, pp 106–110

Brakensiek DL, Rawls, WJ, Stephenson GR (1984) Modifying SCS hydrologic soil groups and curve numbers for rangeland soils. ASAE paper no. PNR-84203, St. Joseph

Briggs LJ, McLane JW (1907) The moisture equivalents of soils. USDA Bureau of soils bulletin, 45. Washington, DC, 23 p

Briggs LJ, Shantz HL (1912) The wilting coefficient for different plants and its indirect determination. USDA Bureau of plant industry Bulletin, 230. U. S. Gov. Printing Office, Washington, DC

Brooks RH, Corey AT (1964) Hydraulic properties of porous media. Hydrology paper no. 3. Colorado State University, Ft. Collins

Bruand A (2004a) Utilizing mineralogical and chemical information in PTFs. In: Pachepsky Y, Rawls WJ (eds) Development of pedotransfer functions in soil hydrology, Developments in soil science, 30. Elsevier, Amsterdam, pp 153–157

Bruand A (2004b) Preliminary grouping of soils. In: Pachepsky Y, Rawls WJ (eds) Development of pedotransfer functions in soil hydrology, Developments in soil science, 30. Elsevier, Amsterdam, pp 159–172

Cagliari J, Veronez MR, Alves ME (2011) Remaining phosphorus estimated by pedotransfer function. Rev Bras Cienc Solo 35:203–212

Campbell GS (1974) A simple method for determining unsaturated conductivity from moisture retention data. Soil Sci 117:311–314

Campbell GS, Shiozawa S (1994) Prediction of hydraulic properties of soils using particle size distribution and bulk density data. In: Van Genuchten MT, Leij FJ, Lund LJ (eds) Indirect methods for estimating the hydraulic properties of unsaturated soils. University of California, Riverside, pp 317–328

Carsel RF, Parrish RS (1988) Developing joint probability distributions of soil water retention characteristics. Water Resour Res 24:755–769

Chagas CS, Carvalho W Jr, Bhering SB, Tanaka AK, Baca JFM (2004) Estrutura e organização do sistema de informações georreferenciadas de solos do brasil (sigsolos – versão 1.0). Rev Bras Cienc Solo 28:865–876

Coelho EF, Conceição MA, Souza VAB (1998) Estimativa dos limites de disponibilidade de água em função da densidade global e da textura do solo. Rev Ceres 45:183–192

Cooper M, Mendes LMS, Da Silva WLC, Sparovek G (2005) A national soil profile database for Brazil available to international scientists. Soil Sci Soc Am J 69:649–652

Correa JC (1984) Características físico hídricas dos solos latossolo amarelo, podzolico vermelho amarelo e podzol hidromórfico do estado do Amazonas. Pesqui Agropecu Bras 20:1317–1322

Cosby BJ, Hornberger GM, Clapp RB, Ginn TR (1984) A statistical exploration of the relationships of soil moisture characteristics to the physical properties of soil. Water Resour Res 20:682–690

Costa A (2012) Retenção e disponibilidade de água em solos de santa catarina: avaliação e geração de funções de pedotransferência. Tese (doutorado) – Centro de Ciências Agroveterinárias// Universidade do Estado de Santa Catarina – UDESC. Lages, 423 p

Da Silva JRL (2010) Caracterização Físico-Hídrica de Três Bacias Experimentais do Estado de Pernambuco para suporte a Modelagem Hidrológica. Dissertação (Mestrado em Engenharia Agrícola) – Universidade Federal Rural de Pernambuco, Departamento Dde Tecnologia Rural, Recife

Da Silva AS, Brito LT, Oliveira CA, Mota AW (1990) Parâmetros de solo em função da umidade na capacidade de campo em áreas irrigáveis do trópico semiárido Brasileiro. Pesqui Agropecu Bras 25:103–116, 19

Da Silva AP, Tormena CA, Fidalski J, Imhoff SS (2008) Funções de pedotransferência para as curvas de retenção de água e de resistência do solo à penetração. Rev Bras Ciênc Solo, Viçosa 32(1):1–10

Dalbianco L (2009) Variabilidade espacial e estimativa da condutividade hidráulica e caracterização físico-hídrica de uma microbacia hidrográfica rural. Universidade Federal de Santa Maria (UFSM), Centro de Ciências Rurais, Dissertação (mestrado), 116 p

Dane JH, Puckett W (1992) Field soil hydraulic properties based on physical and mineralogical information. In: Van Genuchten MT, Leij FJ, Lund LJ (eds) Indirect methods for estimating the hydraulic properties of unsaturated soils. University of California, Riverside, pp 389–403

De Vos B, Van Meirvenne M, Quataert P, Deckers J, Muys B (2005) Predictive quality of pedotransfer functions for estimating bulk density of forest soils. Soil Sci Soc Am J 69:500–510

Demattê JLI (1988) Manejo de Solos ácidos dos Trópico úmidos – Região Amazônica. Fundação Cargill, Campinas

Dos Santos WJR, Curi N, Da Silva SHG, Araújo EF, Marques JJ (2013) Pedotransfer functions for water retention in different soil classes from the center-southern Rio Grande do Sul state. Ciênc Agrotec Lavras 37(1):49–60

Fabian AJ, Ottoni Filho TB (2000) Determinação de capacidade de campo in situ ou através de equações de regressão. Pesqui Agropecu Bras Brasília 35(5):1029–1036

Fazeli M, Shorafa M, Khojasteh DN, Shahri ARP (2010) A fractal approach for estimating soil water retention curve. J Soil Sci Environ Manage 1:177–183

Fidalski J, Tormena CA (2007) Pedotransfer functions for the soil water retention and soil resistance to penetration underground cover management systems in citrus. Ciência Rural Santa Maria 37(5):45–89

Fiorin TT (2008) Estimativa da infiltração de água no solo a partir de pedofunções. Santa Maria, Universidade Federal de Santa Maria, 116 p. (Tese de Doutorado)

França AMS (2011) Função de pedotransferência para estimativa de estoques de carbono em solo de áreas de campo limpo úmido do Distrito Federal. Universidade de Brasília, Instituto de geociências. Brasília, 144 p

Franzmeier DP (1991) Estimation of hydraulic conductivity from effective porosity data for some Indiana soils. Soil Sci Soc Am J 55:1801–1803

Gaiser T, Graef F, Cordeiro JC (2000) Water retention characteristics of soils with contrasting clay mineral composition in semi-arid tropical regions. Aust J Soil Res 38:523–536

Gevaerd R, Freitas S (2006) Estimativa operacional da umidade do solo para iniciação de modelos de previsão numérica da atmosfera. Parte I: descrição da metodologia e validação. Rev Bras Meteorol São Palulo 21(3):1–15

Giarola NFB, Da Silva AP, Imhoff S (2002) Relações entre propriedades físicas e características de solos da Região Sul do Brasil. Rev Bras Cienc Solo 26:885–893

Gonçalves MCPB (1994) Características hidrodinâmicas dos solos: sua determinação e funções de pedotransferência. Tese (Doutorado em Ciência do Solo) – Instituto Superior de Agronomia, Universidade Técnica de Lisboa, Lisboa, 212 p

Gonçalves MC, Pereira LS, Leij FJ (1997) Pedo-transfer functions for estimating unsaturated hydraulic properties of Portuguese soils. Eur J Soil Sci 48:387–400

Gregson K, Hector DJ, McGowan M (1987) A one-parameter model for the soil water characteristic. J Soil Sci 38:483–486

Hall DGM, Reeve MJ, Thomasson AJ, Wright VF (1977) Water retention, porosity and density of field soils, Technical monograph no. 9. Soil Survey of England & Wales, Harpenden, 75 p

Haverkamp R, Parlange JY (1986) Predicting the water retention curve from the particle size distribution 1. Sandy soils without organic matter. Soil Sci 142:325–329

Hodnett MG, Tomasella J (2002) Marked differences between Van Genuchten soil water-retention parameters for temperate and tropical soils: a new water-retention pedo-transfer function developed for tropical soils. Geoderma 108:155–180

Hutson JL, Cass A (1987) A retentivity function for use in soil-water simulation model. J Soil Sci 38:105–113

Imam B, Sorooshian S, Mayr T, Schaap M, Wösten JHM, Scholes B (1999) Comparison of pedotransfer functions to compute water holding capacity using the Van Genuchten model in inorganic soils, IGBP-DIS working paper #22. The International Geosphere-Biosphere Programme, Stockholm

Jagtap SS, Lal U, Jones JW, Gijsman AJ, Ritchie JT (2004) A dynamic nearest-neighbor method for estimating soil water parameters. Trans ASAE 47:1437–1444

Jana RB, Mohanty BP, Springer EP (2007) Multiscale pedotransfer functions for soil water retention. Vadose Zone J 6:868–878

Kroes JG, Van Dam JC, Groenendijk P, Hendriks RFA, Jacobs CMJ (2008) SWAP version 3.2. Theory description and user manual. WUR-Netherlands, Wageningen

Lamorski K, Pachepsky Y, Slawihski C, Walczak RT (2008) Using support vector machines to develop pedotransfer function for water retention of soils in Poland. Soil Sci Soc Am J 72:1243–1247

Lamp J, Kneib W (1981) Zur quantitativen Erfassung und Bewertung von Pedofunktionen. Mitteilungen der Deutschen Bodenkundlichen Gesellschaft 32:695–711

Leenhardt D, Voltz M, Bornand M (1994) Propagation of the error of spatial prediction of soil properties in simulating crop evapotranspiration. Eur J Soil Sci 45:303–311

Leij FJ, Alves WJ, Van Genuchten MTh, Williams JR (1996) The UNSODA unsaturated soil hydraulic database. Rep. EPA/600/R-96/095. USEPA Natl. Risk Manage. Lab., Cincinna,

Leij FJ, Romano N, Palladino M, Schaap MG, Coppola A (2004) Topographical attributes to predict soil hydraulic properties along a hillslope transect. Water Resour Res 40:W02407. doi:10.1029/2002WR001641

Lilly A, Lin H (2004) Using soil morphological attributes and soil structure in pedotransfer functions. In: Pachepsky Y, Rawls WJ (eds) Development of pedotransfer functions in soil hydrology, Developments in soil science, 30. Elsevier, Amsterdam, pp 415–429

Lopes-Assad ML (2006) Funções de pedotransferência para estimativa da disponibilidade de água em diferentes tipos de solos do Brasil. Relatório Final, Centro de Ciências Agrárias da Universidade Federal de São Carlos, 18 p

Lopes-Assad ML, Sans LMA, Assad ED, Zullo J Jr (2001) Relationship between soil water retention and amount of sand in brazilian soils. Rev Bras Agrometeorologia Passo Fundo 9(3):588–596, (Nº Especial: Zoneamento Agrícola)

Marcolin CD, Klein VA (2011) Determinação da densidade relativa do solo por uma função de pedotransferência para a densidade do solo máxima. Acta Scientiarum Agron Maringá 33(2):349–354

Masutti MM (1997) Caracterização da água disponível a partir de parâmetros físico-hídricos em solos da zona da mata do estado de Pernambuco. Universidade Federal Rural de Pernambuco, Recife, 69 p. (Tese de Mestrado)

McBratney AB, Minasny B, Cattle SR, Vervoort RW (2002) From pedotransfer functions to soil inference systems. Geoderma 109:41–73

McBratney AB, Minasny B, Tranter G (2011) Necessary meta-data for pedotransfer functions. Geoderma 160:627–629

McKeague JA, Wang C, Topp GC (1982) Estimating saturated hydraulic conductivity from soil morphology. Soil Sci Soc Am J 46:1239–1244

McKenzie NJ, Jacquier D (1997) Improving the field estimation of saturated hydraulic conductivity in soil survey. Aust J Soil Res 35:803–825

McKenzie NJ, MacLeod DA (1989) Relationships between soil morphology and soil properties relevant to irrigated and dryland agriculture. Aust J Soil Res 27:235–258

Medeiros JC (2012) Funções de pedotransferência em estudos do funcionamento hídrico do solo da região sudeste do estado do Pará. Tese (Doutorado) – Escola Superior de Agricultura "Luiz de Queiroz", USP, 120 p

Mello FFC (2007) Estimativas dos estoques de carbono dos solos nos Estados de Rondônia e Mato Grosso anteriores à intervenção antrópica. Dissertação (Mestrado), Escola Superior de Agricultura Luiz de Queiroz, 88 p

Mello CR, Oliveira GC, Ferreira DF, Lima JM, Lopes D (2005) Modelos para determinação dos parâmetros da equação de Van Genuchten para um Cambissolo. Rev Bras Eng Agríc Ambient Campina Grande 9(1):23–29

Mendonça-Santos ML, Santos HG (2003) Mapeamento Digital de Classes e Atributos de Solos: métodos, paradigmas e novas técnicas. Embrapa Solos, Documentos, Rio de Janeiro, 55, 19 p

Michelon CJ, Carlesson R, Oliveira ZB, Knies AE, Petry MT, Martins JD (2010) Funções de pedotransferência para estimativa da retenção de água em alguns solos do Rio Grande do Sul. Ciênc Rural Santa Maria 41:848–853

Minasny B (2000) Efficient methods for predicting soil hydraulic properties. Thesis (Doctor of Philosophy) – University of Sydney, Sydney, 396 p

Minasny B, Hartemink AE (2011) Predicting soil properties in the tropics. Earth-Sci Rev 106:52–62

Minasny B, McBratney AB (2002) Uncertainty analysis for pedotransfer functions. Eur J Soil Sci 53:417–429

Minasny B, Mcbratney AB, Mendonça-Santos ML, Santos HG (2003) Revisão sobre funções de pedotransferência e novos métodos de predição de classes e atributos do solo.Embrapa Solos, Rio de Janeiro, 50 p (Documentos, 45)

Montenegro S, Ragab R (2012) Impact of possible climate and land use changes in the semi arid regions: a case study from North Eastern Brazil. J Hydrol 55–68:434–435

Naime JM, Vaz CMP, Macedo A (2006) Determinações físicas do solo a partir da granulometria: programa Qualisolo. In: Reunião brasileira de manejo e conservação do solo e da água, 16. Aracaju. Novos desafios do carbono no manejo conservacionista. Resumos... Aracaju: Sociedade Brasileira de Ciência do Solo, 1 CD-ROM, 5 p

Nascimento GB, Anjos LHC, Pereira MG, Fontana A, Santos HG (2010) Funções de pedotransferência do conteúdo de água em Latossolos Amarelos e Argissolos Amarelos. Rev Bras Ciênc Agrárias Recife UFRPE 5(4):560–569

Nebel ALC, Timm LC, Cornelis W, Gabriels D, Reichardt K, Aquino LS, Pauletto EA, Reinert DJ (2010) Pedotransfer functions related to spatial variability of water retention attributes for lowland soils. Rev Bras Cienc Solo 34:669–680

Nemes A, Rawls WJ (2004) Soil texture and particle-size distribution as input to estimate soil hydraulic properties. In: Pachepsky Y, Rawls WJ (eds) Development of pedotransfer functions in soil hydrology, Developments in soil science, 30. Elsevier, Amsterdam, pp 36–50

Nemes A, Schaap MG, Leij FJ, Wösten JHM (2001) Description of the unsaturated soil hydraulic database UNSODA version 2.0. J Hydrol 251:151–162

Nemes A, Rawls WJ, Pachepsky Y (2006) Use of the nonparametric nearest neighbor approach to estimate soil hydraulic properties. Soil Sci Soc Am J 70:327–336

Netto AA (2007) Estimativa da retenção da água no solo a partir do uso de equipamentos não convencionais, rede neurais artificiais e funções de pedotransferência. Tese (Doutorado em Engenharia Ambiental) – Universidade de São Carlos, São Carlos, 176 p

Oliveira LB, Ribeiro MR, Jacomine PKT, Rodrigues JJV, Marques FA (2002) Funções de pedotransfêrencia para predição de umidade retida a potencias específicos em solos do estado de Pernambuco. Rev Bras Cienc Solo 26:315–323

Oliveira SRM, Zurmely HR, Lima Júnior FA, Meirelles MSP, Santos HG, Yamashita HH (2009) Um sistema para organização de informação de solos do Brasil na Internet. In: VII Congresso Brasileiro de Agroinformática, 21 a 25 de setembro de Viçosa, MG

Ottoni MV (2005) Classificação físico-hídrica de solos e determinação da capacidade de campo in situ a partir de testes de infiltração. Dissertação (Mestrado), Departamento de Eng, Civil, Universidade Federal do Rio de Janeiro (UFRJ), 157 p

Pachepsky Y, Rawls WJ (eds) (2004) Development of pedotransfer functions in soil hydrology, Development in soil science 30. Elsevier, Amsterdam

Pachepsky Y, Schaap MG (2004) Data mining and exploration techniques. In: Pachepsky Y, Rawls WJ (eds) Development of pedotransfer functions in soil hydrology, Developments in soil science, 30. Elsevier, Amsterdam, pp 21–32

Pachepsky Y, Mironenko EV, Scherbakov RA (1992) Prediction and use of soil hydraulic properties. In: Van Genuchten MT, Leij FJ, Lund LJ (eds) Proceedings of the international workshop on indirect methods for estimating the hydraulic properties of unsaturated soils. University of California, Riverside, pp 203–212

Pachepsky Y, Timlin D, Várallyay G (1996) Artificial neural networks to estimate soil water retention from easily measurable data. Soil Sci Soc Am J 60:727–733

Pachepsky Y, Crawford JW, Rawls WJ (1999a) Fractals in soil science: preface. Geoderma 88(3–4), 273 p

Pachepsky Y, Rawls WJ, Timlin DJ (1999b) The current status of pedotransfer functions: their accuracy, reliability, and utility in field- and regional-scale modeling. In: Corwin DL, Loague K, Ellsworth TR (eds) Assessment of non-point source pollution in the Vadose zone, Geophysical monograph 108. American Geophysical Union, Washington, DC, pp 223–234

Peraza JES (2003) Retenção de água e pedofunções para solos do Rio Grande do Sul. Tese (Mestrado em Engenharia Agrícola) – Universidade Federal de Santa Maria, Santa Maria, 118 p

Prevedello CL (1999) Programa SPLINTEX para estimar a curva de retenção de água a partir da granulometria (composição) do solo – Versão 1.0

Puckett WE, Dane JH, Hajek BF (1985) Physical and mineralogical data to determine soil hydraulic properties. Soil Sci Soc Am J 49:831–836

Rawls WJ (1983) Estimating soil bulk density from particle size analysis and organic matter content. Soil Sci 135:123–125

Rawls WJ, Brakensiek DL, Saxton KE (1982) Estimation of soil water properties. Trans Am Soc Agric Eng 25:1316–1328

Rawls WJ, Gish TJ, Brakensiek DL (1991) Estimating soil water retention from soil physical properties and characteristics. Adv Agron 16:213–234

Reichardt K (1988) Capacidade de campo. Rev Bras Ciênc Solo 12:211–216

Reichardt K, Timm LC (2004) Solo, planta e atmosfera: conceitos, processos e aplicações. Manole, Barueri, 478 p

Reichert JM, Suzuki LEAS, Reinert DJ, Horn R, Hakansson I (2008) Reference bulk density and critical degree-of-compactness for no-till crop production in subtropical highly weathered soils. Soil Tillage Res 102:242–254

Reichert JM, Albuquerque JA, Kaiser DR, Reinert DJ, Urach FL, Carlesso R (2009) Estimation of water retention under availability in soils of Rio Grande do Sul. Rev Bras Cienc Solo 33:1547–1560

Rodrigues LN, Maia AHN, Da Silva RN (2011) Funções de pedotransferência para estimar capacidade de campo, ponto de murcha permanente e densidade global em solos de uma bacia hidrográfica do bioma cerrado. Anais do XL Congresso Brasileiro de Engenharia Agrícola – Conbea, Cuiabá

Rossato L (2001) Estimativa da capacidade de armazenamento de água no solo do Brasil. São José dos Campos:INPE-(INPE-8915-TDI/809), 145 p

Rossato L, Alvalá RCS, Tomasella J (2004) Variação espaço-temporal da umidade do solo no Brasil: análise das condições médias para o período de 1971–1990. Rev Bras Meteorologia 19(2):113–122

Saxton KE, Rawls WJ (2006) Soil water characteristic estimates by texture and organic matter for hydrologic solutions. Soil Sci Soc Am J 70:1569–1578

Saxton KE, Rawls WJ, Romberger JS, Papendick RI (1986) Estimating generalized soil water characteristics from soil texture. Soil Sci Soc Am J 50:1031–1036

Schaap MG (2004) Accuracy and uncertainty in PTF predictions. In: Pachepsky Y, Rawls WJ (eds) Development of pedotransfer functions in soil hydrology, Developments in soil science, 30. Elsevier, Amsterdam, pp 415–429

Schwertmann U, Taylor RM (1977) Iron oxides. In: Dixon JB (ed) Minerals in soil environments. Soil Science Society of America, Madison. 948 p

Shein EV, Arkhangel'skaya TA (2006) Pedotransfer functions: state of the art, problems, and outlooks. Eurasian Soil Sci 39:1089–1099

Sobieraj J, Elsenbeer H, Vertessy RA (2001) Pedotransfer functions for estimating hydraulic conductivity: implication for modelling storm flow generation. J Hydrol 251:202–220

Souza LD, Reichardt K (1996) Estimativas de capacidade de campo. Rev Bras Cienc Solo 20:183–189

Tietje O, Hennings V (1996) Accuracy of the saturated hydraulic conductivity prediction by pedotransfer functions compared to the variability within FAO textural classes. Geoderma 69:71–84

Tietje O, Tapkenhinrichs M (1993) Evaluation of pedotransfer functions. Soil Sci Soc Am J 57:1088–1095

Tomasella J, Hodnett MG (1996) Soil hydraulic properties and Van Genuchten parameters for an oxisol under pasture in central Amazonia. In: Gash JHC, Nobre CA, Roberts JM, Victoria RL (eds) Amazonian deforestation and climate. Wiley, Chichester, pp 101–124

Tomasella J, Hodnett MG (1997) Estimating unsaturated hydraulic conductivity of Brazilian soils using soil-water retention data. Soil Sci 162:703–712

Tomasella J, Hodnett MG (1998) Estimating soil water retention characteristics from limited data in Brazilian Amazonia. Soil Sci 163:190–202

Tomasella J, Hodnett MG (2004) Pedotransfer functions for tropical soils. In: Pachepsky Y, Rawls WJ (eds) Development of pedotransfer functions in soil hydrology, Developments in soil science, 30. Elsevier, Amsterdam, pp 415–429

Tomasella J, Hodnett MG, Rossato L (2000) Pedotransfer functions for the estimation of soil water retention in Brazilian soils. Soil Sci Soc Am J 64:327–338

Tomasella J, Pachepsky Y, Crestana S, Rawls WJ (2003) Comparison of two techniques to develop pedotransfer functions for water retention. Soil Sci Soc Am J 67:1085–1092. Trans Am Soc Agric Eng, 34:417–422

Tormena CA, Da Silva AP (2002) Incorporação da densidade no ajuste de dois modelos à curva de retenção de água no solo. Rev Bras Cienc Solo 26:305–314

Twarakavi NKC, Simunek J, Schaap MG (2009) Development of pedotransfer functions for estimation of soil hydraulic parameters using support vector machines. Soil Sci Soc Am J 73:1443–1452

Tyler SW, Wheatcraft SW (1989) Application of fractal mathematics to soil water retention estimation. Soil Sci Soc Am J 53:987–996

Urach FL (2007) Estimativa da retenção de água em solos para fins de irrigacao. Tese de Mestrado (Mestrado em Engenharia de Água no solo) – Universidade de Santa Maria, Santa Maria, 79 p

Van den Berg M, Klamt E, Van Reeuwijk LP, Sombroek G (1997) Pedotransfer functions for the estimation of moisture retention characteristics of Ferralsols and related soils. Geoderma 78:161–180

Van Genuchten MT (1980) A closed form equation for predicting the hydraulic conductivity of unsaturated soils. Soil Sci Soc Am J 44:892–989

Vaz CMP, Iossi MF, Naime JM, Macedo A (2004) Validação de modelos para estimativa indireta da capacidade de retenção da água nos solos. Embrapa Instrumentação Agropecuária, São Carlos (Embrapa Instrumentação Agropecuária. Comunicado Técnico, 61)

Vaz CMP, Iossi M, Naime J, Macedo Á, Reichert JM, Reinert DJ, Cooper M (2005) Validation of the Arya and Paris water retention model for Brazilian soils. Soil Sci Soc Am J 69(3):577–583

Veihmeyer FJ, Hendrickson AH (1927) The relation of soil moisture to cultivation and plant growth. In: Proceedings of the 1st international congress of soil science, Washington, 3, pp 498–513

Vereecken H, Maes J, Feyen J, Darius P (1989) Estimating the soil moisture retention characteristic from texture, bulk density, and carbon content. Soil Sci 148:389–403

Vereecken H, Diels J, Vanorshoven J, Feyen J, Bouma J (1992) Functional evaluation of pedotransfer functions for the estimation of soil hydraulic properties. Soil Sci Soc Am J 56:1371–1378

Vereecken H, Weynants M, Javaux M, Pachepsky Y, Schaap YMG, Van Genuchten MT (2010) Using pedotransfer functions to estimate the Van Genuchten–Mualem soil hydraulic properties: a review. Soil Science Society of America. Vadose Zone J 9:795–820

Weynants M, Vereecken H, Javaux M (2009) Revisiting Vereecken pedotransfer functions: introducing a closed-form hydraulic model. Vadose Zone J Madison 8(1):102–123. In: Soils: An Australian Viewpoint, 499–530. Division of Soils, CSIRO, CSIRO: Melbourne

Williams J, Prebble RE, Williams WT, Higgnet CT (1983) The influence of texture, structure and clay mineralogy on the soil moisture characteristic. Aust J Soil Res 21:15–32

Willmott CJ, Ackleson SG, Davis RE, Feddema JJ, Klink KM, Legats DR, O'Donnell J, Rowe CM (1985) Statistics for the evaluation and comparison of models. J Geophys Res Oceans 90:8995–9005

Wösten JHM, Bannink MH, De Gruijter JJ, Bouma J (1986) A procedure to identify different groups of hydraulic-conductivity and moisture-retention curves for soil horizons. J Hydrol 86:133–145

Wösten JHM, Finke PA, Jansen MJW (1995) Comparison of class and continuous pedotransfer functions to generate soil hydraulic characteristics. Geoderma 66:227–237

Wösten JHM, Lilly A, Nemes A, Le Bas C (1999) Development and use of a database of hydraulic properties of European soils. Geoderma 90:169–185

Wösten JHM, Pachepsky Y, Rawls WJ (2001) Pedotransfer functions: bridging the gap between available basic soil data and missing soil hydraulic characteristics. J Hydrol 251:123–150

Zinn YL, Lal R, Resck DVS (2005) Texture and organic carbon relations described by a profile pedotransfer function for Brazilian Cerrado soils. Geoderma 127:168–173

Zinn YL, Guerra AR, Da Silva AC, Marques JJ, Oliveira GC, Curi N (2012) Perfis de carbono orgânico do solo nas regiões sul e serra do espinhaço meridional, Minas Gerais: modelagem em profundidade. Rev Bras Cienc Solo 36:1395–1406

Part III
Different Approaches for the Characterization of Soil Physical Quality Indicators

Chapter 7
An Algorithm for Mapping the Spatial Variability of Soil Physical Quality

Marcos Bacis Ceddia, Sergio Drumond Ventura, Rosane Ferreira de Oliveira, André Luis Oliveira Villela, and Carlos Alberto Alves Varella

Abstract The search for sustainable agricultural production system has required the monitoring of the quality of agricultural soils as well as the generation of modern tools for mapping the spatial variability of factors affecting the productivity of crops. The purpose of this work was to develop and evaluate the performance of an algorithm for mapping the spatial variability of soil physical quality, and develop a user-friendly and free software to perform the algorithm. Basically, the method consisted on: (1) Mapping the spatial variability of Easily Available Water (EAW), Air Filled Porosity (PA) and soil Penetration Resistance (PR), using geostatistics; and (2) Perform the integration of the maps of EAW, PA and PR into a map of Soil Physical Quality classes (SPQ), using Euclidean distance. The algorithm was tested in an agroecological farm with 6.24 ha, and was considered efficient to classify the SPQ. The attributes EAW and PA were the most important to define the SPQ of the study site. The software QUALIFIS met the demand for user-friendly software and can be applied in precision agriculture.

Keywords Digital soil mapping • Soil physical quality • Precision agriculture • Euclidean distance

M.B. Ceddia (✉)
Soils Department, Rural Federal University of Rio de Janeiro, Seropédica, RJ, Brazil
e-mail: marcosceddia@gmail.com

S.D. Ventura • R.F. de Oliveira
Department of Mathematics, Federal Rural University of Rio de Janeiro (UFRRJ), Seropédica, RJ, Brazil

A.L.O. Villela
Technical College of the Federal Rural University of Rio de Janeiro (CTUR-UFRRJ), Seropédica, RJ, Brazil

C.A.A. Varella
Engineering Department, Federal Rural University of Rio de Janeiro (UFRRJ), Seropédica, RJ, Brazil

W.G. Teixeira et al. (eds.), *Application of Soil Physics in Environmental Analyses: Measuring, Modelling and Data Integration*, Progress in Soil Science, DOI 10.1007/978-3-319-06013-2_7,

7.1 Introduction

Soil quality is considered a key element of sustainable agriculture and can be defined as the capacity of a soil-specific, within the limits of the ecosystem (natural or managed), to sustain the productivity of plants and animals, while maintaining or improving the quality of water, air, and support the health and human habitation (Karlen et al. 1997). According to Carter (2002), soil quality has two parts: an intrinsic part covering soil's *inherent* capacity for crop growth and a *dynamic part* that can change over relatively short time periods by the soil user or manager. Moreover, the approaches used in defining soil quality generally use an ecological framework, to evaluate soil quality based on the following sequence: soil functions, soil processes, soil attributes or properties, attribute indicators, and methodology (Carter 2002).

Considering the soil function as the capacity of sustaining the productivity of plants, Letey (1985) pointed out soil water availability, aeration, mechanical resistance and temperature as soil physical properties directly affecting the plant growth. Furthermore, the author stressed that the availability of water is the dominant controlling factor, and that the other three are affected by water content, as schematically illustrated at Fig. 7.1. According to Letey (1985), bulk density and pore size distribution (indirect factors) affect the relationship between water and both aeration and mechanical resistance. The relationship between water and aeration is opposite to that between water and mechanical resistance. Increasing water content decreases aeration, which is undesirable, on the other hand, the same increasing of water decrease the mechanical resistance, which is desirable. The effect of water on both of these attributes (aeration and penetration resistance) is intensified by increase in bulk density and/or presence of small unisize pores.

Taking into account the interrelationships of soil physical factors, Letey (1985) proposed the determination of the Non-Limiting Water range (NLWR), at different soil depth as useful method for physical characterizing of a soil when crop production is being considered. Initially, if a specific soil present a good soil quality, the NLWR will be determined by the range of water content (water availability) which

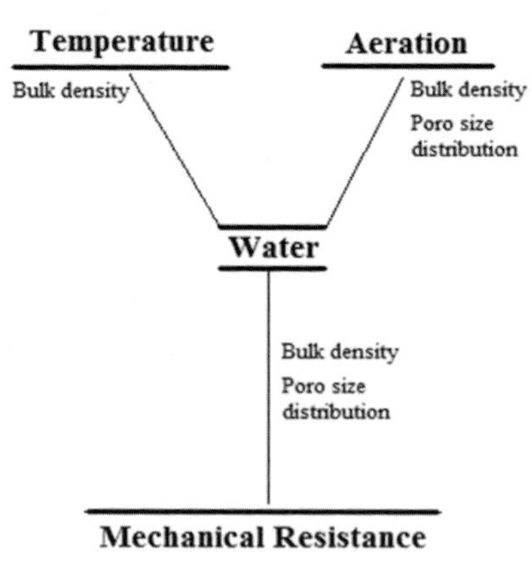

Fig. 7.1 Schematic representation of the relationship between water and other soil physical attributes which directly affect plant growth (Adapted from Letey 1985)

is the interval between the water content at field capacity and permanent wilting point. However, depending on the soil management, the NLWR can be reduced both by decreasing aeration and/or by increasing mechanical resistance.

In 1994, Da Silva et al. proposed the Least Limiting Water Range (LLWR) as an index of soil physical quality for crop growth, since it integrates the effects of soil aeration, resistance to penetration, and water retention on crop growth into a single parameter. The LLWR has been validated as a soil physical quality indicator for a wide variety of soils, crops, and management systems (Leão et al. 2006). However, the use of the LLWR methodology among agronomists, land managers, consultants, and soil scientists has been restricted by: (i) the need for a detailed description of the nonlinear statistical parameter estimation procedures for the θ (ψ) and PR (θ, ρ_b) functions and (ii) the mathematical procedures for calculating the LLWR range and the critical soil bulk density (ρ_{bc}), where the LLWR is equal to zero (Leão et al. 2005). In addition to the restrictions presented above, the assessment of soil physical quality indicators is time consuming, labor-intensive and costly, which represent great barriers to generate spatial variability maps of soil physical quality indicators.

In order to attend the huge demand of society for soil information, digital soil mapping (DSM) techniques have been developed. DSM is defined as "the creation and population of spatial soil information systems by numerical models inferring the spatial and temporal variations of soil types and soil properties from soil observation and expert knowledge and from related environmental variables" (Lagacherie and McBratney 2007). DSM has as its goals the creation, update and quick availability of soil information with lower cost, more accurate and adequate spatial representation. To attend these goals, DSM use many techniques such as data mining, multilinear regression, geostatistics, fuzzy logic, neural networks, decision trees and discriminant analysis (Lagacherie 2008).

Considering the great demand for maps of soil physical quality, and the advantages of the DSM techniques, this work present an approach to integrate the attributes water availability, aeration and penetration resistance into maps of spatial variability of soil physical quality. Specifically, the objectives of this work were: (1) develop and evaluate the performance of an algorithm for mapping the spatial variability of soil physical quality, and (2) develop a user-friendly and free software to perform the algorithm.

7.2 Materials and Methods

7.2.1 The Approach to Integrate the Maps of Soil Physical Quality Indicator

Basically, the method consisted on: (1) Mapping the spatial variability of Easily Available Water (EAW), Air Filled Porosity (PA) and soil Penetration Resistance (PR), using geostatistics techniques; and (2) The spatial variability maps of the soil attributes are integrated into a map of Soil Physical Quality (SPQ) classes, using a standard vector classifier (Euclidean distance). A detailed explanation about the standard vector classifier is presented at Sect. 7.2.2.

In order to formalize the soil physical quality mapping process according to the DSM framework, the Eq. (7.1) represents the SCORPAN model proposed by McBratney et al. (2003). In Eq. (7.2), the approach presented in this work is adapted to the SCORPAN model, taking into account the characteristics of study site.

According to McBratney et al. (2003), the SCORPAN is an empirical quantitative description of the relationship between soil and other spatially referenced factors with a view to using these as soil spatial prediction functions. The term SCORPAN is an acronym composed by the first words of the seven factors interfering in the prediction process.

$$\mathrm{S} = \boldsymbol{f}(\mathrm{s}, \mathrm{c}, \mathrm{o}, \mathrm{r}, \mathrm{p}, \mathrm{a}, \mathrm{n}); \tag{7.1}$$

Where:

S: Soil class (S_c) or Soil attributes (S_a) to be predicted;
f: is the prediction function;
s: soil, other properties of the soil at a point;
c: climate, climatic properties of the environment at a point;
o: organisms, vegetation or fauna or human activity;
r: topography, landscape attributes;
p: parent material, lithology;
a: age, the time factor;
n: space, spatial position (x, y spatial coordinates).

The soil factor (s) is included in the S function (Eq. 7.1) because soil can be predicted from its properties, or soil properties from its class or other properties. The space factor (n) is the implicit spatial coordinates *x, y* of the soil attributes which were collected at the field.

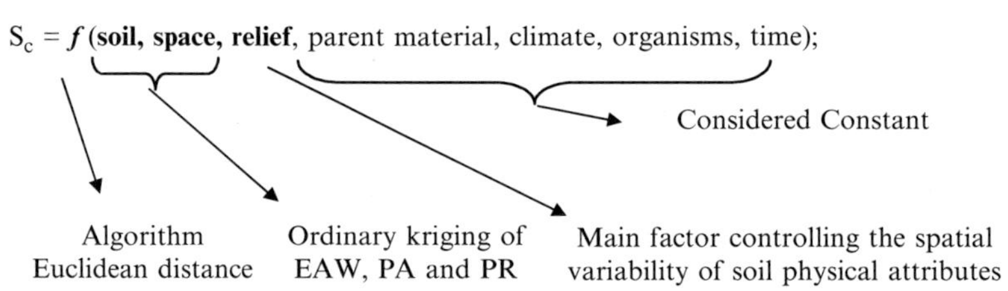

(7.2)

Where:

S_c: is the Soil Physical Quality (SPQ) classes to perform at each pixel of the study site;

f: the algorithm to integrate the soil attributes EAW, PA and PR into SPQ classes (Euclidean distance);

soil: is the values of the soil physical attributes EAW, PA and PR at each pixel of the map generated by ordinary kriging interpolation;
space: is the UTM coordinates of each point of the spatial variability maps of EAW, PA and PR
Parent material: Rocks of acidic character (gneiss)
Climate: Aw, according to Köppen system;
Relief: the main factor controlling the spatial variability of the soil physical attributes (ranging from steep to flat);
Organism: Atlantic Rainforest
Time: Pre Cambrian

The soil formation factors like the parent material, climate, time and organisms are considered constant along the study site. The information about these factors was obtained from Ramos et al. (1973) and is used as an expert knowledge to give support to digital soil mapping of soil physical quality indicators. According to the authors, the relief factor plays an important effect over the spatial variability of the attributes EAW, PA and PR.

7.2.2 The Algorithm to Integrate the Soil Attributes EAW, PA and PR into SPQ Classes

The operation of the algorithm is shown in a schematic form in Fig. 7.2. The algorithm classify the SPQ based on the Euclidean distance between the standard vector of each soil physical quality class and the values of soil physical attributes (EAW, PA and PR) at each pixel of the spatial variability map generated by ordinary kriging. Euclidean distance measures the dissimilarity between individuals, groups and classes. For example, as higher is the Euclidean distance between classes, more different they are. The majority of cluster analysis algorithm use this measurement.

In the context of this work, the Euclidean distance is calculated between the standard vector (standard values of the attributes EAW, PA and PR for each specific SPQ classes) and the values of EAW, PA and PR at a specific pixel of the map. Three SPQ classes are considered and named as Restrictive, Suitable and Optimum. At item 2.3.4, a detailed explanation about these SPQ classes is presented.

The classification process comprises three steps:

1. Define the class interval of EAW, PA and PR for each soil physical class, and organizes a matrix of standard vectors;
2. Calculate the Euclidean distance between the standard vector of each soil physical quality class and the values of soil physical attributes;
3. Allocate the SPQ class, considering the lower Euclidean distance between the standard vector and the values of the soil physical attributes at a specific pixel of the map.

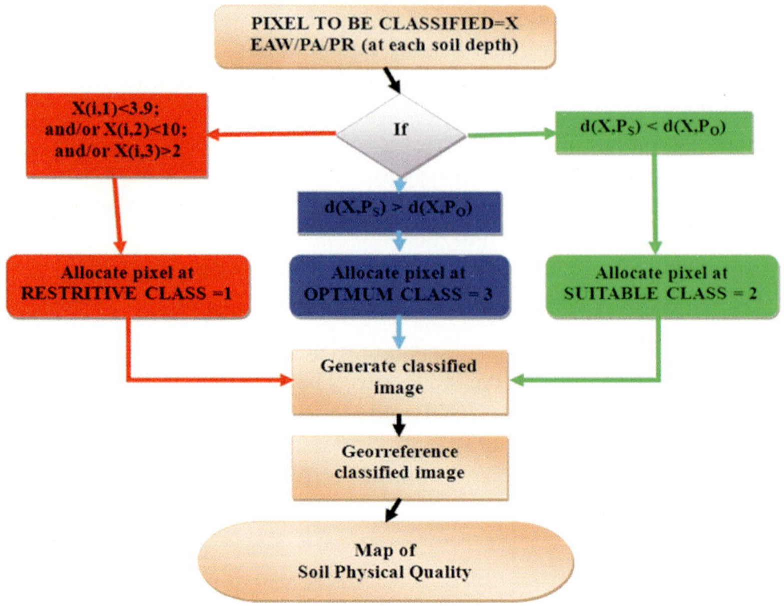

Fig. 7.2 Flowchart of the classificatory system

Three SPQ classes (standard vectors) were create (Restrictive, Suitable and Optimum). Regarding the first step, the interval values of EWA, PA and PR for which of these SPQ classes were labeled as follow:

mEAW*rest – Mean value of the Restrictive interval class for Easily Available Water;
mEAW*suit – Mean value of the Suitable interval class for Easily Available Water;
EAWopti – Minimum value of the Optimum class for Easily Available Water;
mPA*rest – Mean value of the Restrictive interval class for Air Filled Porosity;
mPA*suit – Mean value of the Suitable interval class for Air Filled Porosity;
PAopti – Minimum value of the Optimum class for Air Filled Porosity;
PRrest – Minimum value of the Restrictive class for Penetration Resistance;
mPR*suit – Mean value of the Suitable interval class for Penetration Resistance;
PRopti – Maximum value of the Optimum class for Penetration Resistance;

The asterisk indicate that the parameter is a vector with two positions, while the others (without asterisk) the vector represent a unique value. Thus, for the cases where the parameters have two positions, their mean value is calculated. Consequently, the label of the interval class is preceded by "m". The matrix (3 × 3) of standard vectors is presented below:

$$\mathrm{M} = \begin{bmatrix} mEAW_{rest} & mEAW_{suit} & EAW_{opti} \\ mPA_{rest} & mPA_{suit} & PA_{opti} \\ PR_{rest} & mPR_{suit} & PR_{opti} \end{bmatrix} \tag{7.3}$$

Fig. 7.3 Schematic process of Euclidean distance calculation

Matrix A (m x 3)	Matrix 3 x 3 $\begin{bmatrix} mEAW_{rest} & mEAW_{suit} & EAW_{opti} \\ mPA_{rest} & mPA_{suit} & PA_{opti} \\ PR_{rest} & mPR_{suit} & PR_{opti} \end{bmatrix}$
$\begin{bmatrix} X_{11} & X_{12} & X_{13} \\ X_{21} & X_{22} & X_{23} \\ X_{31} & X_{32} & X_{33} \\ \vdots & \vdots & \vdots \\ \vdots & \vdots & \vdots \\ X_{i1} & X_{i2} & X_{i3} \\ \vdots & \vdots & \vdots \\ \vdots & \vdots & \vdots \\ X_{n1} & X_{n2} & X_{n3} \end{bmatrix}$	$\begin{bmatrix} D_{11} & D_{12} & D_{13} \\ D_{21} & D_{22} & D_{23} \\ D_{31} & D_{32} & D_{33} \\ \vdots & \vdots & \vdots \\ \vdots & \vdots & \vdots \\ D_{i1} & D_{i2} & D_{i3} \\ \vdots & \vdots & \vdots \\ \vdots & \vdots & \vdots \\ D_{n1} & D_{n2} & D_{n3} \end{bmatrix}$ **Matrix of Euclidean distance**

The three columns of the matrix represent the standard vectors of soil physical quality classes (Restrictive – rest, Suitable – suit, and Optimum – opti) and the three rows represent the soil physical attributes used in the classification system (EAW, PA and PR).

The second step is to calculate the Euclidean distance between the standard vector of each soil physical quality class (matrix 3 × 3, Fig. 7.3) and the values of soil physical attributes (matrix A, Fig. 7.3) at each pixel of the spatial variability map generated by ordinary kriging. For this, let A be a matrix of size m × 3, where m is the number of rows and 3 is the number of columns in the input file (data file). Each row of the input file represent a pixel of the map with three columns (the kriged values of EAW – X_{i1}, PA – X_{i2}, and PR X_{i3}), at each soil depth.

The command D = dist (X, M) creates a matrix Dij of size n × 3 (at the bottom right side of the Fig. 7.3), which is the matrix containing the Euclidean distances between the Xij rows of the matrix A and the columns values (M_{i1}, M_{i2}e M_{i3}) of the matrix (3 × 3). The Eqs. 7.4, 7.5 and 7.6, were used to calculate the distance of each pixel of the map in relation to the Restrictive, Suitable and Optimum classes, respectively.

Where:

$$D_{i1} = \sqrt{(X_{i1} - mEAW_{rest})^2 + (X_{i2} - mPA_{rest})^2 + (X_{i3} - PR_{rest})^2} \tag{7.4}$$

$$D_{i2} = \sqrt{(X_{i1} - mEAW_{suit})^2 + (X_{i2} - mPA_{suit})^2 + (X_{i3} - mPR_{suit})^2} \tag{7.5}$$

$$D_{i3} = \sqrt{(X_{i1} - EAW_{opti})^2 + (X_{i2} - PA_{opti})^2 + (X_{i3} - PR_{opti})^2} \tag{7.6}$$

Regarding the third step, it is to perform the allocation of the SPQ class by itself. The algorithm checks the minimum Euclidean distance according to Eq. 7.7 and allocate the SPQ class following the rules 1, 2 and 3, presented below.

$$\mathrm{Di} = \min\{\mathrm{D_{i1}}, \mathrm{D_{i3}}, \mathrm{D_{i3}}\}. \tag{7.7}$$

Where:

Di is the minimum Euclidean distance of a specific pixel (row i) in relation to a standard vector.

1. if $\mathrm{Di} = \mathrm{D_{i1}}$, the row i is classified as "Restrictive".
2. if $\mathrm{Di} = \mathrm{D_{i2}}$, the row i is classified as "Suitable".
3. if $\mathrm{Di} = \mathrm{D_{i3}}$, the row i is classified as "Optimum".

7.2.3 The Study Site

The SIPA (Agro ecological Production and Research Integrated System) is an experimental farm located in the municipality of Seropédica-Rio de Janeiro State/ Brazil. The farm, created in order to develop agro ecological technologies, was implanted in 1993 and encompasses an area of 59 ha, being 6.24 ha used to policultive (Fig. 7.4). The climate of the region is classified as Aw (Köppen classification system), with the domain of high temperatures in the summer and mild in the winter, average annual temperature of 24.5 °C, with rainfall concentrated from November to March and average annual rainfall of 1,213 mm.

The Sampling Procedures

In order to apply geostatistics to investigate the spatial variability of soil physical attributes, the sampling strategy included the definition of a regular square grid with 20 m spacing. As recommended by Trangmar et al. (1985), since spacing between sampling points might affect data modeling, additional soil samples were collected in reduced spacing (1, 5 and 10 m), according to topography and soil classes. The Fig. 7.4 shows the two sampling grids used. The points highlighted in red represent the 169 locations where PR tests were performed, while the blue highlighted points represent the 122 locations where undisturbed soil samples were collected to measure EAW and PA. The coordinates (UTM system) were measured using a GPS with differential correction (DGPS-Trimble-GeoExplorer 3 model), with submetric accuracy.

Soil Physical Analysis

In 122 georeferenced points, undisturbed soil samples were collected at depths 0.0–0.10, 0.10–0.20 and 0.20–0.30 m, for the determination of water retention at 10 kPa (Field Capacity) and 80 kPa (limit of tensiometer reading) and soil bulk

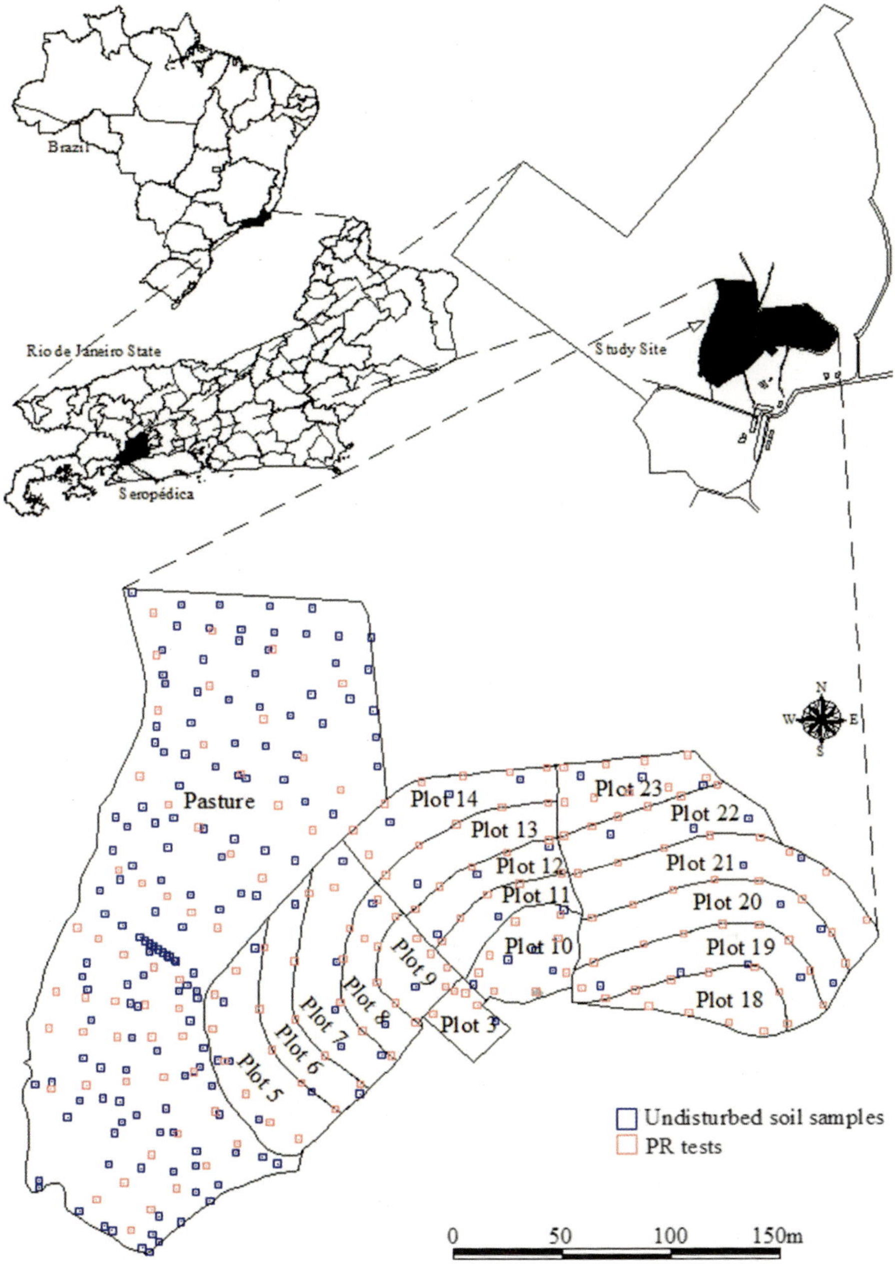

Fig. 7.4 The study site and the grids used to collect soil samples and perform Penetration Resistance tests

density – ρ_b (Embrapa 1997). Thereafter, the soil samples were grinded and air-dried for determination of soil particle density (ρ_s) and soil particle size distribution (Pipette method – Embrapa 1997).

Easily Available Water (EAW) and Air-Filled Porosity (PA) were calculated by the expressions:

$$\mathrm{EAW} = (\theta_{10\mathrm{kPa}} - \theta_{80\mathrm{kPa}}) \tag{7.8}$$

Where:

$\theta_{10\mathrm{kPa}}$ – Volumetric water content at 10 kPa;
$\theta_{80\mathrm{kPa}}$ – Volumetric water content at 80 kPa;

$$\mathrm{PA} = (\alpha - \theta_{10\mathrm{kPa}}) \tag{7.9}$$

Where:

α is the total porosity of the soil, which is calculated by the Eq. 7.10:

$$\left(1 - \frac{\rho_b}{\rho_s}\right) \tag{7.10}$$

ρ_b – Soil bulk density;
ρ_s – Soil particle density;

Penetration resistance (PR) was determined in 169 grid points, at 0.0–0.10, 0.10–0.20 and 0.20–0.30 m soil depth, using a penetrometer of impact developed by Stolf (1991). The determinations were performed 48 h after a rain of 8 mm, when the soil humidity was at about in the field capacity. The PR was calculated by the equation proposed by Stolf (1991) as:

$$\mathrm{PR} = \frac{F}{A} \tag{7.11}$$

Where,

PR = Penetration resistance (kgf.cm^{-2});
F = Strength of the resistance, kgf;
A = The base area of the cone, cm^{-2};

Where,

$$\mathrm{F} = (M + m) \times g + \frac{M}{M + m} \times \frac{Mgh}{x} \tag{7.12}$$

Where,

M = Impact mass (kg);
m = Mass of others components (kg);
g = Gravity acceleration (9.81 $m.s^{-2}$);
h = Falling height (cm);
x = Vertical distance of penetration by impact (cm);

Statistical and Geostatistical Analysis

Descriptive and exploratory analysis were performed using the software XLSTAT 7.5 (Addinsoft 2004). The descriptive analysis consisted on evaluate the existence of outliers and the central tendency and dispersion of data. The coefficients skewness and excess kurtosis of samples were used to evaluate if an specifical attribute present a normal distribution (Webster and Oliver 1990). The software GEOSTAT (Vieira et al. 1983) was used both to determine measures of spatial continuity (experimental semivariograms) and for model fitting to the semivariogram to be further used in the interpolation. The selection of the most proper model was done by cross-validation (jack-knifing).

The experimental semivariogram, $\gamma(h)$, of n spatial observations Z(xi), i = 1,... n, was calculated using the Eq. 7.13:

$$\gamma(h) = \frac{1}{2N(h)} \sum_{i=1}^{N(h)} [Z_{xi} - Z_{xi+h}]^2 \tag{7.13}$$

Where:

$N(h)$ – is the number of observations separated by a distance h;
Z_{xi} – is the soil attribute value measured at a specific point (x_1) of the grid;
Z_{xi+h} – is the soil attribute value measured at a specific neighborhood point apart by a distance h distance (xi + h);

Experimental semivariograms can be fit in to a variety of models that have well known parameters: nugget C_0, sill ($C_0 + C_1$), and range of spatial dependence (a), McBratney and Webster (1986). The fitted model provided information about structure and input parameters for executing kriging interpolation. The kriging method was used because it is recognized as a good linear estimate without bias and uses a moving weighted average (Isaaks and Srivastavas 1989). This estimate was calculated using Eq. 7.14:

$$\hat{Z}(x_0) = \sum_{i=1}^{n} \lambda_i Z(x_i) \tag{7.14}$$

where $\hat{Z}(x_0)$ is the value to be estimated at site x_0 and $Z(x_i)$ is the known value in the site sampled x_i. Weight distribution (λ_i) was not arbitrated, and its calculation was based on the parameters in the fitted variogram models. The sum of the weights is equal to one to ensure that the estimate is not biased and that the error estimate or kriging variance is minimized (Eq. 7.15).

$$\sum_{i=1}^{n} \lambda_i = 1 \tag{7.15}$$

The Class Intervals for Soil Physical Attributes

The class intervals for each one of the soil physical attributes of the study site was generated according to reference values presented at the literature and also considering the kriged values of the EAW, PA e PR at the study site. In fact, the class intervals of each soil physical attributes can be changed by the users, considering specific demands inherent to working conditions, e.g. different soil type, cultures and climate.

The class intervals adopted for the attribute PR and its respective limits values were: Restrictive ($\text{P.R.} > 2$ MPa); Suitable ($2 \geq \text{P.R.} \leq 1.0$ MPa); Optimum ($\text{P.R.} < 1.0$ MPa). The restrictive value (> 2MPa) is frequently presented in the literature as the limiting value of soil resistance (Da Silva et al. 1994). The range values of the suitable and optimum class were settled in accordance to amplitude values of the kriged values observed along the study site.

The EAW class intervals were generated considering the amplitude and the mean of the kriged values of EAW. The class intervals adopted were: Restrictive ($0.0 \leq \text{EAW} \geq 3.9$ mm), Suitable ($3.9 \leq \text{EAW} \geq 5.67$ mm), and Optimum ($\text{EAW} > 5.67$).

For the attribute PA the class intervals adopted were: Restrictive (0–10 %), Suitable (10–20 %) and Optimum (>20 %). The value of 10 % air-filled porosity is presented in the literature as Restrictive for PA, which is believed to be too small for adequate plant aeration (Da Silva et al. 1994; Engelaar and Yoneyama 2000; Da Silva et al. 2004). The range values of the suitable and optimum class were settled in accordance to amplitude values of the kriged values observed along the study site.

7.2.4 *The Software QUALIFIS*

The software QUALIFIS (Qualidade Física Software/Soil Physical Quality Software) was designed with three aims in mind. The first one, geared toward the enduser, is to provide a user-friendly software. The user-friendliness of QUALIFIS mainly comes from its non-obtrusive graphical user interface, and the possibility to directly visualize data sets and results in an interactive environment. The second objective was to design a software whose functionalities could conveniently be augmented. The third objective was to write the software in an open source language.

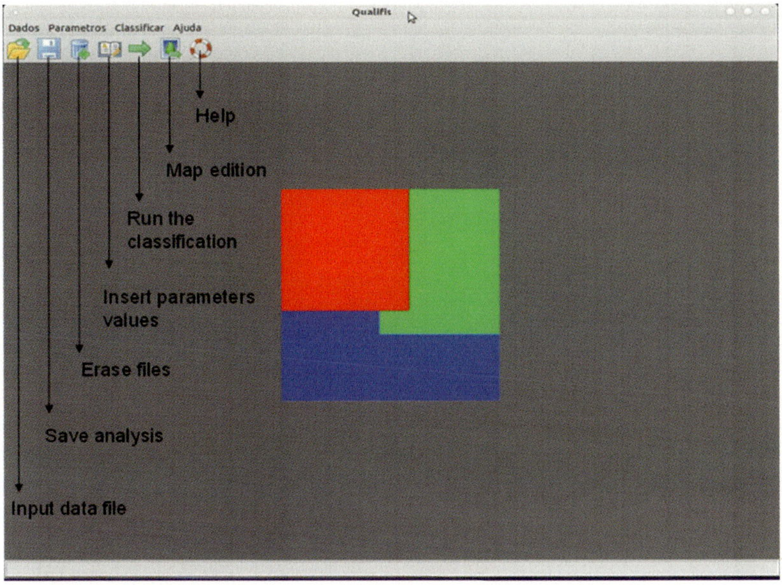

Fig. 7.5 The first screen of the software QUALIFIS

The software QUALIFIS was written in Python language (Linux environment), given its multi-platform and "open-source" character and simplicity. To create and control the Graphical User Interface (GUI), the "GUI toolkit" wxPython was used, which is also multi-platform and "open-source". In relation to the structure, the software code is divided into three classes: libframe (for the GUI), libdata (to work with data) and libpara (working with the parameters).

The first screen of the QUALIFIS software is presented in Fig. 7.5. In the top left side of the software screen, the users will find the following icons: Input data file, Save analysis, Erase files, Insert parameters values, Run the classification, Map edition and Help, respectively.

The input data file is an ASCII file which has the extension ".txt". An example of the data file opened in the software is presented in Fig. 7.6.

The data file used as example presented 11 columns, organized as follows:

X	Y	AFD10	PA10	RP10	AFD20	PA20	RP20	AFD30	PA30	RP30
636106	7483085	3.001	26.23	0.10	3.793	18.57	0.2206	3.828	25.9	0.289
636106	7483087	3.011	25.84	0.10	3.839	18.54	0.2271	3.825	25.1	0.292
636106	7483089	3.021	25.48	0.11	3.892	18.54	0.2371	3.836	24.1	0.293
636106	7483091	3.952	22.17	0.11	3.947	17.77	0.1212	3.861	23.1	0.296
636106	7483093	3.982	21.42	0.11	4.002	17.65	0.1326	3.901	22.0	0.299
636106	7483095	4.012	20.65	0.11	4.064	17.51	0.1216	3.952	20.9	0.305

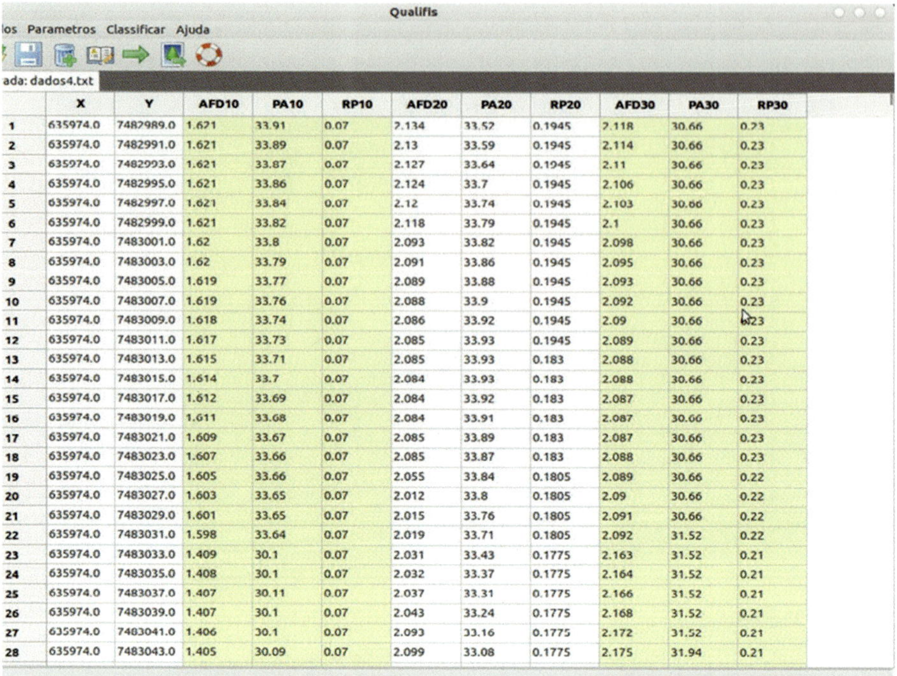

	X	Y	AFD10	PA10	RP10	AFD20	PA20	RP20	AFD30	PA30	RP30
1	635974.0	7482989.0	1.621	33.91	0.07	2.134	33.52	0.1945	2.118	30.66	0.23
2	635974.0	7482991.0	1.621	33.89	0.07	2.13	33.59	0.1945	2.114	30.66	0.23
3	635974.0	7482993.0	1.621	33.87	0.07	2.127	33.64	0.1945	2.11	30.66	0.23
4	635974.0	7482995.0	1.621	33.86	0.07	2.124	33.7	0.1945	2.106	30.66	0.23
5	635974.0	7482997.0	1.621	33.84	0.07	2.12	33.74	0.1945	2.103	30.66	0.23
6	635974.0	7482999.0	1.621	33.82	0.07	2.118	33.79	0.1945	2.1	30.66	0.23
7	635974.0	7483001.0	1.62	33.8	0.07	2.093	33.82	0.1945	2.098	30.66	0.23
8	635974.0	7483003.0	1.62	33.79	0.07	2.091	33.86	0.1945	2.095	30.66	0.23
9	635974.0	7483005.0	1.619	33.77	0.07	2.089	33.88	0.1945	2.093	30.66	0.23
10	635974.0	7483007.0	1.619	33.76	0.07	2.088	33.9	0.1945	2.092	30.66	0.23
11	635974.0	7483009.0	1.618	33.74	0.07	2.086	33.92	0.1945	2.09	30.66	0.23
12	635974.0	7483011.0	1.617	33.73	0.07	2.085	33.93	0.1945	2.089	30.66	0.23
13	635974.0	7483013.0	1.615	33.71	0.07	2.085	33.93	0.183	2.088	30.66	0.23
14	635974.0	7483015.0	1.614	33.7	0.07	2.084	33.93	0.183	2.088	30.66	0.23
15	635974.0	7483017.0	1.612	33.69	0.07	2.084	33.92	0.183	2.087	30.66	0.23
16	635974.0	7483019.0	1.611	33.68	0.07	2.084	33.91	0.183	2.087	30.66	0.23
17	635974.0	7483021.0	1.609	33.67	0.07	2.085	33.89	0.183	2.087	30.66	0.23
18	635974.0	7483023.0	1.607	33.66	0.07	2.085	33.87	0.183	2.088	30.66	0.23
19	635974.0	7483025.0	1.605	33.66	0.07	2.055	33.84	0.1805	2.089	30.66	0.22
20	635974.0	7483027.0	1.603	33.65	0.07	2.012	33.8	0.1805	2.09	30.66	0.22
21	635974.0	7483029.0	1.601	33.65	0.07	2.015	33.76	0.1805	2.091	30.66	0.22
22	635974.0	7483031.0	1.598	33.64	0.07	2.019	33.71	0.1805	2.092	31.52	0.22
23	635974.0	7483033.0	1.409	30.1	0.07	2.031	33.43	0.1775	2.163	31.52	0.21
24	635974.0	7483035.0	1.408	30.1	0.07	2.032	33.37	0.1775	2.164	31.52	0.21
25	635974.0	7483037.0	1.407	30.11	0.07	2.037	33.31	0.1775	2.166	31.52	0.21
26	635974.0	7483039.0	1.407	30.1	0.07	2.043	33.24	0.1775	2.168	31.52	0.21
27	635974.0	7483041.0	1.406	30.1	0.07	2.093	33.16	0.1775	2.172	31.52	0.21
28	635974.0	7483043.0	1.405	30.09	0.07	2.099	33.08	0.1775	2.175	31.94	0.21

Fig. 7.6 The QUALIFIS software with the "Input data file" icon opened

Where:

X and Y are the UTM coordinates of each kriged point (center of the pixel);
AFD10, AFD20 and AFD30 means Easily Available Water (EWA) at 10, 20 and 30 cm soil depth, respectively; PA10, PA20, PA30 means Air Filled Porosity at 10, 20 and 30 cm soil depth, respectively; RP10, RP20 and RP30 means Penetration Resistance (PR) at 10, 20 and 30 cm soil depth, respectively;

To perform the soil quality classification, the user can use the default values presented in Sect. 7.2.3.4 or can change the parameters according to the specific conditions. Figure 7.7 shows an example of the QUALIFIS software screen with the Input parameter icon opened (central panel of the figure) to set the values to perform the soil physical classification. The software classifies the soil physical quality at three soil depths or more, which can change according to each specific study case. In the example of Fig. 7.7, the classification is being performed at the 0–0.10, 0.10–0.20 and 0.20–0.30 m soil layers, according to the default values of input parameters proposed at this work.

As a result, three soil physical quality maps are generated with tree possible classes (restrictive, suitable and optimum).When the user runs the classification, an output file in ASCII format is generated (".txt"). In Fig. 7.8, the output file contains five columns. The first two columns of the output file are the X and Y coordinates of

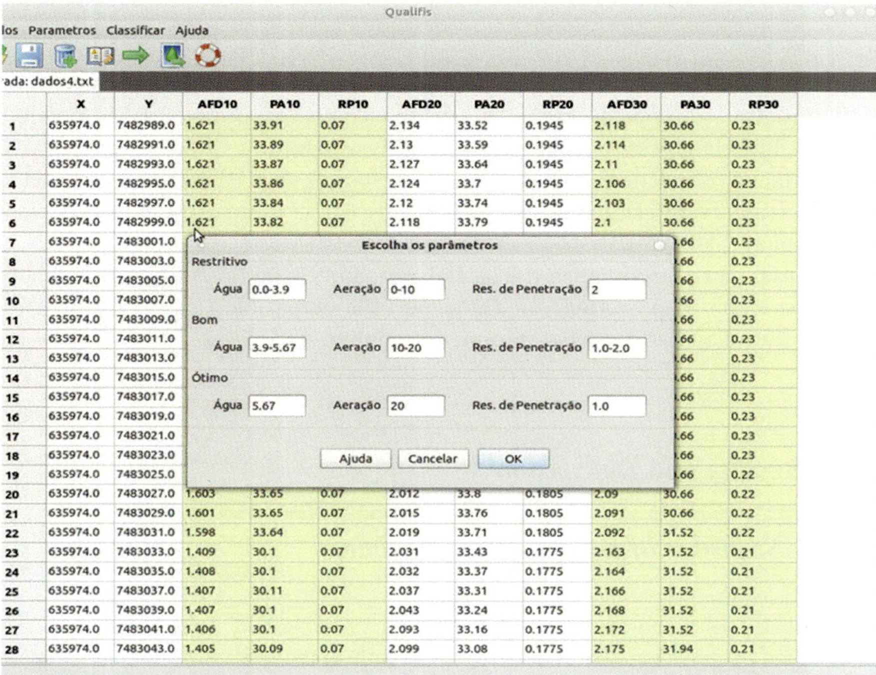

Fig. 7.7 The QUALIFIS software with the "Insert parameters values" icon opened

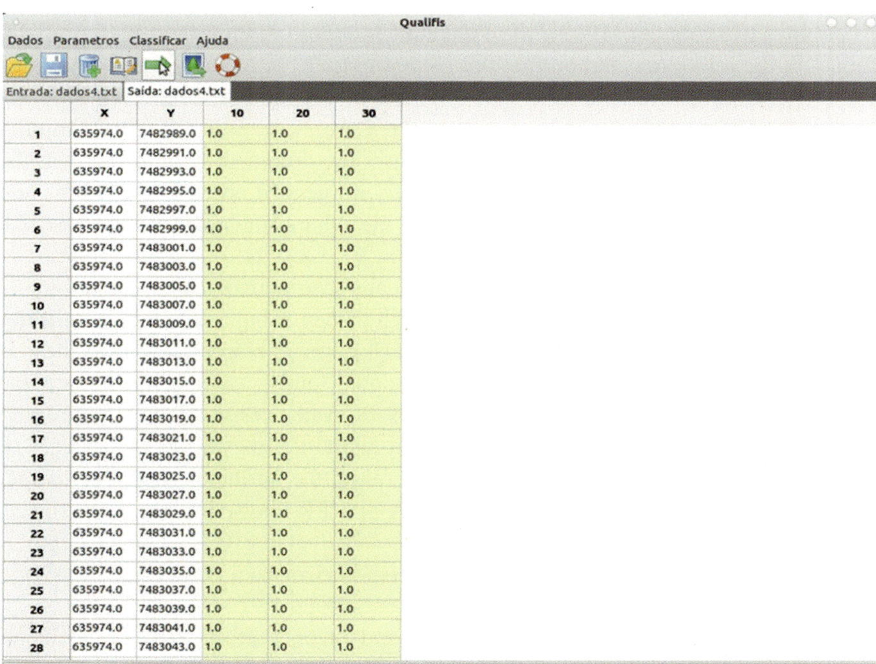

Fig. 7.8 The Output file generated by the QUALIFIS software after running the classification

each pixel of the spatial variability map of SPQ class. The further columns will contain the code values, which represent the soil physical quality at each soil depth (0.0–0.10, 0.10–0.20 and 0.20–0.30 m soil depth layers, respectively). The possible soil physical classification codes are 1, 2 or 3, which means Restrictive, Suitable and Optimum, respectively. In Fig. 7.8, all the points showed in the screen, and at all soil depth, were classified with the code 1, which means restrictive for plant development.

The user can also visualize the spatial variability map of soil physical quality using the icon "Map Edition", or can export the output file (".txt" extension) to other software as Geographical Information Systems.

7.3 Results and Discussion

7.3.1 *Statistics of the Physical Attributes*

The statistics the soil physical attributes are presented in Table 7.1. To better understand the physical characteristics of the soils of the study area, the information about the soil types and its relation to the relief is presented (Fig. 7.9). Samples were collected along a slope, with elevation values ranging from 26 to 39.4 m above the sea level (A.S.L), and the soil classes identified in the area were: (1) Medial over fine clayey Kandiudult, and (2) Sandy over loamy texture Paleudult (Fig. 7.9). According to Embrapa (1999), the Paleudult is more common and preferentially found on the mid-slope and on the foot slope (up to 30 m A.S.L). Medial over fine clayey Kandiudult were the dominant soil class on the top of the hills. Considering the greater occurrence of sandy over loamy texture in the soils, and the sampling restricted to the upper 0.30 m from the surface, the dominant textural class at the three soil depths sampled was sandy. It greatly influenced not only the means and median values of the soil attributes but also the others statistical parameters, like variance, kurtosis and skewness.

The predominance of sand content, basically compounded of quartz mineral (Silva et al. 2001), determine the relatively high values of bulk and particle densities (averaging 1.49–1.61 kg dm^{-3}, and 2.50–2.54 kg dm^{-3}, respectively). Besides, as sandy soils present higher proportion of macropores and lower specific surface, the soil layer up to 0.30 m present low values of water retention and availability (Θ10kPa, Θ80kPa and EAW) and penetration resistance, as well as high values of air-filled porosity (PA). As the soil depth increase, the attributes sand, clay, water retention at Θ10kPa and Θ80kPa, EAW, PA and PR showed a greater variance. Probably, this behavior may result from the fact that the 0.20–0.30 m depth is coincident to the upper bound of the transitional horizon to a clayey horizon (B) in both soil types.

Considering both the coefficient skewness and excess kurtosis, only the attributes particle density, total porosity and EAW, at all soil depths, presented distribution

Table 7.1 Statistics of the physical attributes in three soil depths

0.0–0.10 m	**Sand**	**Silt**	**Clay**	Θ_{10kPa}	Θ_{80kPa}	**ρb**	**ρs**	**α**	**EWA**	**PA**	**PR**
Unity	$g.kg^{-1}$	$g.kg^{-1}$	$g.kg^{-1}$	$cm^3.cm^{-3}$	$cm^3.cm^{-3}$	$kg.dm^{-3}$	$kg.dm^{-3}$	$cm^3.cm^{-3}$	$cm^3.cm^{-3}$	$cm^3.cm^{-3}$	mPa
No. of values	124	124	124	124	124	123	124	123	124	123	168
Minimum	570	10	10	4.1	2.4	1.04	2.26	28.6	0.1	6.5	0.03
Maximum	970	230	340	36.1	31.8	1.73	2.94	59.2	6.1	43.9	0.67
Mean	829	62	110	14.0	10.6	1.49	2.53	41.2	3.4	27.3	0.13
Median	860	60	75	11.5	8.1	1.52	2.51	41.0	3.4	28.6	0.09
Kurtosis[a]	−0.05	4.39	−0.07	0.10	0.42	0.88	0.43	0.48	−0.64	0.27	7.36
Skewness	−0.94	1.38	0.99	0.96	1.09	−0.74	0.21	0.34	−0.20	−0.75	2.48
CV (%)	13	56	81	55	65	9	5	14	44	30	88
Variance	106.3	11.8	78.4	59.1	46.5	0.02	0.01	31.8	2.2	68.4	0.01
SD	10.3	3.4	8.9	7.7	6.8	0.13	0.12	5.6	1.5	8.3	0.11
0.10–0.20 m	**Sand**	**Silt**	**Clay**	Θ_{10kPa}	Θ_{80kPa}	**ρb**	**ρs**	**α**	**EWA**	**PA**	**PR**
No. of values	122	122	122	122	122	104	109	94	122	91	168
Minimum	360	12	13	3.6	1.5	1.22	2.17	27.4	0.5	2.6	0.03
Maximum	970	400	400	34.1	29.4	1.76	2.77	51.2	8.8	37.1	1.33
Mean	819	66	115	13.1	9.6	1.58	2.50	37.4	3.5	25.0	0.35
Median	860	60	75	10.1	6.8	1.59	2.50	37.3	3.6	26.2	0.24
Kurtosis[a]	1.24	18.17	0.04	−0.33	−0.16	1.37	0.13	0.62	−0.10	0.13	1.77
Skewness	−1.20	3.35	1.05	0.93	1.00	−0.93	−0.44	0.32	0.34	−0.80	1.53
CV (%)	15	75	83	63	76	7	5	12	44	30	81
Variance	148.8	25.0	91.1	67.0	53.2	0.01	0.02	19.32	2.38	56.58	0.08
SD	12.2	5.0	9.6	8.2	7.3	0.10	0.13	4.42	1.55	7.56	0.28
0.20–0.30 m	**Sand**	**Silt**	**Clay**	Θ_{10kPa}	Θ_{80kPa}	**ρb**	**ρs**	**α**	**EWA**	**PA**	**PR**
No. of values	120	120	120	118	118	95	111	90	118	89	168
Minimum	470	12	14	3.2	1.4	1.17	2.16	23.1	0.7	3.3	0.03
Maximum	960	170	410	33.9	29.5	1.77	2.86	55.3	8.2	39.5	1.95
Mean	816	60	125	12.7	9.2	1.61	2.54	36.2	3.5	23.4	0.43
Median	875	60	70	9.3	5.7	1.63	2.56	36.3	3.2	24.7	0.28

(continued)

Table 7.1 (continued)

0.0–0.10 m	**Sand**	**Silt**	**Clay**	Θ_{10kPa}	Θ_{80kPa}	**ρb**	**ρs**	**α**	**EWA**	**PA**	**PR**
Unity	$g.kg^{-1}$	$g.kg^{-1}$	$g.kg^{-1}$	$cm^3.cm^{-3}$	$cm^3.cm^{-3}$	$kg.dm^{-3}$	$kg.dm^{-3}$	$cm^3.cm^{-3}$	$cm^3.cm^{-3}$	$cm^3.cm^{-3}$	mPa
Kurtosis[a]	−0.41	0.60	−0.23	−0.42	−0.31	3.19	−0.04	1.44	0.34	−0.93	2.21
Skewness	−0.86	0.59	0.97	0.91	0.95	−1.44	−0.10	0.09	0.80	−0.43	1.56
CV (%)	16	58	86	68	83	6	6	15	46	40	82
Variance	161.6	11.83	114.7	73.59	57.38	0.01	0.02	27.29	2.50	87.80	0.12
SD	12.71	3.44	10.71	8.58	7.58	0.10	0.14	5.22	1.58	9.37	0.35

[a]Excess kurtosis, the standard normal distribution has a kurtosis of zero. A positive kurtosis indicates a "peaked" distribution and negative kurtosis indicates a "flat" distribution. *CV* Coefficient of variation, *SD* Standard deviation, *α* Total porosity

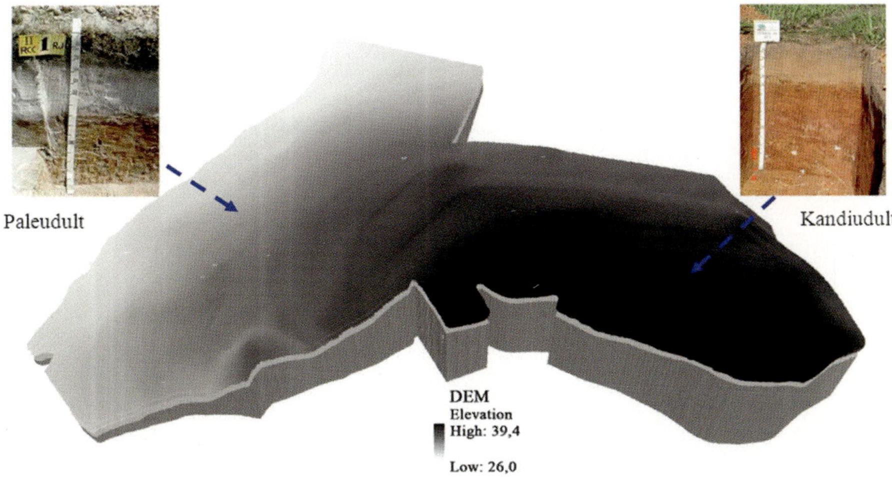

Fig. 7.9 Digital elevation model and the soil classes identified along the study site. The *gray scale* shows the elevation (meters above the sea level)

function similar to normal (skewness and excess kurtosis near to 0). The attributes sand content, PA, and bulk density showed high negative skewness, while the attributes silt and clay content, Θ10kPa, Θ80kPa and PR had high positive skewness at all soil depths. The attributes silt content, PR and ρ_b presented high excess kurtosis ("peaked" distribution) having many observations near the center of the distribution.

7.3.2 *Spatial Variability Analysis*

In order to apply kriging interpolation methods to generate spatial variability maps of EAW, PA and PR, the experimental semivariograms of these attributes were calculated. Despite some soil physical attributes did not present normal distribution function (PA and PR), the experimental semivariograms were calculated without transforming the data. Experimental semivariograms with its respective fitted models and parameters at each soil depth are presented in Fig. 7.10, and Table 7.2, respectively. For all the soil attributes and soil depths, as the lag distance (h) increase, the semi-variance $\gamma(h)$ also increase up to reach a value approximately constant called sill variance (known as the priori variance of the random variable (C_0+C_1), Table 7.2). In general, the sill values of the experimental semivariograms of EAW, PA and PR were very similar to its respective values of variance, at all depths (Table 7.1). For the attributes EAW and PA at all depths, and PR at 0.0–0.10 m, spherical models were fitted, while exponential models were better fitted to PR at 0.10–0.2- and 0.20–0.30 m. According to Isaaks and Srivastava (1989), the type of the model matching the data distribution suggests the spatial continuity of the phenomenon investigated. Comparing to exponential models, the spherical model

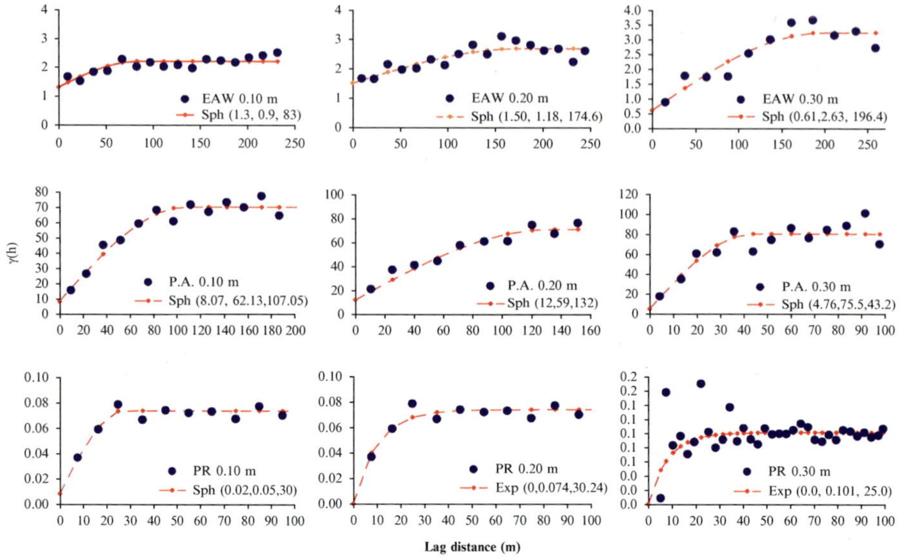

Fig. 7.10 Experimental semivariograms of EAW, PA and PR, and respective fitted models, at 0.0–0.10, 0.10–0.20 and 0.20–0.30 m soil depths

Table. 7.2 Parameters of the semivariograms models

Soil attribute	Soil depth (m)	Model	Co	C_1	Sill C_0+C_1	Range a (m)	$Co/(Co+C_1)*100$
EAW	0.0–0.10	Spherical	1.30	0.90	2.20	83	59.1
EAW	0.10–0.20	Spherical	1.50	1.18	2.68	174.6	55.9
EAW	0.20–0.30	Spherical	0.61	2.63	3.24	196.4	18.8
PA	0.0–0.10	Spherical	8.07	62.13	70.20	107.5	11.5
PA	0.10–0.20	Spherical	12.59	59	71.59	132.0	17.6
PA	0.20–0.30	Spherical	4.76	75.5	80.26	43.2	5.9
PR	0.0–0.10	Spherical	0.02	0.05	0.07	30.2	28.6
PR	0.10–0.20	Exponential	0.0	0.074	0.074	30.2	0.0
PR	0.20–0.30	Exponential	0.0	0.101	0.101	25.0	0.0

Co nugget variance or nugett effect, C_1 structure variance at range distance (a), *a* range distance or spatial dependence distance, $Co/(Co + C1)*100$ Proportion of nugget variance in relation to the sill

describes a more continuous random function. The occurrence of exponential models fitted to the experimental semivariograms suggests a rather random pattern, as it is observed for PR at the depths of 0.10–0.20 and 0.20–0.30 m.

The lag distance at which the variogram reaches its sill is the *range*. This is the spatial dependence, beyond it the variance bears no relation to the separation distance (Webster and Oliver 1990). The spatial dependence of the attribute EAW tended to be higher than the attributes PA and PR, especially at 0.10–0.20 and 0.20–0.30 m soil depth.

Another characteristic features of the semivariograms is the nugget effect (C_0). By definition the semivariance at lag zero is itself zero, however in many circumstances these measured semivariances tend towards some positive value as the lag distance approaches zero. The presence of positive values of C_0 is generally associated to two causes, the first one is due to the sampling grid adopted (the spacing of the grid is too high), and the second by non-controlled errors during the attribute determination. According to Isaaks and Srivastavas (1989), there are two consequences of using a higher value of nugget effect; the first one is that the weight kriging distribution turns into more similar. Increasing the nugget effect makes the estimation procedure become more like a simple averaging of the available data. The second one is that as C_0 increase, the accuracy of the kriging interpolation decrease (higher ordinary kriging variance). Consequently, as the attribute EAW at 0.0–0.10 and 0.10.0.20 soil depths presented higher proportion of C_0 ($Co/(Co+C_1)*100$, Table 7.2), its spatial variability maps will present lower accuracy.

The results of the semivariogram analysis not only allow affirming that EAW, PA and PR, at all soil depths, presented spatial dependence but also that it's possible to take advantage of the kriging interpolation method to generate spatial variability maps.

7.3.3 Spatial Variability Maps of Soil Physical Attributes and Soil Physical Quality

Figure 7.11 shows the spatial variability maps of EAW, PA and PR generated by ordinary kriging. From the left to the right, the maps of each soil physical attributes are organized according to the following sequence of soil depth: 0.0–0.10, 0.10–0.20 and 0.20–0.30 m. These maps give an overview of how the soil physical attributes values vary along the study site, at different soil depths. The area in red represents the restrictive values for the development of plants considering each attribute and in green and blue, the suitable and optimum values, respectively.

Analyzing each soil physical attribute, it is possible to note that PR does not represent a limiting factor for plant development once all the kriged values are lower than the restrictive reference's value of 2 MPa (the three maps at the bottom of the Fig. 7.11). The higher values of EAW are found along the areas with policultive, which present higher clay content (Kandiudult soil class – Fig. 7.9). The clay content can also explain the higher percentage of the areas classified as optimum for EAW (>5.5 mm) at 0.20–0.30 m soil depth. From the layer of 0.20–0.30 m intensifies the increase in clay content of the soil (textural B horizon), consequently is expected to result in higher values EAW. Conversely, the higher values of PA (>20 %, optimum class) are found along the foot slope. The area classified as restrictive (<10 % of PA) occurs in some plots under policultive. However, it occurs only from the layer 0.10 m, becoming higher in depth 0.20–0.30.

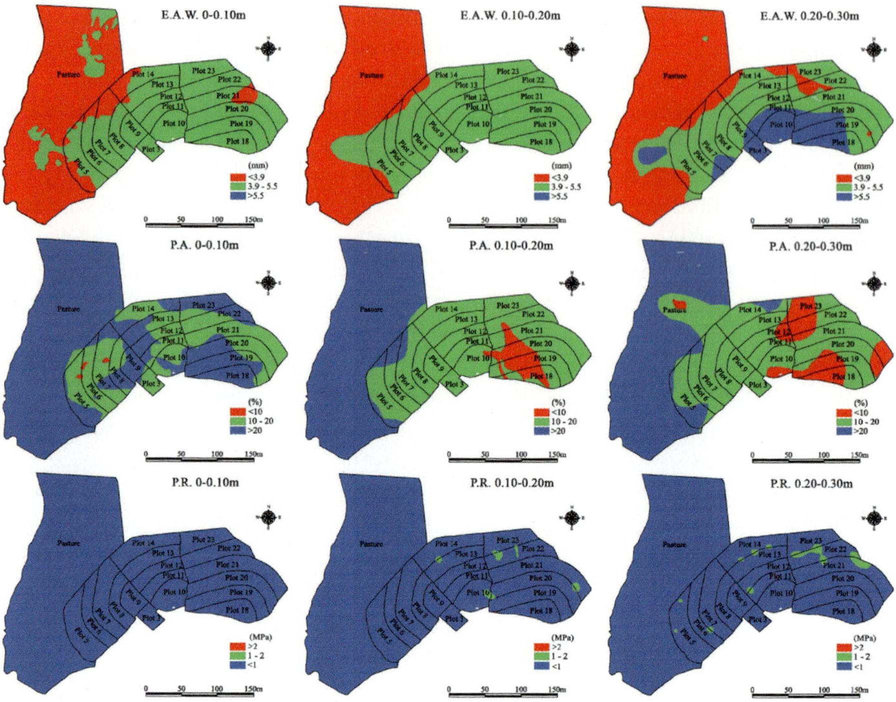

Fig. 7.11 Spatial variability maps of EAW, PA and PR at 0.00–0.10, 0.10–0.20 and 0.20–0.30 m soil depths

Despite the advantage of analyzing the spatial variability maps of each soil physical attributes, a complete evaluation of the soil physical quality can be better performed when all the soil physical attributes are considered simultaneously. Figure 7.12 shows the spatial variability maps of soil physical quality after applying the algorithm "Euclidean distance" using the software QUALIFIS. For each soil depth, the algorithm generated one soil physical quality map integrating the spatial variability of EAW, PA and PR. Considering the total area classified as optimum and suitable, the soil physical quality is higher at the first soil layer (0.0–0.10 m), mainly along the areas with policultive (on the top of the hills). At the 0.0–0.10 m soil depth, the soil physical quality is basically determined by EAW, however, as the soil depth increases, the proportional area classified as optimum decreases while increase the area classified as restrictive, mainly due the reduction of air-filled porosity (PA < 10 %).

Figure 7.13 shows the planimetric data of the soil physical quality classes at each soil depth. The reduction of air-filled porosity is probably a combination of the frequent plowing activity along the plots with policultive and the natural reduction of porosity as a consequence of the increment of clay content in deeper layers of these soils.

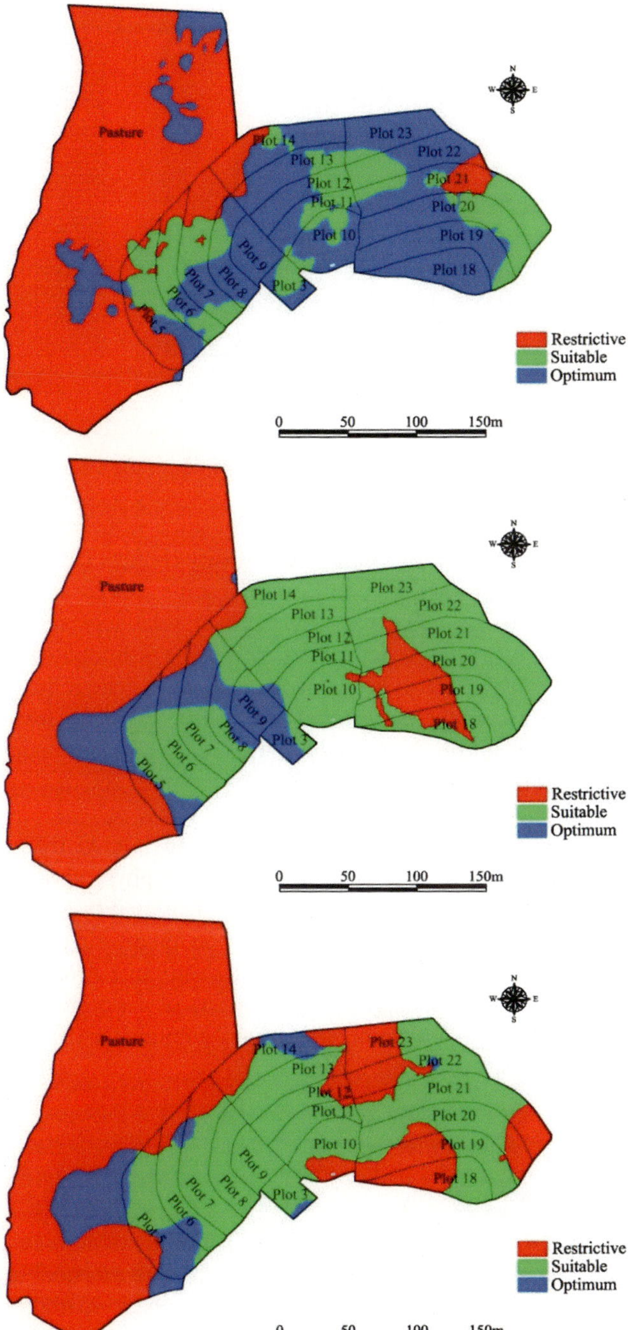

Fig. 7.12 Spatial variability maps of soil physical quality at 0.00–0.10, 0.10–0.20 and 0.20–0.30 m soil depths

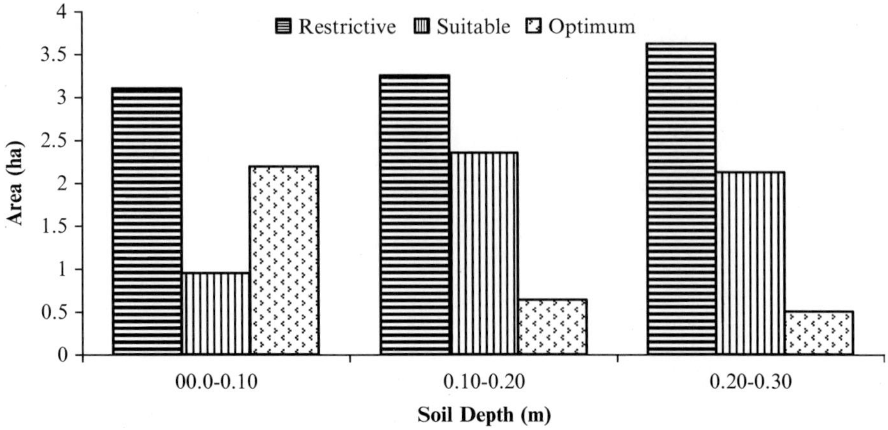

Fig. 7.13 Planimetric data of soil physical quality classes of the study site at 0.0–0.10, 0.10–0.20 and 0.20–0.30 m soil depths

The spatial variability maps of soil physical quality confirm the statement that soil quality has two parts: an intrinsic part covering soil's *inherent* capacity for crop growth and a *dynamic part* that can change over relatively short time periods influenced by the soil user or manager (Carter 2002). The distinction between inherent and dynamic soil quality can also be characterized by the genetic (or static) pedological and management and that are strongly influenced by processes vs. the kinetic (or dynamic) processes in soil as proposed by Richter (1987). The maps presented at Fig. 7.12 give an overview of a mixture of these two parts of soil quality (the *inherent* capacity and *dynamic part*).

For by so doing, the algorithm implemented in the QUALIFIS software was efficient both to integrate the spatial variability of soil physical attributes directly affecting the plant growth and to turn the soil physical quality classification into a more user friendship process. In order to keep on monitoring the soil physical quality at the study site, new maps of spatial variability of EAW, PA and PR must be generated, which remain one of the great barriers of the digital soil mapping of soil physical quality. As already point out by many authors, the development of new soil sensors could allow overcoming this barrier.

7.4 Conclusions

The algorithm proposed was efficient to classify the soil physical quality (SPQ), integrating different soil physical attributes, which present spatial dependence at different soil depth. The attributes EAW and PA were the most important to define the SPQ of the study site. At the soil surface (0.0–0.10 m soil depth) the SPQ is higher and decreases as the soil depth increase. The attribute EAW was the most important factor controlling the SPQ, and the reduction of the air filled porosity (PA) was the main cause of the SPQ reduction at the layers below 0.10 m depth.

The software QUALIFIS attend the demand for user-friendly software and can be applied in precision agriculture. Considering the efficiency of the algorithm and the availability of the Qualifis software, the main barrier to apply this method of mapping the spatial variability of SPQ is the great demand for maps of EAW, PA and PR. The generation of these maps is still time consuming and demand a high number of soil samples in order to generate spatial variability maps with good accuracy.

References

Addinsoft (2004) *XL*STAT-PLS 1.8: statistical software to MS Excel

Carter MC (2002) Soil quality for sustainable land management: organic matter and aggregation interactions that maintain soil functions. Agron J 94:38–47

Da Silva AP, Kay BD, Perfect E (1994) Characterization of the least limiting water range of soils. Soil Sci Soc Am J 58:1775–1781

Da Silva AP, Imhoff S, Kay B (2004) Plant response to mechanical resistance and air-filled porosity of soils under conventional and no-tillage system. Sci Agric (Piracicaba, Brazil) 61(4):451–456

Empresa Brasileira de Pesquisa Agropecuária – Embrapa (1997) Manual de métodos de análise de solo. Embrapa, Rio de Janeiro, p 247

Empresa Brasileira de Pesquisa Agropecuária – Embrapa (1999) Levantamento semidetalhado dos solos da área do Sistema Integrado de Produção Agroecológica (SIPA) – km 47 – Seropédica, RJ. Boletim de Pesquisa n. 5, Embrapa, Rio de Janeiro, 93 p

Engelaar WMHG, Yoneyama T (2000) Combined effects of soil waterlogging and compaction on rice (Oryza sativa L.) growth, soil aeration, soil N transformations and N discrimination. Biol Fertil Soils 32:484–493

Isaaks EH, Srivastava RM (1989) Applied geoestatistics. Oxford University Press, Oxford, 560 p

Karlen DL, Mausbach MJ, Doran JW, Cline RG, Harris RF, Schuman GE (1997) Soil quality: a concept, definition, and framework for evaluation. Soil Sci Soc Am J 61:4–10

Lagacherie P (2008) Chapter 1. Digital soil mapping: state of the art. In: Hartemeink AE, McBratney AB, Mendonça-Santos ML (eds) Digital soil mapping with limited data. Elsevier, Amsterdam, pp 3–14

Lagacherie P, McBratney AB (2007) Chapter 1. Spatial soil information systems and spatial soil inference systems: perspectives for Digital Soil Mapping. In: Lagacherie P, McBratney AB, Voltz M (eds) Digital soil mapping an introductory perspective. Development in soil science, vol 31. Elsevier, Amsterdam, pp 3–24

Leão TP, Silva AP, Perfect E, Tormena CA (2005) An algorithm for calculating the least limiting water range of soils. Agron J 97:1210–1215

Leão TP, Silva AP, Macedo MCM, Imhoff S, Euclides VPB (2006) Least limiting water range: a potential indicator of changes in near-surface soil physical quality after the conversion of Brazilian Savanna into pasture. Soil Tillage Res 88:279–285

Letey J (1985) Relationship between soil physical properties and crop production. Adv Soil Sci 1:277–294

McBratney AB, Webster R (1986) Choosing functions for semi-variograms of soil properties and fitting them to sampling estimates. J Soil Sci 37(3):617–639

McBratney AB, Mendonça Santos ML, Minansny B (2003) On digital soil mapping. Geoderma 117:3–52

Ramos DP, Castro AF, Camargo MN (1973) Levantamento de solos da área da Universidade Federal Rural do Rio de Janeiro. Pesqui Agropecu Bras 8:1–27

Richter J (1987) The soil as a reactor. In: Gregorich EG, Carter MR (eds) Soil quality for crop. Catena Verlag, Cremlingen

Silva MB, Anjos LHC, Pereira MG, Nascimento RAM (2001) Estudo de topossequência da baixada litorânea fluminense: efeitos do material de origem e posição topográfica. Rev Bras Ciênc Solo Viçosa 25:965–976

Stolf R (1991) Teorias e testes experimentais de fórmulas de transformação dos dados de penetrômetro de impacto em resistência do solo. Rev Bras Ciênc Solo (Campinas) 15:229–235

Trangmar BB, Yost RS, Uehara G (1985) Application of geostatistics to spatial studies of soil properties. Adv Agron 38:45–93

Vieira SR, Hatfield JL, Nielsen DR, Biggar JW (1983) Geostatistical theory and application to variability of some agronomical properties. Hilgardia 51:1–75

Webster R, Oliver MA (1990) Statistical methods in soil and land resource survey. Oxford University Press, Oxford, 316p

Chapter 8
A Preliminary Proposal for the Structural Classification of Soil Pore Space

Theophilo Benedicto Ottoni Filho and Marta Vasconcelos Ottoni

Abstract This study presents a preliminary proposal of a structural classification system of soil pore space seeking to group soils with similar air availability curves. The function that relates soil air content to water suction is herein called the air availability curve. The proposed system approach is similar to that of textural classification and allows classifying soil according to types of soil structural families based on nine possible sub-areas (from A to I) represented in a classification triangle, herein called the structural triangle. Besides family types, it is also necessary to predict the family orders according to different ranges of variation for the effective soil porosity. Both family type and order are obtained by parameterization of the moisture retention curve. 467 samples were classified according to the proposed methodology, which confirmed that the structural families grouped soils with similar air availability curves. This result shows the potential usefulness of the proposed methodology for pedologic knowledge and for the development of pedotransfer functions of the soil water properties.

Keywords Air availability • Available water • Brazilian soils • Classification triangle • Effective air • Parameters of van-Genuchten model • Pedotransfer function • Pore size distribution • Residual water • Soil classification • Soil's void space

---

T.B. Ottoni Filho (✉)
Department of Water Resources and Environment, Federal University of Rio de Janeiro, Rio de Janeiro, RJ, Brazil
e-mail: teottoni@poli.ufrj.br

M.V. Ottoni
Department of Hydrology, Companhia de Pesquisa de Recursos Minerais (CPRM), Geological Survey of Brazil, Rio de Janeiro, RJ, Brazil

Laboratório de Recursos Hídricos e Meio Ambiente (LABH2O), Programa de Engenharia Civil, Instituto Alberto Luiz Coimbra de Pós-Graduação e Pesquisa em Engenharia (COPPE), Universidade Federal do Rio de Janeiro (UFRJ), Rio de Janeiro, RJ, Brazil
e-mail: marta.ottoni@cprm.gov.br

W.G. Teixeira et al. (eds.), *Application of Soil Physics in Environmental Analyses: Measuring, Modelling and Data Integration*, Progress in Soil Science,
DOI 10.1007/978-3-319-06013-2_8,

• Structural classification • Structural family • Textural classification • Water retention curve

8.1 Introduction

Soil pore spaces are filled with water by suction *s* (in centimeters of water column), defining a saturation volumetric fraction equal to θ/Φ where θ is the volumetric water content and Φ is the soil porosity. In turn, suction can be understood as being a measure of the size of the hydrated pores, in terms of their equivalent diameter *d*, through the capillarity equation:

$$d = \frac{4\sigma}{\rho g s}, \tag{8.1}$$

where σ and ρ are, respectively, the surface tension and the specific mass of the water, and *g* is the acceleration due to gravity. So, the plot of "$(\Phi - \theta)/\Phi$" versus "s" can be considered a representation of the distribution, as a cumulative function of volume, of the pore size within the soil pore space volume. This distribution is close to zero for the greatest hydrated pore diameter *d* (with the smallest suction s) and increases to the limit of 1.0 with the increase in suction and decrease in size of the pores filled up with water.

The similarity between the construction of the pore size distribution mentioned above and the particle size distribution curve allows the development of a structural classification of the pore space that is similar to the soil textural classification. Because of this similarity this study proposed a structural classification system of soil pore space based on the $(\Phi - \theta)$ curve (and not on the θ curve), herein called the soil air availability curve, as a function of water suction. We expect that the soil grouping by similar air availability curves allowed by the system may contribute to pedologic knowledge and the description of the dynamic processes that occur in soil pore spaces. This classification system is termed as structural because it refers to volume distribution inside the soil void space, which is a direct consequence of soil structure, a classical pedological term that, instead, generally refers to geometric and physical features of the soil solid matrix.

8.2 Material and Methods

8.2.1 Representation of the Pore Size Distribution Curve

For the modeling of the various dynamic processes that occur in soil pore spaces, it is important to obtain an accurate representation of the pore size distribution curve, which, in turn, depends on the representation of the moisture retention curve $\theta(s)$. To this end, we adopted the van Genuchten (1980) model, which relates *s* (cm) to θ (cm^3 cm^{-3}):

$$W(s) = \frac{\theta(s) - \theta_r}{\Phi - \theta_r} = \left[1 + (\alpha s)^{1/(1-m)}\right]^{-m}, \tag{8.2}$$

$$s \geq 0, \theta_r \leq \theta \leq \Phi, 0 < m < 1, \alpha > 0,$$

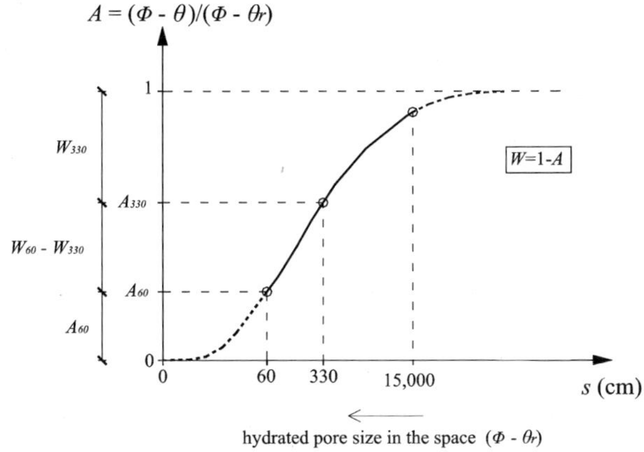

Fig. 8.1 Pore size distribution curve within the effective pore space ($\Phi - \theta_r$). For the structural classification the pore space was arbitrarily divided in three fractions: A_{60}, ($W_{60} - W_{330}$) and W_{330}. In the classification system, the curve was applied only in the suction range 60–15,000 cm (*bold line*)

where θ_r (cm^3 cm^{-3}), α (cm^{-1}) and m (non-dimensional) are considered as fitting parameters, and Φ is the soil saturation moisture (or porosity) determined in laboratory. This model was chosen due to its wide use (Tomasella et al. 2000, 2003; Cornelis et al. 2005; Vereecken et al. 2010) and because its parameters can be used to fit the non-saturated hydraulic conductivity curve (van Genuchten 1980; Schaap and van Genuchten 2006). However, a high accuracy for W ($0 \leq W \leq 1$) is not expected when θ is in a very wet range (close to saturation) or very dry range (close to the residual moisture θ_r), as reported in the literature (van Genuchten 1980; Coppola 2000; Schaap and van Genuchten 2006). Therefore, Eq. 8.2 will only be applied for the intermediate range of suction, 60 cm $\leq s \leq$ 15,000 cm. It also is advantageous that θ_r may be fitted as a negative number (Dourado Neto et al. 2000). However, Ottoni and Ottoni Filho (2011) indicate that there is not a significant improvement in accuracy in the determination of the moisture retention curve if θ_r is more negative than -0.30 cm^{3}cm^{-3}, for which reason we arbitrarily set $\theta_r \geq -0.30$ cm^{3}cm^{-3}.

For the classification of an undisturbed soil sample relative to its pore space structure, the function

$$A(s) = \frac{\Phi - \theta(s)}{\Phi - \theta_r} = 1 - W(s) \tag{8.3}$$

represents the pore size distribution in terms of cumulative volumetric fraction, assuming an effective pore space between $\theta = \Phi$ ($A = 0$) and $\theta = \theta_r$ ($A = 1$) (Fig. 8.1). Effectively, this procedure allows building a parameterized and standardized pore space structural classification system that is similar to the textural classification, as shown later on.

8.2.2 Distribution Curve Parameterization Protocol

To determine function *A(s)* (Eqs. 8.2 and 8.3) with approximately equal accuracy in its entire range (60 cm $\leq s \leq$ 15,000 cm), the three fitting parameters (θ_r, m and α)

are calculated using, correspondingly, only three experimental points of the moisture retention curve, called (s_i, θ_i), $i = 1, 2, 3$, following the protocol below:

- s_1 is the suction the closest to 60 cm, in the range of 30 cm $\leq s \leq$ 80 cm; s_2 is the suction the closest to 330 cm, in the range of $250 \leq s \leq 500$ cm, and s_3 is the suction the closest to 15,000 cm, in the range of 9,000 cm $\leq s \leq$ 18,000 cm.
- Parameters θ_r, m and α will be determined according to the procedure described by Ottoni (2009), which requires that the residuals (observed θ_i – calculated θ_i) be null, $i = 1, 2, 3$. If this is not possible, the RETC software (van Genuchten et al. 1991), with modifications to allow $\theta_r \geq -0.30$ cm^{3}cm^{-3} (Ottoni and Ottoni Filho 2011), will be used for the calculation of the three parameters. This ensures a minimization of the residuals (observed θ_i – calculated θ_i, $i = 1, 2, 3$), which would not be assured if all the water retention data points were used.

8.2.3 Definition of the Structural Families

The air availability curve "$\Phi - \theta(s)$" is the element to be grouped and classified in structural families. The values of A_{60} [$A(s = 60$ cm)], A_{330} and $A_{15,000}$ are calculated using the values of Φ, θ_r, m and α in Eqs. 8.2 and 8.3. Volumetric fractions A_{60}, $A_{330} - A_{60}$ (which is equal to $W_{60} - W_{330}$) and $1 - A_{330}$ (which is equal to W_{330}) add up to 1.0 (Fig. 8.1) and allow the determination of the structural family type of the sample when they are plotted in the pore space structural classification triangle (Fig. 8.2), here called the structural triangle.

By analogy with the textural triangle, fraction A_{60}, called the "effective air", is similar to that of sand, as it tends to be high in sandy soils. The fraction A_{60}, to a certain extent, can be interpreted as being a measure of soil aeration capacity. Fraction W_{330} tends to be high in clayey soils, as these soils usually have great affinity to water and high microporosity. W_{330} is called the "less available water fraction" and it is qualitatively similar to clay in the textural triangle. Finally, $W_{60} - W_{330}$, the "more available water fraction", is similar to the silt fraction. In fact, it is known that silty soils tend to have great water availability for plant use in the low suction range of microporosity, generally being quite appropriate soils in terms of water retention for irrigated agriculture. Although the normalized moisture $W_{15,000}$ is not used for classification purposes, it is useful in the system for the determination of θ_r. Figure 8.2 defines nine types of structural families, from type A to I, relative to plotting areas of fractions A_{60}, ($W_{60} - W_{330}$) and W_{330}. From Eq. 8.2:

$$\left(W_{60}^{-1/m} - 1\right)^{1-m} = (60/330)\left(W_{330}^{-1/m} - 1\right)^{1-m} \tag{8.4}$$

Equation 8.4 allows the calculation of m if W_{60} and W_{330} are known. The representation of the lines of equal m values in the structural triangle is shown in Fig. 8.2, reinforcing the thesis that this triangle can be viewed as a tool of graphical representation of the pore space structure. After determining the value of m,

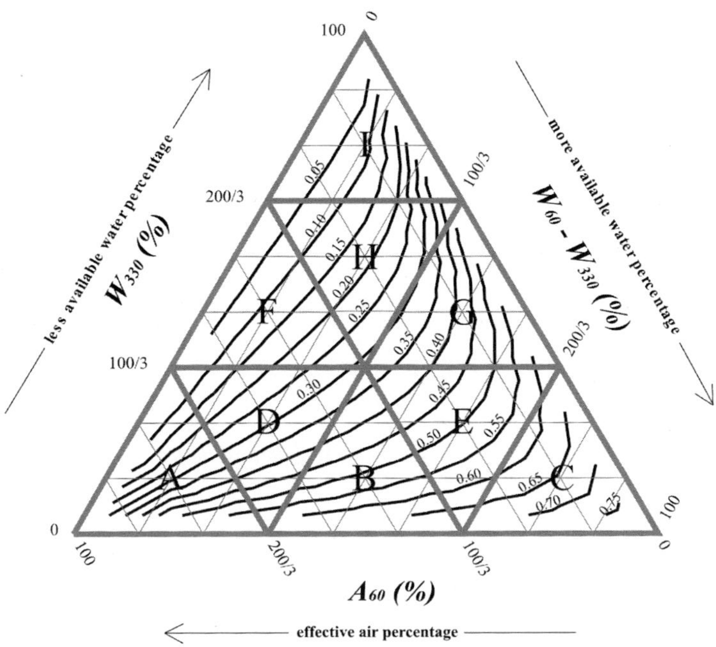

Fig. 8.2 Nine types of structural families (from A to I) defined in the pore space structural classification triangle, or simply, structural triangle. The lines of equal parameter m values are also shown

parameter α is calculated by solving Eq. 8.2 for the known value of W_{60} (or W_{330}). Thus, the pair (W_{60}, W_{330}) determines the pair (m, α).

From Eqs. 8.2 and 8.3, the air availability curve is:

$$\Phi - \theta(s) = (\Phi - \theta_r)\left\{1 - \left[1 + (\alpha s)^{1/(1-m)}\right]^{-m}\right\} = (\Phi - \theta_r)f(s)_{m,\alpha} \quad (8.5)$$

Thus, for the perfect definition of the structural families, it is also necessary to group the soil samples in $(\Phi - \theta_r)$ value ranges. These ranges indicate the various "i" orders of the families. A soil is considered to be in the family of order i (i $\geq$ 1) when

$$0.05\,(\mathrm{i} - 1) < (\Phi - \theta_r) \leq 0.05\,\mathrm{i} \quad (8.6)$$

Equations 8.5 and 8.6 justify the classification of a structural family using the notation iX, where the integer number i (Arabic) represents the family order and X is the family type (from A to I). Order i groups similar values of $(\Phi - \theta_r)$ and type X groups similar (m, α) pair values. Thus, a soil can be graphically regarded in a three-dimensional space as a point with height $(\Phi - \theta_r)$ above the point where the soil is plotted in the structural triangle.

8.2.4 The Studied Soils

Samples (467) of Brazilian soils, most of them classified as Latosols and Argisols, were classified into structural families. The samples were selected by Ottoni (2009) based on the study by Tomasella et al. (2003). The database contains water contents for five suction values (60 cm, 100 cm, 330 cm, 1,000 cm and 15,000 cm) and porosity (or saturation moisture). Soil descriptions and data determination methodology can be found in Tomasella et al. (2003).

8.3 Results and Discussion

8.3.1 Parameterization of the Distribution Curve

For all samples, $s_1 = 60$ cm, $s_2 = 330$ cm and $s_3 = 15{,}000$ cm, and the three parameters, θ_r, m and α, were determined according to Ottoni (2009), by making the residuals (observed θ_i – calculated θ_i) null, $i = 1, 2, 3$. θ_r values were nearly always positive, with the exception of 47 soil samples, for which the minimum θ_r was $-0.183\ \text{cm}^3\text{cm}^{-3}$. The maximum absolute residual in all cases for the validation points ($s = 100$ cm and 1,000 cm) was $0.0044\ \text{cm}^3\text{cm}^{-3}$ (Ottoni 2009), which confirms the good reliability of Eq. 8.2 to represent the water retention curves.

8.3.2 Structural Families

The calculation of fractions A_{60}, $(W_{60} - W_{330})$, and W_{330} (from Φ, θ_r, m and α) led to the plotting of the 467 soil samples in the structural triangle (Fig. 8.3a). The corresponding textural triangle plot is also shown for comparison (Fig. 8.3b). The analysis of the two figures reveals two tendencies: (a) there were few soils in the silty classes (or with silt percentages over 50 %) and few soils with high $W_{60} - W_{330}$ fractions (more available water percentages over 33 %), which confirms the expectation that the materials from those textural classes tend to retain large amount of readily available water for plant use; (b) there was a large number of soils in the heavy clay class and a much smaller number in the "corresponding" class (class I) of Fig. 8.3a. This can be attributed to the "hybrid" tendency of some clayey latosols to behave hydrodynamically as soils of sandy texture (Tomasella et al. 2000), with great "effective air" ($\Phi - \theta_{60}$), which might explain the predominance in the number of samples plotted in classes D and F in Fig. 8.3a. The effective air fractions (A_{60}) of samples in these two classes are greater than those of class I.

A qualitative comparative analysis of Fig. 8.3a, b indicates a correlation between the textural and structural behavior of the soil samples, which is assumed when

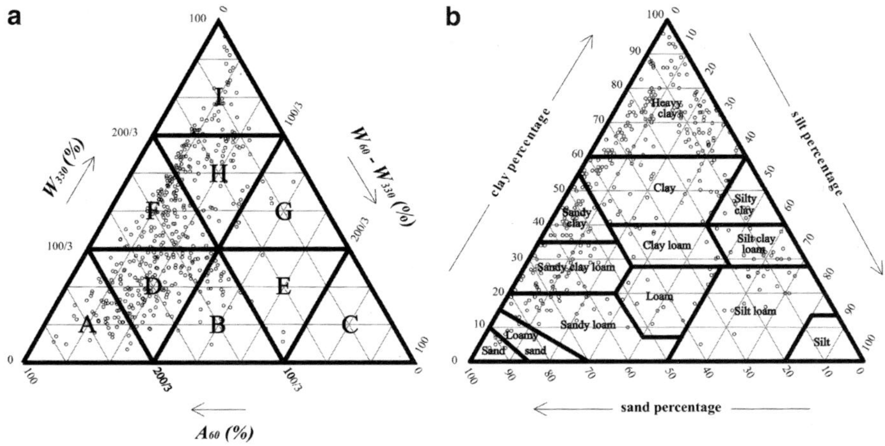

Fig. 8.3 Studied soils plotted in the (**a**) structural and (**b**) textural triangle

one uses pedotransfer functions to estimate water retention from granulometric data (Tomasella et al. 2000, 2003). This result shows the potential use of the proposed pore space classification system as a tool in the construction of pedotransfer functions.

In the soils studied, 13 orders of structural families were identified, from order 3 (0.10 $cm^3cm^{-3} < (\Phi - \theta_r) \leq 0.15\ cm^3cm^{-3}$) to order 15 (0.70 $cm^3cm^{-3} < (\Phi - \theta_r) \leq 0.75\ cm^3cm^{-3}$). Among the family types of highest sample incidence, in decreasing order, were classes D, F and H, totaling 70 % of the samples. The most frequent orders were 6, 7, 5, 8, and 9, in decreasing number of soils, totaling 68 % of the samples. The only family type which did not have any soil sample was C.

8.3.3 Air Availability Curve Grouping

Figure 8.4 shows the $(\Phi - \theta_r)$ curves for soils in families 5A, 9A, 5 F and 9 F, as an example. The classification system was conceived to group similar air availability curves, as shown in Fig. 8.4. The similarity observed among $(\Phi - \theta)$ curves for soils of the same family was confirmed for all families of the 467 samples investigated. One can see that the shape of the curves in 5A and 9A are similar to each other, as their soils correspond to the same class in the structural triangle. What distinguishes the two groups of curves is the scale of values (vertical axis), because they are of different orders. The same analysis applies to the elements of families 5 F and 9 F shown in Fig. 8.4. The curves were plotted only in the suction range allowed by the system (60 cm to 15,000 cm). If they were extrapolated to the zero suction, they would all tend to zero $(\Phi - \theta = 0)$, asymptotically to the horizontal axis. As the elements of class A have a greater effective air fraction (A_{60}) than those of class F,

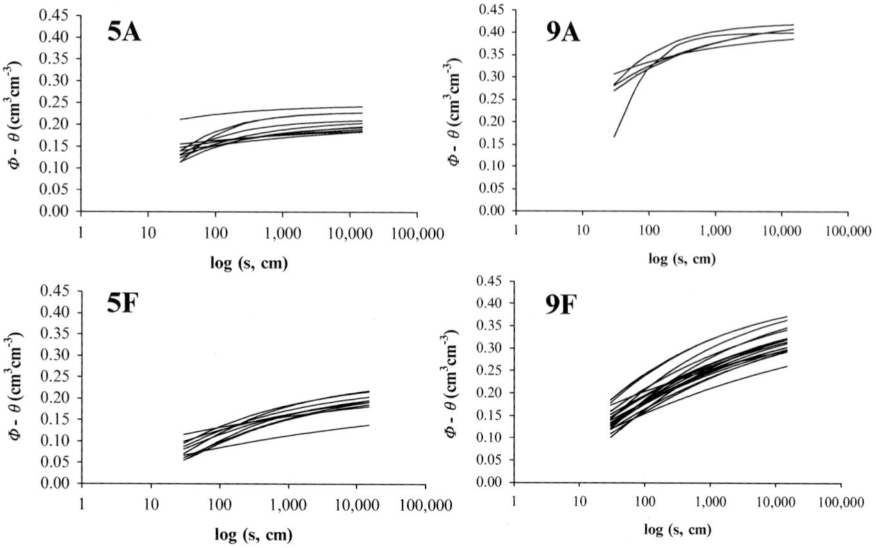

Fig. 8.4 Air availability curves for soil samples of families 5A, 9A, 5F and 9F

the effective air values ($\Phi - \theta_{60}$) of 5A soils tend to be greater than those of 5 F, which is also observed for soils in 9A and 9 F (Fig. 8.4).

8.4 Conclusions

The proposed classification system showed to be able to group soils with similar air availability curves, which lends it to potential use as an instrument of pedologic investigation and for the modeling of processes that occur in pore spaces. The proposed classification system needs to be better evaluated taking into account soils of various pedologic origins. So, we consider that we are presenting a preliminary version of the system, which may be somewhat altered with respect to pragmatic aspects of the structural family definition and nomenclature.

Acknowledgments The authors thank Dr. Javier Tomasella (Instituto Nacional de Pesquisas Espaciais – INPE) for the data used in this study.

References

Coppola A (2000) Unimodal and bimodal descriptions of hydraulic properties for aggregated soils. Soil Sci Soc Am J 64:1252–1262

Cornelis WM, Khlosi M, Hartmann R et al (2005) Comparison of unimodal analytical expressions for the soil-water retention curve. Soil Sci Soc Am J 69:1902–1911

Dourado Neto D, Nielsen DR, Hopmans JW et al (2000) Software to model soil water retention curves (SWRC, version 2.00). Sci Agric 57:191–192

Ottoni MV (2009) Avaliação do ajuste da curva de retenção de água dos solos a partir de três umidades representativas. In: Anais do XVIII SIMPÓSIO BRASILEIRO DE RECURSOS HÍDRICOS, Associação Brasileira de Recursos Hídricos, Campo Grande/MS, 22–26 Novembro

Ottoni MV, Ottoni Filho TB (2011) Avaliação da curva de retenção de umidade considerando a equação de van Genuchten e umidades residuais negativas. In: Anais do XXXIII CONGRESSO BRASILEIRO DE CIÊNCIA DO SOLO, Sociedade Brasileira de Ciência do Solo, Uberlândia/MG, 31 Julho–5 Agosto

Schaap MG, van Genuchten MT (2006) A modified Mualen-van Genuchten formulation for improved description of the hydraulic conductivity near saturation. Vadose Zone J 5:27–34

Tomasella J, Hodnett MG, Rossato L (2000) Pedotransfer functions for the estimation of soil water retention in Brazilian soils. Soil Sci Soc Am J 64:327–338

Tomasella J, Pachepsky YA, Crestana S et al (2003) Comparison of two techniques to develop pedotransfer functions for water retention. Soil Sci Soc Am J 67:1085–1092

Van Genuchten MTh (1980) A closed-form equation for predicting the hydraulic conductivity of unsaturated soils. Soil Sci Soc Am J 44:892–898

Van Genuchten MTh, Leij FJ, Yates SR (1991) The RETC code for quantifying the hydraulic functions of unsaturated soils. In: Report No. EPA/600/2-91/065, RS Kerr Environmental Research Laboratory, US Environmental Protection Agency, Ada, p 85

Vereecken H, Weynants M, Javaux M et al (2010) Using pedotransfer functions to estimate the van Genuchten-Mualem soil hydraulic properties: a review. Vadose Zone J 9:795–820

Chapter 9
Indicators of Soil Physical Quality: From Simplicity to Complexity

Alvaro Pires da Silva, Ary Bruand, Cássio Antônio Tormena, Euzebio Medrado da Silva, Glenio Guimarães Santos, Neyde Fabíola Balarezo Giarola, Rachel Muylaert Locks Guimarães, Robélio Leandro Marchão, and Vilson Antônio Klein

Abstract In working with soil physics, getting new answers to the same questions is a challenge. As soil physicists, we are always hoping to find new ways of understanding such a complex soil science area. In this chapter, we will discuss some of the

A.P. da Silva (✉)
Departamento de Ciência do Solo, Escola Superior de Agricultura Luiz de Queiroz – Universidade de São Paulo, Av. Pádua Dias, 11, Cx.P. 9, 13418-900 Piracicaba, São Paulo, Brazil
e-mail: apisilva@usp.br

A. Bruand
Institut des Sciences de la Terre d'Orléans (ISTO) UMR7327, Université d'Orléans, CNRS/INSU, BRGM, Université François Rabelais – Tours, 1A, Rue de la Férolerie 45071, Orléans, Cedex 2, France

C.A. Tormena
Departamento de Agronomia, Universidade Estadual de Maringá, Av. Colombo 5790, 87020-900 Maringá, Paraná, Brazil

E.M. da Silva • R.L. Marchão
Empresa Brasileira de Pesquisa Agropecuária – Embrapa Cerrados, BR 020, km 18, 73310-970 Planaltina, Distrito Federal, Brazil

G.G. Santos
Escola de Agronomia e Engenharia de Alimentos – Universidade Federal de Goiás, Rodovia Goiânia-Nova Veneza, km 0, Cx.P. 131, 74.001-970 Goiânia, Goiás, Brazil

N.F.B. Giarola
Departamento de Ciência do Solo e Engenharia Agrícola, Universidade Estadual de Ponta Grossa, Av. Gal. Carlos Cavalcanti 4748, 84030-900 Ponta Grossa, Paraná, Brazil

R.M.L. Guimarães
Departamento de Agronomia, Universidade Tecnológica Federal do Paraná, Campus Pato Branco, Via do Conhecimento Km 1, 85503-390 Pato Branco, Paraná, Brazil

V.A. Klein
Faculdade de Agronomia e Medicina Veterinária, Universidade de Passo Fundo, BR 285, São José, 99052-900 Passo Fundo, Rio Grande do Sul, Brazil

W.G. Teixeira et al. (eds.), *Application of Soil Physics in Environmental Analyses: Measuring, Modelling and Data Integration*, Progress in Soil Science, DOI 10.1007/978-3-319-06013-2_9,

ways to assess soil physical quality for crop growth, using ascending complexity classification, from the simplest to the more complex soil physical indicators for crop growth.

Keywords Soil structure • Visual evaluation • Relative compaction • S index • Least limiting water range • Soil compaction

9.1 Introduction

> Student: "Dr. Einstein, aren't these the same questions as last year's [physics] final exam?
> Dr. Einstein: "Yes; But this year, the answers are different."
>
> Albert Einstein

In working with soil physics, getting new answers to the same questions is a challenge. As soil physicists, we are always hoping to find new ways of understanding such a complex soil science area. In this chapter, we will discuss some of the ways to assess soil physical quality for crop growth, using ascending complexity classification, from the simplest to the more complex soil physical indicators for crop growth.

9.2 Visual Evaluation of Soil Structure

The Visual Evaluation of Soil Structure (VESS) method is a combined visual and tactile assessment of soil in terms of structure, root growth, and surface condition, which offers a holistic means of assessing soil physical quality for optimal crop growth. VESS is an evolution of the Peerlkamp test, which was used to visually evaluate soil structure by attributing scores of 1 (worst quality) to 9 (best quality). The scores are related to the soil organic matter and clay content as well as to crop performance (Peerlkamp 1959). It was from this test that Ball et al. (2007) developed a method for evaluating the quality of soil structure, named the "Visual Soil Structure Quality Assessment" (VSSQA). These authors developed a chart to visually assist the user with scoring the soil structure. Further improvements to the method were made by Guimarães et al. (2011). This improved method and its accompanying evaluation chart became called the Visual Evaluation of Soil Structure (VESS).

The VESS method consists of scoring the structural quality of the topsoil by observing attributes such as size, shape, strength and colour of the aggregates, the presence of roots inside and outside aggregates, and the number and size of visible pores. These attributes are compared with a chart (Guimarães et al. 2011) that contains images from different scores (index) of soil structural quality. The final score is computed by averaging the grades, weighted by the thickness of the layer where they occur, and fitting them into the chart.

To conduct an analysis using VESS, one must first select a uniform area of crop, or an area where there is a suspicion of compaction. A minimum of 10 replications

Fig. 9.1 Soil slices after VESS evaluation presenting (**a**) one layer, (**b**) two layers, and (**c**) three layers

should be made. Samples can be taken any time of year, but moisture content is important when conducting the test. The soil cannot be too wet during the evaluation, as the process of extracting the soil slice can compact the soil; if the soil is too dry, however, the pit may become difficult to dig and soil will be too hard to manipulate accurately by hand. Usually, the best time to evaluate is when the soil is friable; to assess friability, the 'worm test' can be performed. For silty soils, roll a 'worm' 10 mm wide × 40 mm long between the palms of your hands (7 mm × 40 mm for clayey soils), if this can be done without the 'worm' cracking the soil is too wet. If the worm cracks when it is 10 mm wide for silty soils and 7 mm wide for clayey soils, the soil is suitable for conducting the test (Shepherd 2009).

To analyze soil quality using VESS, dig out a slice of soil, using a straight spade of 25 cm deep, 20 cm wide, and ~10–15 cm thick. After taking the slice of soil, measure its depth. Using your hands, slowly start to break-up the soil slice, respecting the natural fracture lines between aggregates; do this movement to the sides of the slice to avoid making the slice taller than it is. Look for layers present in soil slices (Fig. 9.1) after fragmentation with different numbers of layers. Gently manipulate the block using both hands to reveal any cohesive layers or clumps of aggregates.

Match the soil with the chart and compare the categories. Size, strength, porosity, roots, and color are some of the parameters used to give the soil structure a score. The scent of the soil after the break-up is an important aspect of the test as well; the presence of anaerobic zones is suggested if the soil presents a rotten egg-like or sulphurous odor. Measure the length of each layer and assign each of them a score. If the category is ambiguous, a break-up of major aggregates can be performed: break larger pieces apart and fragment them until an aggregate size of 1.5–2.0 cm is achieved. Look to the shape, porosity, roots, and size, and compare the latter with the last column of the chart. Scores range from 1 to 5 (good to poor structural quality, respectively); however, the soil may fit between soil structural quality scores if it demonstrates the properties of two categories. The chart can be printed from www.sruc.ac.uk/vess.

To calculate the overall score for the samples above (Fig. 9.1) when two or more layers of distinct structures are present within the soil slice, multiply the score of each layer by its length and divide the product by the overall depth. Repeat the same for the other layers and sum the results. Scores of 1–2 are acceptable, a score of three signals that changes should be made in the long term, and scores of 4–5 require an immediate change of management.

VESS has proven to be one of the simplest methods for the semiquantitative assessment of soil quality, which includes a variety of aspects of soil structure and rooting. With VESS, one is able to distinguish topsoil layers (30 cm) with differing structures, and to evaluate soil layers individually rather than by the weighted average of the total soil sample. These features can improve the choice of management methods to preserve or improve soil quality (Giarola et al. 2013).

VESS enables the evaluation of current soil management by pinpointing specific problems such as compaction, impeded drainage, erosion, and restrictions to roots. It may also be possible to use the VESS method to predict subsoiling requirements, but only in conjunction with other soil physical parameters. A valuable aspect of the methodology is its ability to include observations of unusual features, such as fauna and residues, in the detection of layering.

VESS is a simple, low-cost, reliable, and accurate method, which quickly produces results that are understood by researchers, technical advisors, and farmers (Giarola et al. 2013). However, VESS requires considerable knowledge of pedology and requires field experience.

The VESS method provided the first opportunity for performing small-scale topsoil assessments relevant to agronomy. However, the skill of the operator is important for the successful application of the VESS method. The better the operator's knowledge of soil, especially soil structure, the greater the chance that the scores will be accurate. Standardizing the block-breaking procedure to produce the aggregates is the main difficulty for non-experienced personnel, due to lack of basic knowledge about soil structure and the visualization of weakness planes. Effective teaching is therefore essential, as well as confidence-building and motivational development of the trainee.

The methodology will always require field training and some experience for its effective use. It is of particular value to researchers to understand soil conditions, to identify appropriate locations of soil measurements and sampling, and to understand soil and crop yield variability. In the Brazilian climate, the soil is most often too dry for convenient sampling, so strong laborers are required to dig and collect the samples. For example, one case required approximately 20 h to evaluate 36 sampling points. However, digging small trenches for soil block removal increases the area of disturbed soil. Unfortunately this is unavoidable, as the VESS method requires soil block extraction.

In small experimental plots, it is necessary to take more than one slice to obtain a representative sampling. The requirement for separate access holes and the foot traffic involved in the process of sample extraction causes a significant area of the plot to be destroyed. The score (Sq) is good at revealing the gradient in structure under no-tillage, increasing from Sq 1 at the surface to Sq 3 or 4 (or greater) at the base of the topsoil layer (ca. 25 cm). This gradient arises from the very intensive wetting and drying cycles, biological activity (including roots), tillage by the seeder coulter and unrelieved compaction at depth.

9.3 Relative Compaction

The arrangement of the soil solid particles represents the soil structure and, therefore, the bulk density, which is defined as the ratio of dry soil mass to bulk soil total volume. The bulk density from cultivated soil ranges from 0.9 to 1.8 Mg m^{-3} according to soil texture and organic matter content. Soils with higher clay and organic matter content exhibit lower bulk density. Clayey soils commonly have a large amount of extremely stable soil microaggregates (>1 mm), that do not allow solid particle accommodation, which added to the microaggregates internal porosity is responsible for this lower bulk density.

The bulk density is a soil property that shows a low coefficient of variation (<10 %), which can be attributed to the fact that one sample can blend areas with higher and lower bulk density. The low variability allows a good representation of this soil property with a reduced number of samples. However, these variations in bulk density according to soil type, texture, and organic matter content, make the use of this property difficult as an indicator of physical quality of cultivated soils.

In this context, the Proctor test, by its low cost, ease, and simplicity of realization, is an alternative to determining the maximum soil bulk density (or bulk density reference) for each soil type, which can be used for the evaluation of soil degree of compaction in the field. This test consists of compacting soil samples with a range of water content, through application of 560 kPa of energy. The compaction event can be explained by considering the great influence of the interstitial water on the soil. In the dry range of the compaction curve (Fig. 9.2, Proctor), in which the soil has low humidity, the water is trapped in the pores by the capillary effect, and the water tension tends to cohesively cluster the soil, preventing its disruption and the relative movement of the particles to a new rearrangement. As the soil water content increases, the free water absorbs a considerable portion of the applied compaction energy (Klein 2012).

The soil bulk density values as a function of the gravimetric water content are adjusted by minimizing the sums of squared deviations and obtaining a polynomial equation of the second degree. With this equation, it is possible to mathematically obtain the maximum bulk density and the optimum water content for compaction. For this, the first derivative of this equation is necessary, and gives the point of optimum water content for maximum compaction. Substituting the value of the optimum water content for variable x in the equation, the maximum bulk density for energy applied is obtained. Note that the optimum water content for compaction is actually the level of moisture that is inadequate for performing work with farm machinery, since in this condition changes in soil structure will occur more easily, causing compaction.

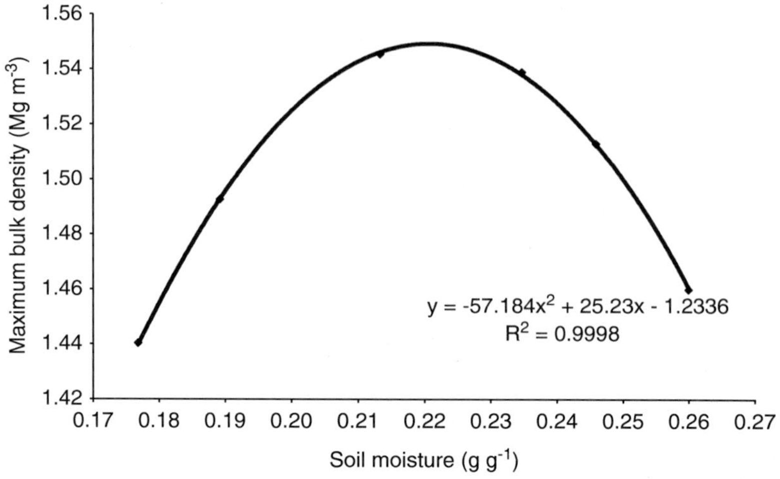

Fig. 9.2 Characteristics of the Proctor compaction curve

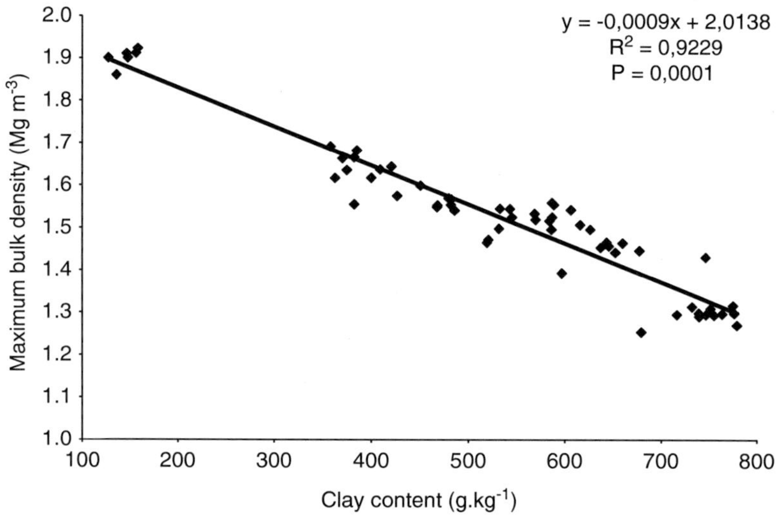

Fig. 9.3 Maximum soil bulk density by Proctor test as a function of clay content, **significant at 0.01 by F test (Marcolin and Klein 2011)

Conducting numerous Proctor tests with soils of different textures permits a negative linear fit of the maximum soil bulk density to be obtained as a function of the increase in clay content, to determine the maximum soil bulk density of the soil as clay content (Fig. 9.3).

The content of organic matter or organic carbon in the soil adversely affects the maximum bulk density, either by its positive effect on the soil structure stability, or by the fact that the organic material has lower density than solid soil particles.

Table 9.1 Reference values for relative compaction from Oxisols under no-tillage conditions in southern Brazil

Value	Compaction level	Observation
<80	Loose	Value below which the plant growth is damaged by excessive porosity aeration and smaller volume of available water
80–90	Not compacted	Range of relative bulk density considered not restrictive to plant growth
90–95	Compacted	Restriction occurs with deformation on the root morphology in moderate levels
>95	Very compacted	Significant deformations in the root system, thickening and deviations in vertical growth, and concentration in the superficial layer

Adapted from Marcolin (2009)

However, for the Oxisols samples studied by Marcolin and Klein (2011), no significant influence was found on the maximum bulk density, once the non-surface layer organic material contents, even in long-term no-tillage, were low (<3 %).

As mentioned earlier, cultivated soils have a bulk density ranging from 0.9 to 1.8 Mg m^{-3} due to their mineralogical and texture characteristics. Soils with the same mineralogical composition and texture may also present a wide range. For example, a clayey soil (≅700 g × kg^{-1} clay) may show a bulk density from 0.92 Mg m^{-3} in natural conditions to 1.3 Mg m^{-3} when intensively tilled (Klein 2012).

These differences in bulk density and the wide range cause difficulties for comparing results, even for other parameters obtained indirectly from the bulk density. The concept of relative bulk density consists of dividing the bulk density from the field by the maximum bulk density obtained by the Proctor test.

$$Relative\ compaction = \frac{soil\ bulk\ density}{maximum\ soil\ bulk\ density} * 100$$

Establishing optimal or critical values for plant development is a constant search. In this sense, through the compilation of research results and the relative compaction in which these studies were conducted, the table below was obtained to provide references values to quantify the physical quality of Oxisols (Table 9.1).

The assessment of Oxisols under no-tillage conditions (Fig. 9.4) demonstrates that the relative compaction in the 0–5 cm layer is below 90; that is, in loose conditions and in deeper layers, the average of most of the samples is above 90, indicating the state of compaction. The challenges ahead are to assess the behavior of plants in different compaction conditions, always considering the soil water content, which minimizes the effects of compaction and draws boundaries and models for soils different from Oxisols.

The opportunity to apply the relative compaction for the evaluation of the soil physical quality, notably in Brazil, is very large and easy to apply. Using equations as shown in Fig. 9.3, the use of relative compaction (RC) is a methodology easily performed, which is inexpensive and conceptually easily to understand; for this reason, it is considered a methodology that will be widely adopted.

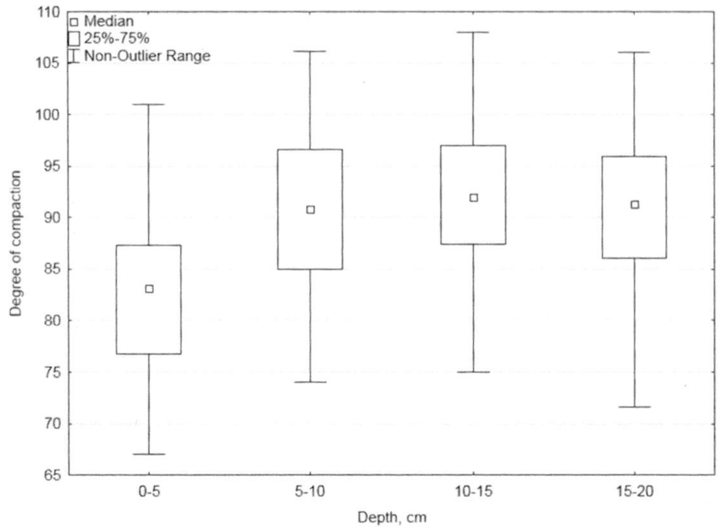

Fig. 9.4 Relative compaction from Oxisol samples (94 per layer) under no-tillage conditions

9.4 *S*-Index

Among indicators of soil quality, the index proposed by Dexter and Bird (2001) and Dexter (2004) enables the physical qualities of soil (workability, permeability, structure stability, etc.) to be investigated, and should be particularly effective for providing information on the soil's hydric functioning. This index is the slope (S) of the soil-water retention curve (SWRC) at its inflection point. It is determined for the SWRC when the gravimetric water content (W), a function of soil-water suction (h) and expressed using the van Genuchten equation, is plotted with the natural logarithm of h. In this study, we use W to denote the gravimetric water content, rather than θ, as in Dexter and Bird (2001), to be more consistent with the literature, since θ usually represents the volumetric water content. As for the van Genuchten equation (van Genuchten 1980), which was written for θ, it remains valid for W.

Dexter (2004) derived the expression of the slope of the SWRC analytically to calculate the value of S, thus leading to the following expression:

$$S = -n(W_s - W_r)\left[1 + \frac{1}{m}\right]^{-(1+m)} \tag{9.1}$$

With m and n, the fitted dimensionless shape parameters of the van Genuchten equation; and W_s and W_r, the saturated and residual gravimetric water contents of the van Genuchten equation, respectively, measured in grams of water per g of oven-dried soil. This characteristic of the SWRC was considered by Dexter (2004) as a physical parameter (S-index) of the physical quality of soil. Dexter (2004) showed that it was related to the texture, bulk density, organic matter content, and root growth of soil. Since its early developments, the *S*-index has been used by many authors (Tormena et al. 2008).

Dexter and Bird (2001), however, noted that there were two possible inflection points depending on whether *W* is plotted against log(*h*) or against *h*. They reported that the two inflection points are in close proximity for soils with a narrow pore-size distribution. This explains why they used the inflection point of curves of *W vs.* log (*h*), believing this was an estimate of air entry into granular materials which were considered in their study (Dexter and Bird 2001). Another point not raised by Dexter and Bird (2001) concerned their choice for computing the slope in a graph *W vs.* ln(*h*) of the *W* curve as a function of *h* according to the van Genuchten equation, instead of the slope of the *W* curve *vs.* ln(*h*), which would have been mathematically more consistent.

In this part of chapter, we discuss the choice of Dexter and Bird (2001) and compare the *S*-index with the slope of the SWRC at its inflection point when it is expressed as a function of the independent variables *h*, ln(*h*), or log(*h*). The equations developed are applied to a non-compacted and compacted soil and the resulting values of the slope are compared to the *S*-index.

9.4.1 *Theory*

Expression of W According to h, ln(h), and log(h)

On the basis of the van Genuchten equation (van Genuchten 1980), *W* can be expressed as:

$$W = \left(W_s - \hat{W}_r\right) \left[1 + (\hat{\alpha}\, h)^{\hat{n}}\right]^{-\hat{m}} + \hat{W}_r \tag{9.2}$$

with *W* the gravimetric soil water content (g g^{-1}); W_s, the measured gravimetric saturated soil water content (g g^{-1}); $\hat{W}_r$, the fitted residual gravimetric soil water content (g g^{-1}); $\hat{\alpha}$, the fitted scaling parameter (kPa^{-1}); and $\hat{n}$ and $\hat{m} = 1 - 1/\hat{n}$, dimensionless fitted shape parameters. In order to facilitate presentation, Eq. 9.2 can be represented as:

$$W = g\left(h \mid W_s,\ \hat{W}_r,\ \hat{n},\ \hat{\alpha},\ \hat{m}\right) \tag{9.3}$$

with *g* being *W* as a function of *h*, given that parameters W_s (g g^{-1}), $\hat{W}_r$ (g g^{-1}), $\hat{n}$ (dimensionless), $\hat{\alpha}$ $\left(\text{hPa}^{-1}\right)$, and $\hat{m}$ (dimensionless) are known. The circumflex on a letter is used to identify a fitted parameter value. $\hat{m}$ can also be fitted, but in this study, it was signed with Mualem (1976) constraint $\left(= 1 - 1/\hat{n}\right)$.

Similarly, W *vs.* ln(h) can be represented by f using Eq. 9.2 as:

$$\begin{aligned}\theta &= f\left(\ln(h) \mid \theta_s,\, \hat{\theta}_{rl},\, \hat{n}_l,\, \hat{\alpha}_l,\, \hat{m}_l\right) W \\ &= f\left(\ln(h) \mid W_{\mathrm{s}},\, \hat{W}_{\mathrm{r}1},\, \hat{n}_1,\, \hat{\alpha}_1,\, \hat{m}_1\right) \text{ for } h \geq 1,\end{aligned} \tag{9.4}$$

with the fitted parameters $\hat{W}_{\mathrm{r}1}$ $(\mathrm{g\,g}^{-1})$, $\hat{n}_1$ (dimensionless), $\hat{\alpha}_1$ $\left(\mathrm{hPa}^{-1}\right)$, and $\hat{m}_1$ (dimensionless), while W *vs.* log(h), can be represented by k, using Eq. 9.2, as:

$$W = k\left(\log(h) \mid W_{\mathrm{s}},\, \hat{W}_{\mathrm{r}1},\, \hat{n}_1,\, \hat{\alpha}_2,\, \hat{m}_1\right) \text{for}\, h \geq 1, \tag{9.5}$$

with $\hat{\alpha}_2$ the only new fitted parameter such as $\hat{\alpha}_2 = \hat{\alpha}_1 \ln 10$, the other fitted parameters being identical to those determined for f.

Derivation of the SWRC to Obtain the Inflection Point

Taking Eqs. (9.3), (9.4), and (9.5) as general representations of the WRC and using Eq. 9.4, we can write the following derivatives:

$$\dot{g} = \frac{dW}{dh} = -\hat{m}\,\hat{n}\left(\hat{\alpha}^{\hat{n}}\right)\left(W_{\mathrm{s}} - \hat{W}_{\mathrm{r}}\right)\left\{\left[1 + (\hat{\alpha}\,h)^{\hat{n}}\right]^{-\hat{m}-1}\right\} h^{\hat{n}-1} \tag{9.6}$$

with $\dot{g}$ the first derivative of W in relation to h,

$$\begin{aligned}\dot{f} &= \frac{dW}{d(\ln(h))} \\ &= -\hat{m}_1\hat{n}_1\left(\hat{\alpha}_1^{\hat{n}_1}\right)\left(W_{\mathrm{s}} - \hat{W}_{\mathrm{r}1}\right)\left\{\left[1 + (\hat{\alpha}_1 \ln(h))^{\hat{n}_1}\right]^{-\hat{m}_1-1}\right\} \ln(h)^{\hat{n}_1-1}\end{aligned} \tag{9.7}$$

with $\dot{f}$ the first derivative of W in relation to ln (h), and:

$$\begin{aligned}\dot{k} &= \frac{dW}{d(\log(h))} \\ &= -\hat{m}_1\hat{n}_1\left(\hat{\alpha}_2^{\hat{n}_1}\right)\left(W_{\mathrm{s}} - \hat{W}_{\mathrm{r}1}\right)\left\{\left[1 + (\hat{\alpha}_2 \log(h))^{\hat{n}_1}\right]^{-\hat{m}_1-1}\right\} \log(h)^{\hat{n}_1-1}\end{aligned} \tag{9.8}$$

with $\dot{k}$ the first derivative of W in relation to log (h).

It is important to state that $\dot{f} = dW/d(\ln(h))$ cannot be computed by simply applying the chain rule from Eq. 9.3, because the parameters determined by fitting either g (Eq. 9.3) or f (Eq. 9.4), subjected to Eq. 9.2, are not necessarily the same. This can be also said for functions g (Eq. 9.3) and k (Eq. 9.5), except that, here, the only difference between $\dot{f}$ and $\dot{k}$ is the magnitude of the scaling parameters $\hat{\alpha}_1$ and $\hat{\alpha}_2$.

It is known that any continuous and differentiable mathematical function has its inflection points located where the second derivative is null throughout its real domain. Thus, at the inflection points for function g, we can set:

$$\ddot{g} = \frac{d^2W}{dh^2} = -\hat{m}\,\hat{n}\,(\hat{\alpha}^{\hat{n}})(W_s - \hat{W}_r)\left\{(-\hat{m}\,\hat{n} - \hat{n})(\hat{\alpha}^{\hat{n}})\left[1+(\hat{\alpha}\,h)^{\hat{n}}\right]^{-\hat{m}-2} h^{2\hat{n}-2} + (\hat{n}-1)(h^{\hat{n}-2})\left[1+(\hat{\alpha}\,h)^{\hat{n}}\right]^{-\hat{m}-1}\right\} = 0$$

$$= \frac{d^2\theta}{dT^2} = -\hat{m}\,\hat{n}\,\hat{\alpha}_{\hat{n}}\,(\theta_s\text{-}\hat{\theta}_r)\left\{(-\hat{m}\,\hat{n} - \hat{n})(\hat{\alpha}^{\hat{n}})\left[1+(\hat{\alpha}\,T)^{\hat{n}}\right]^{-\hat{m}-2} T^{2\hat{n}-2} + (\hat{n}-1)(T^{\hat{n}-2})\left[1+(\hat{\alpha}\,T)^{\hat{n}}\right]^{-\hat{m}-1}\right\} = 0 \tag{9.9}$$

with $\ddot{g}$, the second derivative of W in relation to h. After simplifying Eq. (9.9), we obtain:

$$(-\hat{m}\,\hat{n} - \hat{n})(\hat{\alpha}^{\hat{n}})\left[1+(\hat{\alpha}\,h)^{\hat{n}}\right]^{-\hat{m}-2}(h^{2\hat{n}-2}) + (\hat{n}-1)(h^{\hat{n}-2})\left[1+(\hat{\alpha}\,h)^{\hat{n}}\right]^{-\hat{m}-1} = 0 \tag{9.10}$$

Equation 9.10 can be solved for h to obtain the precise location of its inflection point $(h)_i$ as follows:

$$(h)_i = \frac{1}{\hat{\alpha}}(\hat{m})^{\frac{1}{\hat{n}}} \tag{9.11}$$

Similarly, we can compute the second derivative of Eqs. (9.7) and (9.8) to obtain:

$$(\ln(h))_i = \frac{1}{\hat{\alpha}_1}(\hat{m}_1)^{\frac{1}{\hat{n}_1}} \tag{9.12}$$

with $(\ln(h))_i$, the inflection point of W *vs*. $\ln(h)$, and:

$$(\log(h))_i = \frac{1}{\hat{\alpha}_2}(\hat{m}_1)^{\frac{1}{\hat{n}_1}} \tag{9.13}$$

with $(\log(h))_i$, the inflection point of W *vs*. $\log(h)$.

Calculation of the Slope at the Inflection Point of the SWRC

The slope, S_h, from function g (Eq. 9.3) at its inflection point (Eq. 9.11) is obtained by substituting Eq. 9.11 into Eq. 9.6, yielding:

$$S_h = -(\hat{\alpha})(\hat{n} - 1)(W_s - \hat{W}_r)(\hat{m})^{\hat{m}}(1 + \hat{m})^{-\hat{m}-1}. \tag{9.14}$$

Similarly, the slope, $S_{\ln(h)}$, from function f (Eq. 9.4) at its inflection point (Eq. 9.12), is obtained by substituting Eq. (9.12) into Eq. (9.7):

$$S_{\ln(h)} = -(\hat{\alpha}_1)(\hat{n}_1 - 1)(W_s - \hat{W}_{r1})(\hat{m}_1)^{\hat{m}_1}(1 + \hat{m}_1)^{-\hat{m}_1-1}. \tag{9.15}$$

The slope, $S_{\log(h)}$, from function k (Eq. 9.5) at its inflection point (Eq. 9.13) is obtained by introducing Eq. (9.13) into Eq. (9.8):

$$S_{\log(h)} = -(\hat{\alpha}_2)(\hat{n}_1 - 1)(W_s - \hat{W}_{r1})(\hat{m}_1)^{\hat{m}_1}(1 + \hat{m}_1)^{-\hat{m}_1-1}. \tag{9.16}$$

9.4.2 Application to a Case Study

The equations developed in this study were applied to samples from a cultivated soil where compacted layers were identified (Santos et al. 2011). The soil studied was a clayey Ferralsol (Oxisol) according to the IUSS-WRB (2006) soil classification. It was located on a private farm (latitude 16.493246 S, longitude 49.310337 W, and altitude 776 m), near the "Embrapa Arroz e Feijão Agricultural Research Center", at Santo Antônio de Goiás, GO, Brazil. The native vegetation was a typical Cerrado until 1985. After clearing the land, the soil was occupied by annual crops with conventional tillage for 2 years and then by a pasture of Brachiaria decumbens Stapf cv Basilisk. The soil was managed according to intensive animal grazing without any addition of fertilizer, which led to a compaction of the topsoil. In 2006, soil cores were collected with stainless steel 100 cm^3 cylinders (diameter = 5.1 cm, height = 5.0 cm) in the compacted 0–5 cm and non-compacted 70–75 cm layers (Table 9.2). The higher bulk density found in the 0–5 cm layer is caused by soil compaction, since this type of soil, under native vegetation, exhibits a uniform bulk density profile according to depth, with a bulk density near to 1.0 g cm^{-3} (Santos et al. 2011).

Gravimetric water contents (W in g g^{-1}) at −10, −30, −60, −100, −330, −800, −4,000, −10,000, and −15,000 hPa were determined in triplicate for the two layers studied (Table 9.3), using the centrifuge method (Bruand and Prost 1987). A SWRC was fitted using the van Genuchten equation (van Genuchten 1980) (see Eq. 9.2) to the different water contents measured for the compacted and non-compacted layers, using h, ln(h), or log(h) as the independent variable. The Solver routine embedded in Microsoft Excel was used to obtain the fitting parameters $\hat{W}_r$, $\hat{\alpha}$, $\hat{n}$, and $\hat{m}$ (Table 9.4). During the fitting process, W_s was taken as the mean value of the three saturated water contents measured (Reatto et al. 2008): 0.367 g g^{-1} and 0.544 g g^{-1} for the compacted and non-compacted layer, respectively, and therefore was not adjusted.

At this point, it should be remembered that Dexter and Bird (2001) and Dexter (2004) derived the S-index formulation from the slope of SWRC plotted in a ln scale, and the result was transformed to a log scale by multiplying it by ln 10; this log scale was then used. In order to compare and discuss the location of the inflection point according to the independent variable used, we applied the equations developed here

Table 9.2 Principal physical and chemical properties of the 0–5 cm compacted and 70–75 cm non-compacted layers selected in the studied soil

Soil	Particle size distribution[a]			Organic carbon[a]	Bulk density[b]
	Clay	Silt	Sand		
Compacted	485	71	444	0.70	1.27
Non-compacted	549	72	380	0.16	1.03

[a] $g\ kg^{-1}$
[b] $g\ cm^{-3}$

Table 9.3 Gravimetric soil water content (W g g^{-1}) of the cores originating from the 0–5 cm compacted (C) and 70–75 cm non-compacted (NC) layers according to the suction (hPa)

Suction	W – Compacted layer			W – Non-compacted layer		
(hPa)	Replicate 1	Replicate 2	Replicate 3	Replicate 1	Replicate 2	Replicate 3
0	0.366	0.356	0.380	0.558	0.549	0.523
10	0.356	0.346	0.355	0.540	0.542	0.516
30	0.332	0.324	0.306	0.456	0.455	0.432
60	0.297	0.290	0.281	0.337	0.360	0.333
100	0.277	0.287	0.271	0.287	0.278	0.277
330	0.237	0.242	0.237	0.240	0.231	0.227
800	0.222	0.230	0.226	0.218	0.213	0.214
4,000	0.199	0.206	0.204	0.201	0.195	0.195
10,000	0.185	0.190	0.190	0.193	0.184	0.187
15,000	0.178	0.180	0.181	0.182	0.176	0.175

and those of Dexter and Bird (2001) and Dexter (2004) to water retention properties found for compacted and non-compacted soils (Table 9.4).

The S-index computed using Eq. 9.1 and multiplied by ln 10, according to Dexter (2004), was 0.082 and 0.329 for the compacted and non-compacted soils, respectively. Using Eq. 9.16, the slope at the inflection point of the SWRC, expressed according to log(h) as the independent variable, was 0.081 and 0.326 for the compacted and non-compacted soils, respectively. These values are very close to the S-index computed as described by Dexter (2004). Thus, using an equation of W fitted with h as the independent variable and plotted with log(h) as abscissa, or an equation of W fitted with log(h) as the independent variable and plotted according to log(h), the slopes of the two curves at the inflection point are very similar. This could be expected, since the experimental points remain at the same place in the W - log(h) graph regardless of the independent variable used for the equation to describe the SWRC. Consequently, the slope at the inflection point of the SWRC computed according to Dexter (2004) who leads to the S-index and is used by many authors, would have been similar using Eq. 9.16 instead of Eq. 9.1.

On the other hand, the location of the inflection point of the curve of W *vs.* h, and the slope of the curve at this point, have more physical meaning than the corresponding values computed by Dexter (2004). The value of h at the inflection point can be considered as the "breakthrough" matrix potential at which air

Table 9.4 Fitted parameter values for W *vs.* h, $\ln(h)$, or $\log(h)$, and corresponding inflection points and S-values for the 0–5 cm compacted (C) and 70–75 cm non-compacted (NC) layers

	Independent variable[h]						Dexter (2004)	
	h		$\ln(h)$		$\log(h)$			
Variables	C	NC	C	NC	C	NC	C	NC
$W_s (g\,g^{-1})$	0.367 ± 0.012	0.544 ± 0.018	0.367 ± 0.012	0.544 ± 0.018	0.367 ± 0.012	0.544 ± 0.018	0.367 ± 0.012	0.544 ± 0.018
W_r $(g\,g^{-1})$	0.160 ± 0.010	0.192 ± 0.004	0.147 ± 0.012	0.188 ± 0.004	0.147 ± 0.012	0.188 ± 0.006	0.160 ± 0.010	0.192 ± 0.004
n[a]	1.314 ± 0.045	2.057 ± 0.088	3.182 ± 0.254	6.396 ± 0.364	3.182 ± 0.254	6.396 ± 0.364	1.314 ± 0.045	2.057 ± 0.088
α[b]	0.057 ± 0.009	0.032 ± 0.002	0.227 ± 0.006	0.263 ± 0.003	0.524 ± 0.013	0.606 ± 0.008	0.057 ± 0.009	0.032 ± 0.002
m[1]	0.239 ± 0.025	0.514 ± 0.020	0.686 ± 0.023	0.844 ± 0.008	0.686 ± 0.023	0.844 ± 0.008	0.239 ± 0.025	0.514 ± 0.020
Suction at the inflection point[c]	5.876	22.421	3.948	3.699	1.696	1.606	1.715	1.632
Slope at the inflection point[d]	0.0020	0.0046	0.035	0.142	0.0805	0.3261	0.0816	0.329
Equivalent pore diameter at the inflection point[e]	510	134	60	74	60	74	58	70
Water content at the inflection point[f]	0.300	0.394	0.266	0.373	0.266	0.365	0.266	0.365
RMSE[g]	0.0065	0.0114	0.0065	0.0106	0.0065	0.0106	0.0065	0.0114
R^2	0.987	0.991	0.988	0.992	0.988	0.992	0.987	0.988

[a]Dimensionless

[b]Units in hPa^{-1} for h; ln hPa^{-1} for ln h; and log hPa^{-1} for log h

[c]Units in hPa for h; ln hPa for ln h; and log hPa for log h

[d]Units for S_h $(g\,g^{-1}\,hPa^{-1})$; $S_{\ln h}$ $(g\,g^{-1}\,\ln hPa^{-1})$; or $S_{\log h}$ $(g\,g^{-1}\,\log hPa^{-1})$

[e]Unit in μm

[f]Unit in $g\,g^{-1}$

[g]$RMSE = \sqrt{1/N\sum_{i=1}^{n}\left(\hat{W}_i - W_i\right)^2}$ in $g\,g^{-1}$

[h]The standard errors for W_s were calculated directly from the measured values. Those for W_r, n, α, and m originated from the analysis of variance of errors due to regression when fitting these parameters

penetrates throughout the soil. The slopes at the inflection point of the SWRC, using Eq. 9.14 with h as the independent variable, were 0.0020 and 0.0046 for the compacted and non-compacted soil, respectively. These values are 41 and 72 times smaller than the corresponding S-index values (Table 9.4), respectively. Suction at the corresponding inflection point using Eq. 9.11 was 6 and 22 hPa for the compacted and non-compacted soil, respectively; though, according to Dexter (2004) these values were 52 and 43 hPa (Table 9.3).

Using Jurin's law (Bruand and Prost 1987) we computed the equivalent pore diameter corresponding to the suction at the inflection point of the SWRC (Table 9.4). The results showed a close equivalent pore diameter for compacted and non-compacted soil at the inflection point when the SWRC was plotted with ln(h) or log(h) as the independent variable (60 and 74 μm, respectively) and according to Dexter (2004) (58 and 70 μm, respectively). On the other hand, the equivalent pore diameter at the inflection point of the SWRC was about four times higher for compacted soil (510 μm) than for the non-compacted soil (134 μm) when the SWRC was plotted with h as the independent variable (Table 9.4).

In contrast to what is indicated by the S-index, however, air would penetrate throughout the soil at a smaller suction, and, consequently, for a larger equivalent pore diameter for compacted than for non-compacted soil. This result may appear surprising since compaction leads to smaller porosity with a shift of the inflection point on the SWRC to a larger suction. The effects of compaction on pore geometry are difficult to understand, since they depend on the structure and related pore types prior to compaction, on soil composition and water content, and on the intensity of compaction.

Beneath native vegetation, the soil studied had a weak macrostructure and a pronounced granular structure at the micrometer scale (Balbino et al. 2002). Since the structure of non-compacted soil is considered to be similar to the structure under native vegetation, its theoretical SWRC would be a bimodal curve with two inflection points: (i) one corresponding to pore draining resulting from micro-aggregate assemblage and occurring for a very low suction of several hPa, such as for coarse sandy soils, and (ii) one corresponding to pore draining resulting from the assemblage of elementary particles in micro-aggregates and occurring for values of several hundred hPa. Because of the difficulty to correctly measure water retention of the soils studied at several hPa, only the second inflection point is usually measured (Balbino et al. 2002).

When soil is compacted, the pores resulting from the assemblage of micro-aggregates are transformed into smaller pores. The resulting SWRC contains one inflection point which is related to a continuous distribution of equivalent pore diameters from the smaller pores, which were distorted by compaction, to those resulting from the assemblage of the elementary particles in microaggregates. Figure 9.5, which is based on the results of several studies on similar soils, illustrates how using such a transformation of porosity makes it possible to pass from a SWRC, with a given inflection point and its related equivalent pore diameter for a non-compacted soil, to another SWRC, with an inflection corresponding to a larger equivalent pore diameter for compacted soil.

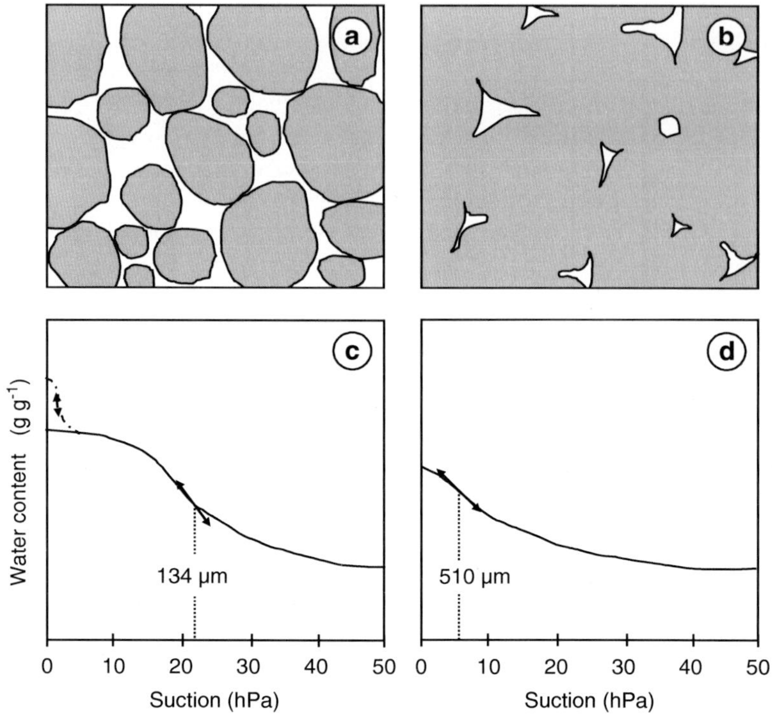

Fig. 9.5 Schematic representation of the structure of the non-compacted (**a**) and compacted soil (**b**), and the soil water retention curve corresponding to the non-compacted soil (**c**). The portion of the curve that relates to the pores resulting from the assemblage of the micro-aggregates (*white* in (**a**) and *dashed curve* in (**c**)) was not measured. The soil water retention curve of the compacted soil (**d**) is shown with the value of the equivalent pore diameter in μm at the inflection point

Finally, our results question the value of S as a possible index to determine the physical quality of soil. The values of h at the inflection point determined for compacted and non-compacted soil are low, thus corresponding to water content close to saturation, which should not be optimal for soil tillage.

The expression of the SWRC according to ln(h) or log(h) as the independent variable, instead of h, leads to different values of the S-index. Computing the S-index when the SWRC is expressed with h as the independent variable is both mathematically and physically consistent. We also show that, independent of the consistency of the approach, the discussion of the physical properties of the soil can be limited according to the independent variable used. For the soil selected, our results, in fact, show that calculation of the S-index when it is expressed with h as the independent variable significantly increases the relevance of the analysis compared to the range of the S-indices when it is expressed as proposed by Dexter and Bird (2001). Further work will aim to determine in which proportion the S-index is affected for a large range of soils and to verify if the use of h as the independent variable effectively increases sensitivity of the analysis.

9.5 Least Limiting Water Range (LLWR)

The term "Least Limiting Water Range" (LLWR) was used (Letey 1985; da Silva et al. 1994) to define the region bounded by the upper and lower soil water content over which water, oxygen, and mechanical resistance become major limitations for root growth. It is delimited by the supply of oxygen to the roots at the wet end, and by the supply of water to the roots and mechanical resistance to root growth at the dry end.

The computation of the LLWR for plant growth requires that functional relationships be obtained between water content (θ) and each of the following properties: water potential, mechanical resistance, and aeration. To compute the LLWR, the matric potential at field capacity and a limiting value of aeration must be defined at the wet end of the range, and a limiting value of mechanical resistance and the matric potential at the permanent wilting point must be defined for the dry end. The water contents are then determined at the limiting values. The smallest range between these limiting water contents is defined as the LLWR. The critical limits defining the LLWR were (1) soil water contents at field capacity ($\psi = -0.01$ MPa) and permanent wilting point ($\psi = -1.5$ MPa); (2) air filled porosity less than 10 %; and (3) soil strength >2.0 MPa.

Undisturbed soil samples are used to determine the functional relationships. The sampling has to encompass the variation of structural conditions, expressed by the soil bulk density, of the soil(s) under study, in order to model properly the water release curve (WRC, θ versus ψ), and the soil resistance curve (SRC, soil strength as a function of soil water content and bulk density). Once the functional relationships have been defined, the LLWR computation is performed, as demonstrated in Fig. 9.6. The critical limits of water contents at field capacity and permanent wilting point are obtained from the WRC (Fig. 9.6a, b) for each soil sample. Similarly, in dry soils, the mechanical resistance may reach limiting values at low values of ψ before the permanent wilting point is reached. In the dry end, depending on the structural condition of the soil, the θ_{WP} can be substituted by the water content at which the soil strength (θ_{SR}) reaches the critical value previously defined (Fig. 9.6d). In soils with degraded structure, the wet end of the LLWR is commonly determined by the θ_{AFP} instead of θ_{FC} (Fig. 9.6e). In these cases, at θ_{FC}, the plant growth is theoretically conditioned by the restricted diffusion of oxygen in the soil. The air-filled porosity is strongly dependent on the soil structural condition, being linear and negatively related with the soil bulk density (ρ_b). Figure 9.3 demonstrates that θ_{AFP} substitutes θ_{FC}, while θ_{SR} substitutes θ_{WP}, starting from values $\rho_b = 1.40$ g cm^{-3}. In this condition, the LLWR is strongly reduced until it reaches a value of zero. The LLWR is zero when the water content values at the wet end and at the dry end are numerically the same. The critical bulk density of the soil (ρ_{bc}) is defined as the value of ρ_{bc} at which LLWR = 0. For the soil illustrated in Fig. 9.6e, the LLWR = 0 when the $\rho_{bc} = 1.55$ g cm^{-3}.

LLWR integrates four static soil characteristics (aeration, field capacity, soil mechanical resistance, and permanent wilting point) into a single variable.

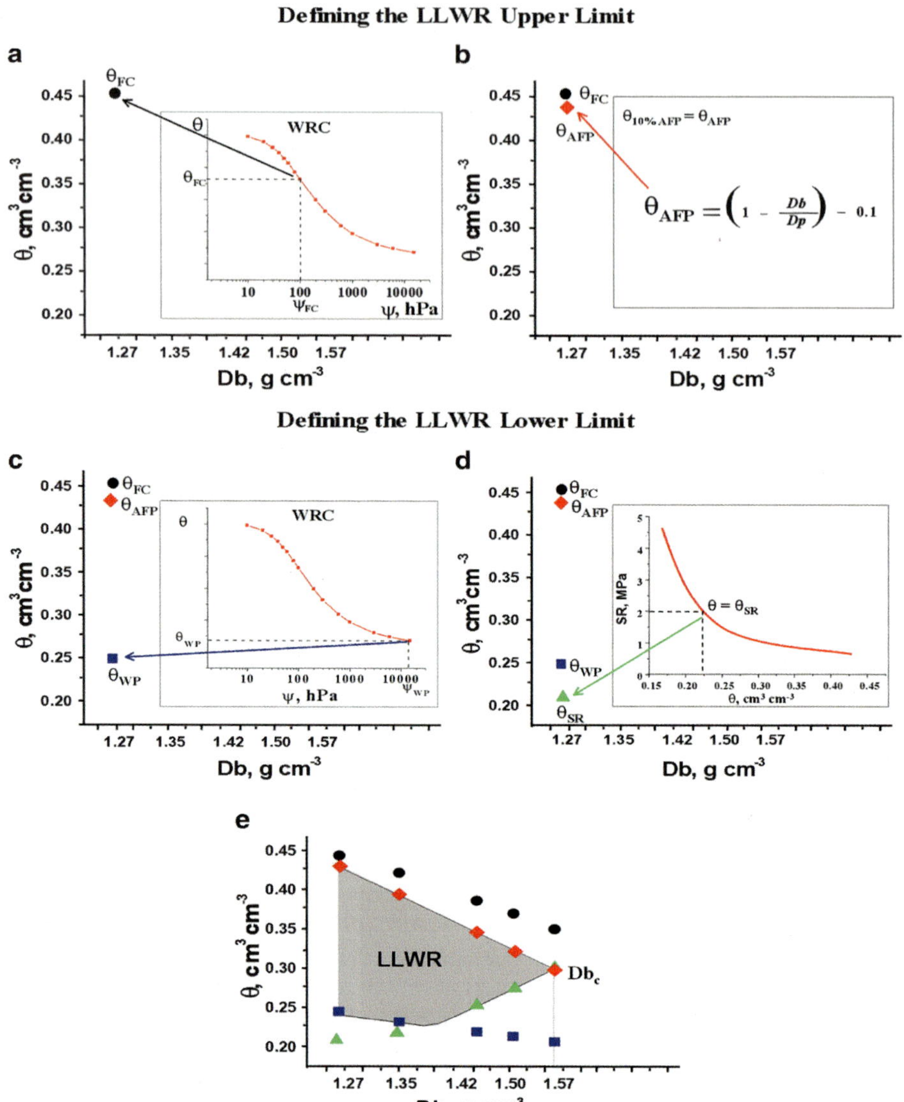

Fig. 9.6 Determination of the critical values of the variables FC, WP, AP, and SR that define the LLWR

Dynamic soil characteristics, such as parameters describing the transport of water or oxygen, are not taken into account. Limiting soil resistance of the soil matrix, and therefore the calculated value of the LLWR, is of limited relevance to compacted soils with an abundance of macropores, because root growth occurs mainly in these macropores, rather than in the soil matrix. Limiting values are time-consuming to measure, and the impact of some characteristics – particularly those of soil

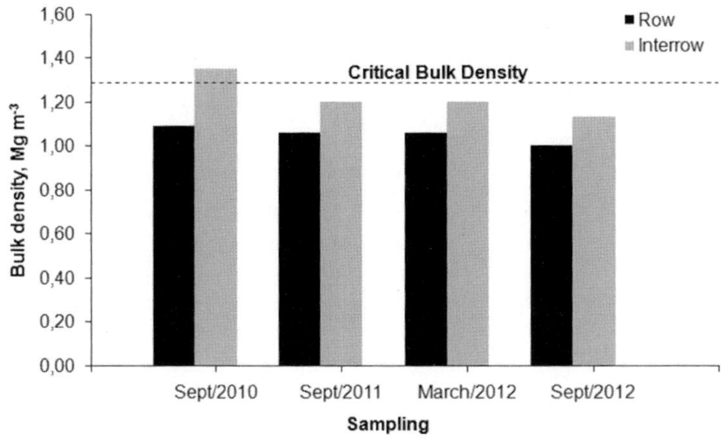

Fig. 9.7 Soil bulk density (ρ_b) in different sampling seasons in an Oxisol under no-tillage. The *dotted line* indicates the critical bulk density equal to 1.29 Mg m^{-3}

resistance and aeration - on the extent of rate reduction of those physiological processes that have the greatest influence on yield is not well-defined. The variation in this reduction among different crops is largely unexplored. Although interaction between limiting characteristics (e.g., the effect of reduced aeration on limiting values of soil resistance) could be accounted for in the concept of the LLWR, this interaction is generally not considered.

An example of an application of the LLWR for the control and monitoring of the soil physical quality is shown in Fig. 9.7. This study was conducted on a commercial grain production farm in northwestern Parana State, in the municipality of Maringa (23°30′40″ S, 51°59′48″ W), in a soil classified as Oxisol (Rhodic Ferralsol), in the clay textural class (800 g kg^{-1} clay content). The land is under a no-tillage system since 1979 with a crop rotation including soybeans, corn, wheat, and oats. Corn and soybeans were sown using furrow opener drills, while oats and wheat were sown with double-disc coulters. The LLWR was evaluated as described by da Silva et al. (1994) and Betioli Junior et al. (2012). The critical bulk density value (ρ_{bc}, the soil bulk density [ρ_b] value in which LLWR = 0) was 1.29 Mg m^{-3}. Sampling for monitoring surface soil physical quality was performed immediately after the harvest of corn in September 2010, September 2011, and September 2012, and after the soybean harvest in March 2012, at 40 locations along a transect crossing 40 crop rows. At each location (40 in rows and 40 in the adjacent inter-row positions), at 0–10 cm depth, undisturbed soil cores (50 mm high × 70 mm in diameter) were collected. We sampled this layer because it is the most biologically active zone, and where soil disturbance associated to furrow opening during seeding occurs under no-tillage.

The critical bulk density estimated from LLWR can be used as a reference for the monitoring of soil physical quality and only requires the measurement of soil ρ_b. The determination of ρ_b is simple, requires minimal instrumentation, is relatively inexpensive, and can be performed at the farm level. Increasing values of ρ_b over ρ_{bc} values are indicative of greater physical limitations of the soil, since excessive soil resistance to root penetration or reduced aeration may limit plant growth due to the natural variability of soil water content. As shown in Fig. 9.7, in no-till there is a consistent and systematic difference between inter-row and row conditions, with soil physical quality being more favorable in the row area. The favorable soil physical conditions within the rows may be attributed to the loosening action of the chisel coulters of the no-till seeder, which open the soil to a depth of 0.12 m. The average values of ρ_b indicated that only in September 2010 did ρ_b exceed ρ_{bc} in the inter-row positions, yet this value was not high enough to require remedial action. In all subsequent sampling, ρ_{bc} systematically exceeded ρ_b, suggesting that a reduction in ρ_b within rows may due to soil's physical resilience associated with wetting and drying cycles, effects of root crops, and the residual effect of the disturbance of the soil by chisel coulters.

To revisit Einstein´s quote, these are the answers for this year's "exam". For sure, soil physics will continue to attract new people, and much improved answers will be available for the same questions in the near future.

References

Balbino LC, Bruand A, Brossard M et al (2002) Changes in porosity and microaggregation in clayey Ferralsols of the Brazilian Cerrado on clearing for pasture. Eur J Soil Sci 53:219–230

Ball BC, Batey T, Munkholm L (2007) Field assessment of soil structural quality – A development of the Peerlkamp test. Soil Use Manage 23:329–337

Bruand A, Prost R (1987) Effect of water content on the fabric of a soil material: an experimental approach. J Soil Sci 38:461–472

Betioli Júnior E, Moreira WH, Tormena CA, Ferreira CJB, da Silva AP, Giarola NFB (2012) Intervalo hídrico ótimo e grau de compactação de um Latossolo Vermelho após 30 anos sob plantio direto. Revista Brasileira de Ciência do Solo 36:971–982. doi:10.1590/S0100-06832012000300027

da Silva AP, Kay BD, Perfect E (1994) Characterization of the least limiting water range of soils. Soil Sci Soc Am J 58:1775–1781

Dexter AR (2004) Soil physical quality: part I, theory, effects of soil texture, density, and organic matter, and effects on root growth. Geoderma 120:201–214

Dexter AR, Bird NRA (2001) Methods for predicting the optimum and the range of soil water contents for tillage based on the water retention curve. Soil Tillage Res 57:203–212

Giarola NFB, Silva AP, Tormena CA, Guimarães RML, Ball BC (2013) On the visual evaluation of soil structure: the Brazilian experience in oxisols under no-tillage. Soil Tillage Res 127:60–64

Guimarães RML, Ball BC, Tormena CA (2011) Improvements in the visual evaluation of soil structure. Soil Use Manage 27:395–403

International Union of Soil Science Working Group (2006) World reference base for soil resources 2006: a framework for international classification, correlation and communication. FAO, Rome

Klein VA (2012) Física do Solo. EDIUPF, 2ª Ed. 240 p, Pelotas, Brazil

Letey J (1985) Relationship between soil physical properties and crop production. Adv Soil Sci 1:277–294

Marcolin CD (2009) Indicadores da qualidade física de solos sob plantio direto obtidos por funções de pedotransferência. Tese (Doutorado em Agronomia) – FAMV-UPF, 183 p

Marcolin CD, Klein VA (2011) Determinação da densidade relativa do solo por uma função de pedotransferência da densidade do solo máxima. Acta Sci Argon 33:349–354

Mualem Y (1976) A new model predicting the hydraulic conductivity of unsaturated porous media. Water Resour Res 12:513–522

Peerlkamp PK (1959) A visual method of soil structure evaluation. Medelingen von de Landbouwhogeschool en Opzoekingstations van de Staat te Gent Deel XXIV 24:216–221

Reatto A, Silva EM, Bruand A et al (2008) Validity of the centrifuge method for determining the water retention properties of tropical soils. Soil Sci Soc Am J 72:1547–1553

Santos GG, Marchão RL, Silva EM et al (2011) Qualidade física do solo sob sistemas de integração lavoura-pecuária. Pesq Agropec Bras 46:1339–1348

Shepherd TG (2009) Visual soil assessment, vol 1. Field guide for cropping and pastoral grazing on flat to rolling country, 2nd edn. Horizons Regional Council, Palmerston North

Tormena CA, Silva AP, Imhoff SC, Dexter AR (2008) Quantification of the soil physical quality of a tropical oxisol using S index. Scientia Agricola 65:56–60

van Genuchten MT (1980) A closed-form equation for predicting the hydraulic conductivity of unsaturated soils. Soil Sci Soc Am J 44:892–898

Chapter 10
Monitoring Soil Water in Irrigated Soils of the Brazilian Semi-arid Region: An Opportunity to Improve Water Use

Luis Henrique Bassoi

Abstract This chapter discusses the Brazilian semi-arid region, which is briefly characterized according to climate, vegetation, geology, soil types, and some soil formation processes. The soil water content in deeper layers of an intensively irrigated area of this region has been monitored at several sites. Excess water for crops and slow drainage caused by dense subsurface layers and crystalline rocks have contributed to water storage at deeper soil depths. Several horticulture crops cultivated in the Brazilian semi-arid region have root systems capable of reaching soil depths of at least 1 m and are drought tolerant. Deficit irrigation should be considered as an opportunity for reduced water application by irrigation systems to enable crops to use the water stored in deep layers of irrigated soils. The results of soil water monitoring in irrigated soils are discussed and suggestions for future research are presented.

Keywords Semi-arid • Crystalline rock • Irrigation • Soil water retention curve • Crop evapotranspiration • Root system • Dense subsurface soil layer • Plinthitic mottle • Soil water content • Microsprinkler • Drip • Deficit irrigation • Plant water stress • Water use

10.1 The Brazilian Semi-arid Region

The Brazilian semi-arid region covers an area of 969,589.4 km^2, with 1.133 municipios within nine states, including Alagoas, Bahia, Ceará, Minas Gerais, Paraíba, Pernambuco, Piauí, Rio Grande do Norte, and Sergipe. The majority of its area lies within the northeast region, with the remainder in the southeast region (north of Minas Gerais State) (Brazil 2005).

L.H. Bassoi (✉)
Brazilian Agricultural Research Corporation (EMBRAPA) – Embrapa Semiárido,
P. O. Box 23, 56302-970 Petrolina, PE, Brazil
e-mail: luis.bassoi@embrapa.br

W.G. Teixeira et al. (eds.), *Application of Soil Physics in Environmental Analyses: Measuring, Modelling and Data Integration*, Progress in Soil Science,
DOI 10.1007/978-3-319-06013-2_10,

10.2 Climate, Vegetation, Geology, and Soil

Rainfall in the Brazilian semi-arid region is marked by inter-annual variability associated with low values of total annual rainfall, which is a major contributing factor in drought occurrence. Average annual temperatures vary between 23 °C and 27 °C. The average relative humidity is 50 %, and the period of insolation reaches values of 2,800 h year^{-1}. The potential evapotranspiration ranges between 1,500 mm year^{-1} and 2,000 mm year^{-1}. The predominant vegetation in the semi-arid region is Caatinga, a unique biome with endemic species (Silva et al. 2010).

The geology in this semi-arid environment is quite variable, but has a predominance of crystalline rocks followed by sedimentary areas. In part of that region, crystalline rocks are covered by thick sandy or sandy–clayey sediments. The topography varies from flat to undulating. The main soil types covering the semi-arid surface according to Brazilian Soil Classification – and their corresponding types in U. S. Soil Taxonomys, include Latossolos – Oxisols (21 %), Argissolos – Ultisols (15 %), Luvissolos – Alfisols (10 %), Planossolos – Alfisols (11 %) , Neossolos – Entisols (34 %), Cambissolos – Inceptisols (4 %), and Vertissolos – Vertisols (1.5 %); respective percentages of soil surfaces are given in parentheses. Other types with minor surface coverage include Chernossolos – Molisols, Plintossolo – Albaquults, and Gleissolo – Entisols (Cunha et al. 2010).

10.3 Irrigated Soils

Initially, the alluvial soils in the Brazilian semi-arid region were used for irrigation purposes; however, these soils became saline because of inappropriate soil and water management. The need for expansion of irrigated areas has led to a demand for deep soils without permeability restrictions. Latossolos (Oxisols) and Argissolos (Ultisols) overlying crystalline rocks and currently represent the largest areas of irrigated soils in the semi-arid region. Several soils with good permeability along the soil profile, which has been irrigated with good quality (C1S1) water from the São Francisco River, show secondary salinization and elevated water tables (Amaral 2011). In many irrigated soils, problems related to drainage have also been reported; these may be linked to the presence of dense subsurface layers, shallow to medium-depth soils, and low permeability (Ribeiro 2010). The Brazilian System for the Classification of Land for Irrigation, with a focus on the semi-arid region (Amaral 2011), has presented an interesting overview of soil classes with and without drainage restrictions. Approximately 40 % of the land has some type of drainage limitation, whereas 47 % has high permeability.

10.4 Dense Subsurface Soil Layers

Fragipans, which are dense subsurface soil layers that severely restrict water and root penetration, occur worldwide. Although this type of soil is considered to be a pedogenic horizon, its precise origin is not well understood. In numerous soils, the restricting layer has a geogenic origin due to limited weathering or lithological discontinuities. The pedogenic accumulation of clay, carbonates, gypsum, iron, silica, or a combination of these can also lead to the formation of hard, dense horizons such as petrocalcic and petrogypsic horizons, duripan, ortstein, placichorizon, plinthite, and fragipan. The formation of these horizons and layers is fairly well understood, with the exception of fragipan. Earth flows triggered by earthquakes, lithologic discontinuities, or lithic–paralithic contacts and repetitive wetting and drying cycles that enable both dispersion and precipitation of clay on mineral grains are among the explanations for fragipan formation (Bockheim and Hartemink 2013).

The Low Plateaus of Pernambuco State is a geoenvironmental unit characterized by a cover layer of gneissic rock from the Precambrian period, with sandy–clayey sediments from the Tertiary–Quaternary. It is a flat surface occupied by soils of moderate to imperfect drainage and is restricted by acidity, low fertility, and physical limitations caused by the occurrence of dense subsurface horizons (Burgos and Cavalcanti 1990). The pedogenic process that contributes to the differentiation of the characteristics of surface horizons in relation to subsurface compaction is known as clay eluviation–illuviation (Silva et al. 2002). The iron concretions found in some soils may have been inherited from the parent materials. During pedogenesis, these soils may have undergone physical and chemical degradation that turned into plinthitic mottles, which were later dissipated in the soil mass. The iron forms were dispelled by elluviation–illuviation along the soil profile or were lost with the drainage water (Santos et al. 2002). The strong presence of iron, together with quartz sand grains in mottles, demonstrates its contribution to the formation of dense soil layers (Silva et al. 2007).

An additional important feature of Brazilian semi-arid soil is the high value of soil bulk density. Mean values of this physical soil attribute range from 1.46 Mg m^{-3} to 1.50 Mg m^{-3} for the A horizon with loamy sand texture (0.15–0.18 m layer) and the B horizon with sand loamy texture (0.17–0.30 m layer), as reported by Silva et al. (2002). Similarly, Fante Júnior et al. (2002) reported that mean values range from 1.54 Mg m^{-3} to 1.65 Mg m^{-3} for the A horizon with loamy sand texture (0–0.45 m layer), whereas Timm et al. (2010) suggested a range of 1.450 Mg m^{-3} to 1.446 Mg m^{-3} for both the 0–0.20 m and 0.20–0.40 m soil layers. Soil particle packing caused by clay eluviation causes clay particles to occupy the pore spaces between coarse particles, which leads to increases in soil bulk density (Dantas et al. 1998).

10.5 Soil Water Dynamics Throughout Growing Seasons of Irrigated Crops

Solar radiation is abundant throughout the year in the Brazilian semi-arid region, which leads to high occurrences of evapotranspiration. Cropping of mainly horticultural species is possible during any month of the year because of the practice of irrigation. Hence, two growing seasons per year are possible for some crops, and two case studies on soil-water dynamics and root distribution during three consecutive growing seasons illustrate such conditions. One case study relates to a banana crop from January 1999 to September 2001 (Bassoi et al. 2004a, b) and the other to a guava crop from March 1999 to December 2001 (Bassoi et al. 2002b, 2007a, b) at Bebedouro Irrigation Scheme in Petrolina, Pernambuco State.

The soil water matric potential was measured at various depths and soil water content (θ, $m^3\ m^{-3}$) was obtained by analyzing soil water retention curves of a medium-texture Latossolo or Oxisol (82 % sand, 6 % silt, 12 % clay) cropped with banana (Fig. 10.1) and guava (Fig. 10.2). Both crops were irrigated by microsprinkler.

For the banana crops, the greatest variation of soil moisture was observed at 0.2 m, 0.4 m, and 0.6 m depths because the effective rooting depth of these crops was 0.4 m and 0.6 m until 9 and 12 months after planting, respectively (Bassoi et al. 2004a). Conversely, the guava crop presented most of its roots at 0.4 m and 0.8 m until 3 and 12 months after planting, respectively (Bassoi et al. 2002b, 2007b).

In the banana crop, the presence of downward soil water flux was observed at a depth of 0.4 m owing to frequent irrigation by microsprinklers, and the presence of upward flux was observed at depths of 0.6 m and 0.8 m. These results demonstrate that these soil depths contributed to the evapotranspiration process because the roots reached a depth of 1.0 m. Within 783–787 days after planting during the third growing season, in the absence of rainfall or irrigation, a great change in soil water content was observed to a depth of 0.6 m and at a distance of up to 0.6 m from the pseudostem as a result of the previously mentioned root presence (Bassoi et al. 2004a).

In trenches dug for root distribution data collection and analysis in both crops (Bassoi et al. 2002b, 2004a, 2007b), the presence of plinthitic mottles was observed at an approximate depth of 1.0 m, indicating the presence of slow drainage or water table elevation due to continuous irrigation throughout the year for several years. Consequently, the soil moisture was higher at greater depths (i.e., at 0.8 m and 1.0 m) in both crops during the majority of the monitoring time (32 and 33 months for banana and guava crops, respectively).

In an additional case study (presented in Figs. 10.3 and 10.4; Nascimento et al. 2012), the soil-water content (θ, $m^3\ m^{-3}$) in a Neossolo Quartzarênico or Arenosol (92 % sand, 1 % silt, 7 % clay) was measured by a neutron probe. In 2008, monitoring was conducted in a microsprinkler-irrigated vineyard throughout two consecutive table grape growing seasons (January to May and July to November) at Senador Nilo Coelho Irrigation Scheme in Petrolina, Pernambuco State. In the first season, θ values in the upper layers (Fig. 10.3, top) were smaller than those in the

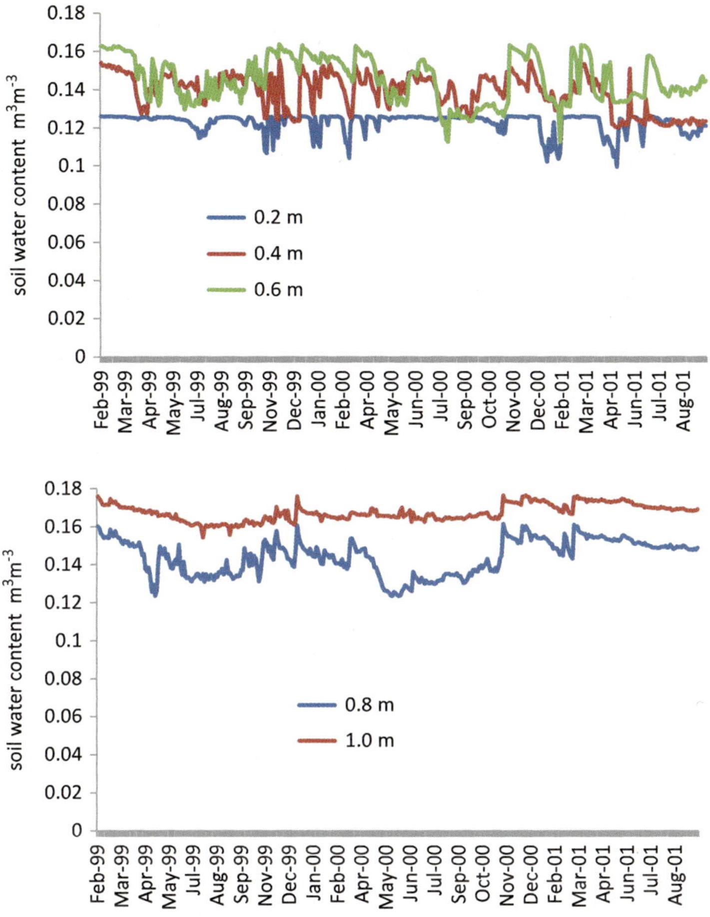

Fig. 10.1 Soil water content from depths of 0.2 m to 1.0 m during three consecutive growing seasons of banana crop (Bassoi et al. 2004a)

lower layers (Fig. 10.3, bottom) during the majority of the monitoring time. However, differences in θ were greatest at depths of 0.15 m, 0.30 m, 0.45 m, and 0.60 m, likely as a result of the contribution of these soil layers to the evapotranspiration process and owing to rainfall of 199 mm. In lower soil layers (from 0.75 m to 1.2 m), θ values were higher for most of the period considered, and the magnitude of θ variation was low because of the lower contribution to the evapotranspiration process. The presence of plinthitic mottles observed in a trench at a depth of 1.2 m indicated the presence of slow drainage in that orchard.

In the second growing season of 2008, a spreadsheet proposed by Nascimento et al. (2012) to adjust the irrigation depth calculated on the basis of crop evapotranspiration was adopted by the irrigation manager. If irrigation depth was higher than the amount of water calculated to reach the field capacity in the effective root zone of

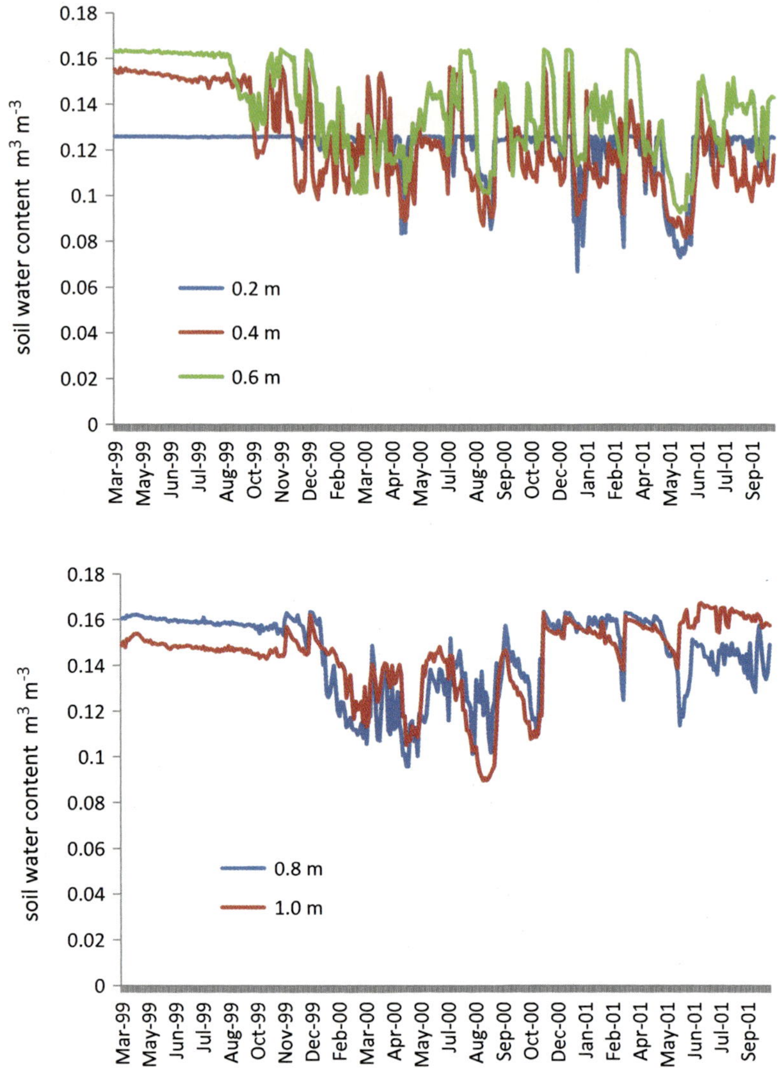

Fig. 10.2 Soil water content from depths of 0.2 m to 1.0 m during three consecutive growing seasons of guava crop (Bassoi et al. 2002b, 2007a)

vines, considered to be at a depth of 0.6 m (Bassoi et al. 2002a, 2003), irrigation was postponed until the following day or the irrigation depth was reduced on the basis of the amount indicated by the spreadsheet to avoid excessive water application. The θ values from the upper layers (Fig. 10.4, top) were smaller than those of the lower layers (Fig. 10.4, bottom). Such differences were more evident than those during the first growing season because only 1 mm of rainfall was recorded during the second growing season. Even so, continuity of soil moisture was present at the lower layers,

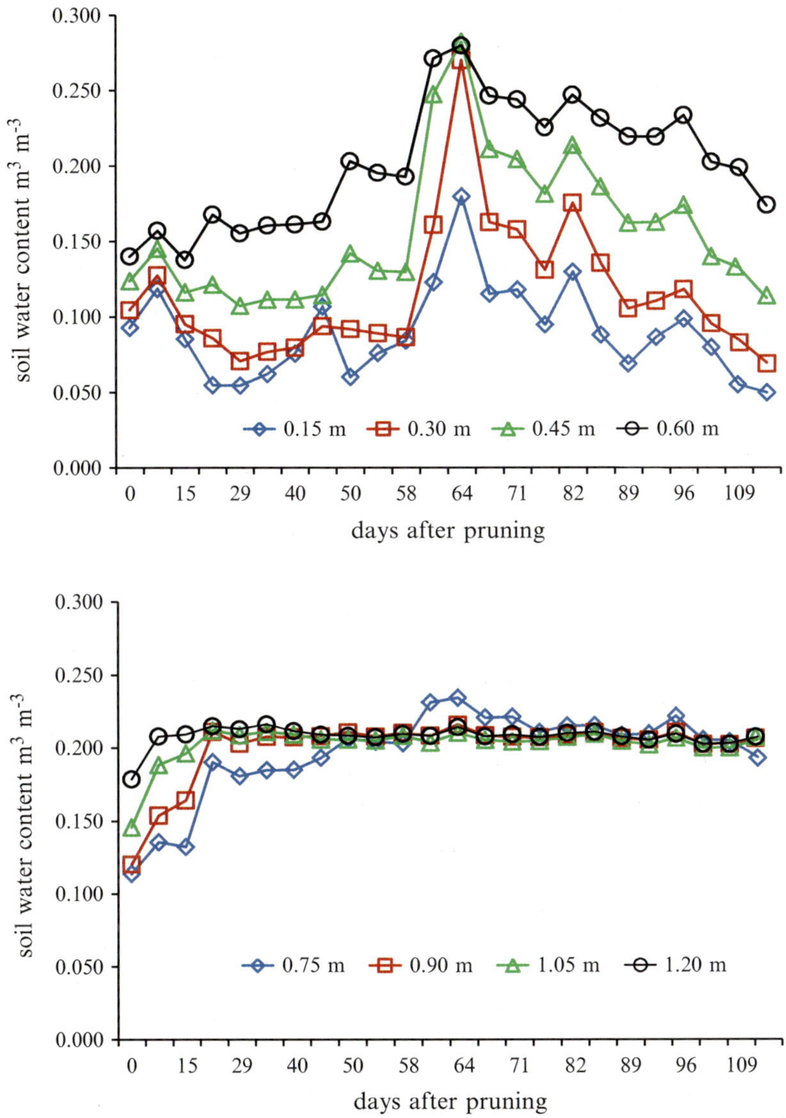

Fig. 10.3 Soil-water content from 0.15 m to 1.2 m depths during the first growing season in 2008 for a table grapevine orchard irrigated by microsprinkler (Nascimento et al. 2012)

with little reduction in the θ values by the end of the season, when irrigation was reduced due to the lower water requirement of the vines.

The application of irrigation depths greater than the amount required by plants can contribute to the formation of a water table, even in sandy soils such as Neossolo Quartzarênico or Arenosol (Amaral 2011).

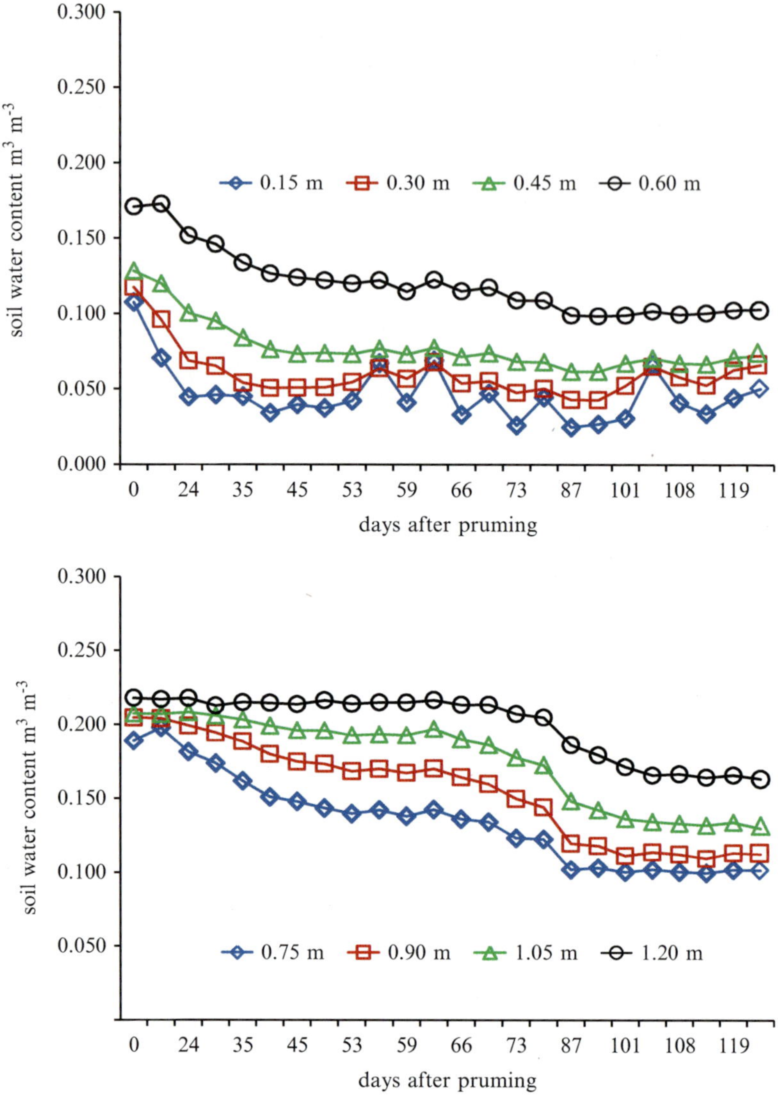

Fig. 10.4 Soil-water content from 0.15 m to 1.2 m depths during the second growing season in 2008 for a table grapevine orchard irrigated by microsprinkler (Nascimento et al. 2012)

10.6 The Opportunity for Deficit Irrigation Practice

Deficit irrigation can be defined as water application to a crop in a lesser amount than that which would completely meet the water requirement for the plant. Among the reasons for using this strategic irrigation method are water shortages, inadequate design of distribution systems and water application, and excessive increases in water usage beyond sustainability in irrigated areas (Ferreres and Soriano 2007).

Table 10.1 Yield of wine grapes and water use efficiency (WUE) during the three consecutive growing seasons (April to August 2010, November 2010 to February 2011, and May to September 2011) as a function of full irrigation (FI), regulated deficit irrigation (RDI), and deficit irrigation (DI) in Petrolina, Pernambuco State

Yield (kg vine^{-1})	FI	RDI	DI
1st season[1]	2.16 a	1.70 a	1.91 a
2nd season[2]	1.47 a	1.52 a	1.49 a
3rd season[3]	1.98 a	1.54 b	1.20 c
WUE (kg m^{-3})	FI	RDI	DI
1st season[1]	2.30 b	2.61 b	4.44 a
2nd season[2]	1.27 b	1.98 a	2.02 a
3rd season[3]	3.02 b	4.93 a	4.64 a

Sources: [1]Bassoi et al. (2011); [2]Correia (2012); [3]Silva (2013). Values followed by the same letter do not differ at the 5 % level for the Tukey test

Regulated deficit irrigation (RDI) is a strategy originally developed to control vegetative growth in orchards through high-density planting and to optimize fruit set, fruit size, and quality. Moderate water stress applied during the slow growth period of the fruit can control excessive vegetative growth to increase the yield and efficiency of irrigation water usage. The period of RDI application, root distribution within the soil, and soil layer moisture levels are important factors for the success of this strategy (Goodwin and Boland 2002).

The practice of deficit irrigation strategies for wine grapevines was evaluated in a medium-texture Argissolo or Ultisol (74 % sand, 10 % silt, 16 % clay) at Bebedouro Irrigation Scheme in Petrolina, Pernambuco State. The full irrigation (FI) strategy consisted of water application by drip irrigation during the entire growing season based on crop evapotranspiration, whereas irrigation was practiced until the phenological phase of closure cluster in the RDI and deficit irrigation (DI) strategies. The difference between the latter two irrigation strategies is that water was not applied until harvesting in DI, whereas it was applied at various times during RDI, based on soil water monitoring by neutron probe in the effective rooting zone (0.6 m depth), to increase the soil water content in that soil layer. The results for three consecutive growing seasons are reported in Table 10.1. Despite a small reduction in the yield of clusters per vine in some seasons, water use efficiency was higher when some type of water deficit irrigation strategy was applied.

The soil profile of that Argissolo (Ultisol) also presented slow drainage and plinthitic mottles in the lower layers, i.e., at a depth of approximately 1.0 m. The dynamics of soil water content were monitored by neutron probe readings. The results from the third growing season of May to September 2011 are presented in Figs. 10.5 and 10.6 (Silva 2013) and exhibit the same pattern observed in the previous two growing seasons.

Soil water content markedly decreased until the 0.6 m depth when irrigation was interrupted in the RDI and DI strategies, although some decreases occurred at the 0.75 m and 0.9 m depths. However, very little change in θ was observed.

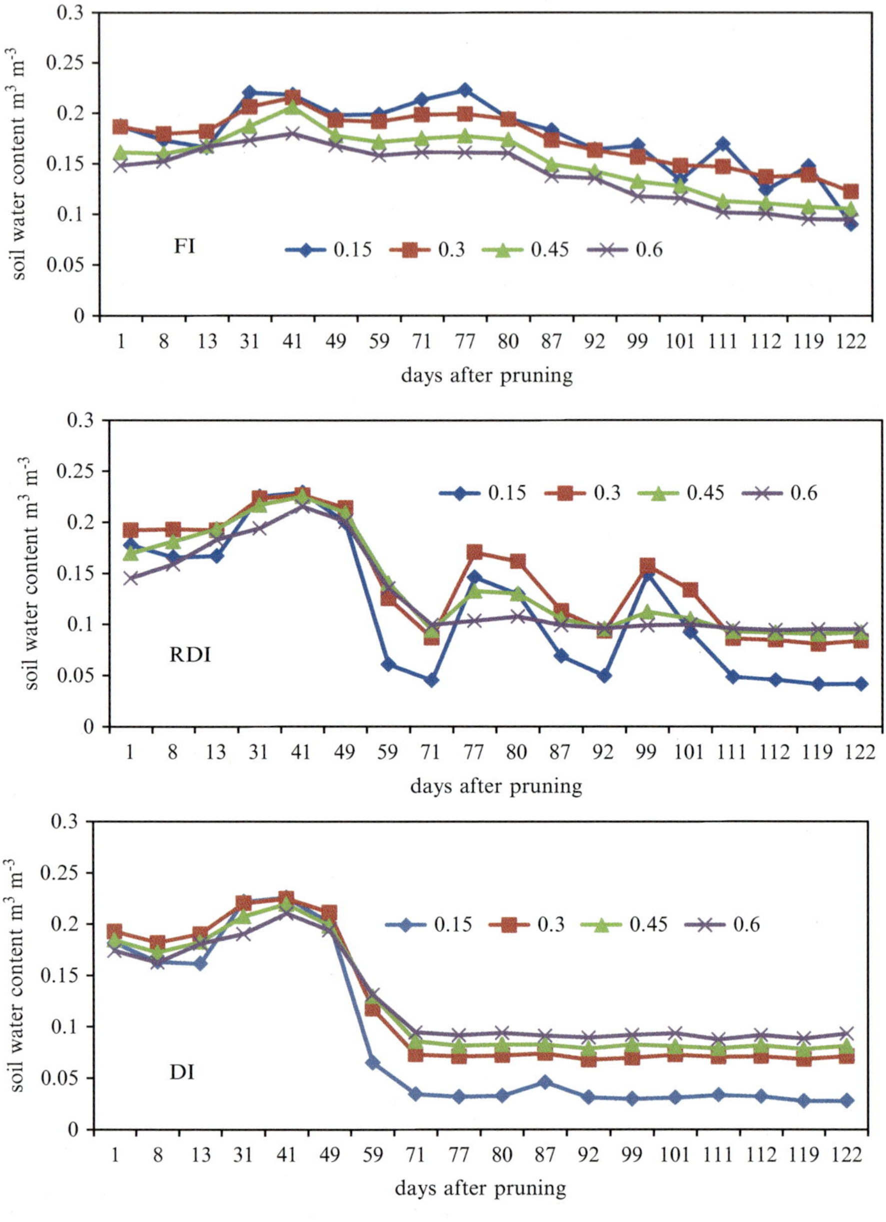

Fig. 10.5 Soil-water content from 0.15 m to 0.6 m depths in the third growing season of wine grapevines irrigated according to full irrigation (FI), regulated deficit irrigation (RDI), and deficit irrigation (DI) strategies (Silva 2013)

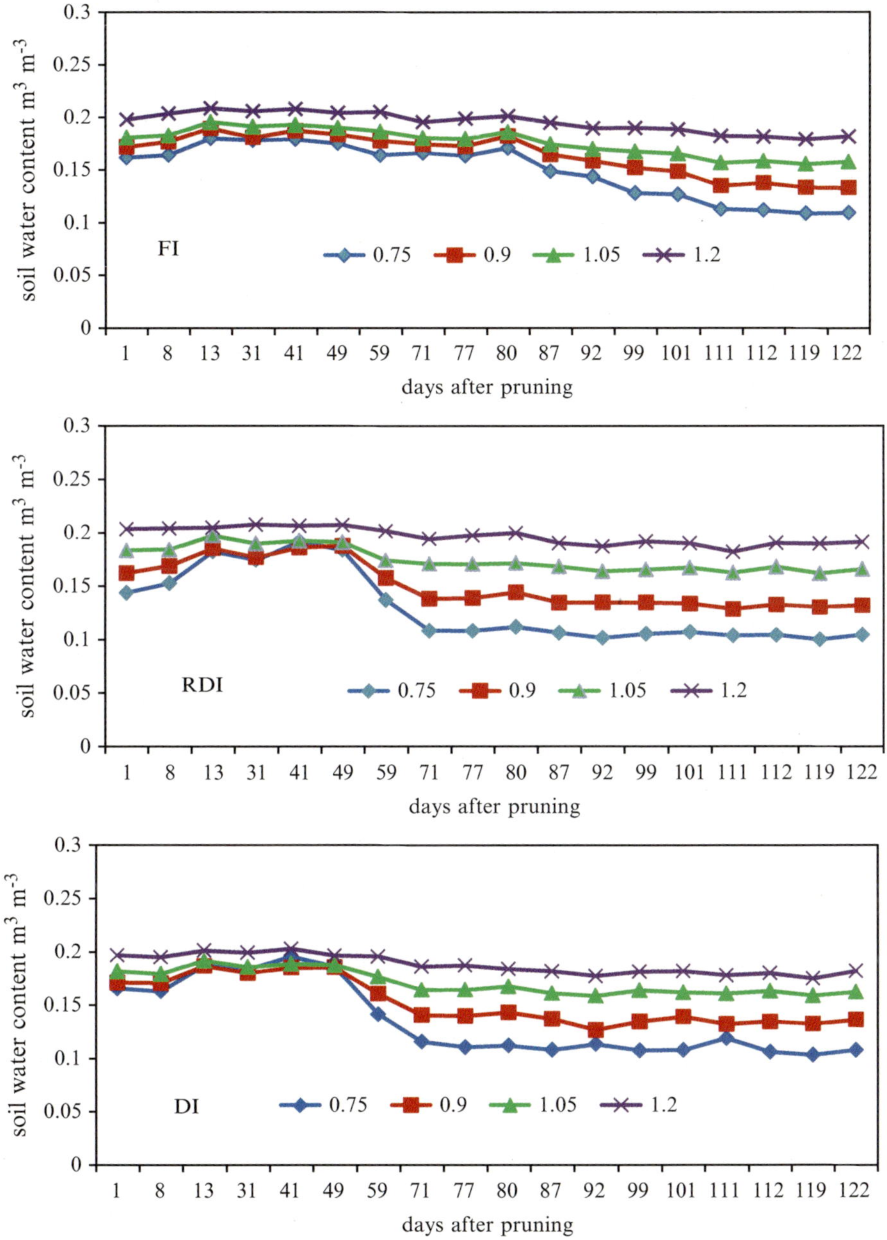

Fig. 10.6 Soil-water content from 0.75 m to 1.20 m depths in the third growing season of wine grapevines irrigated according to full irrigation (FI), regulated deficit irrigation (RDI), and deficit irrigation (DI) strategies (Silva 2013)

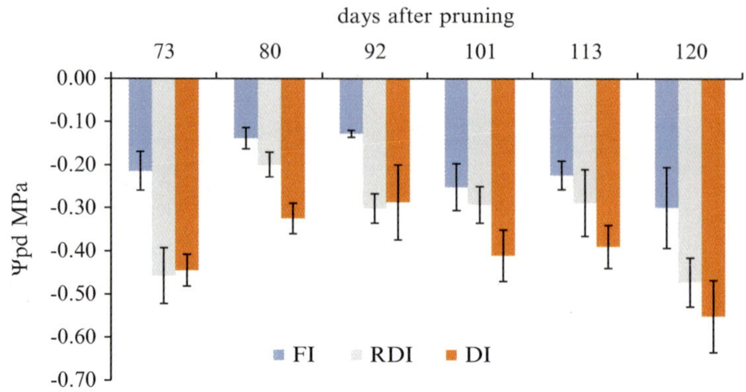

Fig. 10.7 Pre-dawn leaf water potential (Ψ_{pd}) on vines irrigated according to full irrigation (FI), regulated deficit irrigation (RDI), and deficit irrigation (DI) strategies (Silva 2013)

Vines exhibited differences in leaf water potential measured at pre-dawn among irrigation strategies in some cases (Fig. 10.7). However, these differences were insufficient for the characterization of a severe plant water stress (Carbonneau 1998) owing to soil water availability and the root system present at a depth of 1.0 m. The effective rooting depth for vines in such soil conditions is generally 0.6 m; however, some roots can reach a depth of 1.0 m (Bassoi et al. 2002a, 2003).

10.7 Final Remarks

In the Brazilian semi-arid region, various soils under irrigation over long periods of time store water in their deeper layers owing to several reasons, as discussed in this paper. Research efforts should be focused on horticulture species with deep root systems and drought tolerance, and on agricultural practices such as deficit irrigation strategies, that can harness the water stored in the soil profile.

Therefore, the knowledge of soil type and its attributes together the soil water monitoring is essential to help growers to make decision about deficit irrigation practices in the Brazilian semi-arid region. Depending on the crop, benefits such as water use efficiency should increase in each unit of cropped area, without significant prejudice of biomass production.

References

Amaral FCS (2011) Sistema brasileiro de classificação de terras para classificação: enfoque na região semi-árida. Embrapa Solos, Rio de Janeiro

Bassoi LH, Grangeiro LC, Silva JAM et al (2002a) Root distribution of irrigated grapevine rootstocks in a coarse texture soil of the São Francisco Valley, Brazil. Rev Bras Frutic 24:35–38

Bassoi LH, Teixeira AHC, Silva JAM et al (2002b) Parâmetros para o manejo de irrigação da goiabeira no Vale do São Francisco. Paper presented at the 31 congresso brasileiro de engenharia agrícola, SBEA, Salvador, 29 July–2 August 2002

Bassoi LH, Hopmans JW, Jorge LAC et al (2003) Grapevine root distribution in drip and microsprinkler irrigation. Sci Agric 60:377–387

Bassoi LH, Silva JAM, Silva EEG et al (2004a) Guidelines for irrigation scheduling of banana crop in São Francisco Valley, Brazil. I – Root distribution and activity. Rev Bras Frutic 26:459–463

Bassoi LH, Teixeira AHC, Lima Filho JMP et al (2004b) Guidelines for irrigation scheduling of banana crop in São Francisco Valley, Brazil. Rev Bras Frutic 26:464–467

Bassoi LH, Ferreira MNL, Coelho RD et al (2007a) Goiabeira irrigada: consumo de água e coeficiente de cultura no Vale do São Francisco. Paper presented at the 36 congresso brasileiro de engenharia agrícola, SBEA/Embrapa Agropecuária Oeste, Bonito, 30 July–2 August 2007

Bassoi LH, Ferreira MNL, Coelho RD et al (2007b) Distribuição radicular da goiabeira irrigada no Vale do São Francisco. Paper presented at the 36 congresso brasileiro de engenharia agrícola, SBEA/Embrapa Agropecuária Oeste, Bonito, 30 July–2 August 2007

Bassoi LH, Gonçalves SO, Santos ARL et al (2011) Influência de manejo de irrigação sobre aspectos de ecofisiologia e produção da videira cv. Syrah Irriga 16:395–402

Bockheim JG, Hartemink AE (2013) Soils with fragipans in the USA. Catena 104:233–242

Brazil (2005) Relatório final do grupo de trabalho interministerial para a redelimitação do Semi-Árido nordestino e do polígono das secas. Ministério da Integração Nacional, Brasília

Burgos N, Cavalcanti AC (1990) Levantamento detalhado de solos da área de sequeiro do CPATSA, Petrolina-PE, escala 1:7500. Embrapa, Brasilia (Boletim de Pesquisa, 38)

Carbonneau A (1998) Aspects qualitatifs. In: Tiercelin JR (ed) Traité d'irrigation. Tec et Doc Lavoisier, Paris, pp 258–276

Correia JS (2012) Irrigação deficitária em videira de vinho cv. Syrah durante o período chuvoso no Semiárido. Dissertation, Universidade Estadual Paulista Júlio de Mesquita Filho

Cunha TJF, Petrere VG, Silva DJ et al (2010) Principais solos do Semiárido tropical brasileiro: caracterização, potencialidades, limitações, fertilidade e manejo. In: Sá IB, Silva PCG (eds) Semiárido Brasileiro. Pesquisa, Desenvolvimento e Inovação, 1st edn. Embrapa Semiárido, Petrolina, pp 49–87

Dantas JA, Santos MC, Heck RJ (1998) Caracterização de Podzólicos Amarelos irrigados e não irrigados do Submédio São Francisco. Rev Bras Cienc Solo 22:761–771

Fante Júnior L, Oliveira JCM, Bassoi LH et al (2002) Tomografia computadorizada na avaliação da densidade de um solo do semi-árido brasileiro. Rev Bras Cienc Solo 26:835–842

Ferreres E, Soriano MA (2007) Deficit irrigation for reducing agricultural water use. J Exp Bot 58:147–159

Goodwin I, Boland AM (2002) Scheduling deficit irrigation of fruit trees for optimizing water use efficiency. In: FAO (ed) Deficit irrigation practices. FAO, Rome, pp 67–78 (FAO Report 22)

Nascimento PS, Bassoi LH, Paz VPS (2012) Planilha eletrônica para auxilio à tomada de decisão em manejo de irrigação. Irriga 17:1–15

Ribeiro MR (2010) Manejo do solo e da água em perímetros irrigados da região Nordeste do Brasil. In: Prado RB, Turetta APD, Andrade AG (eds) Manejo e Conservação do Solo e da Água no Contexto das Mudanças Ambientais. Embrapa Solos, Rio de Janeiro, pp 171–180

Santos MC, Silva MSL, Almeida BG et al (2002) Caracterização micromorfológica e considerações sobre a gênese de solos de tabuleiro do semi-árido do Brasil. Rev Bras Cienc Solo 26:1031–1046

Silva JA (2013) Estratégias de irrigação para economia de água em videira cv. Syrah no Vale do Submédio São Francisco. Dissertation, Universidade Federal do Vale do São Francisco

Silva MSL, Klamt E, Cavalcante AC et al (2002) Adensamento subsuperficial em solos do semi-árido: processos geológicos e/ou pedogenéticos. Rev Bras Eng Agríc Ambient 6:314–320

Silva MSL, Chaves VC, Cavalcanti AC et al (2007) Adensamento e compactação de solos irrigáveis da zona semi-árida do Nordeste brasileiro. Embrapa Solos, Recife (Circular Técnica, 38)

Silva PCG, Moura MSB, Kiill LHP et al (2010) Caracterização do Semiárido brasileiro: fatores naturais e humanos. In: Sá IB, Silva PCG (eds) Semiárido Brasileiro. Pesquisa, Desenvolvimento e Inovação, 1st edn. Embrapa Semiárido, Petrolina, pp 17–48

Timm LC, Reckziegel NL, Aquino LS et al (2010) Spatial variability of physical and hydraulic attributes of a sandy soil in the Northeast Brazil. Suelos Ecuatoriales 40:139–151

Part IV
Sensors and Monitoring in Soil Physics

Chapter 11
Sensor Network Data Assimilation in Soil Water Flow Modeling

Yakov Pachepsky, Feng Pan, and Gonzalo Martinez

Abstract Data assimilation is the model performance improvement methodology that relies in correcting (updating) modeling results based on observations, and using updated values as new initial conditions that will be used in modeling until the next measurement will be available and the update will be in order. The central problem of data assimilation is the accounting for the uncertainty in both monitoring data and modeling results. Such accounting for soil water sensor networks is described in this book chapter. With soil system specifics in mind, two most promising data assimilation methods are characterized, the statistical model of the uncertainty in soil data is presented, and a methodology to assess uncertainty in modeling results is proposed. Research needs and outlook are offered.

Keywords Soil water monitoring • Ensemble Kalman filter • Particle filter • Pedotransfer functions • Soil water flow modeling • Soil water sensors • Spatio-temporal variability

Y. Pachepsky (✉)
Environmental Microbial and Food Safety Laboratory, USDA-ARS Beltsville Agricultural Research Center, 10300 Baltimore Ave., Bldg. 173, Beltsville, MD 20705, USA
e-mail: yakov.pachepsky@ars.usda.gov

F. Pan
Energy and Geoscience Institute, University of Utah, Salt Lake City, UT, USA

G. Martinez
Department of Natural Resources and Agroecology, Andalusian Institute of Agricultural Research and Training, Fisheries, Food and Organic Production (IFAPA), Seville, Spain

W.G. Teixeira et al. (eds.), *Application of Soil Physics in Environmental Analyses: Measuring, Modelling and Data Integration*, Progress in Soil Science, DOI 10.1007/978-3-319-06013-2_11,

11.1 Introduction

A large number of soil water flow and storage models have been developed for applications in hydrology, meteorology, agronomy, contaminant hydrology, and other fields (Ranatunga et al. 2008; Köhne et al. 2009). Each of these models is based on simplifications formulated as a set of assumptions about the mechanisms, processes, and parameters of water retention and flow. It is often not possible to predict whether a particular set of assumptions will be applicable for a specific site. Therefore, errors in soil water modeling predictions arise that result from both conceptual uncertainty and the lack of detailed knowledge about spatial distributions and temporal changes of model parameters. If a model is selected, then there are two basic ways to improve its performance – to find better parameters or to correct model predictions from time to time and continue modeling using corrected values as initial conditions. These two approaches can be combined, and the correction can be applied both to simulated soil water contents and to model parameters. Soil water monitoring data are needed to improve the model performance.

Improvement in model parameter values consists of monitoring soil water content for a long period of time and subsequently calibrating the model with data over the calibration period. This is usually a nontrivial task, given the high nonlinearity of realistic soil water flow models, layering that may require separate parameter sets and observations within each layer, the uncertainty in plant water uptake for specific soil water content profiles and their history, and the need to accumulate observations of a substantial number of both very wet and very dry conditions (Scott et al. 2000; de Jong van Lier et al. 2013; Morbidelli et al. 2013).

Using monitoring data to periodically correct modeling results consists of updating simulated values, i.e. replacing simulated values of environmental variables with values that are closer to the measured ones. This operation is called data assimilation (DA). It has become a common approach in modeling atmospheric, oceanic, and hydrologic systems (Park and Xu 2009; Lahoz et al. 2010). There are two basic approaches to data assimilation: *sequential* assimilation, that only considers observation made in the past until the time of analysis, and *non-sequential*, or *retrospective* assimilation, where observation, after the time of analysis, can be used. Sequential assimilation will be discussed in this chapter.

Applications of DA in soil water flow and storage modeling have a substantial history. First applications focused on modeling water storage in irrigated soils with soil water balance computed for the whole soil profile (Aboitiz et al. 1986; Or and Hanks 1992). Aboitiz et al. (1986) developed daily soil water content estimates that were based on both the model and the most recent field measurements according to the explicit weighting scheme using estimates of their relative errors. With this approach, extended forecasts could be sequentially generated given current estimates and explicitly evaluated error covariances. In his work, the model was much less accurate than soil moisture measurements at sprinkler irrigated fields, while the opposite was true for surface irrigated fields. DA effectively reflected these differences in terms of the relative accuracy of the model and data. Or and Hanks (1992) applied DA to explicitly evaluate the evolution of the spatial variability in soil water content data and to include it in their estimates of soil water contents derived by combining measurements and simulations. They used neutron probe measurements, to correct the simulated total soil water storage in 1.5-m deep soil profiles, under sprinkler irrigation. Wendroth et al. (1999) were the first to use the soil water sensor networks, in this case tensiometers, to apply the DA with explicit account of soil layering. A simple Darcy-based mass balance equation was applied in each layer. They showed that DA in soil water modeling could be efficient for a soil water model that includes three layers. Moreover, they demonstrated that the DA can be useful if applied only in a part of soil profile which is the most affected by errors in measurements.

A large volume of research was devoted to assimilating remote sensing data, on surface soil moisture, to infer distributions of soil water contents in soil profiles. Originally, coupled heat transport and water equations were corrected with DA (Entekhabi et al. 1994). In later works, the soil water flow model given by Richards equation was used in remote sensing DA under the assumption that the remote sensing retrieval algorithm provided values directly applicable for the surface soil water content update (Heatman et al. 2003; Das and Mohanty 2006). Lately, remote sensing surface soil water content data were assimilated with the semiempirical soil water flow and storage models, a small number of vertical compartments used for coarser spatial scales (Crow and van Loon 2006; Huang et al. 2008).

Two different approaches to the update step were attempted in soil water DA. The first approach, termed *direct insertion*, assumed that measured data can be used for the update without any corrections. The second approach postulates that data-assimilation based correction of modeling results should use information on both uncertainty in data and uncertainty in modeling results. Simulated values should be changed to values very close to measured ones if the uncertainty in data is much less than the uncertainty in modeling results. On the other hand, there is no reason to substantially change simulated values if the uncertainty in modeling results is much less than the uncertainty in data. Although the direct insertion method has been applied from time to time (Houser et al. 1998; Heatman et al. 2003), it was eventually recognized that taking into account uncertainties in modeling results and in data, leads to better results. Walker et al. (2001) and Crow and Wood (2003) made this clear using a synthetic example with a Darcy-based model of a layered soil, and with a land surface model with real-world independent observations, respectively.

The rest of this chapter describes the DA approaches in detail and outlines the features that are specific for the network sensor data in soil hydrology.

11.2 Data Assimilation Methods

The DA approaches currently used in soil hydrology recognize the fact that both simulation results and measurements are actually uncertain – they are random values sampled from some distribution. The fundamental difference between the approaches consists in the number of moments of this distribution that is used to decide on how to use measurements to update simulations. Using only the first moment is equivalent to the direct insertion – no information about the uncertainty is used. The first and the second moments, i.e. mean and variance, are used in data assimilation with the Kalman filter (more details on Kalman filtering below).

The assumption that two moments are sufficient to define the distribution means that the distribution can be defined with two parameters. Kalman filter users typically assume that both data and modeling results are normally distributed values. If a two-parametric distribution is far from reality, then empirical distributions are used to characterize uncertainties in both measured and simulated values. DA in this case is using so-called particle filters.

The uncertainty-based approach to DA in soil water modeling was originally implemented using the statistical technique called the Kalman filter (Aboitiz et al. 1986; Or and Hanks 1992). However, the application of the Kalman filter requires not only the model of soil water content in time but also the model uncertainty in soil water contents in time. Developing the latter model appeared to be hardly tractable not only in soil hydrology but also in many other fields. To overcome this difficulty, Evensen (1994) proposed the ensemble Kalman filter (EnKF). The EnKF is a sequential DA method, which uses an ensemble of model states to represent the error statistics of the model estimation. The idea is to start an ensemble of (many) simulations by varying model parameters, initial state variables, and forcing within feasible ranges. The variation in modeling results within the ensemble, at the time of state variable update, is used to define the uncertainty in modeling results. Vereecken et al. (2008) noted that the conceptual simplicity, relative ease of implementation, and computational efficiency of the EnKF make the method an attractive option for DA in vadose zone hydrology. The EnKF has been proven to be an efficient approach to correct Richards equation-based soil water content estimation by assimilating surface soil moisture (Das and Mohanty 2006; Pan et al. 2012).

11.2.1 Ensemble Kalman Filter

The Kalman Filter is an implementation of the Bayesian update method. Given a probability density function (PDF) of the state of the modeled system (the prior) and the probability distribution function of data, the Bayes theorem is used to obtain the PDF after the data has been taken into account (the posterior). The Bayesian update incorporates new data when they become available and model advances in time from one update to another. The following description is based on the work of Mandell (2007).

The Kalman filter relies on normal distributions of data and modeling results. Let the model for any simulated time generate N state variables $x_1, x_2, \ldots x_{n.}$. The probability distribution function $p(x)$ of the vector of simulation results $\mathbf{x} = \{x_1, x_2, \ldots x_{n.}\}$ is

$$p(\mathbf{x}) = A_1 \exp\left(-\frac{1}{2}(\mathbf{x}-\boldsymbol{\mu})^T \mathbf{Q}^{-1}(\mathbf{x}-\boldsymbol{\mu})\right) \tag{11.1}$$

where $\boldsymbol{\mu}$ is the vector of mean values of variables x_1, x_2, x_n, $\mathbf{Q}$ is the covariance matrix, A_1 as well as A_2, A_3, and A_4 in equations below, are scaling multipliers to have the integral of probability distribution function equal to one. The function $p(\mathbf{x})$

is the prior probability distribution on the moment the state has to be updated to account for data. The vector of data values **d** is also assumed to be normally distributed with the mean $\overline{\mathbf{d}}$ and covariance matrix **R**. It is assumed that the mean data vector $\overline{\mathbf{d}}$ is related to state variables **x** via matrix **H** as

$$\overline{\mathbf{d}} = \mathbf{Hx} \tag{11.2}$$

The value **Hx** represents the value of the data for the state **x** in the absence of data errors. Then the probability density $p(\mathbf{d}|\mathbf{x})$ of the data **d** conditional of the system state **x**, is

$$p(\mathbf{d}|\mathbf{x}) = A_2 \exp\left(-\frac{1}{2}(\mathbf{d}-\mathbf{Hx})^T\mathbf{R}^{-1}(\mathbf{d}-\mathbf{Hx})\right) \tag{11.3}$$

For the update purposes, one needs the probability density of states conditioned on data $p(\mathbf{x}|\mathbf{d})$ rather than the probability density of data conditioned on states $p(\mathbf{d}|\mathbf{x})$. The conversion of $p(\mathbf{d}|\mathbf{x})$ to $p(\mathbf{x}|d)$ can be done using the Bayes theorem:

$$p(\mathbf{x}|\mathbf{d}) = A_3 p(\mathbf{d}|\mathbf{x})p(\mathbf{x}) \tag{11.4}$$

States conditioned on data, i.e. $\mathbf{x}|\mathbf{d}$, are posterior states; they are referred below as x^{P}. When Eqs. 11.1 and 11.3 are used to compute the right-hand side of Eq. 11.4, the expression for $p(\mathbf{x}^{\mathrm{P}})$ is obtained in the form:

$$p(\mathbf{x}^p) = A_3 \exp\left(-\frac{1}{2}(\mathbf{x}^p-\boldsymbol{\mu}^p)^T\mathbf{Q}^{p-1}(\mathbf{x}^p-\boldsymbol{\mu}^p)\right) \tag{11.5}$$

The posterior mean $\boldsymbol{\mu}^{\mathrm{P}}$ and posterior covariance Q^{P} in Eq. 11.5 are given by the Kalman update formulas:

$$\begin{aligned} \boldsymbol{\mu}^{\mathbf{p}} &= \boldsymbol{\mu} + \mathbf{K}(\mathbf{d} - \mathbf{H}\boldsymbol{\mu}) \\ \mathbf{Q}^{\mathbf{P}} &= (\mathbf{I} - \mathbf{KH})\mathbf{Q} \end{aligned} \tag{11.6}$$

where

$$\mathbf{K} = \mathbf{QH}^{\mathbf{T}}\left(\mathbf{HQH}^{\mathbf{T}} + \mathbf{R}\right)^{-1} \tag{11.7}$$

is the Kalman update matrix. The Kalman update changes state variables taking into account (a) data available at the moment when predictions have been obtained, (b) the accuracy of those data, and (c) variability of state variables. An important feature of the Kalman filter is that the number of elements (measurements) in the data vector **d** is usually much smaller than the number of state variables – elements of the vector **x**.

The ensemble Kalman filter (EnKF) has been developed to overcome the difficulty of using the original Kalman filter in cases when the dependence of the covariance matrix **Q** on time is difficult to find. The EnKF estimates the covariance matrix **Q** as the sample covariance computed from the ensemble simulation results. The ensemble is composed from randomly generated equiprobable realizations of the studied model. The randomness may apply to initial conditions, model parameters, and boundary conditions or forcing.

Let the ensemble consists of N models and each model predicts n state variables. Let the predictions of the ith model form the vector $\mathbf{x}_i$ that has n elements $x_i,\ i = 1,2,\ldots n$, which are predicted values of state variables. The $n \times N$ matrix $\mathbf{X} = [\mathbf{x}_1, \mathbf{x}_2, \ldots \mathbf{x}_N]$ is the prior ensemble. The goal is to correct the predictions at each of the preset update times by changing this matrix to the posterior ensemble $\mathbf{X}^p = \left[\mathbf{x}_1^p, \mathbf{x}_2^p, \ldots \mathbf{x}_N^p\right]$. It is assumed that the data is comprised of the vector **d** that has m elements. The vector $\boldsymbol{\varepsilon}$ is the random error in data characterized by the $m \times m$ error covariance matrix **R**.

The ensemble Kalman filter (or EnKF) update consists of four basic steps.

1. Find the $n \times N$ covariance matrix **C** of ensemble predictions $\mathbf{x}_\mathrm{i}$
2. Generate representative random data separately for each ensemble member: $\mathbf{d}_1 = \mathbf{d} + \boldsymbol{\varepsilon}_1$, $\mathbf{d}_2 = \mathbf{d} + \boldsymbol{\varepsilon}_2$, …, $\mathbf{d}_\mathrm{N} = \mathbf{d} + \boldsymbol{\varepsilon}_\mathrm{N}$, where the random vector $\boldsymbol{\varepsilon}$ belongs to the n-dimensional normal distribution N(0,R).
3. Collect the random data in the $m \times N$ matrix $\mathbf{D} = [\mathbf{d}_1, \mathbf{d}_2, \ldots \mathbf{d}_\mathrm{N}]$
4. Find the corrected predictions as:

$$\mathbf{X}^\mathrm{p} = \mathbf{X} + \mathbf{K}(\mathbf{D} - \mathbf{H}\mathbf{x}) \tag{11.8}$$

where the Kalman gain matrix K relates the variability in predictions and the data accuracy and is estimated as

$$\mathbf{K} = \mathbf{C}\mathbf{H}^\mathrm{T}\left(\mathbf{H}\mathbf{C}\mathbf{H}^\mathrm{T} + \mathbf{R}\right)^{-1} \tag{11.9}$$

The one-dimensional case gives a general idea of how the ensemble Kalman filter works. Consider the case $n = 1$ and $m = 1$ when there is only one model-predicted state variable and its value is measured. All matrices then will become scalars, and **H** will be equal to 1. Let $x_i = \mu + \xi_i$, ξi belongs to $N(0,\ \sigma_x^2)$, and $d_i = d + \varepsilon_i$, ε_i belongs to $N(0,\ \sigma_d^2)$. The gain K will be

$$\mathrm{K} = \frac{\sigma_x^2}{\sigma_x^2 + \sigma_d^2} \tag{11.10}$$

and

$$x_i^p = (1 - \mathrm{K})(\mu + \xi_i) + \mathrm{K}(d + \varepsilon_i) \tag{11.11}$$

Values of K are between 0 and 1. The value of x_i^p is close to x_i when K is close to zero, i.e. $\sigma_d^2 >> \sigma_x^2$ and accuracy in data is much lower compared with the variability in predictions. On the contrary, the value of x_i^p is close to d_i when K is close to one, i.e. $\sigma_x^2 >> \sigma_d^2$ and accuracy in data is much higher than the variability in predictions.

11.2.2 Particle Filter

The particle filter (PF) is known also as a bootstrap filter, the condensation algorithm, sequential Monte Carlo sampling, or interacting particle approximations. The PF is another implementation of the Bayesian update approach. Given a probability density function (PDF) of the state of the modeled system (the prior) and the probability distribution function of data, the Bayes theorem is used to obtain the PDF after the data has been taken into account (the posterior). PF works with the set of random samples, or ensemble, as EnKF does. Members of the ensemble are called particles. Three main conceptual differences between EnKF and PF are (Moradkhani 2008):

1. the assumption about the normal distributions of data and modeling results is applied in EnKF but is not made in PF. Empirical distributions for measurements and simulation results are used in PF;
2. for the updating step, the updated ensemble members are changed in EnKF but are kept the same in PF. Instead of changes in ensemble members as in EnKF, values of the weights assigned to ensemble members are updated in PF;
3. all originally selected ensemble members continue to be included in simulations after update in EnKF. The PF samples the distribution of ensemble members, and the probability for an ensemble member to be included in simulations after the update is equal to its weight.

The posterior weight of the ensemble member "i" w_i^p is the normalized likelihood function computed for this member with priors. In case of the Gaussian likelihood Eq. 11.3, weights will be

$$w_i^p = b\exp\left(-\frac{1}{2}(\mathbf{d}_i - \mathbf{H}\mathbf{x}_i)^T\mathbf{R}^{-1}(\mathbf{d}_i - \mathbf{H}\mathbf{x}_i)\right) \tag{11.12}$$

The normalization, i.e. selection of the value for "b", is achieved by requiring the sum of all weights to be equal to one. The probability distribution to use for sampling is

$$p(\mathrm{x}|d) = \sum_{i=1}^{N} w_i^p\delta(\mathrm{x} - \mathrm{x}^p) \tag{11.13}$$

In case of the Gaussian likelihood there is a possibility of the ensemble degeneration, i.e. ensemble collapsing to a single member, as those particles that are closer to the measurement get higher weights while others are discarded. To avoid this, resampling techniques are used (Plaza et al. 2012). Some of the sampling techniques used in particle filtering are the Sequential Importance Sampling (SIS), Sequential Importance Resampling or Sampling Importance Resampling (SIR) and Regularized Sampling (Ristic et al. 2004). The most important property of the particle filter is its ability to handle complex, multi-modal (non-Gaussian) posterior distributions. This may require replacing the simple relationship Eq. 11.3 with a heuristically defined relationship between model predictions and measured data.

11.3 Statistical Modeling of the Spatio-temporal Variability in Soil Water Contents from Sensor Network Data

It is assumed in this section that the soil water content model predicts the average water content across the area served with the sensor network. The uncertainty metrics for this average needs to be derived for the same measurements.

Sensor network data characterize both temporal and spatial variability of soil water contents. Published soil water sensor network data invariably demonstrate temporal stability of soil water contents. The concept of temporal stability in soils was first introduced by Vachaud et al. (1985). They observed that well defined locations exist where soil was consistently wetter or consistently dryer than average across the field. They also noted that some locations represented well the average soil water contents across the area of study. The temporal stability in soil water contents (TS SWC) has been also reported both at scales finer than the field, e.g. within a plot (Rolston et al. 1991; Pachepsky et al. 2005; Herbst et al. 2009), and at scales coarser than the watershed, e.g. regional scale (Martinez-Fernández and Ceballos 2003).

The statistical spatio-temporal variability model for two-dimensional soil moisture fields, including temporal stability, was proposed by Vanderlinden et al. (2012). The authors assume that soil water content, $\theta(x, y; t)$ in the location, with lateral coordinates x and y, at the time t includes the deterministic component $M(x, y; t)$, and the random stationary zero-mean component $\epsilon(x, y; t)$

$$\theta(x, y; t) = M(x, y; t) + \epsilon(x, y; t) \tag{11.14}$$

Assume that the deterministic component has independent temporal W and spatial V components

$$M(x, y; t) = m + V(x, y) + W(t) \tag{11.15}$$

Here m is the average of $M(x, y; t)$ across the observation area A over all the observation time from 0 to N_t, $V(x, y)$ is the deviation of the soil water content at location (x, y) relative to the average across the area A measured at any moment of observation, $W(t)$ is the deviation of the average soil water content across the area A on time t from the average water content over all the period of observations. Both V and W may have both positive and negative values and average values of zero.

The estimation of *m*, *V*, and *W* is straightforward:

$$m = \frac{1}{N_t N_s} \sum_{j=1}^{N_t} \sum_{i=1}^{N_s} \theta(x_i, y_i; t_j) \tag{11.16}$$

$$V(x, y) = \sum_{j=1}^{N_t} \frac{\left(\theta(x_i, y_i; t_j) - m\right)}{N_t} \tag{11.17}$$

$$W(t) = \sum_{i=1}^{N_s} \frac{\left(\theta(x_i, y_i; t_j) - m\right)}{N_s} \tag{11.18}$$

where N_s and N_t are the number of locations in space and the number of observations in time, respectively, θ is the measured soil water content.

If the model (11.15) is applicable, the statistical modeling of the average water content, without accounting for the spatial component, i.e. as the deterministic time series plus random component, can lead to gross overestimation of the standard error of the average water content.

This results from the fact that the variance of water content found from Eqs. 11.14 and 11.15, formally computed, and includes values of *V*:

$$\iint (\theta - E < \theta >)^2 dA = \iint (V + \varepsilon)^2 dA$$

However, the random component in the average water content for any given observation time is just ε, and its statistical properties have to be used to compute the covariance matrix **R** in Eq. 11.3.

The need in removing biases in measurement locations, to properly estimate the covariance matrix of data, exists also in the cases when the profile distributions of soil water contents, rather than areal distributions are recorded with the network sensors. Pan et al. (2012) presented an example of data assimilation, for the temporally stable soil moisture field, derived from the sensor network in soil profile. The network is shown in Fig. 11.1. In brief, the experimental field was located in Bekkevoort, Belgium, at the bottom of a gentle slope and was covered with a meadow. The soil was classified as Eutric Regosol (FAO 1975). A trench, 1.2 m deep and 8 m long, was dug at the field site. The grass cover was removed from the experimental area. A plastic sheet to isolate the disturbed trench zone covered one side of the trench. Volumetric water content was measured with time domain reflectometry (TDR). Sixty two-rod TDR probes (25 cm long, 0.5 cm rod diameter, 2.5 cm rod spacing) were installed along the 5.5 m of the trench at 12 locations at five depths of 15, 35, 55, 75, and 95 cm (Fig. 11.1).

The observed time series of soil water contents were previously analyzed in the work of Pachepsky et al. (2005). Substantial temporal stability was found, manifesting itself in the similarity of soil water content time series shapes in different locations at the same depth, and shifts of the time series graphs relative to each other along the water content axis (Fig. 11.2).

Because the time series at the same depths were correlated, the 'naive' computation of the covariance matrix of data errors **D** under the assumption of independence

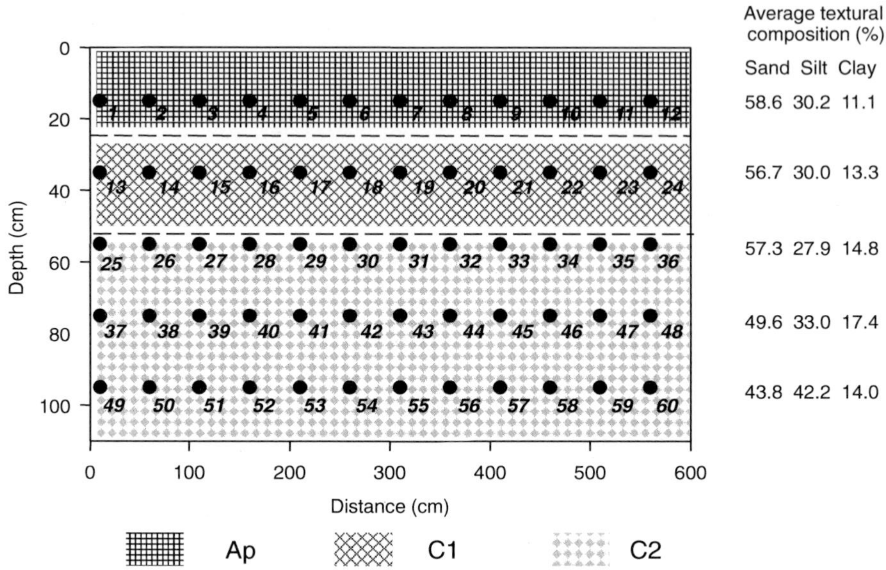

Fig. 11.1 Time domain reflectometry probe placement (●) at the trench wall. Locations 1–12 denote 12 positions along the trench where sets of five sensors – one for each measurement depths were installed. *Filled rectangles* show Ap, C1, and C2 horizons top to bottom. *Dashed lines* show the average position of the horizon boundary, and *white bands* show the observed range of horizon boundary depths. Average values of clay, silt, and sand content are given for the probe installation depths (Adapted from Pachepsky et al. (2005) with permission)

of data in different locations at the same depth could result in large inaccuracies (Wigley et al. 1984) since correlated observations result in inflated type 1 errors (Quinn and Keough 2002). Therefore, the statistical model of the data was assumed in the form (Jacques et al. 2001):

$$\theta_{i,j}(t) = \Theta_i(t) + \beta_{i,j} + \varepsilon_{i,j} \tag{11.19}$$

where i is the subscript to denote depth, $i = 15, 35, 55, 75, 95$ cm, j is the subscript to denote location across the trench, $j = 1,2,\ldots,12$, Θ_i is the average water content at the depth "i", $\beta_{i,j}$ is the bias of the measurement in location j at the depth "i" relative to the average water content at this depth, and η_{ij} is the random component that is used to define the covariance matrix. The bias values were derived by fitting Eq. 11.8 to the whole observed time series. Figure 11.3 shows that values of β have a good connectivity in the studied 2D domain along the trench. Two well-defined isolated regions of negative bias, i.e. low water contents compared to the areal average at observation depths are well expressed and can be traced vertically at two positions along the trench. Using the information about the temporal stability of soil water content measurements was essential for the data assimilation procedure. The "naïve" standard errors of soil water contents at the five observation depths (15, 35, 55, 75 and 95 cm) were in the ranges 0.022–0.030, 0.017–0.023,

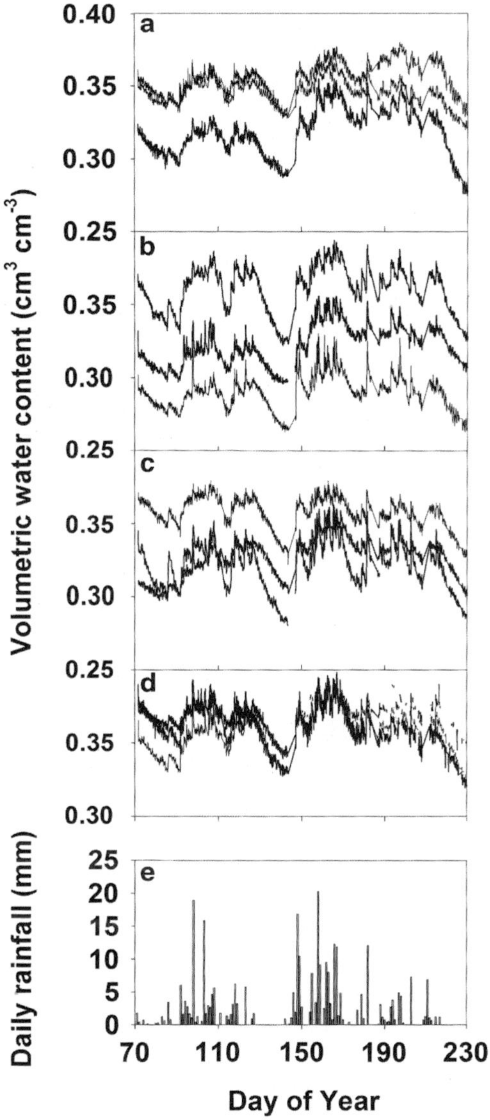

Fig. 11.2 Time series of time domain reflectometry (TDR)–measured water contents at the 15-cm depth (**a**–**d**) and precipitation (**e**). Location numbers top to bottom: (**a**) 2, 3, 1; (**b**) 5, 6, 4; (**c**) 7, 8, 9; and (**d**) 11, 12, 10. Location numbering is shown in Fig. 11.1 (Adapted from Pachepsky et al. (2005) with permission)

0.013–0.019, 0.019–0.023, and 0.020–0.028, respectively. The standard errors of the noise values in Eq. 11.2 for the same depths in the same order ranged from 0.004 to 0.011, 0.003 to 0.012, 0.003 to 0.009, 0.004 to 0.012, and 0.003 to 0.007.

Use of sensor networks facilitates the determination of the temporal stability in soil water contents and removal of local biases $\beta_{i,j}$. This demonstrates one of advantages of using soil water sensor networks that can radically change the accuracy of data assimilation. Expert judgment on the data errors need to be used in absence of the laterally distributed sensor network. For example, Wu and

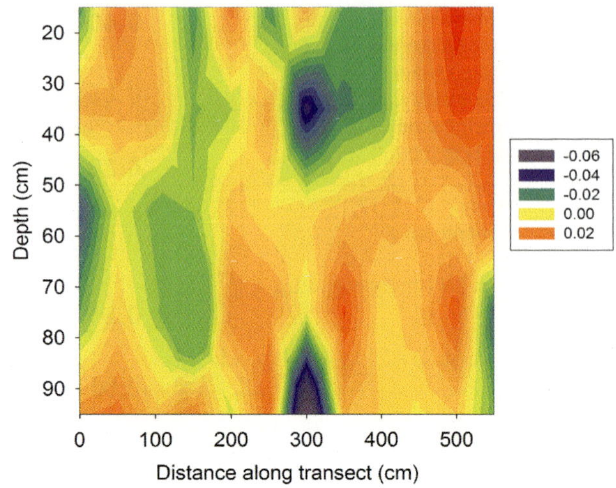

Fig. 11.3 Map of the sensor bias β ($cm^3\ cm^{-3}$) along the trench at the Bekkevoort study site

Margulis (2013) worked with a setup in which only vertical distribution of sensors was present, and assumed unbiased errors with the standard deviation of error equal to 0.01 $cm^3\ cm^{-3}$.

11.4 Using Pedotransfer Functions to Develop an Ensemble for Data Assimilation

Ensemble components to perform data assimilation have to be realistic. In DA applications, these components are obtained randomly, and their properties and parameters are drawn from empirical distributions. The most popular parameterization of soil flow models is the van Genuchten-Mualem parameterization, which, in its generic form, is based on models of water retention and unsaturated hydraulic conductivity

$$\theta = \theta_r + \frac{\theta_s - \theta_r}{\left[1 + (\alpha|\psi|)^n\right]^{1-1/n}} \tag{11.20}$$

$$k = K_{sat} S^L \left\{1 - \left[1 - S^{n/(n-1)}\right]^{1-1/n}\right\}^2 \tag{11.21}$$

Here θ is the volumetric water content, θ_s is the saturated water content, θ_r is the residual water content, ψ is soil water matric potential, S is the soil water saturation, $S = (\theta - \theta_r)/(\theta_s - \theta_r)$. Other soil hydrology parameterizations in use include Brooks-Corey parameter sets, Campbell parameter sets, Vereecken parameter sets, parameters sets including macroporosity, etc.

Current approaches to building an ensemble of water flow models are summarized in Fig. 11.4. All of them have been so far based on pedotransfer functions

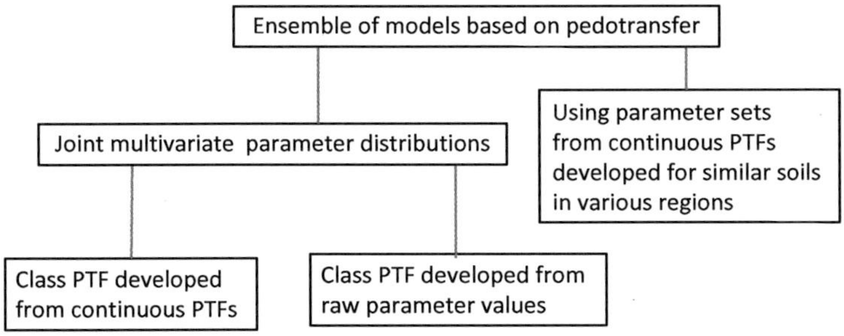

Fig. 11.4 Methods of model ensemble development for network sensor data assimilation for soil water flow modeling

(PTFs), i.e. on the empirical relationships between soil hydraulic parameters θ_r, α, m, n, L, and K_{sat} in Eqs. 11.20 and 11.21 (or their analogues from other soil hydrology parameterizations) and soil basic properties that are easier to obtain either by querying soil survey databases or from site specific measurements. One approach consists in random drawing parameter set values from joint multivariate distributions. The reason for using multidimensional distributions is the evidence, that correlation between soil hydraulic parameters can be fairly strong (Carsel and Parrish 1988; Meyer et al. 1997; Faulkner et al. 2003). So far it has been assumed that the multidimensional parameters distributions need be defined for each of 12 soil textural classes defined by the USDA textural classification. Figure 11.4 lists two methods of obtaining those parameter distributions. One of them has been introduced by Carsel and Parrish (1988). They (a) selected continuous PTFs, i.e. regressions of soil basic properties vs. hydraulic parameters, (b) developed sets of combinations of basic properties for each of textural classes, (c) applied the regressions to obtain hydraulic parameter sets within each class, (d) found transformations of hydraulic parameter values to assure normal distributions of transformed values, (e) constructed joint multivariate normal distributions or each textural class. These distributions have been used, e.g. by Wu and Margulis (2013) in data assimilation work with soil water content and salinity sensors as well as by Scharnagl et al. (2011) in a parameter estimation project. The continuous pedotransfer functions, for this work, were developed from the US data base. The international database UNSODA was used for the same purposes by Faulkner et al. (2003). Figure 11.5 based on the data from this work demonstrates that correlations between parameters can be very strong and are textural class specific.

One feature of the above methodology is that its results depend on pedotransfer relationships not only within a given textural class but also on pedotransfer relationships within other textural classes since the pedotransfer relationships are developed across the database containing data for several textural classes (Carsel and Parrish 1988). Therefore, the joint multivariate parameter distributions for a specific class may not be sufficiently accurate. The alternative is to use the dataset on parameters obtained from raw data for soils from a given textural class and to

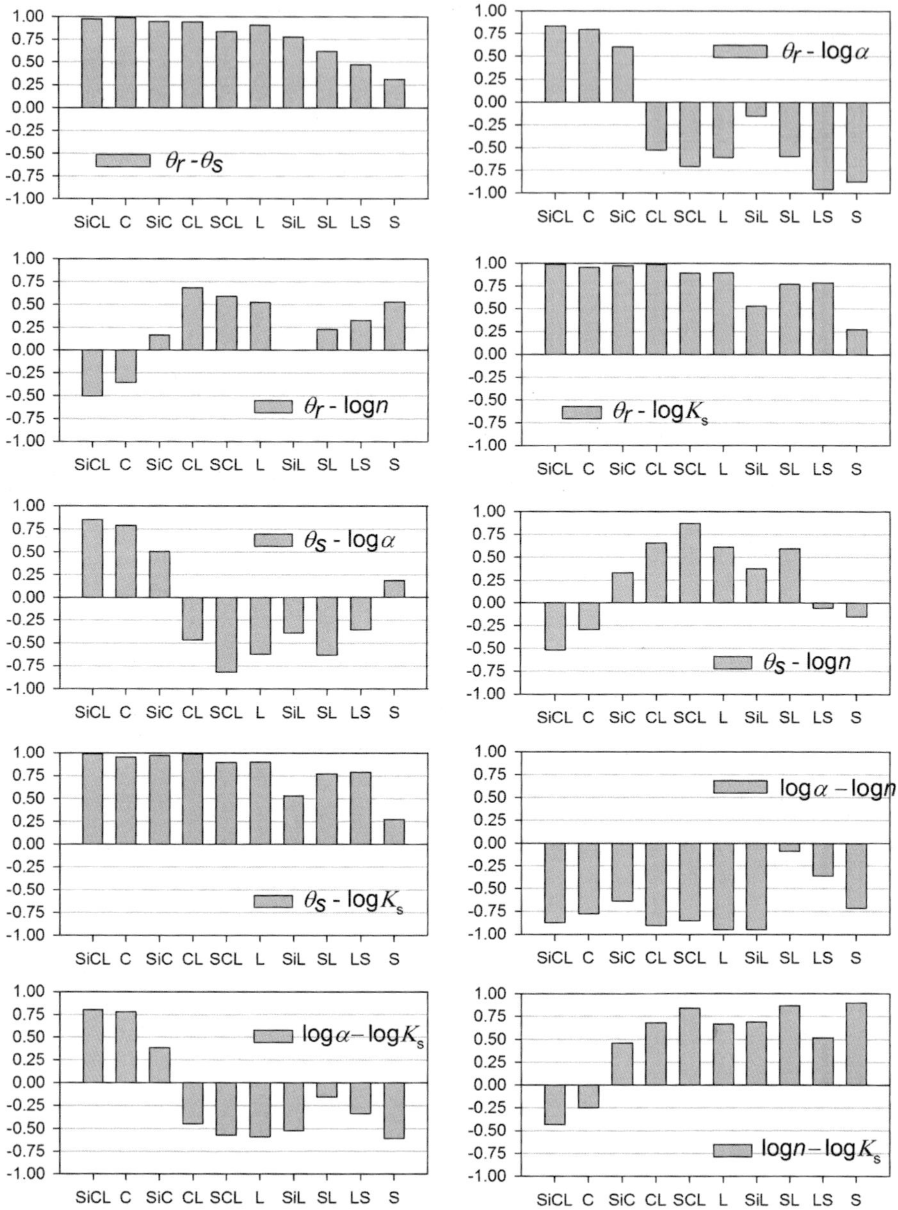

Fig. 11.5 Correlations between soil hydraulic parameters by textural classes developed by Faulkner et al. (2003). The soil textural class symbols are *SiCL* silty clay loam, *C* clay, *SiC* silty clay, *CL* clay loam, *SCL* sandy clay loam, *SiL* silt loam, *LS* loamy sand, *S* sand

use this dataset in developing the joint multivariate parameter distributions specifically for this class. More class-specific distributions are expected to be developed in this way.

The development of joint multivariate distributions relies on pedotransfer functions developed from specific database or from a specific dataset. The dependence of pedotransfer results on the database has been amply demonstrated (e.g., Schaap and Leij 1998). To address this uncertainty, Guber et al. (2006), proposed to build the ensemble of pedotransfer functions for a specific site using a set of pedotransfer functions developed in different regions. This method is referred to in Fig. 11.4 as "using parameter sets from continuous PTFs developed for similar soils in various regions." Guber et al. (2006) noted that, essentially, using the PTF ensemble is the utilization of generic information available before the site study. They argued that although predictions with a single PTF depend on the database used for the PTF development, an ensemble prediction will be less dependent on the individual PTFs if the number of individual PTFs is sufficient enough to represent natural variability in soil water retention. Pan et al. (2012) successfully applied this method in the TDR network data assimilation study. The set of PTFs used in their work and the code to apply them to generate the ensemble can be found in Guber and Pachepsky (2010).

11.5 Data Assimilation and Interpolation

One important feature of the Ensemble Kalman filter applications is that the number of elements (measurements) in the data vector **d** is usually much smaller than the number of state variables – elements of the vector **x**. Updates based on a small number of observation points appear to be efficient across the entire represented domain. In case of soil water flow modeling, updating at several points in a soil profile creates corrections for the whole profile (Das and Mohanty 2006). Pan et al. (2012) described TDR data assimilation for the flow in the domain shown in Fig. 11.1. They found that soil water content data assimilation with the Ensemble Kalman filter provided good results across the whole soil profile even when the data from one depth or from two depths were used for the assimilation in the HYDRUS 1D model. They attributed this to the fact that the assumptions of the soil water flow model were applicable to the site conditions during the observation period which did not include events likely to result in preferential flow or long dry spells when Richards equation might not be fully applicable. Satisfactory results across the whole soil profile were obtained from the assimilation of water contents at the top of the profile. This is in line with results and conclusion of works that have used the ensemble Kalman filter to assimilate the remotely sensed data at the soil surface to reproduce the water contents in soil profile (Das and Mohanty 2006). If the mechanism is correctly described by the model, then it is able to capture the process, if the boundary conditions are corrected. Similar results were reported by Wu and Margulis (2013) who investigated a monitoring and modeling system assimilating embedded sensor data streams into HYDRUS 1D, to improve real-time soil water state estimates over the whole soil profile. They carried out spatial data withholding experiments that demonstrated improvements in water content predictions in unobserved areas, if the soil system was relatively homogeneous and effects from erroneous irrigation on forcing were reduced.

In case of using small number of observation points, an interpolation procedure may be beneficial. The data assimilation update can be done only at observation locations, and update in other numerical grid points can be done by interpolation. This can make computations more efficient, but results may then depend on the interpolation procedure. Martinez et al. (2013) described interpolation of updated states across the soil profile assuming linear dependences of logarithms of pressure heads on depths between interpolation points. The top and the bottom pressure heads obtained from simulations were preserved to perform the linear interpolation outside of the range of measurement depths at all assimilation times. The assimilation depths set included the measurement depths set. The initial water contents for simulations were set at assimilation depths. The initial distribution of water contents across the profile was constructed by linear interpolation, between the initial water content depths and linear extrapolation outside of the range of initial water content depths.

11.6 Research Needs and Conclusions

Soil moisture data assimilation, from sources other than remote sensing of surface soil moisture, received little attention so far. At the same time, soil water content or soil matric potential sensors have become the important and wide-spread source of data on water contents in deep soil layers (Vereecken et al. 2008). A number of methodological issues need to be addressed to provide guidance for the soil water sensor network accumulation.

Up until the present, particle filters have not been applied in soil water sensor accumulation. Research is needed to uncover whether the use of this method or of the hybrid methods, combining EnKF and PF (Stordal et al. 2011), may be beneficial. Sabater et al. (2007) compared several methods of data assimilation for a soil-vegetation-atmosphere model with two soil layers and concluded that the ensemble Kalman filter was one of the best to use. However, it is not clear if this conclusion is specific to the used model type. The need in using other than EnKF data assimilation methods may be due to model-specific violations of the EnKF assumptions. Specifically, EnKF requires the normality of model and data errors distributions (Eqs. 11.2 and 11.4). The distributions of water contents, simulated with the PTF generated model ensemble, conformed to the normality hypothesis in the majority of cases in the work of Pan et al. (2012). However, in some cases simulated soil water contents were not normally distributed. This percentage may be different in other soils and with other weather conditions.

In principle, the EnKF allows for the estimating states and parameters simultaneously. The data assimilation to simultaneously correct both the parameters and state variables, i.e. hydraulic conductivity and soil water contents, may be a way to develop a reliable soil water model for a specific site (e.g., Montaldo and Albertson 2003). Systematic procedures for such dual estimation in hydrologic models were introduced by Moradkhani et al. (2005a, b) and Vrugt et al. (2005). Data assimilation, to determine parameters of the Richards equation, was recently demonstrated with a synthetic dataset (Montzka et al. 2011). However, in real world applications, since the parameters become more and more accurate over time, covariances of parameter values decreases. This can result in filter divergence, because reduction in covariance of parameters leads to a narrowing ensemble spread in model states and, therefore, new measurements will have limited ability to correct the model trajectory (Wu and Margulis 2011). Estimating parameters "on the go" is a very attractive possibility, to avoid the need of accumulating long time series for model calibration. However, it remains questionable if and how the representation of the different parts of possible water content range may affect the dual estimation of parameters and states in the course of data assimilation.

Data error estimation is the promising topic for exploration. Using information about the temporal stability of soil water content measurements appears essential for the data assimilation procedure. On average, the standard errors of noise were about 30 % of the "naïve" standard errors in the work of Pan et al. (2012). Using errors of noise, instead of naïve errors in this work, made the uncertainty in soil water content data about one order of magnitude smaller than the uncertainty in modeling results and caused the updated modeling results to be close to measurements. These results are similar to results of Starr (2005) who worked at the coarser scale and found that the temporal stability model explained 47 % of the observed variability in soil water content whereas additional 20 % of the variability was attributed to random measurement errors. Applying results, of the temporal stability analysis, is independent from the necessary accounting for the errors in sensor measurements *per se*.

Appropriateness of the temporal stability models (11.14, 11.15) or (11.19) depends on location and number of sensors. If sensor locations are such that the applicability temporal stability model is limited, the bias values may be inaccurate, the noise values may be excessively large and updates will not lead to significant changes in simulated water contents. Sensor placement may also affect the value of the average water content that is used in updates with the Kalman gain matrix (Eq. 11.7). Although the selection of locations where soil water content sensors should be placed, to record values of water content close to the average over the study area, have been suggested (Grayson and Western 1998; Jacobs et al. 2004; Rivera et al. 2012), there are still no general recommendations on selection of such sensor locations. The additional difficulty is that locations for sensors representing the average over the entire study area may be different for different soil depths (Guber et al. 2008). Finding the environmental factors that may indicate probable locations of representative soil moisture measurements will help to decrease the number of sensors and improve the effectiveness of the soil water content data assimilation.

The building of ensembles of models for water sensor data assimilation needs more research. Crow and van Loon (2006) discussed the remote sensing data assimilation and noted that relatively little guidance exists concerning strategies for selecting the appropriate magnitude and/or type of introduced model noise. They used the example of coarse-scale soil water model to demonstrate that inappropriate model error assumptions can worsen the performance of a model. In case of Richards equation-based soil water modeling, feasible ranges of initial soil water contents can be established for a specific case, but establishing a feasible ensemble of soil water flow parameter sets is far from trivial. Existing approaches shown in Fig. 11.4 need to be evaluated and compared. They need to be coupled with proper random sampling of the soils basic property space. The number of models in ensemble is known to affect the accuracy of assimilation results (Houtekamer and Mitchell 1998), and some indications need to be developed on the number of models in ensembles for soil water sensor data assimilation.

The effect of observation location placement and the number of sensors in soil profile on the efficiency of data assimilation is currently not clear. In essence, a single soil water sensor or tensiometer can provide enough information to correct the performance of a physically-based soil hydraulic model for the whole soil profile. The top part of the profile was the preferable location of the sensors for assimilation in the work of Pan et al. (2012). However, the site had no vegetation and soil water dynamics was not affected by shallow groundwater or by intensive evaporation in their work. For a scenario with different soil water dynamics, a site-specific exploratory research, on synthetic cases, would be needed to establish preferable locations of soil water content sensors.

Interpolation of updates in observation locations also requires investigations. Methods of interpolating soil water contents in soil profiles have been developed (Fernández-Gálvez et al. 2006), and their comparative efficiency in data assimilation algorithms needs to be established. This is especially important when the mass balance needs to be preserved, for example in modeling chemical transport in soils. Temporal frequency of updates needs to be determined for the extremes – large infiltration events and dry downs.

Evaluation of the data assimilation efficiency needs to be put in the context of the measurement and modeling purpose of specific types of projects, which may give very different soil moisture network monitoring requirements. For example, capacitance sensors have been used in irrigation scheduling (Fares et al. 2006), estimating soil hydraulic properties (Kelleners et al. 2005), evaluating tree water uptake (Schaffner 1998), upscaling soil water contents (Guber et al. 2009) and many other applications. Different monitoring goals dictate the focus of measurement of particular soil water regime features and events. Studies of contaminant transport in soils have focus on understanding and estimating fluxes, whereas studies of plant productivity are geared towards accurate water storage estimates. Metrics for the data assimilation evaluation have to be in line with the "monitoring + modeling" project goals.

Little is known about the meaning of data assimilation corrections. The changes in modeling results due to updates in data assimilation are a currently untapped source of information about the model and the observed system. It may be beneficial to evaluate soil water flow model by the quantitative and qualitative attributes of model update time series.

Overall, existing, albeit not numerous, examples show that assimilation of soil water content sensor data appeared to be very effective in correcting soil water content profiles simulated with the Richards equation based model; a small number of sensors can be sufficient to correct the simulated profile. Bringing together developments in pedotransfer functions, temporal stability of soil water patterns, and soil water content sensors can create a new source of data to improve modeling results in soil hydrology and related fields.

References

Aboitiz M, Labadie JW, Heermann DF (1986) Stochastic soil moisture estimation and forecasting for irrigated fields. Water Resour Res 22:180–190

Carsel RF, Parrish RS (1988) Developing joint probability distributions of soil water retention characteristics. Water Resour Res 24(5):755–769

Crow WT, Van Loon E (2006) The impact of incorrect model error assumptions on the sequential assimilation of remotely sensed surface soil moisture. J Hydrometeorol 8:421–431

Crow WT, Wood EF (2003) The assimilation of remotely sensed soil brightness temperature imagery into a land surface model using ensemble Kalman filtering: a case study based on ESTAR measurements during SGP97. Adv Water Resour 26(2):137–149

Das NN, Mohanty BP (2006) Root zone soil moisture assessment using remote sensing and vadose zone modeling. Vadose Zone J 5:296–307

de Jong van Lier Q, van Dam JC, Durigon A, dos Santos MA, Metselaar K (2013) Modeling water potentials and flows in the soil–plant system comparing hydraulic resistances and transpiration reduction functions. Vadose Zone J 12(3). doi:10.2136/vzj2013.02.0039

Entekhabi D, Nakamura H, Njoku EG (1994) Solving the inverse-problem for soil moisture and temperature profiles by sequential assimilation of multifrequency remotely sensed observations. IEEE Trans Geosci Remote Sens 32(2):438–448

Evensen G (1994) Sequential data assimilation with a non-linear quasi-geostrophic model using Monte Carlo methods to forecast error statistics. J Geophys Res 99(C5):10143–10162

Fares A, Hamdhani H, Polyakov V, Dogan A, Valenzuela H (2006) Real-time soil water monitoring for optimum water management. J Am Water Resour Assoc 42:1527–1535

Faulkner BR, Lyon WG, Khan FA, Chattopadhyay S (2003) Modeling leaching of viruses by the Monte Carlo method. Water Res 37(19):4719–4729. doi:10.1016/S0043-1354(03)00419-6

Fernández-Gálvez J, Simmonds LP, Barahona E (2006) Estimating detailed soil water profile records from point measurements. Eur J Soil Sci 57(5):708–718. doi:10.1111/j.1365-2389.2005.00761.x

Food and Agriculture Organization of the United Nations (FAO) (1975) Soil map of the world at 1:5,000,000, vol 1. Europe. UNESCO, Paris

Grayson RB, Western AW (1998) Towards areal estimation of soil water content from point measurements: time and space stability of mean response. J Hydrol 207:68–82

Guber AK, Pachepsky YA (2010) Multimodeling with pedotransfer functions. Documentation and user manual for PTF calculator (CalcPTF). Software and user manual. Available at: http://www.ars.usda.gov/research/docs.htm?docid=21090

Guber AK, Pachepsky YA, van Genuchten MT, Rawls WJ, Simunek J, Jacques D, Nicholson TJ, Cady RE (2006) Field-scale water flow simulations using ensembles of pedotransfer functions for soil water retention. Vadose Zone J 5(1):234. doi:10.2136/vzj2005.0111

Guber AK, Gish TJ, Pachepsky YA, van Genuchten MT, Daughtry CS, Nicholson TJ, Cady RE (2008) Temporal stability of soil water content and soil water flux across agricultural fields. Catena 73:125–133

Guber AK, Pachepsky YA, van Genuchten MT, Simunek J, Jacques D, Nemes A, Nicholson TJ, Cady RE (2009) Multimodel simulation of water flow in a field soil using pedotransfer functions. Vadose Zone J 8:1–10

Heatman GC, Starks PJ, Ahuja LR, Jackson TJ (2003) Assimilation of surface soil moisture to estimate profile soil water content. J Hydrol 279:1–17

Herbst M, Prolingheuer N, Graf A, Huisman JA, Weihermüller L, Vanderborght J (2009) Characterization and understanding of bare soil respiration spatial variability at plot scale. Vadoze Zone J 8:762–771

Houser PR, Shuttleworth WJ, Famiglietti JS, Gupta HV, Syed KH, Goodrich DC (1998) Integration of soil moisture remote sensing and hydrologic modeling using data assimilation. Water Resour Res 34(12):3405–3420

Houtekamer PL, Mitchell HL (1998) Data assimilation using an ensemble Kalman filter technique. Mon Weather Rev 126:796–811

Huang C, Li X, Lu L, Gu J (2008) Experiments of one-dimensional soil moisture assimilation system based on ensemble Kalman filter. Remote Sens Environ 113(3):888–900

Jacobs JM, Mohanty BP, Hsu EC, Miller D (2004) SMEX02: field scale variability, time stability and similarity of soil water. Remote Sens Environ 92:436–446

Jacques D, Mohanty B, Timmerman A, Feyen J (2001) Study of time dependency of factors affecting the spatial distribution of soil water content in a field plot. Phys Chem Earth B Hydrol Oceans Atmos 26:629–634. doi:10.1016/S1464-1909(01)00060-0

Kelleners TJ, Soppe RWO, Ayars JE, Im Nek J, Skaggs TH (2005) Inverse analysis of upward water flow in a groundwater table Lysimeter. Vadose Zone J 4:558–572

Köhne JM, Köhne S, Simůnek J (2009) A review of model applications for structured soils: a) Water flow and tracer transport. J Contam Hydrol 104(1–4):4–35. doi:10.1016/j.jconhyd.2008.10.002

Lahoz WA, Khattatov B, Ménard R (2010) Data assimilation making sense of observations. Springer, Heidelberg/London

Mandell J (2007) A brief tutorial on the ensemble Kalman filter. Cornell University. http://arxiv.org/PS_cache/arxiv/pdf/0901/0901.3725v1.pdf. Accessed 28 Jan 2014

Martinez-Fernández J, Ceballos A (2003) Temporal stability of soil moisture in a large-field experiment in Spain. Soil Sci Soc Am J 67:1647–1656

Martinez G, Feng P, Pachepsky YA (2013) SSDA code to apply data assimilation in soil water flow modeling: documentation and user manual. Software and user manual. Available at: http://www.ars.usda.gov/research/docs.htm?docid=21090

Meyer P, Rockhold M, Gee G, Nicholson T (1997) Uncertainty analyses of infiltration and subsurface flow and transport for SDMP sites Nureg/CR – 6565. U. S. Nuclear Regulatory Commission, Washington, DC

Montaldo N, Albertson JD (2003) Multi-scale assimilation of surface soil moisture data for robust root zone moisture prediction. Adv Water Resour 26:33–44

Montzka C, Moradkhani CH, Weihermüller L, Hendricks Franssen HJ, Canty M, Vereecken H (2011) Hydraulic parameter estimation by remotely-sensed top soil moisture observations with the particle filter. J Hydrol 399:410–421

Moradkhani H (2008) Hydrologic remote sensing and land surface data assimilation. Sensors 8(5):2986–3004

Moradkhani H, Sorooshian S, Gupta HV, Houser P (2005a) Dual state-parameter estimation of hydrological models using ensemble Kalman filter. Adv Water Resour 28:135–147

Moradkhani H, Hsu K, Gupta HV, Sorooshian S (2005b) Uncertainty assessment of hydrologic model states and parameters: sequential data assimilation using particle filter. Water Resour Res 41, W05012. doi:10.1029/2004WR003604

Morbidelli R, Saltalippi C, Flammini A, Rossi E, Corradini C (2013) Soil water content vertical profiles under natural conditions: matching of experiments and simulations by a conceptual model. Hydrol Process n/a–n/a. doi:10.1002/hyp.9973

Or D, Hanks RJ (1992) Spatial and temporal soil water estimation considering soil variability and evapotranspiration uncertainty. Water Resour Res 28:803–814

Pachepsky YA, Guber AK, Jacques D (2005) Temporal persistence in vertical distributions of soil moisture contents. Soil Sci Soc Am J 69:347–352

Pan F, Pachepsky Y, Jacques D, Guber A, Hill RL (2012) Data assimilation with soil water content sensors and pedotransfer functions in soil water flow modeling. Soil Sci Soc Am J 76(3):829–844. doi:10.2136/sssaj2011.0090

Park SK, Xu L (eds) (2009) Data assimilation for atmospheric, oceanic and hydrologic applications. Springer, Berlin

Plaza DA, De Keyser R, De Lannoy GJM, Giustarini L, Matgen P, Pauwels VR (2012) The importance of parameter resampling for soil moisture data assimilation into hydrologic models using the particle filter. Hydrol Earth Syst Sci 16:375–390. doi:10.5194/hess-16-375-2012

Quinn GP, Keough MJ (eds) (2002) Experimental design and data analysis for biologists. Cambridge University Press, New York

Ranatunga K, Nation E, Barratt D (2008) Review of soil water models and their applications in Australia. Environ Model Software 23(9):1182–1206

Ristic B, Arulampalam S, Gordon N (2004) Beyond the Kalman filter: particle filters for tracking applications. Artech House, London

Rivera D, Granda S, Arumí JL, Sandoval M, Billie M (2012) A methodology to identify representative configurations of sensors for monitoring soil moisture. Environ Monit Assess 184(11):6563–6574. doi:10.1007/s10661-011-2441-8

Rolston DE, Biggar JW, Nightingale HI (1991) Temporal persistence of spatial soil-water patterns under trickle irrigation. Irrig Sci 12:181–186

Sabater JM, Jarlan L, Calvet JC, Bouyssel F, de Rosnay P (2007) From near-surface to root-zone soil moisture using different assimilation techniques. J Hydrometeorol 8:194–206

Schaap M, Leij FJ (1998) Database-related accuracy and uncertainty of pedotransfer functions. Soil Sci 163(10):765–779

Schaffner B (1998) Flooding responses and water-use efficiency of subtropical and tropical fruit trees in an environmentally sensitive wetland. Ann Bot (London) 81:475–481

Scharnagl B, Vrugt JA, Vereecken H, Herbst M (2011) Inverse modelling of in situ soil water dynamics: investigating the effect of different prior distributions of the soil hydraulic parameters. Hydrol Earth Syst Sci 15(10):3043–3059

Scott RL, Shuttleworth WJ, Keefer TO, Warrick AW (2000) Modeling multiyear observations of soil moisture recharge in the semiarid American Southwest. Water Resour Res 36(8):2233–2247

Starr GC (2005) Assessing temporal stability and spatial variability of soil water patterns with implications for precision water management. Agric Water Manage 72:223–243

Stordal A, Karlsen H, Naevdal G, Skaug H, Valles B (2011) Bridging the ensemble Kalman filter and particle filters: the adaptive Gaussian mixture filter. Comput Geosci 15(2):293–305

Vachaud G, Passerat De Silans A, Balabanis P, Vauclin M (1985) Temporal stability of spatially measured soil water probability density function. Soil Sci Soc Am J 49:822–828

Vanderlinden K, Vereecken H, Hardelauf H, Herbst M, Martínez G, Cosh MH, Pachepsky YA (2012) Temporal stability of soil water contents: a review of data and analyses. Vadose Zone J 11(4). doi:10.2136/vzj2011.0178

Vereecken H, Huisman JA, Bogena H, Vanderborght J, Vrugt JA, Hopmans JW (2008) On the value of soil moisture measurements in vadose zone hydrology: a review. Water Resour Res 44:W00D06. doi:10.1029/2008WR006829

Vrugt JA, Diks CGH, Gupta HV, Bouten W, Verstraten JM (2005) Improved treatment of uncertainty in hydrologic modeling: combining the strengths of global optimization and data assimilation. Water Resour Res 41, W01017. doi:10.1029/2004WR003059

Walker JP, Willgoose GR, Kalma JD (2001) One-dimensional soil moisture profile retrieval by assimilation of near-surface measurements: a simplified soil moisture model and field application. J Hydrometeorol 2:356–373

Wendroth O, Rogasik H, Koszinski S, Ritsema CJ (1999) State-space prediction of field-scale soil water content time series in a sandy loam. Soil Tillage Res 50:85–93

Wigley TML, Briffa KR, Jones PD (1984) On the average value of correlated time series with applications in dendroclimatology and hydrometeorology. J Clim Appl Meteorol 23:201–213

Wu CC, Margulis SA (2011) Feasibility of real-time soil state and flux characterization for wastewater reuse using an embedded sensor network data assimilation approach. J Hydrol 399:313–325

Wu CC, Margulis SA (2013) Real-time soil moisture and salinity profile estimation using assimilation of embedded sensor datastreams. Vadose Zone J 12(1). doi:10.2136/vzj2011.0176

Chapter 12
New Perspectives for the Application of High-Resolution Benchtop X-Ray MicroCT for Quantifying Void, Solid and Liquid Phases in Soils

Carlos Manoel Pedro Vaz, Markus Tuller, Paulo Renato Orlandi Lasso, and Sílvio Crestana

Abstract Recent technological advances of imaging techniques such as X-ray computed tomography (CT) provide powerful means for visualization of the soil interior to evaluate pore topological properties and structure and study dynamic processes in a noninvasive way. The objective of this chapter is to present new opportunities for the application of advanced benchtop X-ray MicroCT scanners in soil science and vadose zone research and to discuss advantages and limitations associated with the MicroCT technique. Instrumentation, spatial resolution, image contrast and other aspects that directly affect the ability to discriminate and quantify various soil phases are discussed. Analysis of nearly 400 published papers regarding the application of computed tomography in soil and porous media research indicate a surge of applications of benchtop MicroCT systems over the last few years. These instruments are becoming more affordable and popular and have great potential to significantly enhance the ability to characterize and better understand dynamic processes in soils and provide important insights for the development of advanced mathematical pore network and fluid dynamics models. New advanced benchtop MicroCT systems are very flexible in terms of sample size and image spatial resolution and provide high quality images, comparable to advanced synchrotron systems. Results of soil samples scanned at different spatial resolutions are presented and their air phase (porosity) and solid phase are evaluated and quantified. Examples of liquid phase quantification using sodium iodide as dopant are also presented for glass bead and undisturbed soil samples.

C.M.P. Vaz (✉) • P.R.O. Lasso • S. Crestana
Embrapa Agricultural Instrumentation, Brazilian Agricultural Research Corporation-EMBRAPA, P.O. Box 741, 13560-970 São Carlos, SP, Brazil
e-mail: carlos.vaz@embrapa.br

M. Tuller
Department of Soil, Water and Environmental Science, The University of Arizona, Tucson, AZ 85721, USA

W.G. Teixeira et al. (eds.), *Application of Soil Physics in Environmental Analyses: Measuring, Modelling and Data Integration*, Progress in Soil Science,
DOI 10.1007/978-3-319-06013-2_12,

Keywords Microtomography • Soil porosity • Soil structure • CT instrumentation

12.1 Introduction

Recent technological advances of imaging techniques such as X-ray computed tomography (CT) provide powerful means for visualization of the soil interior to evaluate pore topological properties and structure and study dynamic processes in a noninvasive way. The idea of looking inside an opaque object or body without cutting or dissecting it was considered impossible a century ago. However, because of the groundbreaking discoveries about ionizing radiation by Wilhelm Röntgen, Henri Becquerel, and Pierre and Marie Currie and the development of mathematical theory by Johann Radon, Allan Cormack and many others scientists and engineers, it was possible to design the first commercial computed tomography scanner for non-invasive medical diagnosis that was fabricated in 1972 by a team led by Godfrey Hounsfield (Hounsfield 1973). Some years after the first computer-assisted tomography systems (CT) were installed in hospitals, other scientific disciplines such as archeology (Heinemann 1978), material science (Vainberg et al. 1982), soil (Petrovic et al. 1982) and plant sciences (Onoe et al. 1983) started to utilize this new powerful technology.

In soil science, X-ray CT was first recognized for its great potential to contribute to the noninvasive characterization of the soil structure and to studies of dynamic processes such as water infiltration, seed germination, root water uptake, soil shrinking, compaction, earthworm ecology, only to name a few (Petrovic et al. 1982; Hainsworth and Aylmore 1983; Crestana et al. 1985).

The objective of this chapter is to present new opportunities for the application of advanced benchtop microCT systems in soil science and vadose zone research and to discuss advantages and limitations associated with the microCT technique. A number of excellent reviews and book chapters describing the basic theory and concepts, most relevant applications in soils science and anticipated future developments have been recently published (Helliwell et al. 2013; Cnudde and Boone 2013; Wildenschild and Sheppard 2013; Singh and Niven 2013; Peth 2010). So, rather than presenting a general overview covering all relevant aspects of CT in soil science, the present chapter focuses on the discussion of instrumentation, image spatial resolution, image contrast and other aspects which directly affect the ability to discriminate and quantify various different soil phases (air/porosity, solid/minerals-root, liquid/soil solution). Results of soil samples scanned at different spatial resolutions are presented and their air phase (porosity) and solid phase are evaluated and quantified. Examples of liquid phase quantification are also presented using glass beads and undisturbed soil samples.

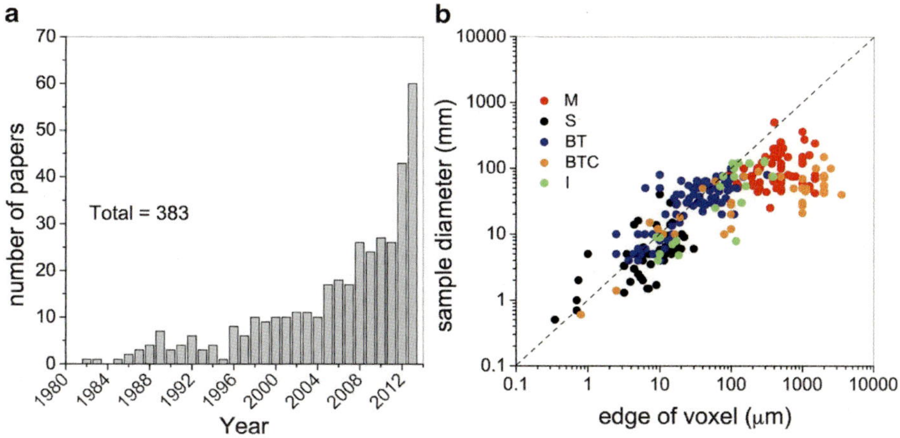

Fig. 12.1 Number of papers about computed tomography applications in soil science published annually (**a**); and tradeoff between sample size (diameter) and spatial resolution (edge of voxel size) for different CT scanner systems (**b**). *M* medical scanners, *S* synchrotron-based scanners, *BT* benchtop scanners, *BTC* customized benchtop scanners, *I* industrial scanners

12.2 Latest Trends in CT Instrumentation Applied in Soil Science Research

In recent years CT has been extensively applied to many different fields of soil science and a comprehensive search of scientific publication databases reveals that there were nearly 400 papers about the application of CT to soils and other porous materials such as packed glass beads, sands or sediments published over the last three decades (1982–2013) (Fig. 12.1a). The average number of papers published in the first 10 years between 1982 and 1991 was about 3 papers/year, increasing to 7 in the period from 1992 to 2001 and to 19 in the period of from 2002 to 2011. However, in 2012 alone, 43 papers were published and 60 in 2013, demonstrating the surge in applications of CT in soil science over the last few years.

Table 12.1 shows the number of published papers grouped in instrument classes that include medical X-ray CT scanners (M), synchrotron based X-ray MicroCT scanners (S), commercial benchtop scanners (BT), customized benchtop systems (BTC), and industrial CT scanners (I), in periods of 4 years from 1982 to 2013. The significant increase in scientific publications occurred after 2005–2006 and this was in great part due to the release and use of new commercial benchtop CT scanners (also called desktop or cabinet-type systems) and the intensification of applications of synchrotron X-ray based microtomography systems (Table 12.1).

The use of medical CT systems was predominant in the first 20 years, but practically stabilized over the last 10 years, being replaced by new and more adequate benchtop and synchrotron MicroCT systems. Therefore, it is evident that the relatively restricted use of CT in soil science in the first two decades, after the first paper came out in 1982 and the remarkable increase over the last decade, mainly in the last

Table 12.1 Number of papers published in 4 years periods from 1982 to 2013 for different groups of CT scanner systems

	Number of published papers					
Period	M	S	BT	BTC	I	Total
1982–1985	3	0	0	0	0	3
1986–1989	9	0	0	4	0	13
1990–1993	11	0	0	2	0	13
1994–1997	12	4	0	1	1	18
1998–2001	22	2	1	8	5	38
2002–2005	23	8	3	12	4	50
2006–2009	20	13	33	9	7	82
2010–2013	27	24	66	17	5	139
Total	127	51	103	53	22	356

M medical scanners, *S* synchrotron-based scanners, *BT* benchtop scanners, *BTC* customized benchtop scanners, *I* industrial scanners

few years, was driven by easier access to scanning facilities and improvements of image spatial resolution, acquisition time and computation for image analysis. It is expected that the actual expansion observed in terms of new CT instruments and facilities (mainly BT and S-CT systems) will further popularize the CT technique in soil science and other fields such as material and biomaterial science, wood and seed technology, paleontology, archeology, entomology and botany, and lead to the development of sophisticated new analytical procedures, image analysis software and probably yield new standard or reference methods.

A detailed analysis of publications summarized in Table 12.1 (complete reference list available at www.cnpdia.embrapa.br/downloads/microCT/reference_dataset.pdf) allows discussion of important aspects of specific instruments, image spatial resolution, drawbacks and future trends and challenges for application of X-ray CT in soil science as discussed in the following.

One of the advantages of medical CT systems is the opportunity to scan large soil columns within a short period of time. However, the image spatial resolution obtained with such systems is relatively low (about 0.5 mm on average), only allowing characterization and quantification of the spatial distribution of larger pores, roots and earthworm burrows, cracks and larger structures. Synchrotron-based systems on the other hand can feature at spatial resolutions of few micrometers, providing means to explore the intra-aggregate porosity and pore-particle interfaces among other features of interest, but only for soil samples of a few millimeters in size. The constraint or tradeoff between sample size and spatial resolution (voxel size) is imposed by the CT scanner detector characteristics (detector size and individual pixel size), magnification and other geometrical and instrument characteristics.

In Fig. 12.1b this tradeoff between the image spatial resolution, represented by the edge of voxel size in μm and sample size (diameter) in mm is illustrated based on information provided in papers evaluated in Table 12.1 for different classes of CT systems (detailed data available at www.cnpdia.embrapa.br/downloads/microCT/reference_table.pdf). In general, the relation between sample diameter and edge of voxel lies close to the 1:1 line for synchrotron and benchtop systems, but drops below this line for medical, industrial and customized systems. This is

Table 12.2 Statistics describing the relation between sample and voxel sizes and sample/voxel ratios achievable with various discussed classes of CT systems applied in soil science studies

		Average	Min.	Max.	SD	CV%
Sample diameter (mm)	M	111	25	360	60	54
	S	7	0.5	40	6	95
	BT	39	4	100	23	60
	BTC	47	0.6	150	28	60
	I	61	4	130	50	82
Edge voxel (μm)	M	549	100	2,000	455	83
	S	10	0.4	30	6	61
	BT	44	2.5	320	42	96
	BTC	953	0.8	3,500	826	87
	I	121	9	387	109	90
Ratio sample/voxel (mm/μm)	M	0.31	0.02	0.89	0.19	61
	S	0.89	0.19	5	0.92	103
	BT	1.28	0.18	8	1.22	95
	BTC	0.26	0.01	2.05	0.43	165
	I	0.58	0.07	1.14	0.29	50

Min. minimum, *Max.* maximum, *SD* standard deviation, *CV* coefficient of variation

due to the more advanced and flexible configuration of BT and S-CT systems with regard to image magnification and detector length pixel size ratio. For instance, some new BT systems employ rectangular detectors as large as 4,000 × 2,000 pixels (Vaz et al. 2011), or even higher, with pixel sizes below 1 μm, while medical systems normally employ circular array detectors with spatial resolutions in the millimeter range.

Statistics about the relation between common sample sizes and image spatial resolutions used for soil studies and the ratio between these two parameters are presented in Table 12.2 (for the dataset evaluated for Table 12.1). The average voxel size was about 1 mm for customized, 0.5 mm for medical, 120 μm for industrial, 40 μm for commercial benchtop and 10 μm for synchrotron-based systems, while the average sample diameters were about 11 cm for medical, 6 cm for industrial, 5 cm for customized, 4 cm for commercial benchtop and 0.7 cm for synchrotron-based systems.

The different CT classes significantly differ in terms of observable sample size and achievable spatial resolution (Fig. 12.2). The BTC-CT systems that include self-fabricated instruments exhibit spatial resolutions ranging from very low as is the case for gamma ray systems (1–3 mm pixel size) to very high-resolution of X-ray MicroCT systems (about 1 μm pixel size), but most of them are in the millimeter range with an average edge voxel size of 953 μm. A recent review (Pires et al. 2010) summarize several interesting applications of the gamma ray CT in soil physics, including homemade instrumentation developments at the Brazilian Embrapa Instrumentation group (Crestana et al. 1986; Vaz et al. 1989; Cruvinel et al. 1990; Macedo et al. 1998).

Synchrotron-based CT instrumentation commonly only allow characterization of very small samples (7 mm on average). Examples of available synchrotron MicroCT facilities are listed in Table 12.3. BT-CT systems can achieve spatial

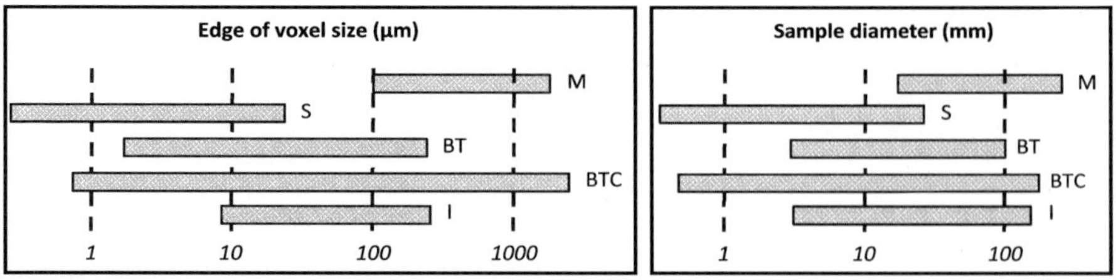

Fig. 12.2 Illustration of the different ranges of edge voxel sizes and sample diameters obtained with different CT-systems as presented in Table 12.2

Table 12.3 List of synchrotron MicroCT facilities, and common benchtop and medical CT-systems used for soil science research

Synchrotron	BenchTop	Medical
APS – USA	X-Tek HMX (Metris) – Nikon	Somatom Plus – Siemens
SLS – Switzerland	Phoenix Nanotom – GE	Prospeed SX Advantage – GE
NSLS – USA	Phoenix vltomelx – GE	Mx8000 IDT 16 – Philips
SSRF – China	SkyScan 1172 – Bruker	CT Pace – GE
HASYLAB – Germany	SkyScan 1072 – Bruker	CT 5005 – EMI
ESRF – France	MS 8X-130 – GE	Aquilon – Toshiba
DLS – UK	SMX-225CT, Shimadzu	Lightspeed VTC – GE
ALS – USA	Toscanner 32250 – Toshiba	W2000 – Hitachi
LSU – USA	µCT 40; 80 – Scanco	HD350 – Universal Systems
Spring 8 – Japan	Xtreme – Scanco	Technicare Deltascan 2020-G

resolutions ranging from the same order as synchrotron systems (few micrometers) to lower resolutions similar to that achievable with the most advanced medical systems (around 100 µm). The most commonly employed commercial BT and medical systems used in these studies are also listed in Table 12.3. The industrial systems have more restricted use in soil science because they are in general large and expensive instruments that are better suited for evaluation of highly dense and large objects.

One evident advantage of synchrotron-based CT systems is the monoenergetic nature of the X-rays, which are modulated with monochromators and the high photon flux at the beam line, compared to the much less intense polychromatic radiation generated with X-ray tubes employed for medical, benchtop, customized and industrial systems. However, the high costs for construction and maintenance of synchrotron accelerators, the small number of available systems and the limited beam time allocated per user are considerable restrictions for a significant increase of S-CT system applications. On the other hand, high-resolution benchtop CT systems are becoming more affordable and popular with an increasing number of new systems, models and brands commercially available. Because of these reasons we project a great expansion and popularization of benchtop MicroCT instrumentation in various fields of materials research, including soil science, in the coming years.

12.3 Quantification of Void and Solid Phases

Soils are very heterogeneous porous materials in terms of composition, particle size distribution and shape, resulting in very complex three-dimensional structures.

The actual MicroCT image spatial resolution of about 1–2 μm allows observation of the full range of macropores (>75 μm), mesopores (30–75 μm) and micropores (5–30 μm) and part of the ultramicropores (0.1–5 μm), according to the classification of Cameron and Buchan (2006). Although some recently released nanoCT scanners and new CT Synchrotron facilities (e.g. SLS-Switzerland, APS-USA) allow sub-micron spatial resolutions, only samples smaller than about 1 mm can be imaged (Bailey et al. 2013; Rozenbaum et al. 2012; Sorensen et al. 2012), which generally is not representative for the whole natural soil system. Therefore, for soils containing significant amounts of clay-sized particles (i.e. diameters lower than 2 μm) a large percentage of the total porosity cannot be accessed with MicroCT, even at the highest possible spatial resolutions achievable today. This is illustrated in Fig. 12.3, where image edge of voxel size is plotted against CT derived porosities obtained from the dataset of published papers in soil science. Despite the large variation of evaluated soil types and utilized CT systems, a clear trend of increasing CT-derived porosity with increasing image spatial resolution is evident. The trend line shows that for edge voxel sizes larger than 30 μm, less than 10 % of the total soil porosity generally can be extracted depending on soil type and the pore size distribution. About 30–40 % of the porosity can be accessed when high-resolution MicroCT systems with voxel sizes of few micrometers are applied.

The uniqueness of the computed tomography technique is its non-invasiveness and the possibility to quantify internal structures based on the linear attenuation coefficients (μ), which depend on the chemical composition and physical density of

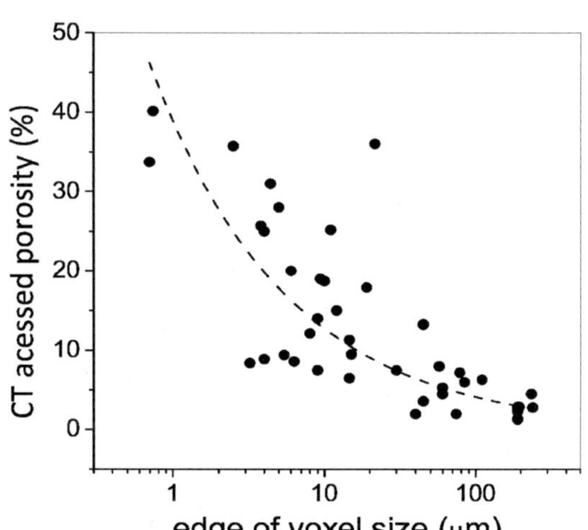

Fig. 12.3 CT-derived porosities as a function of edge voxel size obtained from published papers

Table 12.4 Basic soil physical properties of investigated Brazilian soils

		Clay	Silt	Sand	ρ_b	ρ_p	ϕ
Soil	Class	$g\ g^{-1}$			$g\ cm^{-3}$		$cm^3 cm^{-3}$
S1	Rhodic Kandiudalf	0.395	0.123	0.487	1.61	2.93	0.390
S2	Typic Hapludox	0.292	0.074	0.634	1.27	2.74	0.470
S3	Albaqualf	0.196	0.243	0.561	1.37	2.66	0.416
S4	Rhodic Hapludox	0.732	0.147	0.121	1.15	2.78	0.502
S5	Typic Hapludult	0.793	0.092	0.115	1.13	2.81	0.541

ρ_b: soil bulk density, ρ_p: soil particle density, ϕ: total porosity

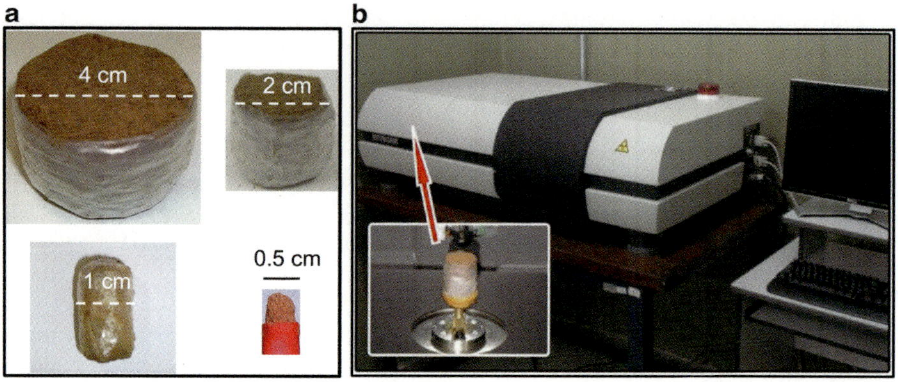

Fig. 12.4 Undisturbed soil samples of different sizes (**a**) scanned at different spatial resolutions with a benchtop SkyScan 1172 MicroCT system (**b**)

the analyzed materials (Wildenschild et al. 2002). Therefore, internal soil structures with sufficient contrast due to their differences in linear attenuation coefficients can be detected and quantified. The relatively high attenuation coefficients of soil mineral particles when compared to air yield good contrast and allow discrimination of solid and void phases, which also depends on the X-ray energy, image spatial resolution and average soil pore size or distribution (Peth 2010).

To evaluate the ability to discriminate void and solid phases and quantify structural features of soil porous media, five oven-dried undisturbed soils were scanned at four different sample sizes (4, 2, 1 and 0.5 cm) and image spatial resolutions (30, 12, 6 and 3 μm, respectively), using a high-resolution benchtop CT-system (SkyScan 1172, Bruker). These soils cover a wide range of soil textures and some selected soil physical properties are presented in Table 12.4. Details on the soil physical analysis, sample preparation for CT analysis, scanning and reconstruction parameters can be found in Vaz et al. (2011). Examples of soil samples (A) analyzed with the benchtop the SkyScan 1172 MicroCT (B) are shown in Fig. 12.4.

Grayscale MicroCT cross-sectional images of the scanned soil samples are presented in Fig. 12.5, where white represents very dense particles (e.g. iron oxides), black represents the air-filled pore space and the light to dark gray areas

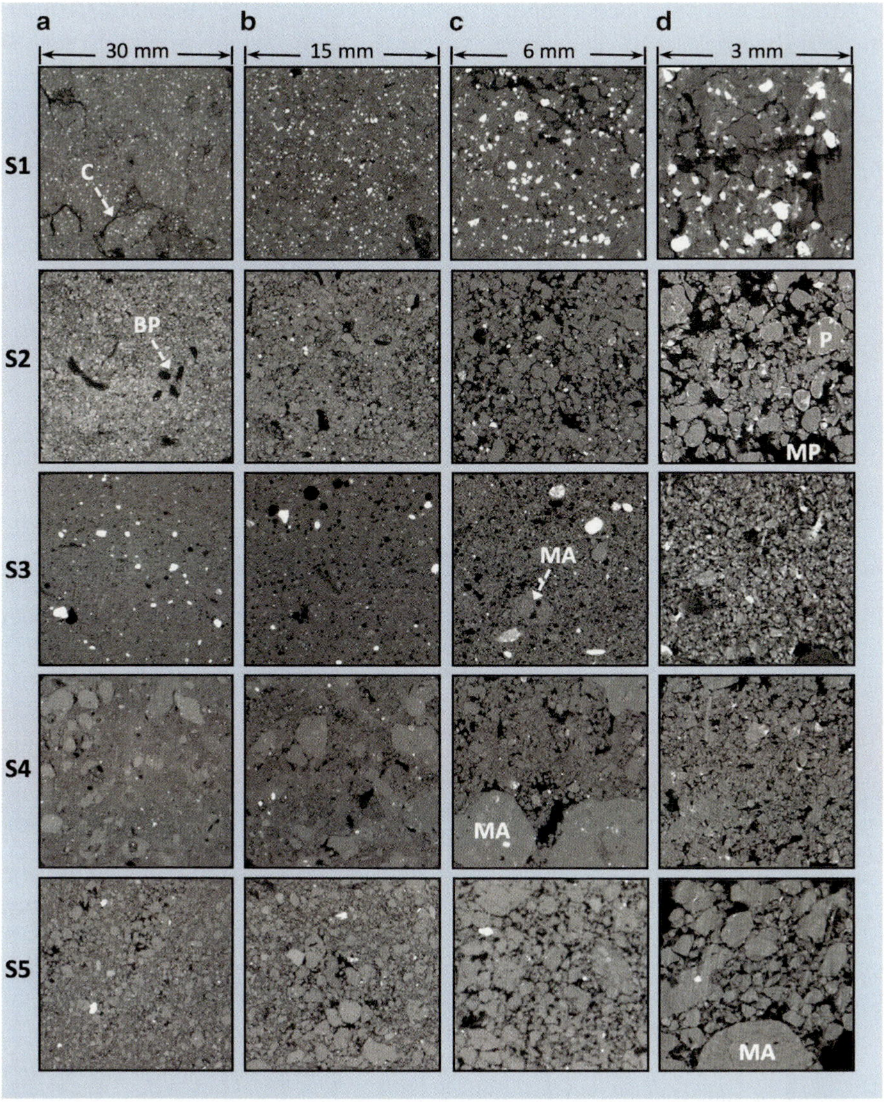

Fig. 12.5 Grayscale cross-sectional MicroCT images of soils S1 to S5 (Table 12.4) for different sample and voxel sizes (A ≈ 4-cm diameter/30 μm edge voxel size; B ≈ 2 cm/12 μm; C ≈ 1 cm/6 μm; D ≈ 0.5 cm/3.5 μm). *Black* represents air and *light to dark gray* the solid phase (particles and aggregates). *White spots* are very high attenuating solid particles. *BP* biopores, *P* particles, *MP* macropores, *MA* microaggregates

represent less dense particles or aggregates. Scanning of soil samples at different resolutions, which is only possible with very flexible benchtop MicroCT systems, provide valuable and complimentary information about large biopores and smaller pores between individual soil particles, inter-aggregate and intra-aggregate pores.

At the largest evaluated sample size (4 cm diameter) large biopores (BP) and cracks (C) with different geometries are clearly visible, mainly in soils S1 and S2 (more round and elongated in soil S2 and cracks and more heterogeneous patterns in soil S1). The large amount of sand in soil S2 (63 %) allows visualization of individual sand particles (P) and large and very well connected macropores (MP) in the smallest scanned sample (0.5 mm in diameter). Soil S3 exhibits the highest amount of silt (24 %) and fine sand particles and shows very well defined round pores, observed mainly in the largest samples (lowest image spatial resolution). These round-shaped macropores are probably formed by entrapped air bubbles, since soil S3 is a flooded Albaqualf. Relatively small silt sized particles (2–50 μm) and very small microaggregates (MA) can be visualized in the smallest investigated soil samples images (1 and 0.5 cm in diameter) in soil S3. Soils S4 and S5 are heavy clayey soils (clay contents higher than 70 %) characterized by large amounts of microaggregates, which on average are larger for soil S4 than soil S5. It is evident that the analysis of MicroCT images for varying sample size/image spatial resolution allows identification of interesting structural and textural features of soil porous systems such as the pore size distribution, shape and connectivity, micro-aggregate distribution and very high dense mineral particles.

Quantification of void and solid phases from MicroCT images are performed after image binarization or segmentation, based on the analysis of the frequency distribution (histogram) of linear attenuation coefficients (global thresholding) or by means of more sophisticated locally-adaptive or dynamic thresholding techniques (Iassonov et al. 2009; Kulkarni et al. 2012; Helliwell et al. 2013; Tuller et al. 2013). Global thresholding generally performs well when image histograms exhibit distinct separation between air (pore) and solid peaks (Peth 2010) and a threshold value for separating the two phases can be chosen as the minimum value or the half distance between these two peaks (Peth 2010; Capowiez et al. 1998).

The linear attenuation coefficient histograms are presented in Fig. 12.6 for soils S1–S5 for investigated sample size/image voxel ratios (4 cm/30 μm; 2 cm/12 μm; 1 cm/6 μm; 0.5 cm/3.5 μm). In order to visualize the "air" or void peak, the main solid peak and other very high attenuating or dense mineral particles, histograms are displayed in two parts (main graphs and inserts) with adequate voxel count number scales to fit the different peaks.

In general, air, solid and very dense particle peaks are better defined as sample size decreases and image spatial resolution increases, due to decreasing influence of partial volume effects (Clausnitzer and Hopmans 1999) for both ranges of linear attenuation coefficients. Variations among histograms for a given soil also occur due to local small-scale variability of different soil sample diameters. Differences in the main solid peak between investigated soils are due to differences in bulk and particle densities, with the more dense soil (S1) having the highest linear attenuation coefficients.

Binarized cross-sectional images are presented in Fig. 12.7 for soils S1–S5 for various investigated sample size/voxel size ratios. A clear increase of CT-derived porosity (black) can be observed as image spatial resolution increases. Figure 12.8 displays the relative CT-derived porosity as function of image spatial resolution,

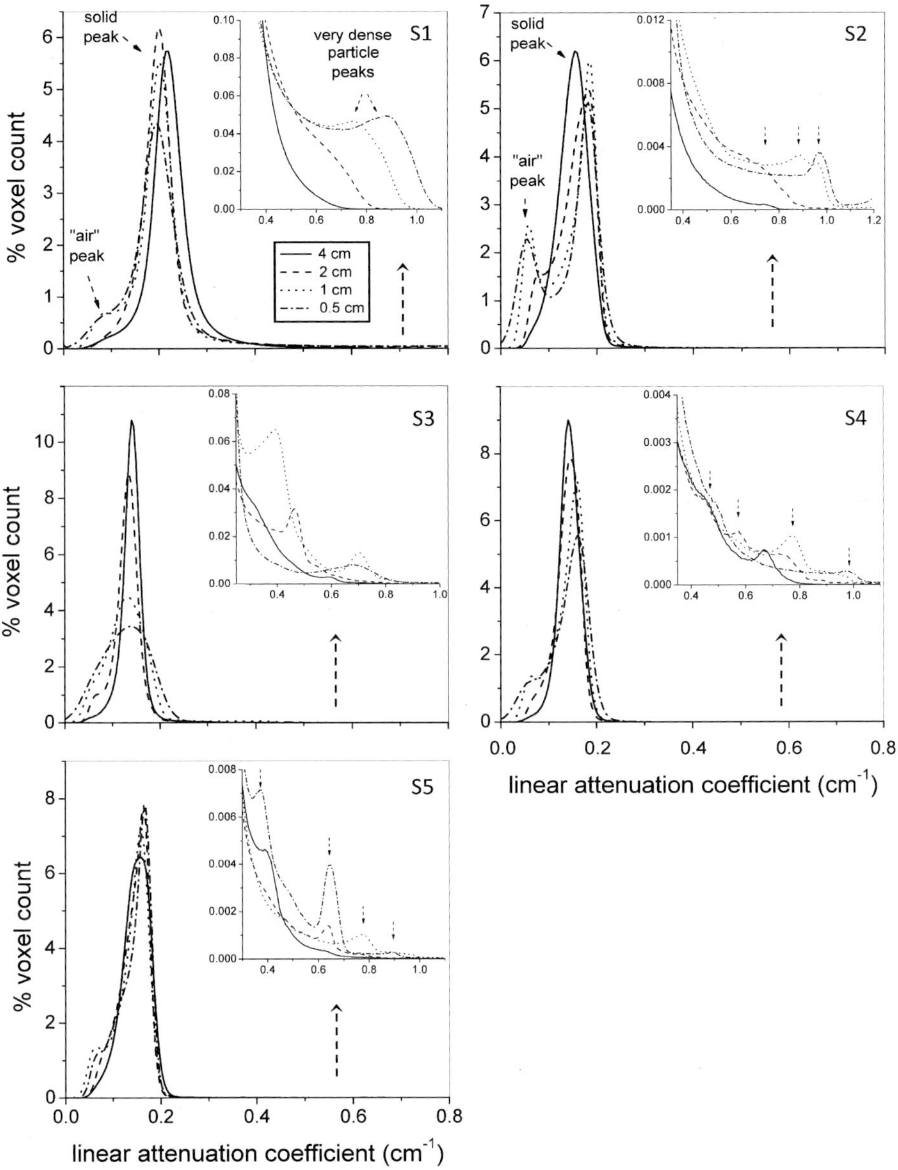

Fig. 12.6 Linear attenuation coefficient histograms of MicroCT gray scale images (Fig. 12.5) obtained for soils S1 to S5. The main graphs on *left side* depict the air peak and the solid phase peaks and *insert* graphs show the high attenuating and very dense soil particles

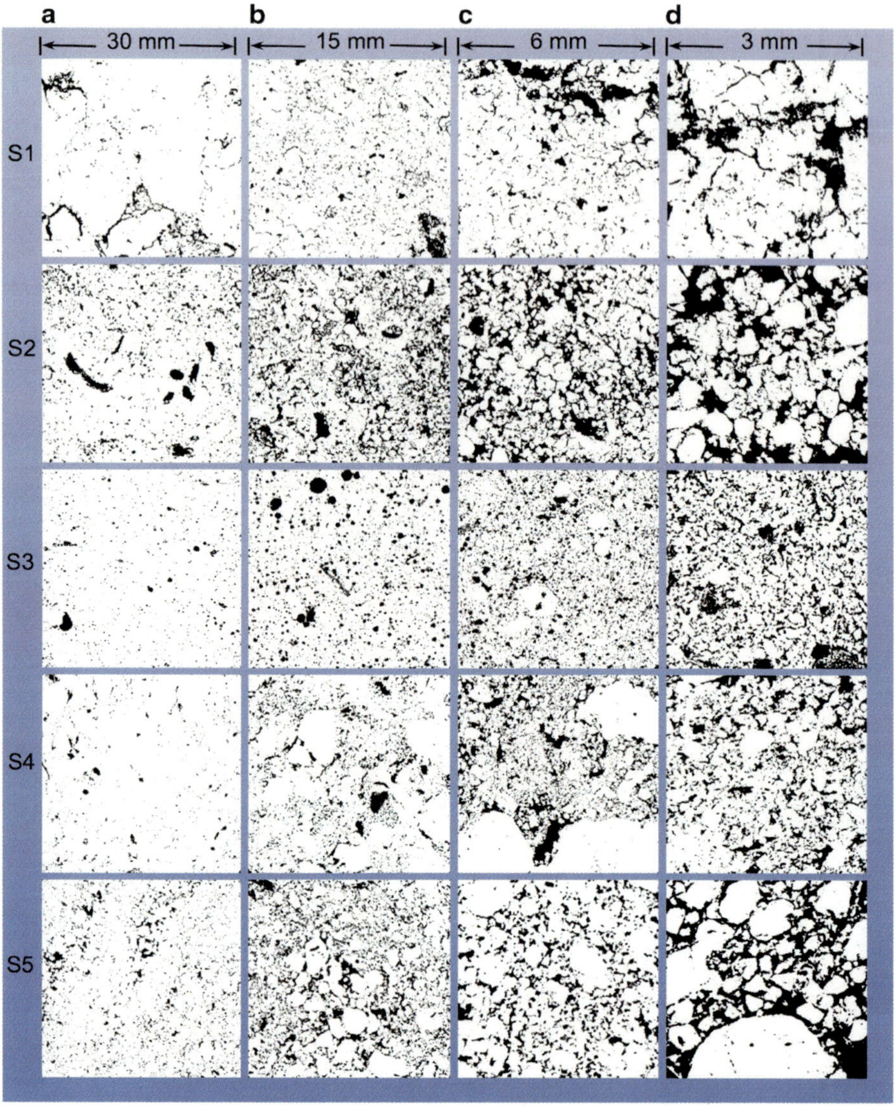

Fig. 12.7 Binarized cross-sectional MicroCT images of soils S1 to S5 (Table 12.4) for investigated sample and voxel sizes (A $\approx$ 4 cm diameter/30 μm edge voxel size; B $\approx$ 2 cm/12 μm; C $\approx$ 1 cm/6 μm; D $\approx$ 0.5 cm/3.5 μm)

which exhibits similar behavior as the previously published data depicted in Fig. 12.3. Scanning soil samples of various sizes at the highest spatial resolution as performed with the SkyScan 1172 benchtop MicroCT system is a useful approach for quantitative analysis of pore-scale structure.

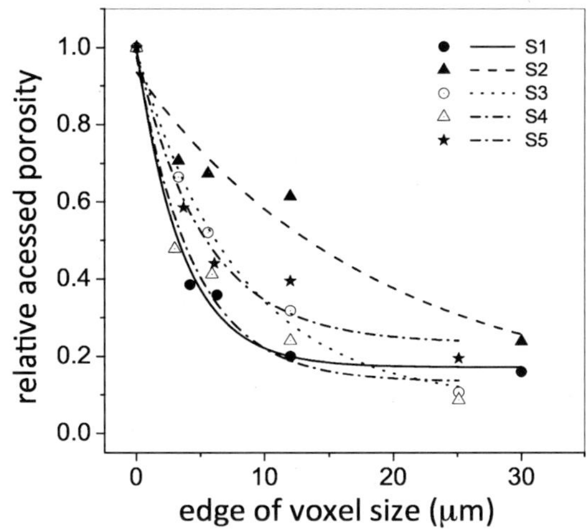

Fig. 12.8 CT-derived porosities for soils S1 to S5 scanned at different sample/voxel sizes, normalized by physically measured total porosities (Table 12.4)

12.4 Quantification of Very Dense Mineral Particles

The main interest in the analysis of soil MicroCT images still lies in visualization of the pore system and quantification of parameters such as pore size distribution, connectivity, tortuosity and related hydrological, agricultural and ecological processes (Helliwell et al. 2013; Peth 2010). However, CT soil images also provide valuable information on undisturbed, three-dimensional and spatial distribution of mineral particles (Ketcham 2005) and microaggregates, especially when high-resolution MicroCT systems are applied.

While discrimination of individual components of the complex soil solid phase or minerals with similar linear attenuation coefficients are difficult (Sorensen et al. 2012) there are opportunities to differentiate between highly dense mineral particles (e.g., iron, titanium and manganese oxides) and other less dense materials such as quartz and aluminum silicate minerals (Vaz et al. 2011). For instance, Taina et al. (2010) applied a thresholding approach to distinguish soil voids and Fe-Mn oxide segregations and to quantify their size distribution and shape. Bellato et al. (2013), who studied a soil-cement material, were also able to distinguish different solid phase components as packed clay inclusions, small organism exoskeletons and very highly absorbing particles (metal-bearing minerals).

The peaks shown in the inserts of Fig. 12.6 are assigned to high attenuating and very dense minerals, which appear in images shown in Fig. 12.5 as white spots. In general, these peaks situate at linear attenuation coefficient higher than 0.3 cm^{-1} for the applied specific scanning conditions (SkyScan 1172, 100 kV and 100 μA, Al/Cu filter). The presence of two or three peaks in the high linear attenuation coefficient range indicates the presence of significant amounts of iron, manganese and titanium oxides, which are common in Brazilian Oxisols. Therefore, assuming a global manual threshold value of 0.35 cm^{-1} (Vaz et al. 2011) between the more intense solid peak and the high attenuation peaks allows to segment the very dense

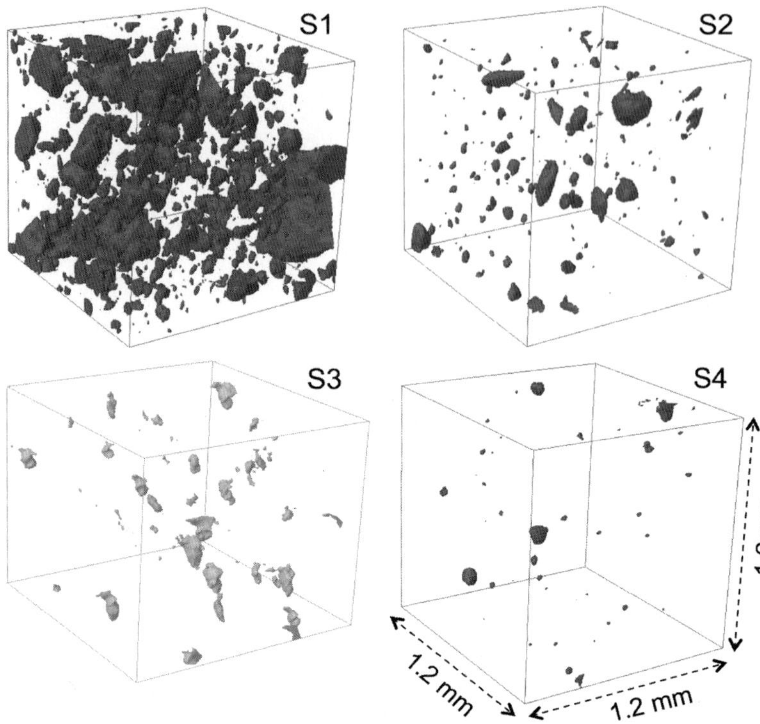

Fig. 12.9 Segmented high dense solid particles of Fe, Ti and Mn oxides of cube-shaped sub volumes (300 × 300 × 300 voxels) for soils S1 to S4

mineral particles which are associated with the presence of Fe, Ti and Mn oxide particles. Figure 12.9 shows reconstructed 3-D views of the five soils scanned at the highest resolution of about 3.5 μm (column D in Fig. 12.5). The cube-shaped sub volumes (300 × 300 × 300 voxels) were segmented for the high attenuating and very dense particles.

MicroCT-extracted volume percentages of the highly dense mineral solid particles were 9.2, 0.65, 0.38 and 0.10 % for soils S1, S2, S3 and S4, respectively. These values are in good agreement with iron and titanium oxides amounts commonly found in Oxisols (Costa et al. 2002). Particle size distributions of these high dense particles varied from 20 to 800 μm, with most frequent diameters of about 100 μm (Fig. 12.10).

Three-dimensional visualization of the solids and void phases (300 × 300 × 300 voxel cubes at the highest image spatial resolution-column D in Fig. 12.5) are depicted in Fig. 12.11 for soils S1 and S2, where light gray and white represent the solid phase (white are the very dense mineral particles, i.e. Fe, Ti, Mn oxides and light gray are the less dense particles and aggregates) and dark gray is the pore space.

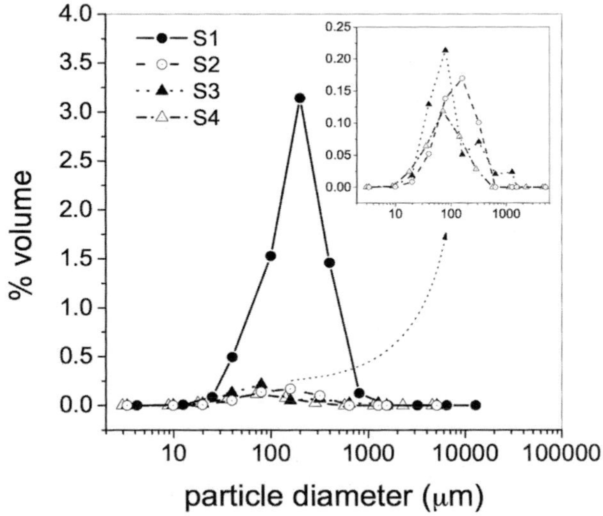

Fig. 12.10 Particle size distribution of high dense mineral particles obtained from MicroCT images of the largest soil samples (column A in Fig. 12.5). Insert graph shows data for soils S2, S3 and S4 zoomed for better visualization

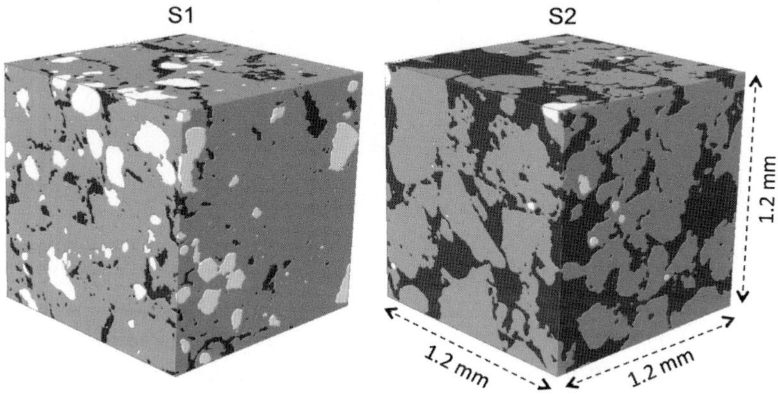

Fig. 12.11 Segmented solid phase (*white* – very dense minerals; *light gray* – less dense minerals and aggregates) and pore or void phase (*dark gray*) of cube-shaped sub volumes (300 × 300 × 300 voxels) for soils S1 and S2

12.5 Quantification of the Liquid Phase

Differentiation between the liquid phase and the other soil components at the microscale can be attained with the MicroCT technique, but it depends on several instrument features such as the X-ray energy spectrum, image spatial resolution and also some characteristics of the evaluated porous media (e.g., texture and pore size distribution). In general, there is not sufficient contrast between the liquid and the solid phase and some chemical dopant has to be added to the liquid phase to enhance contrast and to allow segmentation of the liquid phase (Wildenschild et al. 2002).

Table 12.5 Materials, container diameter and scanning parameters for experiments 1–3. Samples were scanned in saturated state (with 1 % NaI), partially saturated and dried

	Porous media	Voxel size (μm)	Container diameter (mm)	Filter
Exp 1	GB diam. 0.5 mm	5.7	20	Al, Al + Cu, "no filter"
Exp 2	GB diam. 1 mm	6.4	12	Al + Cu
Exp 3	Sandy soil	2.5	12	Al + Cu

GB glass beads, *Al* 0.5 mm thick aluminum filter, *Al+Cu* 0.5 mm aluminum + 0.04 mm Cu thick filter, *no filter* scanned without filter

Fig. 12.12 Experimental setup to saturate glass bead, sand and soil samples for MicroCT analysis from saturation to dry conditions

To illustrate the influence of these parameters on the phase contrast and consequent differentiation of the liquid phase, three experiments were conducted as summarized in Table 12.5. In all studies a 1 % sodium iodide solution (NaI) was used to saturate the porous materials, using a hanging column as shown in Fig. 12.12. In experiment 1 (Exp 1) different filters and X-ray energies were employed to evaluate the effect of different X-ray spectra on the liquid phase for pure water and 1 % NaI solution. The X-ray tube operational parameters were 100 kV/100 μA when an aluminum/copper filter (0.5 mm/0.004 mm) was used, 89 kV/112 μA for an aluminum filter (0.5 mm), and 59 kV/167 μA when no filter was deployed. In Exp 2 and 3, glass beads (1 mm diameter) and a sandy soil (S2) were scanned at 1 % NaI solution saturation, intermediate wet and dry conditions. These samples were dried using a 60 W heater positioned close to the sample surface, covered with a PVC film and equilibrated over 48 h, prior to scanning.

The grayscale histograms of images obtained in experiment 1 (Exp 1) with Al, Al/Cu filters and without filter at saturation with deionized water, 1 % NaI solution and dry glass beads are presented in Fig. 12.13a–c. The liquid phase peaks are much better defined when employing the metal filters, especially the Al/Cu filter, as compared with no filter. This is because metal filters block low energy or soft X-rays that cause beam-hardening image artifacts. Additionally, the use of 1 % NaI solution instead of deionized water displaces the liquid peak to the right (higher linear attenuation coefficient or grayscale values) improving liquid phase detectability.

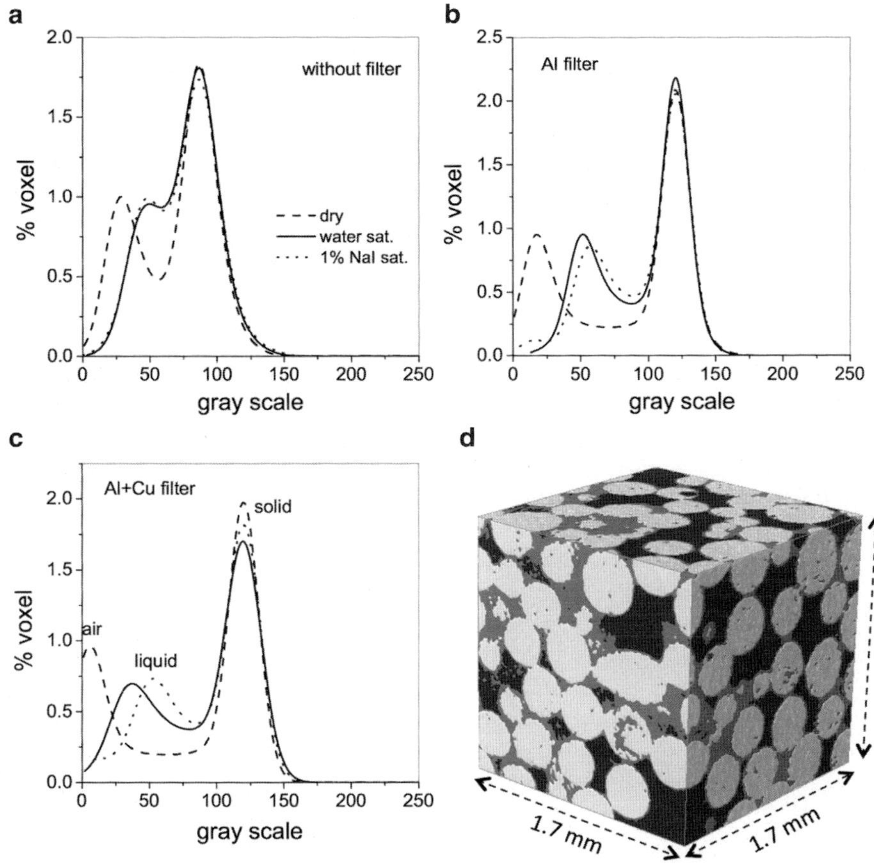

Fig. 12.13 MicroCT image grayscale histograms of 0.5 mm diameter glass beads at dry conditions and saturated with pure water and 1 % NaI, scanned with an aluminum/copper filter (**c**), an aluminum filter (**b**) and without filter (**a**). Segmented cube sub volume (300 × 300 × 300 voxels) of a partially saturated sample (*dark gray* is air, *medium gray* is 1 % NaI solution and *light gray* is the solid phase) (**d**)

In general, 1 % potassium iodide (KI) has been used to enhance detectability of the liquid phase (Anderson et al. 1992; Luo and Lin 2009) or as a marker for macropore network and preferential flow studies (Heijs et al. 1996) due to the iodine absorption edge at about 34 keV (Wildenschild et al. 2002). Although mass attenuation coefficient of sodium (0.4 cm^2g^{-1}) is about three times lower than potassium (1.5 cm^2g^{-1}) at 40 keV, the combined use of the diluted NaI with metal filters and adequate X-ray voltage and current yielded a very good contrast for discriminating the liquid phase from the void (air) and the solid phases in experiment 1 as commonly obtained when using diluted KI solutions (Wildenschild et al. 2002). The good peak separation verified in Fig. 12.13a, c can be even further improved by optimizing the X-ray voltage and current and using other metal filters.

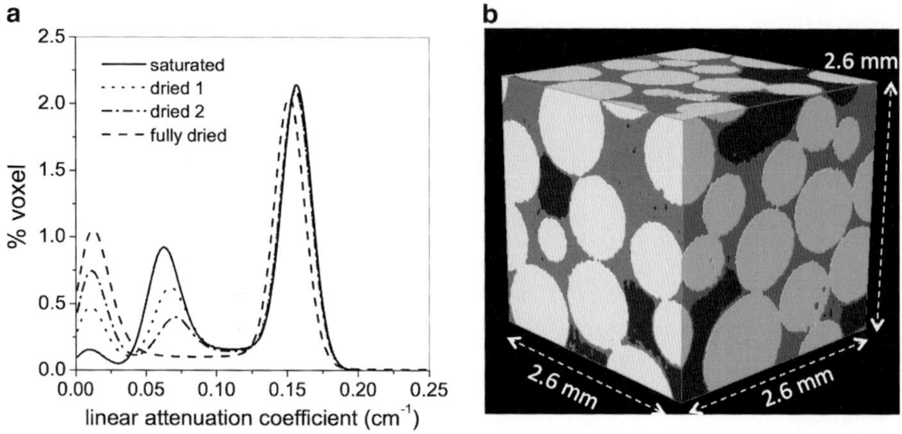

Fig. 12.14 Linear attenuation histograms (**a**) of 1 mm diameter glass bead sample at saturation (1 % NaI solution), partially dried (dried 1 and 2) and fully dried conditions and cube-shaped sub volumes (400 × 400 × 400 voxels) showing void (*dark gray*), liquid (*medium gray*) and solid (*light gray*) phases (**b**)

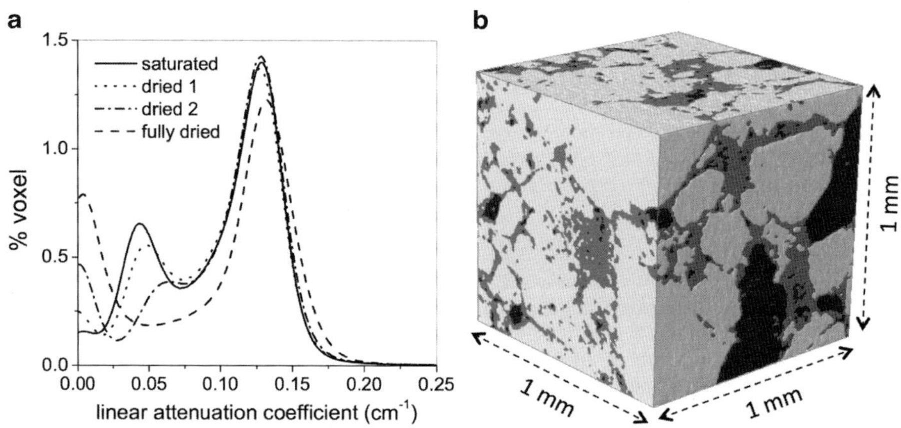

Fig. 12.15 Linear attenuation histograms (**a**) of a sandy soil at saturation (1 % NaI solution), partially dried (dried 1 and 2) and fully dried conditions and cube shaped sub volumes (400 × 400 × 400 voxels) showing void (*dark gray*), liquid (*medium gray*) and solid (*light gray*) phases (**b**)

Figure 12.13d shows a three-dimensional representation of a subsample (300 × 300 × 300 voxels) for the same glass bead sample (Exp 1), scanned with the Al/Cu filter, but partially saturated, clearly exhibiting the air, liquid (1 % NaI) and solid phases.

Figures 12.14a and 12.15a depict histograms of MicroCT images for 1 mm glass beads and a sandy soil (Exp 2 and 3), respectively, at saturated, partially saturated and dried conditions. Very well defined peaks for the air, liquid and solid phases allow segmentation of each phase, which are represented in three-dimensional sub volumes (400 × 400 × 400 voxels) in Figs. 12.14b and 12.15b.

12.6 Conclusions

Analysis of a set of nearly 400 published papers regarding the application of computed tomography in soil and porous media research indicate a surge of applications of benchtop MicroCT systems over the last few years. These instruments are becoming more affordable and popular and have great potential to significantly enhance the ability to characterize and better understand dynamic processes in soils and provide important insights for the development of advanced mathematical pore network and fluid dynamics models. New advanced benchtop MicroCT systems are very flexible in terms of sample size and image spatial resolution and provide high quality images, comparable to advanced synchrotron systems. Lab-based benchtop MicroCT systems, in contrast to large and expensive synchrotron or industrial CT facilities, where user time is a constraint, allow long term and continuous experimentation and provide opportunities to test different filters, adjust and optimize X-ray energy and acquisition parameters for best image quality and phase contrast.

Results obtained with a commercial MicroCT benchtop system showed the value of scanning undisturbed soil samples of various sizes at different spatial resolutions to account for the micrometer to the millimeter range soil structural features and pore systems. This approach allows extracting and quantifying large biopores and cracks, microaggregates and a significant portion of the microporosity. With regard the soil solid phase it was demonstrated that examining and segmenting the highest values of the linear attenuation coefficient histogram (out of the main solid peak) allows quantifying very dense minerals such as Fe, Ti and Mn oxides. Finally, a very good discrimination of the liquid phase from other soil components (void and solid phases) was achieved with 1 % NaI solution in artificially packed glass beads and undisturbed soil samples.

References

Anderson SH, Peyton RL, Wigger JW, Gantzer CJ (1992) Influence of aggregate size on solute transport as measured using computed-tomography. Geoderma 53(3–4):387–398

Bailey VL, McCue LA, Fansler SJ, Boyanov MI, DeCarlo F, Kemner KM, Konopka A (2013) Micrometer-scale physical structure and microbial composition of soil macroaggregates. Soil Biol Biochem 65:60–68

Bellato D, Simonini P, Coste AD (2013) Microstructural and mineralogical evaluation of the effectiveness of mixing treatments in stabilized clays. Geotech Test J 36(5):742–754

Cameron KC, Buchan GD (2006) Porosity and pore size distribution. In: Lal R (ed) Encyclopedia of soil science. CRC Press, Boca Raton, pp 1350–1353

Capowiez Y, Pierret A, Daniel O, Monestiez P, Kretzschmar A (1998) 3D skeleton reconstructions of natural earthworm burrow systems using CAT scan images of soil cores. Biol Fert Soils 27(1):51–59

Clausnitzer V, Hopmans JW (1999) Determination of phase-volume fractions from tomographic measurements in two-phase systems. Adv Water Resour 22(6):577–584

Cnudde V, Boone MN (2013) High-resolution X-ray computed tomography in geosciences: a review of the current technology and applications. Earth-Sci Rev 123:1–17

Costa ACS, Almeida VC, Lenzi E, Nozaki J (2002) Determination of copper, aluminum and iron in basaltic soils by sequential extractions. Quim Nova 25:548–552

Crestana S, Mascarenhas S, Pozzi-Mucelli RS (1985) Static and dynamic 3-dimensional studies of water in soil using computed tomographic scanning. Soil Sci 140(5):326–332

Crestana S, Cesareo R, Mascarenhas S (1986) Using a computed-tomography miniscanner in soil science. Soil Sci 142(1):56–61

Cruvinel PE, Cesareo R, Crestana S, Mascarenhas S (1990) X-rays and gamma-ray computerized minitomograph scanner for soil Science. IEEE T Instrum Meas 39(5):745–750

Hainsworth JM, Aylmore LAG (1983) The use of computer-assisted tomography to determine spatial-distribution of soil-water content. Aust J Soil Res 21(4):435–443

Heijs AWJ, Ritsema CJ, Dekker LW (1996) Three-dimensional visualization of preferential flow patterns in two soils. Geoderma 70(2–4):101–116

Heinemann S (1978) Computerized axial tomography-application to archeological material. Anthropol J Canada 16(3):13–16

Helliwell JR, Sturrock CJ, Grayling KM, Tracy SR, Flavel RJ, Young IM, Whalley WR, Mooney SJ (2013) Applications of X-ray computed tomography for examining biophysical interactions and structural development in soil systems: a review. Eur J Soil Sci 64(3):279–297

Hounsfield GN (1973) Computerized transverse axial scanning (tomography) 1. Description of system. Brit J Radiol 46(552):1016–1022

Iassonov P, Gebrenegus T, Tuller M (2009) Segmentation of X-ray computed tomography images of porous materials: a crucial step for characterization and quantitative analysis of pore structures. Water Resour Res 45(9):W09415

Ketcham RA (2005) Three-dimensional grain fabric measurements using high-resolution X-ray computed tomography. J Struct Geol 27:1217–1228

Kulkarni R, Tuller M, Fink W, Wildenschild D (2012) Three-dimensional multiphase segmentation of X-ray CT data of porous materials using a Bayesian Markov Random Field framework. Vadose Zone J 11. doi:10.2136/vzj2011.0082

Luo L, Lin H (2009) Lacunarity and fractal analyses of soil macropores and preferential transport using micro-X-ray computed tomography. Vadose Zone J 8(1):233–241

Macedo A, Crestana S, Vaz CMP (1998) X-ray microtomography to investigate thin layers of soil clod. Soil Tillage Res 49(3):249–253

Onoe M, Tsao JW, Yamada H, Nakamura H, Kogure J, Kawamura H, Yoshimatsu M (1983) Computed-tomography for use on live trees. Mater Eval 41(6):748–749

Peth S (2010) Applications of microtomography in soils and sediments. Dev Soil Sci 34:73–101

Petrovic AM, Siebert JE, Rieke PE (1982) Soil bulk-density analysis in 3 dimensions by computed tomographic scanning. Soil Sci Soc Am J 46(3):445–450

Pires LF, Borges JAR, Bacchi OOS, Reichardt K (2010) Twenty-five years of computed tomography in soil physics: a literature review of the Brazilian contribution. Soil Till Res 110(2):197–210

Rozenbaum O, Bruand A, Trong EL (2012) Soil porosity resulting from the assemblage of silt grains with a clay phase: new perspectives related to utilization of X-ray synchrotron computed microtomography. CR Geosci 344(10):516–525

Singh K, Niven RK (2013) Non-aqueous phase liquid spills in freezing and thawing soils: critical analysis of pore-scale processes. Crit Rev Env Sci 43(6):551–597

Sorensen HO, Hakim SS, Pedersen S, Christiansen BC, Balogh ZI, Hem CP, Pasarin IS, Schmidt S, Olsen UL, Oddershede J, Frandsen C, Feidenhans'L R, Stipp SLS (2012) Non-destructive identification of micrometer-scale minerals and their position within a bulk sample. Can Mineral 50(2):501–509

Taina IA, Heck RJ, Elliot TR (2010) Micromorphological and X-ray mu CT study of Orthic Humic Gleysols under different management conditions. Geoderma 158(3–4):110–119

Tuller M, Kulkarni R, Fink W (2013) Segmentation of X-ray CT data of porous materials: a review of global and locally adaptive algorithms. In: Anderson SA, Hopmans JW (eds) Tomography and imaging of soil-water-root processes, 2nd edn, SSSA special publication 61. Soil Science Society of America, Madison, pp 157–182. ISBN 978-0-89118-958-9

Vainberg EI, Kazak IA, Kurozaek VP et al (1982) Materials and method of certification of standard specimens for checking computerized X-ray tomographs. Sov J Nondestr Test+ 18(9):672–677

Vaz CMP, Crestana S, Mascarenhas S, Cruvinel PE, Reichardt R, Stolf R (1989) Using a computed tomography miniscanner for studying tillage induced soil compaction. Soil Technol 2(3):313–321

Vaz CMP, de Maria IC, Lasso PO, Tuller M (2011) Evaluation of an advanced benchtop micro-computed tomography system for quantifying porosities and pore-size distributions of two Brazilian oxisols. Soil Sci Soc Am J 75(3):832–841

Wildenschild D, Sheppard AP (2013) X-ray imaging and analysis techniques for quantifying pore-scale structure and processes in subsurface porous medium systems. Adv Water Resour 51:217–246

Wildenschild D, Hopmans JW, Vaz CMP, Rivers ML (2002) Using X-ray computed tomography in hydrology: systems, resolutions, and limitations. J Hydrol 267(3–4):285–297

Chapter 13
Unsaturated Soil Hydraulic Conductivity in the Central Amazon: Field Evaluations

Wenceslau Geraldes Teixeira, Götz Schroth, Jean Dalmo Marques, and Bernd Huwe

Abstract Determination of hydraulic properties may be required to solve many questions in agriculture and environmental research. Hydraulic properties are spatially and temporally variable, consequently, reliable soil hydraulic characterization is complex and time-consuming. Furthermore, many hydraulic properties are better represented by functions (i.e., unsaturated hydraulic conductivity and soil water retentivity) rather than by the mean values. The variability due to the presence of macropores or soil cracks commonly found in some soil classes in the tropics can enhance the naturally high soil hydraulic variability. A tension infiltrometer is an equipment that can be used to evaluate near-saturated hydraulic conductivity and sorptivity. In structured soil, small changes in soil tension, near the saturation, lead to dramatic changes of the infiltration rates. In this chapter, firstly, a discussion about soil hydraulic parameters and methods more appropriate to evaluate them on tropical soils is provided. Specific sources of errors and procedures to avoid or alleviate them are pointed out. Secondly, unsaturated hydraulic conductivity measurements carried out using a tension infiltrometer are used to compare different land use systems in the Central Amazon. The scaling theory and the statistical techniques of piecewise continuous regression are briefly described and used to analyze the results. Tension infiltrometers show to be relatively cheap, robust and a simple field method to

W.G. Teixeira (✉)
Soil Physics, Embrapa Soils, Rua Jardim Botânico, 1026, 22460-00 Rio de Janeiro, RJ, Brazil
e-mail: wenceslau.teixeira@embrapa.br

G. Schroth
Sustainable Agriculture Department, Rainforest Alliance, Wageningen, The Netherlands

J.D. Marques
Federal Institute of Education, Science and Technology of the Amazonas – IFAM, Campus Center of Manaus, Manaus, Brazil

B. Huwe
Soil Physics Department, University of Bayreuth, Bayreuth, Germany

W.G. Teixeira et al. (eds.), *Application of Soil Physics in Environmental Analyses: Measuring, Modelling and Data Integration*, Progress in Soil Science,
DOI 10.1007/978-3-319-06013-2_13,

evaluate unsaturated hydraulic conductivity in tropical soils. Since K near saturation is highly variable, the data were better analyzed using the relative hydraulic conductivity and the scaling theory. These small values of the tension in the breakpoint that divide the fluxes dominated by gravity reflect the importance to evaluate unsaturated hydraulic onductivities near saturation.

Keywords Saturated hydraulic conductivity • Unsaturated hydraulic conductivity • Infiltration • Amazon • Ferralsol • Tension infiltrometer • Scaling • Piecewise equation • Soil water content • Agroforestry system • Peach palm • Bactris gassipaes • Theobroma grandiflorum

13.1 Introduction

Determination of hydraulic properties may be required to solve many questions in agriculture and environmental research. Studies involving water fluxes and chemical transport through the soil profile, changes in soil water storage, water balance and soil-plant-water relations need accurate estimations of soil hydraulic properties. Hydraulic properties are spatially and temporally variable, consequently, reliable soil hydraulic characterization is complex and time-consuming. Furthermore, many hydraulic properties are better represented by functions (i.e., unsaturated hydraulic conductivity and soil water retentivity) rather than by the mean values. Besides, the results are dependent on the methods used and soil and climate conditions at the time of determination.

Soil hydraulic properties are easily altered by natural events such as rainfall, faunal and microbial activity, and by management practices such as harrowing, ploughing, cultivating, draining, fertilization and use of cover crops. The change of cohesive forces between the soil constituents, determined by alteration of characteristics such as pH, CEC, iron and aluminum oxides and organic matter, can affect the mechanical and consequently the hydraulic properties of the soil (El-Swaify 1980; Assouline et al. 1997). Other factors that can also interfere in the water transport processes into the soil vadose zone are the mineralogy of the solid particles (Iwata et al. 1995), the nature of the ions present (Kutílek 1973), and initial volumetric water content (Hillel 1998).

Temperature effects are normally considered too small to be of importance, especially in comparison with the spatial and temporal variability encountered in the field. However, the effect of temperature on the hydraulic conductivity in function of pressure K (h) was analysed by Stoffregen et al. (1999) who show that the temperature dependence is independent of water pressure head, and can be described by an exponential function. However, the experimental data did not agree with the expected values of the theory, and the hydraulic conductivity can be doubled by varying the temperature by about 15 °C, when all other factors remain equal. Mathieu and Pieltain (1998) presented factors of correction to permit comparison between infiltration tests using water at different temperatures. The magnitude of the temperature effect on the hydraulic properties is not completely understood, and similar conditions should be maintained during the evaluation process to permit comparisons.

In general, most undisturbed tropical forest soils are characterized by a rich organic matter layer on the surface with considerable biotic activity which opens large channels in the soil and allows for rapid infiltration of water. In addition, the forest canopy and litter effectively intercept raindrops, slowing their kinetic energy during high intensity thunderstorms, which are frequent in tropical climates.

The removal of vegetation and introduction of mechanized tillage operations result in disturbance and exposure of soil to the direct impact of raindrops and more intensive cycles of wetting and drying. Deforestation frequently cause a rapid decline in the infiltration rate (Lal et al. 1979; Lal 1989; Corrêa 1985), which could lead to runoff and erosion processes, especially in hill slope agriculture fields. The ability of a soil to transmit water at a rate comparable to that of rainfall intensity is an important property that maintains an adequate soil water balance and controls runoff and erosion. Land use systems with perennial crops have a potential for erosion control by providing a permanent soil cover of tree canopies, litter and eventually cover crops, in addition to the role of trees as runoff barriers and the beneficial aspects in the maintenance of a stable structure with high inputs of organic matter (Young 1997). Although these positive effects of perennial land use systems have been verified in many experiments, we still have to demonstrate this function under other climate conditions, soil types and management regimes. The hypothesis that land use systems with perennial crops maintain favorable infiltration rates due to bio-channels and better soil structure than annual food-crop-systems was proven in a long-term agroforestry experiment established on an Alfisol in western Nigeria. The decline in the infiltration rate, evaluated with a double ring infiltrometer, was less in the agroforestry than in the food-crop-systems with either plow-till or no-till treatments (Lal 1989). Hedgerow intercropping with *Senna spectabilis* on a Typic Kandiudult in southern Cameroon increased infiltration rates measured with the double ring infiltrometer (Hululgalle and Ndi 1993). In northern Zambia, on an Orthic Ferralsol, an evaluation of the infiltration rate using the inverse falling head method showed high values in alley cropping with *Leucaena leucocephala* (Dalland et al. 1993).

The variability due the presence of macropores or soil cracks commonly found in some soil classes in the tropics can enhance the naturally high soil hydraulic variability (Beven and Germann 1982; Bouma et al. 1982; Teixeira 2001). Other sources of inaccuracy in the evaluation of hydraulic properties arise from the measurement technique itself (Marques et al. 2008), such as soil disturbance during installation of the equipment or irregular hole diameters for the auger (Amoozegar and Warrick 1986). Because of the high variability of soil hydraulic properties, it is better for many purposes to have several points with smaller precision than just a few highly accurate data. Certainly, both properties are desirable but not easily obtained. The variability of the hydraulic properties in the field needs to be taken into account when choosing the measurement method and the number of replications. If they are insufficient, straightforward treatment effects can be misinterpreted or not detected.

13.1.1 Measuring Soil Hydraulic Properties in Tropical Soils

The term "infiltration rate" refers to the vertical entry of freely available water into a soil surface. It should not be confused with hydraulic conductivity,[1] which is a measure of the ability of a soil to transmit water in a three-dimensional system. The infiltration rate tends to be numerically equal to the saturated hydraulic conductivity if the hydraulic gradient of the soil is unity. This condition is frequently approximated for homogenous and isotropic soils; it is discussed further in the section about the instantaneous profile method. For many practical problems of large-scale significance, the mean flow path over a sufficiently large area is approximately one-dimensional in the vertical direction, especially in relatively homogeneous soils (Stephens 1996).

In tropical climates, the exposure of the soil surface frequently leads to the formation of a superficial soil crust. To verify if such a thin, superficial layer of low permeability controls the infiltration rate, measurements should be made both under natural surface conditions and after the surface layer has been carefully removed. Worms tend to move upwards and crawl out of the soil when the surface is covered with water (Bouwer 1986). The resulting open wormholes can greatly increase infiltration rates, particularly if the test is carried out over a long time. Prewetting the site before the measurements alleviates this trouble frequently encountered in the tropics. The temperature and chemical composition of the water used for infiltration measurements should be the same as those of the soil water to avoid dissolution of soil air in the infiltrating water (Bouwer 1986), and changes of the flocculation status of the clay particles (Hillel 1998). If possible, rainwater from the site should be used to avoid such problems.

Most of the water movements above the water table in the field, including water and nutrient fluxes to plant roots, rainfall infiltration and leaching of nutrients through the soil profile, occur while the soil is unsaturated and are thus controlled by unsaturated hydraulic conductivity. Field measurements of unsaturated hydraulic conductivity are generally preferred to laboratory measurements if the site is sufficiently accessible, has reasonable homogeneity and level topography, is not too stony, and has predominantly vertical flow during drainage (Green et al. 1986).

A tension infiltrometer or disc permeameter consists of a porous baseplate, which establishes hydraulic continuity with the soil through a nylon membrane. It is connected to a Mariotte reservoir that allows supplying the soil surface with water at a constant and regulated tension. The soil water potential under the infiltrometer is controlled by air entry into the mariotte flask through tubes of different lengths, with tension infiltrometers normally operating in the range of 0 to −20 cm. The infiltration rate is measured as water flows into the soil and the water level falls in the

[1] The term permeability that is sometimes found as synonym for hydraulic conductivity is not correct. Permeability is an intrinsic property of soil and has the dimension L^2 while hydraulic conductivity has the dimension LT^{-1} (Jury et al. 1991).

bubble tower. The data can be recorded manually or automatically (Ankeny 1992; Elrick and Reynolds 1992). Tension infiltrometers have been used to characterize near-saturated hydraulic properties, sorptivity, macroscopic capillary length, characteristic pore size (Ankeny 1992; Reynolds 1993; Stephens 1996), soil structural conditions at the soil surface as a function of short-term variations in weather conditions (White and Perroux 1989), and effects of tillage practices (Ankeny et al. 1991; Messing and Jarvis 1993). The near-saturated hydraulic conductivity and sorptivity obtained using tension infiltrometry have been used to distinguish infiltration through macropores from that through the soil matrix (Watson and Luxmoore 1986; Wilson and Luxmoore 1988; Messing and Jarvis 1993). The method involves supplying water to the soil surface at different potentials to exclude a range of macropores from the flow whose contribution to infiltration can thus be quantified. A good soil surface preparation and hence good contact between the infiltration disc and the soil is essential. The hydraulic conductivity of the contact material should be greater than or equal to the hydraulic conductivity of the soil over the range of pressure heads set on the tension infiltrometer. The pore water pressure head at which the contact material spontaneously saturates should be less than the minimum pressure head set on the infiltrometer membrane. The supply membrane must be visible during the infiltration to permit examination for air leaks. Perroux and White (1988) concluded that 3–5 mm thickness of single grain contact material having a fine sand texture and a hydraulic conductivity of about 10^{-5} m s^{-1} should be adequate for most agricultural soils and the usual range of tension infiltrometer pressure heads (e.g., 0 to -150 mm water tension).

For measurements near saturation the infiltration disc must be leveled, otherwise the potential varies across the supply surface. Close to saturation, small changes in soil tension lead to dramatic changes of the infiltration rate and sometimes to higher variability in structured soils. The contact material can have a large influence on the pore water pressure head and hydraulic head gradient at the soil surface, and this can have a substantial impact on the validity of tension infiltrometer results (Reynolds and Zebchuk 1996). Solar heating of the headspace in the Mariotte reservoir should be avoided by shading. According to Reynolds (1993), the main factors affecting the accuracy of tension infiltrometer measurements are soil heterogeneity, soil collapse under the infiltrometer during the measurement, inadequate hydraulic contact between infiltrometer and soil, and the hydraulic properties and thickness of the contact material. A shortcoming of the technique is the time necessary to reach steady flow at low tensions in clayey soils, which can make manual recording impracticable. In this situation, an automated infiltrometer should be used.

In this chapter, firstly, a discussion about soil hydraulic parameters and methods more appropriate to evaluate them on tropical soils is provided. Specific sources of errors and procedures to avoid or alleviate them are pointed out. Secondly, results about unsaturated hydraulic conductivity are shown. The measurements were carried out using a tension infiltrometer to compare unsaturated hydraulic conductivities of different land use systems in the Central Amazon. The scaling theory and the statistical techniques of piecewise continuous regression are briefly described and used to analyze the results.

13.2 Material and Methods

The most part of the studies presented in this chapter were carried out on the research station of EMBRAPA – Amazônia Ocidental, Fig. 13.1, near the city of Manaus in the central Brazilian Amazon (3° 8′ S, 59° 52′ W, 40–50 m a.s.l.).

13.2.1 The Soils in the Central Amazon

The predominant soils covering sediments in the central Amazon result from the weathering of tertiary fluvio-lacustrine deposits, which belong to the so-called "Alter do Chão" formation. It consists of cross-bedded sandstone with interspersed argilites, with predominance of quartz and kaolinite and with feldspar and muscovite as minor components. The soil formed on the plateaux and on the upper part of the slopes are clayey, deeply weathered Ferralsols (Chauvel 1982; Teixeira et al. 2010). This has as a consequence that the soil is normally very poor in nutrients (e.g., phosphorus, calcium, magnesium, and potassium). The soil in the experiment was classified as Xanthic Ferralsol according to WRB (2007). It is classified as a Xanthic Hapludox following the Soil Taxonomy (Soil Survey Staff 1997) or as Latossolo Amarelo muito argiloso distrófico using the Brazilian classification (EMBRAPA 2013). The clayey Ferralsol is representative of large areas in the central Amazon (Teixeira et al. 2010), see Fig. 13.2.

The soil physical and chemical characteristics evaluated in this study show a high clay content on the surface, with 600–750 kg m^{-3} of clay in the first 30 cm; and very clayey further down, with 750–850 kg m^{-3} of clay. The clay fraction is dominated by kaolinite with some gibbsite. It shows a strong microaggregation (Tomasella and Hodnett 1996; Teixeira 2001), possibly as a consequence of the strong binding between positively charged iron oxyhydroxides and negatively charged kaolinite and organic matter (Sanchez 1976).

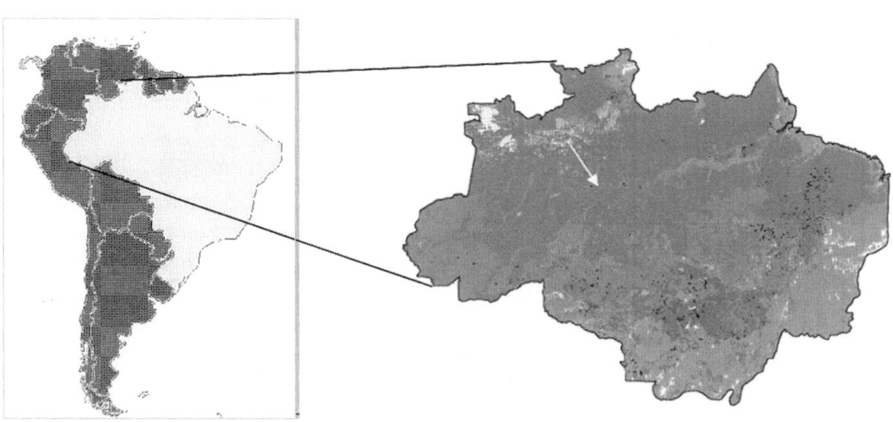

Fig. 13.1 Map of Brazil in South America and the Brazilian Amazon. The *arrow* points to the location of the study area, near the city of Manaus

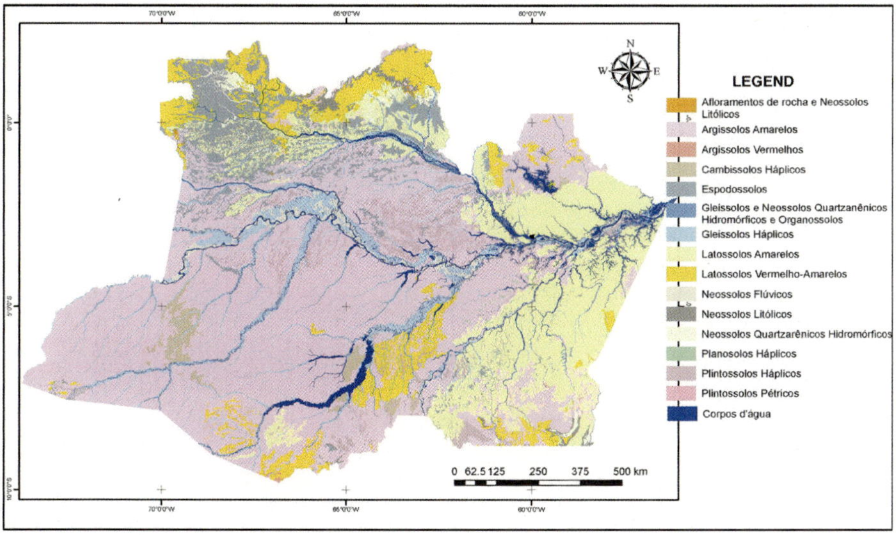

Fig. 13.2 Soil map of the state of Amazonas in Brazil. The *legend* shows the soil classification in the Brazilian system

Land Use Systems and Experimental Design

The study site was first cleared from primary forest in 1980, then an experiment with rubber trees (*Hevea brasiliensis*) was established, being abandoned in 1986. Until preparation of the area for the evaluated experiment used in this study, secondary vegetation was growing together with the remaining rubber trees. The developing secondary forest was manually cleared in 1992 and the vegetation was burnt on the site. The experimental plots studied were planted in 1993 and evaluated in 1997. The three land use systems investigated were: (i) two monocultures of peach palm (*Bactris gasipaes*), for production of fruits and palm heart, respectively; (ii) a monoculture of cupuaçu (*Theobroma grandiflorum*); and (iii) a complex agroforestry system with peach palm, cupuaçu, Brazil nut (*Bertholletia excelsa*) and annatto (*Bixa orellana*). The soil was covered by *Pueraria phaseoloides* in all systems except the peach palm monoculture, which tolerates little ground vegetation. Adjacent areas of primary and secondary forest were investigated for comparison. A more detailed description of the experimental design and the plant species are given in Teixeira (2001) and Schroth et al. (1999, 2000, 2002).

Evaluations Using Tension Infiltrometer

The field evaluations of unsaturated hydraulic water conductivity were carried out with a tension infiltrometer (TI) at ≈40 cm from the trunk of the respective trees which were grown either in monoculture or in association, as well as under the leguminous cover crop, *Pueraria phaseoloides*. Samples were also collected

Fig. 13.3 Scheme of tension infiltrometer used in the evaluations

close to the dominant species in the adjacent secondary forest, *Vismia spp*. A double ring tension infiltrometer constructed by Umwelt-Geräte-Technick GmbH – Müncheberg – Germany was used to evaluate the unsaturated hydraulic conductivity. This disk infiltrometer has a special design (Fig. 13.3) where the infiltrometer was constructed separately from the water tower, making it more stable and accurate even under windy conditions, also avoiding the collapse of the soil structure near saturation.

Moreover, it allows to perform, in addition to the three-dimensional analysis in unsaturated and saturated hydraulic conditions, also the one-dimensional analysis for saturated hydraulic conductivity, when the tension infiltrometer is used like a simple double ring infiltrometer with the outer ring functioning as a small buffer system. The base radii are 77.5 mm and 124 mm for the inner and for the outer ring, respectively. At each infiltration site, an area approximately with the same diameter as the ring was cleared and leveled with help of water levels. Minimal compaction and disturbances are avoided with careful preparations and the cutting of roots with scissors. To delimit the surface infiltration area and to prevent lateral surface flow of pounded water, a sharpened steel ring with 25-cm diameter and 1.0 cm high was pushed 0.5 cm into the soil.

The evaluation was started when bubbling was observed in the Mariotte reservoir. The time to approach steady-state infiltration rates (Q) [L^3 T^{-1} – cm^3 h^{-1}] varied from ca. 2 h at near 50 mm of tensions (h) to less than 5 min near saturation. Data were collected manually reading the water levels in the Mariotte reservoirs when the values were stable (not varied for three consecutive readings), then it was recorded as a pair value of Q and water pressure head (h) [L – cm]. The water pressure head or tension was evaluated with a double U water-manometer (see details in Fig. 13.3).

A broadcloth-covered 400-mesh nylon filter, (with a bubbling point ≈25 cm of water) was used as porous membrane for hydraulic contact with the soil and covered the base of the tension infiltrometer. A thin layer of fine sand was applied and levelled to the soil surface to provide a smoother contact between the membrane on the base of tension infiltrometer and the soil surface. Fine sand, with diameter between 0.5 and 0.02 mm – K minimum about 5,000 cm day^{-1}, was used as a contact material. A layer of maximal 5 mm thickness was applied on the soil surface and levelled using a special equipment, then the disc was installed and the level was checked by using water balances. Unsaturated hydraulic conductivity measurements at approximately 50, 40, 30, 20, 10 mm of h were followed by measurement of saturated hydraulic conductivity at 0 h. An ascending sequence of supply pressure was adopted, since a descending sequence may enhance hysteresis with progressive drainage occurring close to the disk while wetting continues at the infiltration front (Reynolds and Elrich 1991). The measurement were performed under a tent to protect the Mariotte reservoir from the direct sunbeam and also to reduce temperature variations.

The paired K and h values were obtained from Q at successive h using the theory outlined by Ankeny et al. (1991) with modification of Jarvis and Messing (1995). This solution is based on the Wording's solution for a three-dimensional infiltration from a disc source with a constant pressure head at the soil surface (Wooding 1968), it is also assumed that the relation K (h) can be described by the exponential function of Gardner (Eq. 13.1).

$$K = K_{fs}\exp(\alpha h) \tag{13.1}$$

Where K_{fs} is the field-saturated hydraulic conductivity [L T^{-1} – cm day $^{-1}$] and α is the empirical fitting exponential slope [L – cm^{-1}]. The α parameter is calculated by measuring the steady-state flow rate at two different supply potentials (such h_i and h_{i+1}) and solving Eq. 13.2.

$$\alpha_{i+1/2} = \frac{\ln \frac{Q_i}{Q_{i+1}}}{h_i - h_{i+1}} \quad i = 1, \ldots, n-1 \tag{13.2}$$

Where Q_i is the steady-state infiltration rate [L^3 T^1 – cm^3 day $^{-1}$] measured at the tension h [L – cm] and Q_{i+1} is the subsequent evaluation rate at the tension h_{i+1}. The notation ½ denotes that the estimated value of α at the midpoint between successive supply pressure heads h.

Finally, hydraulic conductivity is calculated using Eq. 13.3, where r [L – cm] is the radius of the disc infiltrometer.

$$K_{i+1/2} = \frac{\sqrt{Qi \times Q_{i+1}}}{\left(1 + \frac{4}{\pi r \alpha_{i+1/2}}\right)} \quad i = 1, \ldots, n-1 \tag{13.3}$$

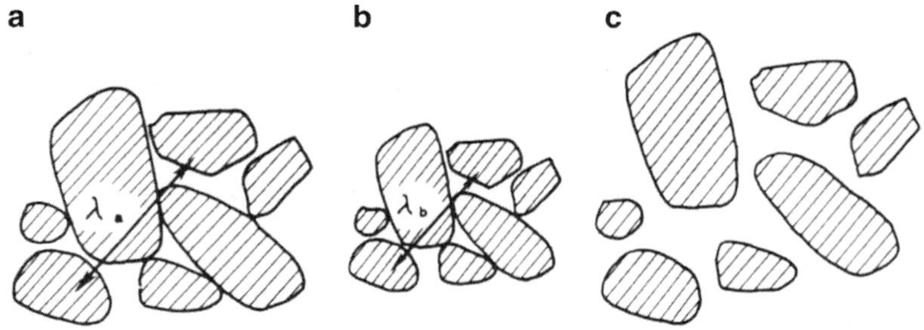

Fig. 13.4 Concept of similitude of Miller and Miller (1956). The porous media (**a**) and (**b**) are similar (Adapted from Miyazaki 1993)

The denominator term in the Eq. 13.3 represents the geometric mean of infiltration rates (Qi and Q_{i+1}) between adjacent supply potentials $[h_i + h_{i+1}]/2$. Two additional paired values of K and h were obtained presuming that the largest and smallest supply potentials, respectively, equal to $\alpha_{3/2}$ and $\alpha_{n-½}$ respectively (Jarvis and Messing 1995).

Scaling Theory

Miller and Miller (1956) introduced scaling factors derived from similitude analysis in soil science. Their original geometric similitude concept for scaling hydraulic properties is based on capillary laws and viscous flow laws, and it presumes that for different locations, the geometry of the pores is the same with differences only in a characteristic length. Therefore, it can be characterized completely by scale factors (Fig. 13.4). Hence, it implies a constant porosity that limits the use of the original scaling concept for many natural soils (Warrick et al. 1977). The original scaling concept was changed, and nowadays scaling methods are based on similarities between soil hydraulic functions instead of similarities between respective geometries (Simmons et al. 1979; Tillotson and Nielsen 1984; Jury et al. 1987; Pachepsky et al. 2003). Scale factors are calculated by regression methods, and scaling methods based on regression analysis are called "*functional normalisation techniques*" (Simmons et al. 1979; Tillotson and Nielsen 1984), further they do not require similar media. Although these scale factors are not directly related to Miller scaling and have no explicit physical meaning, they are useful to express the variability by a single parameter. The purpose of scaling is to reduce the number of parameters needed to characterize the variability of soil properties.

In this study, the usefulness of the scaling theory was investigated in combining with piecewise-regression in order to describe and reduce the spatial variability of hydraulic conductivity functions. The scaling theory was used in a "*hybrid media scaling*" combining the original concept of Miller and the functional normalization. It was based on the scaling technique proposed by Shouse and Mohanty (1998), however some modifications are explained below. Similar approaches were used for analysing tension infiltrometer data by Messing and Jarvis (1993) and Mohanty (1999).

In this the paired data $(K_i, h_i)_{i,\ldots,N}$ obtained by tension infiltrometry under the different positions within the land use systems were scaled. Application of the scaling method requires a set of reference parameters. There are several methods for determining the reference parameters, such as using the field-average parameter values, using parameters for an arbitrary site within the transect, measuring the parameters at a site independent of the transect, or using previously measured data (Shouse et al. 1992).

To define the reference model, measured data $[K_r(h), h]$ were arbitrarily chosen, and Eq. 13.7 was fitted by non linear regression, giving hp = 21 mm, and $\alpha_1 = -0.1342$ and $\alpha_2 = -0.0803$. Paired $[K_r(h), h)]_{i=1,\ldots,N,\, x=1,\ldots,x}]$ data points were subsequently scaled to the reference curve using the regression analysis. The scale factors α_x for a certain infiltration sequence at position [x], consisting on N data points (i.e., water pressure heads steps) was then found by minimizing the sum of squared differences (SS) between the reference relative hydraulic conductivity curve and the scaled data point (Eq. 13.4).

$$SS = \sum_{i=1}^{i=N} \left[\alpha_x h_i - h_i^{ref}\right]^2 \tag{13.4}$$

A computational code was developed to calculate α_x by computing the Eqs. 13.5 and 13.6

$$\alpha_{i,x} = \frac{h_{i,x}^{ref}}{h_{i,x}} \quad i = 1, \ldots, N \tag{13.5}$$

$$\alpha_x = \frac{1}{N}\sum_{i=1}^{i=N} \alpha_{i,x} \tag{13.6}$$

Here, h^{ref} is the water pressure head at the reference curve; α_x is the scaling factor for a certain infiltration sequence at a position [x] consisting of N steps. Once α_x for a position x is known, scaled values $h^*_{i,x}$ $(= \alpha_x h_{i,x})$ are calculated for all water pressure head steps (i = 1, ..., N).

Piecewise Continuous Regression for Two Segments

Previous results show that the clayey Ferralsols have a bimodal pore size distribution instead of a conceptualized soil with a continuous unimodal pore size distribution (Teixeira 2001). Furthermore, the results show a discontinuity in the functional relationship between K and h. Therefore, it may be appropriate to consider the function K (h) in discrete intervals and to express the relation by a two-line regression model. Therefore, the paired data (K, h) obtained from each sequence of infiltration were then fitted by means of a piecewise continuous regression (with two segments). A difficulty in fitting a piecewise continuous regression is to determine the break point or joint point, which separates the two

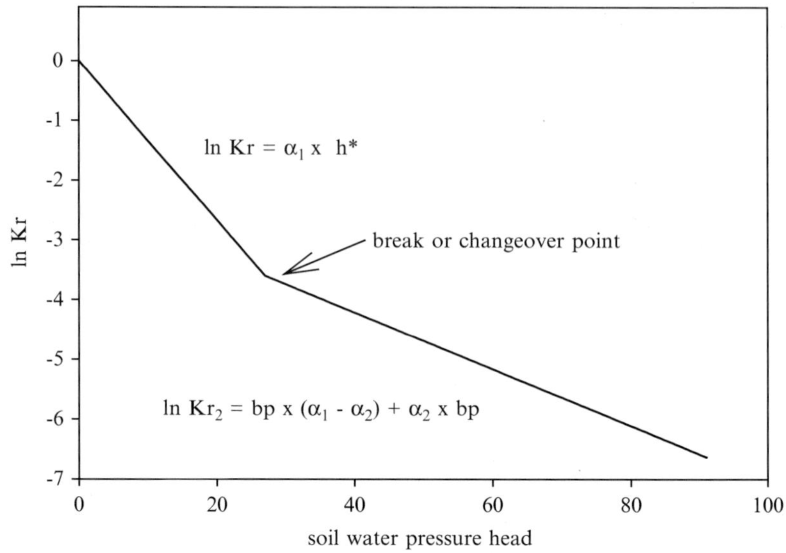

Fig. 13.5 Scheme of a piecewise continuous regression

domains. Since its position is unknown, it has to be estimated by interactive methods of nonlinear squares. The piecewise regression analysis was used by Ahuja et al. (1980) in an Instantaneous Profile Method (IPM) experiment, and it has been used to fit data from tension infiltrometer measurements in a soil with multi-domains fluxes (Keng and Lin 1982; Messing and Jarvis 1993; Jarvis and Messing 1995; Mohanty 1999). The exponential model of Gardner (Eq. 13.1) was linearized and rearranged so that Kr = ln (K Kfs^{-1}). The slopes, the intercept for the second segment, and the water pressure head at the break point (hp) were determined by non linear least square techniques, see Eq. 13.7 A conceptualized piecewise continuous regression is illustrated in Fig. 13.5. A simple computational protocol was developed to estimate the unknown parameter (α_1, α_2 and hp) using the non-linear least square method of Levenberg – Marquardt.

$$Kr = \begin{cases} K_1 \ \ if \ \ h \leq hp \\ K_2 \ \ if \ \ h \geq hp \end{cases} \tag{13.7}$$

where

$K_1(h) = \alpha_1 h$ and $K_2(h) = h_p(\alpha_1 - \alpha_2) + \alpha_2 \cdot h$

Here α_1 and α_2 are the slopes of segment 1 and 2 respectively. The function $K_1(h)$ summarizes the unsaturated hydraulic function governed by the gravitational process, and K_2 (h) segment the relation which is basically governed by the flux in capillary pores. The value of hp can be interpreted as a matric pressure head that divides operationally the two flow domains, i.e., the critical h where flow changes from gravity dominated to capillary dominated flow or vice versa (Keng and Lin 1982; Shouse and Mohanty 1998). The mathematically and statistically simpler piecewise equation, is not a smooth function, but may have advantages over other empirical power or exponential equations normally used to fit hydraulic parameters. Furthermore, the segments, and the break point in the piecewise continuous regression can be statistically compared (Keng and Lin 1982; Draper and Smith 1998).

13.3 Results and Discussion

The scope of measurements carried out with Tension Infiltrometer (TI) are shown in Table 13.1. The measurements with TI were carried out using very small increments near saturation ranging between four to eight tensions for each position. However, the K(h) values for tensions h >10 mm should be interpreted in a qualitative manner, because of possible inaccuracies in the applied h. Also, a very intense bubbling in the chamber reservoir made it very difficult to read accurately the water level. Moreover, in the structured soils as the Ferralsol, variations of millimeters near saturation change the infiltration substantially. Measurements with values of h around 70 mm or higher are very difficult to obtain in the clayey Ferralsols by TI measurements, because they are very time-consuming. Consequently, the likelihood of precipitation events, and changes in soil structure above the disc (because of macrofauna movements) make these measurements not feasible. The measurement of K(h) is very helpful in terms of supplying a matching factor (Luckner et al. 1989), rather than the use of saturated K to transform relative values of K into absolute ones when using indirect methods of estimation of hydraulic properties (Teixeira 2001). This is because saturated K of macroporous

Table 13.1 Scope of inflitration measurements conducted with the tension infiltrometer

Number	Land use system	Common name	Replicate infiltration sequence	Tension steps	Total data Q (h) for K analysis
1	Agroforestry	Cupuaçu	3	4–7	17
2	Agroforestry	Annatto	3	4–7	17
3	Agroforestry	Brazil nut	2	6	12
4	Agroforestry	Peach palm – F	2	7	14
5	Agroforestry	Peach palm – P	3	6–7	20
6	Agroforestry	Pueraria	3	6–8	21
7	Monoculture	Cupuaçu	3	6–7	20
8	Monoculture	Grasses and pueraria	3	4–7	17
9	Monoculture	Peach palm – F(m)	3	6–7	20
10	Monoculture	Peach palm – F(ma)	3	6–7	19
11	Monoculture	Peach palm – P(m)	3	5–7	18
12	Monoculture	Peach palm – P(ma)	3	6–7	20
13	Secondary forest	Vismia	3	7	21
		Total	37	–	237

m monoculture, *a* among the palms, *F* Peach palm for fruits production, *P* Peach palm – for palm heart production

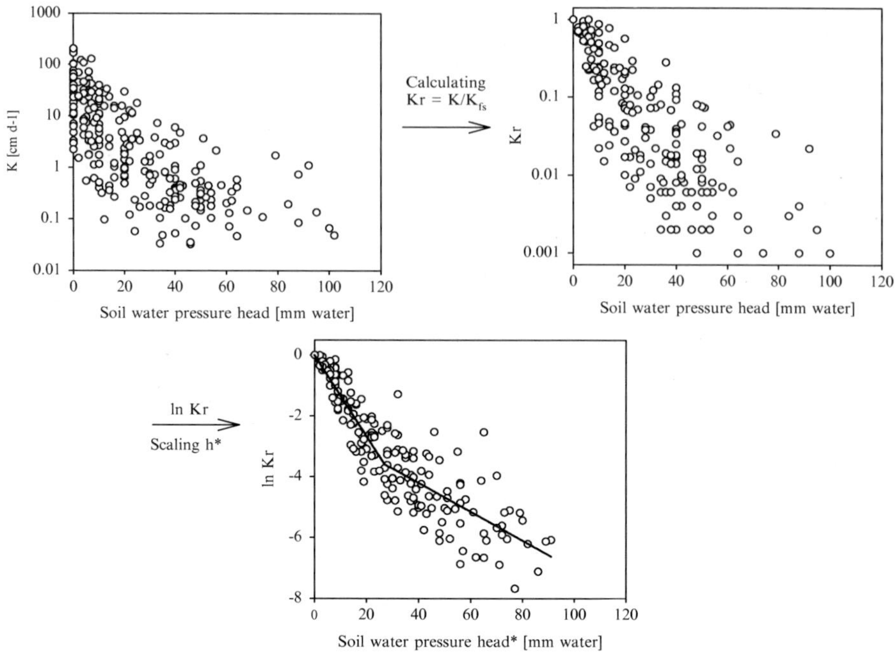

Fig. 13.6 Scatter of unsaturated hydraulic conductivity in function of tension – K(h) evaluated in different land use systems on a clayey Ferralsol in the central Amazon. The figures show also the effect of use of relative hydraulic conductivity (Kr) and scaling theory in the reduction of variability

soils may be unrelated to the unsaturated K in the region near saturation (Clothier and Smettem 1990). The accuracy of field determinations with TI is affected primarily by soil heterogeneity, soil collapse under the infiltrometer during the measurement, inadequate hydraulic contact material (Reynolds and Zebchuk 1996) or by the solar heating of the head space in the Mariotte reservoirs (White et al. 1992; Elrick and Reynolds 1992).

The K(h) values vary by 3–4 orders of magnitude between 0 and 70 mm of (h) (Fig. 13.6). The logarithmic values of the relative hydraulic conductivity Kr [Kr = K(h)/Ks] were calculated and it alleviates the influence of gravity dominated macropore flow near saturation, which has little or no influence on water flow under unsaturated conditions (Shouse and Mohanty 1998; Mohanty 1999). Moreover, assuming geometric similarity among the locations, h could be successfully scaled (Fig. 13.6). In an overall analysis of the K data, a breakpoint that operationally divides the flux is located near a tension of 20 mm. Wilson and Luxmoore (1988) found also a division ≈20 mm of h to separate macropore from micropore flow and Jarvis and Messing (1995) around ≈40 mm.

The coefficients of variations (CV) of the scale factors ranged from 10 % to 51 % (Table 13.2), much smaller than those previously reported by Warrick et al. (1997) but in the same range as those found by Hopmans (1987) and Messing and

Table 13.2 Coefficient of variation of the scale factors in the different land use system

Location	Mean	Coefficient of variation (%)
Agroforestry		
Cupuaçu	1.348	29
Annato	1.316	39
Brazil nut	0.883	51
Peach palm F	1.037	35
Peach palm – PH	1.128	11
Pueraria	1.438	43
Monocultures		
Cupuaçu	1.592	10
Grasses	1.184	22
Peach palm – F	1.816	33
between Peach palm F	1.893	21
Peach palm – PH	1.391	24
between Peach palm – PH	1.148	14
Vismia	1.265	20

F fruit, *PH* palm heart

Table 13.3 Coefficients and statistical parameters of unsaturated soil hydraulic conductivity (K) in function of pressure (h) fitted with piecewise continuous regression in different land use systems in Central Amazon

Specie	n	α_1	α_2	β_2	bp	s	r^2
Cupuaçu[a]	39	−0.1361	−0.0319	−3.4488	33	0.5459	0.93
Brazil nut	12	−0.1522	−0.0016	−3.0120	20	0.4851	0.89
Annatto	17	−0.1481	−0.0531	−1.6150	17	0.4754	0.96
Pueraria	21	−0.1567	−0.0440	−2.0416	18	0.4744	0.94
Grasses	21	−0.1342	−0.0803	−1.1662	22	0.5144	0.96
Peach palm – fruit[a]	35	−0.1539	−0.0618	−1.4312	15	0.7477	0.88
Peach palm – palm heart[a]	38	−0.1282	−0.0740	−1.3417	25	0.6854	0.90
Peach palm among trunks[a]	39	−0.1345	−0.0060	−4.0514	31	0.7412	0.86
Vismia	21	−0.1250	−0.0039	−5.4034	45	0.4871	0.96

[a]Pooled data

Jarvis (1993). The rather small to moderate degree of spatial variability in the scale factors probably reflects the fact that a Representative Elementar Volume (REV) was encompassed by the relative large disc radius (Jarvis and Messing 1995), as well as a reduced disturbance of the soil during preparation of the site where the disc was installed.

The coefficients and statistical parameters of the piecewise functions K(h) for soil surface near different species growing in different land use systems and positions are shown in Table 13.3. Although measurements of K(h) with TI have shown great spatial variability in concordance with the studies of Logsdon and Jaynes (1993), Messing and Jarvis (1993), Mohanty et al. (1994), Angulo-Jaramillo et al. (1997) and Logsdon (1997), trends can nevertheless be observed and discussed in the present study.

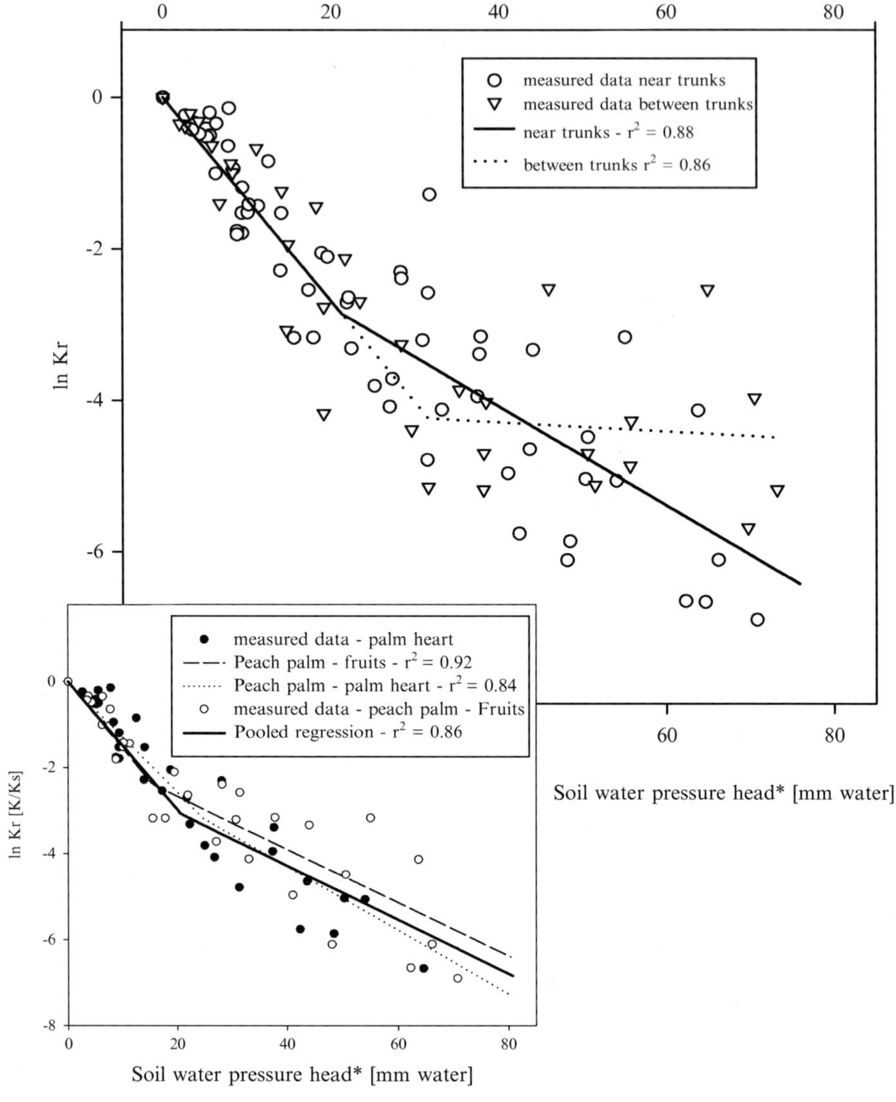

Fig. 13.7 Unsaturated hydraulic conductivity in function of tension – K(h) of the soil surface evaluated at 0.40 and 1.20 m from the trunks of peach palms growing in monoculture for fruit production and palm heart production in the Central Amazon. The *small figure* shows the similarity of the unsaturated hydraulic functions measured at 1.20 m from the trunks of peach palm fruits and peach palm for palm heart

A comparison between the results from the monoculture of peach palm for fruits and palm heart did not show relevant differences in the form of the curve. Therefore, the land use systems of peach palm plantations were pooled and analyzed together (Fig. 13.7). The piecewise equations fitted for peach palm for fruits, palm heart and with all data pooled are shown in Table 13.4. The $h \approx 30$ mm seems to be the

Table 13.4 Coefficients and statistical parameters of unsaturated soil hydraulic conductivity in function of tension – K(h) in peach palm plantations for palm hearts and fruits in the central Amazon

Treatment	n	α_1	α_2	β_2	bp	s	r^2
Peach palm – fruits	19	−0.1672	−0.0084	−4.3036	27	0.6220	0.92
Peach palm – palm heart	20	−0.1182	−0.0001	−3.7913	32	0.6925	0.84
Pooled regression	39	−0.1345	−0.0060	−4.0514	31	0.7412	0.86

breakpoint that divided the flux dominated by macropore and meso and micropores. These small values of tension in the breakpoint that divide the fluxes dominated by gravity and capillarity reinforce the importance to evaluated water content in small tension when is planned to derived K(h) from soil water retention curve.

In the monoculture of peach palm, the K(h) near the trunks ≈40 cm and at ≈120 cm show similar behavior near saturation, but this similarity changes for h greater than ≈20 mm with the position near the trunk showing a more significant reduction of K (Fig. 13.7). This partly agrees with a greater concentration of micropores near these palms (Teixeira 2001). The compaction of the soil between the peach palms, while reducing total porosity and specifically reducing macroporosity (Teixeira 2001) actually increased the number of small pores. Therefore, positions under the influence of the peach palm roots show a reduced K in the capillary range. Moreover, the lower inclination of the K function for h in the range of 30–70 mm in the position between the plants showed that pore radii in the range of 0.5–0.21 mm practically did not exist (Teixeira 2001).

A high infiltration rate near the peach palms was expected because of the large amount of water that arrives from stemflow (e.g., 100 l in a 24 mm rainfall – Schroth et al. 1999), which does not remain on the soil surface and probably infiltrates along both dead root channels and fauna tunnels. These macropores, although occupying a small percent of the pore volume, are very effective in conducting water at high potentials. Some intermittent ponding conditions were observed between the palms in monoculture after strong rainfall events. Therefore, the monitoring of the K functions in this land use system can be an important factor for management decisions.

The K(h) evaluations between the positions within monoculture of cupuaçu, were done near the trunks and between trunks, the soil between the cupuaçu trees was covered by grasses, are shown in Fig. 13.8. Near the soil saturation, $h > 20$ mm, the soil covered by grasses shows lower values of K(h) than the values found near the trunks of cupuaçu. Although, both values are greater in magnitude, cupuaçu roots appears to create a higher near saturated hydraulic conductivity near the trunks, probably due to an expressive concentration of fine roots near the soil surface. For comparisons, in the Fig. 13.8, is also shown the K(h) function for soil covered by pueraria (originally evaluated in the agroforestry system). The results show higher values of K(h) in soil covered by pueraria than grasses. Pueraria is normally recommended as cover crop for monocultures in the tropics in south America, Africa and Asia, its seems to have an effect on the maintenance and enhancement of the soil

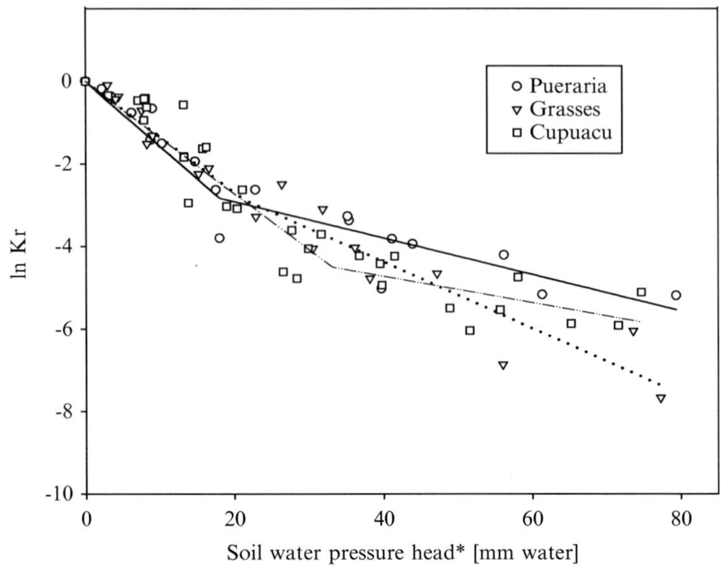

Fig. 13.8 Unsaturated soil hydraulic conductivity (K) in function of tension (h) evaluated near different species of plants growing in a clayey Ferralsol in the Central Amazon. The *dotted line* represents the K(h) for soil covered by grass; *solid line* the K(h) by soil covered pueraria and *dashed line* the K(h) by soil near the trunks of cupuaçu

mesoporosity (Teixeira 2001) what reinforces Pueraria as good option for cover plants in addition of being a leguminous with capacity of nitrogen fixation.

The scatter plot of all K and h data measured within the agroforestry system at different positions are shown in Fig. 13.9. In spite of the spatial variability of the surface hydraulic properties, the influence of intrinsic factors (e.g. soil porosity, pore continuity) and deterministic factors (e.g., plant species, pattern in throughfall and stemflow, specific fertilisers and different rate for each species within the agroforestry system) could be tentatively identified. Similar results were found for the evaluation of K near the trunks of peach palms growing in monoculture and agroforestry. The K(h) values between the peach palms in the monocultures for fruits and palm hearts were similar and were also pooled in a unique function in the agroforestry land use system. The h values ranged from 15 to 45 mm with a mean around 20 mm (Fig. 13.9). However, the results showed that using a pooled function with all K(h) values to characterize the agroforestry land use system will subestimate the water fluxes in some positions and overestimate in others (Fig. 13.9), because some K(h) functions near a specific species are below and some functions are above the pooled K(h) function.

For instance, for h $\approx$ 60 mm the K is about 2 orders of magnitude greater in the soil under the cover crop pueraria than in the soil near the trunks of peach palms. Mallants et al. (1997) demonstrated that the use of no optimal K(h) functions under predicted the cumulative drainage by more than 30 % for a macropore soil.

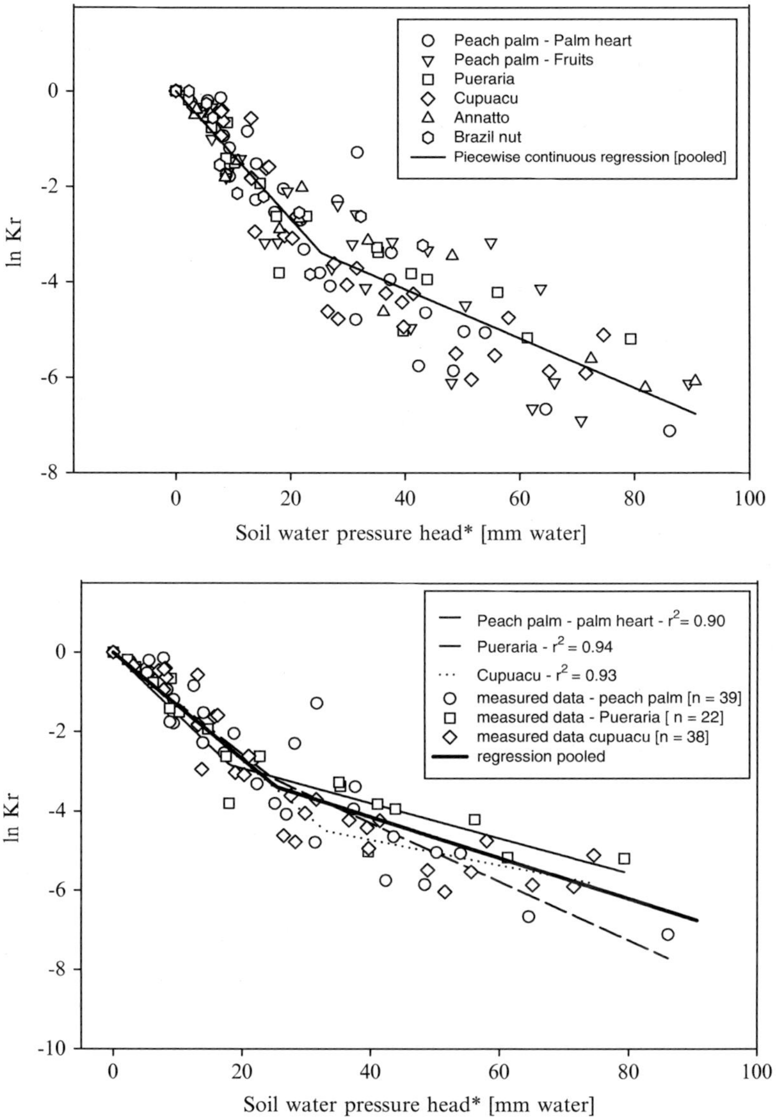

Fig. 13.9 Unsaturated hydraulic conductivity in function of tension – K(h) – evaluated near different species of plant growing in an agroforestry system on a clayey Ferralsol in the Central Amazon

It should be noticed that when the measurements were carried out under the central Amazonian conditions of sunny and hot weather conditions, the protection of the Mariotte reservoirs from the direct sun is a fundamental factor for reliable measurements. Moreover, a good preparation of the soil surface which allows using only a fine layer of the contact material was important. Further, the installation of the disc in leveled position is fundamental for the measurements at small h. Roots should be carefully cut, otherwise the membrane is easily damaged and the soil is disturbed.

13.4 Final Remarks

13.4.1 Evaluation of Unsaturated Hydraulic Conductivity (K)

Tension infiltrometry shows to be a relatively cheap, robust and simple method to evaluate unsaturated hydraulic conductivity in clayey Ferralsols. Since K near saturation is highly variable, the data were better analyzed using the relative hydraulic conductivity and the scaling theory.

The use of piecewise continuous regression analysis allowed the estimation of a changeover point, located between 15 and 40 mm of h, which operationally separates two domains that govern the hydraulic conductivity of the clayey Ferralsol.

K changes by 3–4 orders of magnitude between 0 and 70 mm of water pressure head. Effect of land use systems and different species of plants on unsaturated hydraulic functions were identified, these differences if not taken into account could be a source of error in water and solute transport studies.

These small values of tension in the breakpoint that divide the fluxes dominated by gravity and capillarity reinforce the importance to evaluated water content in small tension when is planned to derived K(h) from soil water retention curve.

A limitation of the tension infiltrometers is the small range in which it operates. However, data from TI can be combined with data derived from other methods to obtain an overview of a large unsaturated range. Very small discs in TI can lead to very high variability because of sampling in those TI are less than Representative Elementar Volume (REV) for unsaturated conductivities in almost all structured soils.

References

Ahuja LR, Green RE, Chong SK, Nielsen DR (1980) A simplified functions approach for determining soil hydraulic conductivities and water characteristics *in situ*. Water Resour Res 16:947–953

Amoozegar A, Warrick AW (1986) Hydraulic conductivity of saturated soils: field methods. In: Klute A (ed) Methods of soil analysis. Part 1. Physical and mineralogical methods. 2. American Society of Agronomy & Soil Science Society of America, Madison, pp 735–770

Angulo-Jaramillo R, Moreno F, Clothier BE, Thony J, Vachaud G, Fernadez-Boy E, Cayuela JA (1997) Seasonal Variation of hydraulic properties of soils measured using a tension disk infiltrometer. Soil Sci Soc Am J 61:27–32

Ankeny MD (1992) Methods and theory for unconfined infiltrometer measurements. In: Topp GC, Reynolds WD, Green RE (eds) Advances in measurements of soil physical properties: bringing theory into practice. SSSA, Madison, pp 123–141

Ankeny MD, Ahmed M, Kaspar TC, Horton R (1991) Simple field method for determining unsaturated infiltration measurement. Soil Sci Soc Am J 55:467–470

Assouline S, Tavares Filho J, Tessier D (1997) Effect of compaction on soil physical and hydraulic properties: experimental results and modeling. Soil Sci Soc Am J 61:390–398

Beven K, Germann P (1982) Macropores and water flow in soils. Water Resour Res 18:1311–1325

Bouma J, Belmans C, Dekker LW (1982) Water infiltration and redistribution in a silt loam subsoil with vertical worm channels. Soil Sci Soc Am J 46:917–921

Bouwer H (1986) Intake rate: cylinder infiltrometer. In: Klute A (ed) Methods of soil analysis. Part 1-Physical and mineralogical methods. 2. American Society of Agronomy, Inc., Madison, pp 825–844

Chauvel A (1982) Os Latossolos amarelos, álicos, argilosos dentro dos ecossistemas das bacias experimentais do INPA e da região vizinha. Acta Amazonica 12:47–60

Clothier BE, Smettem KRJ (1990) Combining laboratory and field measurements to define the hydraulic properties of soil. Soil Sci Soc Am J 54:299–304

Corrêa JC (1985) Características físicas de um latossolo amarelo muito argiloso (typic Acrorthox) do Estado do Amazonas, sob diferentes sistemas de preparo do solo. Pesquisa Agropecuária brasileira 20:1381–1387

Dalland A, Våje PI, Matthews RB, Singh BR (1993) The potential of alley cropping in improvement of cultivation systems in the high rainfall areas of Zambia. III. Effects on soil chemical and physical properties. Agrofor Syst 21:117–132

Draper NR, Smith H (1998) Applied regression analysis, 3rd edn. Wiley, New York, 820 pp

Elrick DE, Reynolds WD (1992) Methods for analyzing constant head well permeameter data. Soil Sci Soc Am J 56:320–323

El-Swaify SA (1980) Physical and mechanical properties of oxisols. In: Theng BKG (ed) Soils with variable charge. New Zealand Society of Soil Science, Lower Hutt, pp 303–324

EMBRAPA (2013) Sistema Brasileiro de Classificação de Solos. EMBRAPA, Brasilia, 353 pp

Green RE, Ahuja LR, Chong SK (1986) Hydraulic conductivity, diffusivity, and sorptivity of unsaturated soils: field methods. In: Klute A (ed) Methods of soil analysis. Part 1: Physical and mineralogical methods. 2. American Society of Agronomy, Madison, pp 771–798

Hillel D (1998) Environmental soil physics. Academic Press, San Diego, 771 pp

Hopmans JW (1987) A comparison of various methods to scale soil hydraulic properties. J Hydrol 93:241–256

Hulugalle NR, Ndi JN (1993) Effects of hedgerow species in alley croping systems on surface soil physical properties of an Oxic Paleustalf in south-western Nigeria. Agroforest Syst 22:207–220

Iwata S, Tabuchi T, Warkentin BP (1995) Soil-water interactions, 2nd edn. Marcel Dekker, New York, 440 pp

Jarvis NJ, Messing I (1995) Near-saturated hydraulic conductivity in soils of contrasting texture measured by tension infiltrometers. Soil Sci Soc Am J 59:34

Jury WA, Russo D, Sposito G (1987) The spatial variability of water and solute transport properties in unsaturated soil. II. Scaling models of water transport. Hilgardia 55:33–56

Jury WA, Gardner WR, Gardner WH (1991) Soil physics, 5th edn. Wiley, New York, 328 pp

Keng JCW, Lin CS (1982) A two-line approximation of hydraulic conductivity for structured soils. Can Agric Eng 24:77–80

Kutílek M (1973) The influence of clay minerals and exchangeable cations on soil moisture potential. In: Hadas A, Swartzendruber D, Ritjema PE, Fuchs M, Yaron B (eds) Physical aspects of soil water and salts in ecosystems. Springer, Berlin, pp 153–160

Lal R (1989) Agroforestry systems and soil surface management of a tropical alfisol: IV. Effects on soil physical and mechanical properties. Agrofor Syst 8:197–215

Lal R, Wilson GF, Okigbo BN (1979) Changes in properties of an alfisol produced by various cover crops. Soil Sci 127:377–382

Logsdon SD (1997) Transient variation in the infiltration rate during measurement with tension infiltrometer. Soil Sci 162:233–241

Logsdon SD, Jaynes DB (1993) Methodology for determining hydraulic conductivity with tension infiltrometers. Soil Sci Soc Am J 57:1426–1431

Luckner LMT, van der Ploeg RR, Nielsen DR (1989) A consistent set of parametric models for the two-phase flow of immiscible fluids in the subsurface. Water Resour Res 25:2187–2193

Mallants D, Tseng PH, Toride N, Timmerman A, Feyen J (1997) Evaluation of multimodal hydraulic functions in characterizing a heterogeneous field soil. J Hydrol 195:172–199

Marques JD, Teixeira WG, Reis AM, Cruz OF, Martins GC (2008) Avaliação da condutividade hidráulica do solo saturada utilizando dois métodos de laboratório numa topossequência com diferentes coberturas vegetais no baixo Amazonas. Acta Amazonica 38:193–206

Mathieu C, Pieltain F (1998) Analyse physique des sols. Lavoisier, Paris, 275 pp

Messing I, Jarvis NJ (1993) Temporal variation in the hydraulic conductivity of a tilled clay soil as measured by tension infiltrometers. J Soil Sci 44:11–24

Miller EE, Miller RD (1956) Physical theory for capillary flow phenomena. J Appl Phys 27:324–332

Miyazaki T (1993) The water flows in soils. Marcel Dekker, New York, 296 pp

Mohanty BP (1999) Scaling hydraulic properties of a macroporous soil. Water Resour Res 35:1927–1931

Mohanty BP, Kanwar RS, Everts CJ (1994) Comparison of saturated hydraulic conductivity measurement methods for a glacial-till soil. Soil Sci Soc Am J 58:672–677

Pachepsky Y, Radclife DE, Selim HM (2003) Scaling methods in soil physics. CRC Press, Boca Raton, 456 pp

Perroux KM, White I (1988) Designs for disc permeameters. Soil Sci Soc Am J 52:1205–1215

Reynolds WD (1993) Unsaturated hydraulic conductivity: field measurement. In: Carter MR (ed) Soil sampling and methods of analysis. Lewis, Boca Raton, pp 633–644

Reynolds WD, Elrich DE (1991) Determination of hydraulic conductivity using a tension infiltrometer. Soil Sci Soc Am J 55:633–639

Reynolds WD, Zebchuk WD (1996) Use of contact material in tension infiltrometer measurements. Soil Tech 9:141–159

Sanchez PA (1976) Properties and management of soil in the tropics. Wiley, New York, 618 pp

Schroth G, da Silva LF, Wolf MA, Teixeira WG, Zech W (1999) Distribution of throughfall and stemflow in multi-strata agroforestry, perennial monoculture, fallow and primary forest in central Amazonia, Brazil. Hydrol Process 13:1426–1436

Schroth G, Teixeira WG, Seixas R, da Silva LF, Schaller M, Macêdo JLV, Zech W (2000) Effect of five tree crops and a cover crop in multi-strata agroforestry at two fertilization levels on soil fertility and soil solution chemistry in central Amazonia. Plant and Soil 221:143–156

Schroth G, D'Angelo SA, Teixeira WG, Haag D, Lieberei R (2002) Conversion of secondary forest into agroforestry and monocultures plantations in Amazonia: consequences for biomass, litter and soil carbon stocks after 7 years. For Ecol Manage 163:131–150

Shouse PJ, Mohanty BP (1998) Scaling of near-saturated hydraulic conductivity measured using disc infiltrometers. Water Resour Res 34:1195–1205

Shouse PJ, Sisson JB, Ellsworth TR, Jobes JA (1992) Estimating in situ unsaturated hydraulic properties of vertically heterogeneous soils. Soil Sci Soc Am J 56:1673–1679

Simmons CS, Nielsen DR, Biggar JW (1979) Scaling of field-measured soil water properties. I. Methodology II. Hydraulic conductivity and flux. Hilgardia 47:74–173

Soil Survey Staff (1997) Keys to soil taxonomy, 7th edn. Pocahontas Press, Blacksburg, 545 pp

Stephens DB (1996) Vadose zone hydrology. Lewis, Boca Raton, 339 pp

Stoffregen H, Wessolek G, Renger M (1999) Effect of temperature on hydraulic properties. In: van Genuchten MTH, Leij FJ, Wu L (eds) Workshop on indirect methods for estimating the hydraulic properties of unsaturated soils, 2, 1997. University of California, Riverside, pp 497–506

Teixeira WG (2001) Land use effects on soil physical and hydraulic properties of a clayey Ferralsol in the Central Amazon, vol 72. Bayreuther Bodenkundliche Berichte, Bayreuth, pp 1–255

Teixeira WG, Arruda WC, Shinzato E, Macedo RS, Martins GC, Lima HN, Rodrigues TE (2010) Solos do Amazonas. In: Maia MAM, Marmos JL (eds) Geodiversidade do estado do Amazonas. CPRM, Manaus – Rio de Janeiro, pp 71–86

Tillotson PM, Nielsen DR (1984) Scale factors in soil science. Soil Sci Soc Am J 48:953–959

Tomasella J, Hodnett MG (1996) Soil hydraulic properties and van Genuchten parameters for an Oxisol under pasture in Central Amazonia. In: Gash JHC, Nobre CA, Roberts JM, Victoria RL (eds) Amazonian deforestation and climate. Wiley, West Sussex, pp 101–124

Warrick AW, Mullen GJ, Nielsen DR (1977) Scaling field-measurement soil hydraulic properties using a similar media concept. Water Resour Res 13:3550362

Watson KW, Luxmoore RJ (1986) Estimating macroporostiy in a forest watershed by use of a tension infiltrometer. Soil Sci Soc Am J 50:578–582

White I, Perroux KM (1989) Estimation fo unsaturated hydraulic conductivity from field sorptivity measurements. Soil Sci Soc Am J 51:1093–1101

White I, Sully MJ, Perroux KM (1992) Measurement of surface soil hydraulic properties: disc permeameters, tension infiltrometers and other techniques. In: Topp WD, Reynolds WD, Green RE (eds) Advances in measurements of soil physical properties: bringing theory into practice. SSSA, Madison, pp 69–104

Wilson GV, Luxmoore RJ (1988) Infiltration, macroporosity, and mesoporosity distributions on two forested watersheds. Soil Sci Soc Am J 52:329–335

Wooding RA (1968) Steady infiltration from a shallow circular pond. Water Resour Res 4:1259–1273

WRB. IUSS Working Group. World reference base for soil resources 2006, first update 2007. World soil resources reports no. 103. FAO, Rome

Young A (1997) Agroforestry for soil management, 2nd edn. CAB, Wallingford, 320 pp

Chapter 14
Soil Moisture and Soil Water Storage Using Hydrological Modeling and Remote Sensing

Otto Corrêa Rotunno Filho, Afonso Augusto Magalhães de Araujo, Luciano Nóbrega Rodrigues Xavier, Daniel Medeiros Moreira, Rafael Carneiro Di Bello, Adilson Elias Xavier, and Lígia Maria Nascimento de Araujo

Abstract Soil moisture and soil water storage play a significant role in lumped and distributed hydrological simulation, both for model initialization and in later time steps to control and to correct model performance. On the other side, rainfall-runoff models still need to be improved to simulate reasonably well the vertical exchanges of heat and water between the soil and the atmosphere, that may result in inconsistent soil moisture fields. Therefore, many issues remain to be adequately addressed, such as how to include new data sources as well as how to improve methods for calibration, validation, parameterization and upscaling of hydrological models.

O.C. Rotunno Filho (✉)
Laboratório de Recursos Hídricos e Meio Ambiente (LABH2O), Programa de Engenharia Civil, Instituto Alberto Luiz Coimbra de Pós-Graduação e Pesquisa em Engenharia – COPPE, Universidade Federal do Rio de Janeiro, Campus Ilha do Fundão – Cidade Universitária, Caixa Postal 68540, CEP 21.945-970 Rio de Janeiro, RJ, Brazil
e-mail: otto@coc.ufrj.br

A.A.M. de Araujo
Departamento de Recursos Hídricos e Meio Ambiente, Escola Politécnica de Engenharia, Universidade Federal do Rio de Janeiro, Campus Ilha do Fundão – Cidade Universitária, Caixa Postal 68540, CEP 21.945-970 Rio de Janeiro, RJ, Brazil
e-mail: afonsoaraujo@poli.ufrj.br

L.N.R. Xavier
Centro de Pesquisa de Energia Elétrica (CEPEL), Avenida Horácio de Macedo, 354, Cidade Universitária, CEP 21941-911 Rio de Janeiro, RJ, Brazil
e-mail: lnrxavier@gmail.com

D.M. Moreira
Laboratório de Recursos Hídricos e Meio Ambiente (LABH2O), Programa de Engenharia Civil, Instituto Alberto Luiz Coimbra de Pós-Graduação e Pesquisa em Engenharia – COPPE, Universidade Federal do Rio de Janeiro, Campus Ilha do Fundão – Cidade Universitária, Caixa Postal 68540, CEP 21.945-970 Rio de Janeiro, RJ, Brazil

Divisão de Hidrologia Aplicada, Serviço Geológico do Brasil, CPRM, Av. Pasteur, 404 – URCA, CEP 22.290-240 Rio de Janeiro, RJ, Brazil
e-mail: daniel.moreira@cprm.gov.br

W.G. Teixeira et al. (eds.), *Application of Soil Physics in Environmental Analyses: Measuring, Modelling and Data Integration*, Progress in Soil Science, DOI 10.1007/978-3-319-06013-2_14, © Springer International Publishing Switzerland 2014

This work focuses on relatively new data sources, addressing the supply of soil moisture (Soil Moisture Experiment 2003 – SMEX03) and soil water storage information (GRACE – *Gravity Recovery and Climate Experiment*) to a typical lumped rainfall-runoff model (SMAP – *Soil Moisture Accounting Procedure* running at daily and monthly steps) from *in situ* measurements and remotely sensed imagery for the São Francisco and Amazon basins in Brazil. In particular, a sensitivity analysis is conducted to evaluate jointly both hydrological simulations and data collected and acquired for the studied areas highlighting soundly based and good results and also pointing out some of the challenges to be faced in the near future. We should mention that much more work on soil physics is still necessary for applications regarding rainfall-runoff models to work properly at the watershed scale.

Keywords Soil moisture • Soil water storage • Soil physics • Hydrological modeling • Remote sensing • Soil moisture experiment 2003 – SMEX03 • Grace • São Francisco watershed • Amazon watershed

14.1 Introduction

The broad objective of this chapter is to improve the knowledge about the evaluation of the input of soil moisture and soil water storage to lumped and distributed hydrological models. Under this general framework, we briefly investigate the use of *in situ* measurements and remotely sensed data for detecting and explaining soil moisture variability occurring at the surface layer within a watershed jointly with soil water storage with the support of a lumped hydrological model allowing us to

R.C. Di Bello
Tribunal de Contas da União – TCU, Setor de Administração Federal Sul – SAFS, Quadra 4, Lote 1, CEP 70.042-900 Brasília, DF, Brazil
e-mail: rcdibello@gmail.com

A.E. Xavier
Programa de Engenharia Sistemas e Computação, Instituto Alberto Luiz Coimbra de Pós-Graduação e Pesquisa em Engenharia – COPPE, Universidade Federal do Rio de Janeiro, Campus Ilha do Fundão – Cidade Universitária, Caixa Postal 68540, CEP 21.945-970 Rio de Janeiro, RJ, Brazil
e-mail: adilson@cos.ufrj.br

L.M.N. de Araujo
Laboratório de Recursos Hídricos e Meio Ambiente (LABH2O), Programa de Engenharia Civil, Instituto Alberto Luiz Coimbra de Pós-Graduação e Pesquisa em Engenharia – COPPE, Universidade Federal do Rio de Janeiro, Campus Ilha do Fundão – Cidade Universitária, Caixa Postal 68540, CEP 21.945-970 Rio de Janeiro, RJ, Brazil

Agência Nacional de Águas – ANA, Setor Policial Sul – SPS, Área 5, Quadra 3, Blocos "B", "L", "M" e "T", CEP 70.610-200 Brasília, DF, Brazil
e-mail: ligiamna@gmail.com

envisage and to provide reflections and comments on future developments in hydrology and on a proposed agenda for hydrological research.

Therefore this work intends to emphasize the role of soil moisture and soil water storage in rainfall-runoff models performance, where, frequently, the initial conditions are unknown and its evolution inside the model is not followed. The soil moisture and soil water storage behavior in SMAP (*Soil Moisture Accounting Procedure*) rainfall-runoff lumped model was observed. The SMAP daily standard algorithm version proposed initially by Lopes et al. (1981) works with three linear conceptual reservoirs (surface runoff, soil surface zone and groundwater flow), whereas its monthly version operates with two linear conceptual reservoirs (surface runoff and groundwater flow). Two different study cases are shown herein.

In the first one, the adopted SMAP daily version uses a hyperbolic smoothing routine (Xavier 2001; Xavier and Oliveira 2005), which makes possible the application of powerful minimization algorithms. The case study was the Barreiras region in the Grande River watershed, which is a portion of the São Francisco basin (northeastern Brazil), where a soil moisture field campaign was conducted in December 2003 (Sano and Assad 2004; Sano et al. 2004), making available a database of soil moisture measurements (SMEX03 – *Soil Moisture Experiment* in Brazil 2003). The influence of soil moisture was revealed for the SMAP model through a sensitivity analysis. A set of computational experiments was conducted aiming at illustrating both the reliability and the efficiency of the proposed calibration approach through the hyperbolic smoothing method. In addition, simulated and field measured data were compared.

In the second case study, a SMAP monthly version was developed for Amazon basin hydrologic modeling. Understanding large scale water fluxes is a great challenge to be met when addressing climate modeling. Data from GRACE (*Gravity Recovery and Climate Experiment*) mission has been derived since 2002 for allowing a more precise determination of the Earth gravity field (Tapley et al. 2004). Cazenave and Chen (2010) presented an in depth revision of GRACE data main applications in many research fields. Over the continents, main gravity changes taking place at the seasonal scale are due to the water cycle. This application aims at discussing the use of TWS (*Total Water Storage*) series derived from GRACE mission suitability as an additional data source for large scale hydrological modeling. Analysis of TWS spatiotemporal variability over Amazon watersheds showed that GRACE TWS series could represent main water cycle characteristics at the seasonal and interannual scales (Xavier et al. 2010). The suitability of integrating GRACE TWS data in hydrologic modeling has been addressed within a simple approach applied to the Negro River watershed, one of the most important sub-basins of Amazon basin. From the obtained results many of the limitations in using these data for hydrological modeling purposes could be discussed, and an improved analysis framework could be proposed.

14.2 General Background

Hydrology deals principally with the movement, distribution, and storage of moisture. From a global perspective, three systems are involved: (1) the atmospheric system, (2) the land system, and (3) the ocean system. The domain of hydrology is confined to the land system but may include those aspects of atmospheric and oceanic systems that directly affect the movement of moisture in the earth system. Thus a better understanding of the land phase of the hydrological cycle is a very important issue in hydrology. In recent years, great effort has been made to pursue this goal through experimental work and mathematical modeling. The present chapter fits well in this framework, since its major focus is on the supply of soil moisture and soil water storage information to rainfall-runoff models using remotely sensed data.

Soil moisture information and soil water storage are used in such models to determine infiltration. This is the term applied to the process of water entry into the soil, generally by downward flow through all or part of the soil surface. The rate of this process is dependent on other processes such as precipitation, surface runoff and evaporation. Knowledge of the infiltration process, as it is affected by the soil's properties and transient conditions, and by the mode of water supply, is therefore a prerequisite for efficient water and soil management as well as a prerequisite for the development of sound physically-based hydrological models.

The infiltration process has been studied extensively in the literature at the continuum or hydrodynamic scale, or at what may be called the microscale. However, hydrologists in engineering practice have to deal with problems at much larger space and time scales. At large scales it is an almost impossible task if one were to model the hydrologic response using the continuum theory.

As Dooge (1986) says, unfortunately most of the problems arising in catchment hydrology do not fall within the scope either of analytical mechanics or of statistical mechanics. In reality catchment hydrology falls in the category of complex systems with some degree of organization.

Hydrologic processes can be analyzed on the basis of the equations of hydraulics and soil physics, but the high degree of spatial variability in a catchment of any size poses serious problems of parameter specification and parameter estimation.

Perhaps the most important property of a mathematical model is the degree of representation given to the spatial and temporal variations in the input and output parameters. One can differentiate between a lumped-system model and a distributed-system model. While in the lumped models the parameters and variables are assumed to change only in time, in the distributed models the parameters and variables are assumed to change in time as well as in space.

In the past, lumped models of a quasi-physical/semi-empirical nature have been most common. They have not necessarily increased the understanding of the systems being modelled, nor have the model parameters had a sound physical basis. However, they have played, and still play, an important role in engineering practice.

More recently, the focus has been changed towards spatially distributed hydrological models. There are at least five areas for which physically-based spatially distributed models can fulfill the needs of practical applications: flood forecasting, catchment changes, spatially variable inputs and outputs, the movement of pollutants and sediment through a catchment and forecasting the hydrological response of ungauged catchments.

On the other hand, distributed hydrological models require much more data when compared to lumped models. These data may be provided through remotely sensed images. Before remote sensing, the hydrologist was limited to the extraction of information from sparse conventional field data using mathematical techniques. The possibility of overcoming this limitation by coupling remote sensing with conventional data sets has been shown to be feasible in the literature, although there are significant drawbacks to the use of remote sensing data. The main difficulty usually lies in calibrating and understanding the electromagnetic signals in hydrological terms as well as data availability.

In the middle of the 90s of the last century, the state of the art in the use of remote sensing as input to distributed hydrological models could be summarized as: (a) the use of visible and infrared part of the electromagnetic spectrum applied to determination of land cover in a watershed; (b) the use of microwave energy applied to the identification of water, including snow, on or near the land surface.

Presently, the state of the art in the use of remote sensing as input to lumped and distributed hydrological models can be summarized as: (a) the use of visible and infrared part of the electromagnetic spectrum applied to determination of soil use and land cover in a watershed and corresponding vegetation and built-up area indices; (b) the use of the thermal part of the electromagnetic spectrum for determination of brightness temperature and land surface temperature; (c) the use of microwave energy applied to the identification of water, including rainfall, water levels with the support of altimetry datasets and snow, on or near the land surface; (d) the measurement of the total continental water storage including surface water, soil moisture and groundwater using gravimetric remotely sensed data and (e) the use of microwave energy and laser technology for providing digital elevation models (DEM) and digital terrain models (DTM) jointly with GPS measurements for altimetry levelling at the watershed scale.

Related study areas and hydrological datasets are:

- temporal and spatial rainfall fields (e.g. rain gauges, TRMM – *Tropical Rainfall Measuring Mission*, GPCP – *Global Precipitation Climatology Program*, meteorological radar datasets);
- energy balance at the soil surface with the determination of net radiation, latent heat, sensible heat and soil heating components for modeling the evaporation process (hydrometeorological ground stations, visible and infrared satellite images of different resolutions, such as Landsat and MODIS – *Moderate-Resolution Imaging Spectroradiometer*);
- infiltration component and its corresponding upscaling spatial analysis applied to distributed hydrological models (surface and profile soil properties and soil retention curves by means of infiltrometers and laboratory permeameters, satellite vegetation indices, satellite land use and land cover);

- total water storage from the precisely defined variations of the Earth gravity field (GRACE mission);
- digital elevation model and digital terrain model (contour levels, GPS measurements, SRTM images, LIDAR – *Light Detection and Range* – technology);
- river stage variability at monthly and submonthly scale (e.g. TOPEX-POSEIDON, ENVISAT, ERS-1, ERS-2, ICESAT, JASON-1, JASON-2).

In the next sections, we develop a brief review on rainfall-runoff models and on remote sensing technology to provide support to better understanding some of the key issues for stimulating new reflections on hydrology and soil physics given the applications herein presented.

14.3 Rainfall-Runoff Models

Hydrological simulation models have become an integral part of almost all aspects of hydrology. For example, current and forecasted weather conditions are input to models to estimate future flood discharges for emergency operations, flood control, and reservoir regulation. Models are used in planning studies to evaluate the benefits and impacts of alternative land use plans or flood control measures. In design studies, models are used to assist in developing specifications for engineering structures to solve specific flood control, navigation, or water supply problems.

Due to the complexity of the physical processes that occur in nature, suggestions are made to simplify the representation of these processes in a hydrological model, trying to achieve a model behavior as close as possible to nature behavior. By identifying, isolating, and quantifying sources of modeling uncertainty and their effects, a proper quantitative view of the usefulness and reliability of basin modeling will emerge.

14.3.1 Calibration of a Hydrological Model

According to Kuczera (1983), the literature has identified the following five major applications of catchment models:

(a) extension of streamflow records;
(b) generation of runoff statistics;
(c) assessment of the effects of land use changes;
(d) prediction at ungauged catchments;
(e) prediction of the effects of land use change on the hydrologic regime.

We could still add to the above list the use of rainfall-runoff models as boundary conditions applied to global circulation models for assessment and prediction of the effects of climate change. It is generally accepted that the first two applications have been successfully implemented. To succeed in the first two applications, it suffices to identify and fit a model capable of satisfactorily converting hydrologic inputs such as rainfall into outputs such as runoff.

However, implementation of the remaining applications is considerably more difficult. Basically the problem arises when regression relationships between optimized parameters and catchment characteristics are attempted. Magette et al. (1976) developed regression relationships between six fitted parameters of the Kentucky watershed model and 15 measurable catchment characteristics. Independent tests revealed that errors in prediction of mean annual flows ranged from less than 1–860 %.

One of the possible reasons for this limited success is that the adjusted parameters have not been correctly identified through the calibration process. Model calibration is one of the greatest difficulties in hydrological modeling. This is due to the fact that the response surface generated by such a model is very complex, usually made up of multiple local minima and many times with the occurrence of valleys and saddle points. On the other hand, the available mathematical tools are only able to solve problems efficiently when the function associated to the model is well-behaved.

In general, calibration can be classified in two types: manual calibration by the method of trial and error and automatic calibration. In the second approach, a mathematical optimization method is programmed to run on a digital computer, and it is used to adjust the values of the unknown parameters based on changes in the values of a pre-specified objective function. The latter introduces higher speed in the calibration process as well as less subjectivity. Lumb et al. (1994) proposed the use of manual calibration by expert system technology.

Calibrating a model aims at obtaining the set of parameters commonly named the optimum set of parameters. According to Troutman (1985a, b), there are two ways of defining the optimum set of parameters. The first one is to define the set of parameters assuming values physically compatible with nature and, more precisely, with the catchment under study. The second possible definition is that in which a certain mathematical criterion, also named objective function, is established. In this case, it is attempted to obtain a set of parameters where the generated flows are as close as possible to the observed flows.

In reality, these two definitions approach each other. It is expected that accurate forecasted flows results in a low objective function value in a minimization problem. At the same time, it is expected that the found parameters are to be physically realistic. If there was a model that could represent nature exactly, the optimum set of parameters would satisfy both definitions. In practice, there is no perfect model, what leads to the definition of an optimum set of parameters that do not assume physical values unless constraints are used. Problems in this area have been recognized, among others, by Johnston and Pilgrim (1976), Diskin and Simon (1977), Canedo (1979), Sorooshian and Dracup (1980), Moore and Clarke (1981), Sorooshian and Arfi (1982), Sorooshian et al. (1983), Gupta and Sorooshian (1983, 1985), and Duan et al. (1992), Rotunno Filho (1995), and Kavetski and Kuczera (2007).

14.4 Remote Sensing Technology Applied to Hydrology

Application of lumped and distributed physically-based hydrological models requires the evaluation of a relatively large number of soil, vegetation and topographic parameters on a spatially distributed basis. Two particular problems concern the number of measurements implied and the means of obtaining effective parameter values, representative at the grid scale. Efforts to resolve these problems will spur developments in field sampling strategies and the use of remote sensing technology.

The basic principle involved in remote sensing methods is that in different wavelength ranges of the electromagnetic spectrum, each type of object reflects or emits a certain intensity of light, which is dependent upon the physical or compositional attributes of the object. Thus, using information from one or more wavelength ranges, it may be possible to differentiate between different types of objects (e.g. dry soil, vegetation, limonitic area, ice, etc.), and map their distribution on the ground.

In hydrology, remote sensing is being recognized as a viable approach to handle spatial variability of watershed properties, because the basic data are spatial in nature. Despite all the progress achieved by studies dealing with remotely sensed data, there is a limited number of hydrological models that have been used successfully in conjunction with satellite remote sensing.

Jackson et al. (1977) used land cover information acquired from Landsat data to estimate runoff coefficients for the STORM model (U.S. Army Corps of Engineers 1976). The authors conclude that, except on very small watersheds, the agreement between the hydrologic simulations produced by a conventionally defined model and those obtained from a Landsat-based model should be within a range acceptable for watershed planning studies. In addition, significant reductions in the costs and man-hours associated with the development of land cover and parameter estimates for hydrologic planning models can be achieved using the Landsat approach. Ragan and Jackson (1980) and Harvey and Solomon (1984) have both used remotely sensed land cover data to estimate runoff curve numbers for further input into the Soil Conservation Service (SCS) runoff model. In both studies, Landsat data was used for the estimation of the land cover and subsequently, the runoff curve numbers, time to peak and peak flow. In all three cases the results obtained from the remote sensing land cover were acceptable when compared with those having curve numbers defined by Soil Conservation Service land use/cover maps (Soil Conservation Service 1972).

Another study by Webb et al. (1980) showed that land use classification percentages derived from Landsat data were well within acceptable levels to be used in hydrological modeling. Although the classifications obtained were not as detailed when using the Landsat estimates as compared to the conventional estimates, acceptable results were still obtained. In a study by Rango et al. (1983), conventional and Landsat land use data were input into the HEC-1 model (U.S. Army Corps of Engineers 1981) for generating discharge frequency curves.

The differences between the Landsat and conventionally derived results were found to be insignificant. Also, it was found that for basins larger than 26 km^2, the Landsat approach for extracting data was more cost efficient.

Fortin et al. (1986) gave an example of using land-cover information for daily flood forecasting in their CEQUEAU model. Simulations made with the Landsat data gave results similar to or better than simulations made with map data. The authors also stressed in the same work the capabilities of the CEQUEAU model in handling weather radar data and snow cover parameters. More recently, Tao and Kouwen (1989) compared two alternatives, with and without Landsat data, using the WATFLOOD model (Kouwen 1988) applied to the Upper Grand River basin using a 10 km grid size. Hourly rainy events were used. The model was run in two modes. Without Landsat data the model was a lumped model, while with Landsat data it was a distributed model. The results indicated that an improvement of predicting flood peaks and total runoff gained by using Landsat data is at the 10 % level of significance.

Rotunno Filho et al. (1996) presented results obtained for the relationship between soil moisture in 28 pasture fields and corresponding backscatter in eight ERS-1 SAR images (C-band, VV polarization) for a 1,500 km^2 area of the Grand River watershed, Ontario, Canada. Across-scene regression results were much better than within-scene results. The authors also mentioned that the recovery of soil moisture in the basin scale has received little attention due to the heterogeneity of the types of land use coverage. In addition, Leconte et al. (2004) emphasized that soil moisture has a direct impact on streamflow forecasting, highlighting the difficulties of moisture mapping near the soil surface at a regional scale. In that work, seven RADARSAT-1 SAR C_{HH} band images were acquired over the northern portion of the Châteauguay River watershed in southwestern Québec, Canada. This area is dominated by herbaceous and agricultural fields. The authors obtained a soil surface roughness map from SAR images, inverting an empirical model based on the return signal or backscattering with known or arbitrated soil moisture. The resulting map was then employed in the backscattering model developed by Dubois et al. (1995) in order to rescue the moisture content close to the surface soil for other SAR images. Field campaigns were conducted in parallel with the acquisition of SAR images, for purposes of measuring soil moisture and surface roughness in 24 regions on an area of approximately 200 km^2. Good agreement was observed between soil moisture values at the watershed scale and the average values for all sampled fields with theta-probe instrument, the same equipment employed in the SMEX03 experiment in Barreiras-BA, which is based on impedance probes. The correlation coefficient between the values of soil moisture obtained by RADARSAT-1 and the averages of field measurements was 0.96 ($R^2 = 0.92$), and the mean square error was 2.2 %, demonstrating the potential use of RADARSAT-1 for soil moisture estimation at the basin scale, covering areas with a few hundred square kilometers (km^2).

Another approach that has been widely used to perform the mapping of soil moisture is based on passive microwave radiometer technique, such as the sensor AMSR -E (*Advanced Microwave Scanning Radiometer*). According to Njoku et al. (2003), the AMSR -E operates in a sun-synchronous orbit, crossing the Equator at approximately 13:30 h. This sensor was developed by JAXA (*Japan Aerospace Exploration Agency*) and it is onboard the EOS Aqua satellite, launched on May 4, 2002. The AMSR-E measures the brightness temperature at six frequencies (6.92, 10.65, 18.7, 23.8, 36.5 and 89.0 GHz), with vertical and horizontal polarization for each band, for a total of 12 channels, and a comprehensive daily coverage. The spatial resolution of AMSR-E ranges from 60 km corresponding to 6.92 GHz until 5 km corresponding to 89.0 GHz. One of the AMSR-E products is a soil moisture global map.

McCabe et al. (2008) evaluated the consistency of remotely sensed data with respect to the various components of the hydrologic cycle, namely: AMSR-E (soil moisture), TRMM (precipitation) and MODIS data, ranging from visible to thermal infrared for the surface energy balance using the SEBS (*Surface Energy Balance System*) algorithm. These authors found that the spatial distribution of precipitation based on TRMM showed good agreement with data on soil moisture obtained from AMSR-E. Moreover, they also observed that the spatial distribution of the sensible heat flux was well represented by means of the relationship soil moisture – precipitation. Thus McCabe et al. (2008) suggested that an approach using multiple sensors associated to atmospheric observations could provide a consistent characterization of the hydrological cycle.

Most hydrological, environmental and agricultural applications require information on soil moisture in the root zone. As estimates of water content from passive microwave technique refers to soil layer up to approximately 10 cm, this technique alone is not enough to provide information on the moisture in the root zone. Even admitting that in many crops or in native vegetation the roots reach a depth of 100 cm or more, the conditions of soil moisture at this depth can present a completely independent behavior of near surface soil moisture. Another limitation in the estimation of soil moisture from passive sensor data in the microwave range is the spatial resolution of radiometers which can vary from 50 to 150 km (Scott et al. 2003).

More recently, Lopes et al. (2011) used NDVI and surface temperature data for mapping soil moisture through a soil moisture index (IUS) in a region covering municipalities located between the states of Paraíba and Rio Grande do Norte in northeastern Brazil. The results showed that the moisture content of the soil varied with the landscape indicating a direct relationship with NDVI and highlighting that there is an intense degradation in neighboring counties, resulting in an increased area of desertification in the region of Seridó.

Lucena et al. (2013) used NDVI maps to better identify and evaluate the formation and presence of heat islands in the metropolitan area of Rio de Janeiro (MARJ) over the last four decades through the use of Landsat images. Additionally, it was evaluated the use of NDVI maps as input to the Penn State-NCAR fifth-generation Mesoscale Model (MM5) over the MARJ region. Viana et al. (2013), in turn, explores the NDVI from the perspective of supplying information and assessment of the SMAP rainfall-runoff model focused on the estimation of evaporation at the river basin scale and on the determination of model parameters. This last research work also seeks to evaluate the potential use of the SMAP model for streamflow regionalization applied to the Preto River watershed, which is part of the Paraíba do Sul River basin and a natural boundary for the states of Minas Gerais and Rio de Janeiro in Brazil. The performance indices showed that the regionalization methods studied using the SMAP model could be used to estimate water availability in watersheds without appropriate flow gauge monitoring since the soil use and land cover are similar.

The surface energy balance is also a related approach to estimate soil moisture. This approach is based on the assumption that the partitioning energy at the surface is a reflection of water content in the soil. Some research has been dedicated to understanding and estimation of the components of energy balance (Bastiaanssen 1995; Bastiaanssen 2000; Trezza 2002; Ahmad and Bastiaanssen 2003; Hemakumara et al. 2003; Scott et al. 2003; Tasumi 2003; Allen et al. 2005; Tasumi et al. 2005a, b; Allen et al. 2007) by means of remotely sensed images in the visible to thermal infrared bands and also by combining the optical and microwave ranges of the electromagnetic spectrum (Bastiaanssen et al. 2012).

We should emphasize that studies devoted to understanding the estimated soil moisture in the root zone based on the partitioning of the energy balance in conjunction with hydrological models and remotely sensed images are being developed under different frameworks (Vereecken et al. 2008), but they are still limited. Thus, given the importance of soil moisture and soil water storage in applications both in hydrology and in modeling of weather and climate, there is an evident need for studies to enable a better understanding of this variable on a regional scale, mainly covering watersheds prominent in the social and economical context for the development of a country, such as the São Francisco River and Amazonia basins in Brazil, which are addressed under the proposed framework for this chapter.

14.5 Daily Hydrological Modeling: São Francisco Basin

As previously highlighted, there has been a great effort in mathematical modeling coupled with experimental work during the last four decades in order to better understand the behavior of nature. Despite the spatial variability and corresponding scale issues at the basin scale, lumped models are very useful in water resources engineering, especially in areas with poor streamflow records and scarce precipitation data or forecasting problems. The low structural complexity, less data set demand and calibration facilities are advantages of these models, despite the progress in the development of distributed models.

On the other hand, soil moisture information is used in physically-based hydrological model to determine infiltration. There are two important ways to use soil moisture in hydrological models. The first one would be to use it in the initialization of a hydrological model. Initial soil moisture is recognized to play an important role in defining the hydrograph for rainfall events. The second way would be to use soil moisture in different time steps to control and correct the performance of the model. This would require an updating procedure. In hydrologic modeling, soil moisture is usually obtained as a result of running a model. It is considered simply a step to make the model work and it does not mean that the values are reflecting what is happening in nature.

An unsolved problem is the definition of the parameters associated with the physical processes, such as infiltration, used in hydrological models. Because of the great variability of many of the variables involved, for example, rainfall, topography, soil properties, and vegetation, hydrologists usually resort to using models of the overall system without reference to the detailed structure of the components of the hydrological processes involved. The nature of the system operation is inferred from the input and output observations, achieved through a model-fitting process.

Therefore, application of such models requires the identification of proper values for the parameters which govern the functions that describe the underlying physical system. Calibration is the stage in the simulation process where the parameters should be identified and estimated. Although the literature has devoted much attention to this stage, there is still a clear need for improving the analysis on parameter optimization and on detection of model structural identifiability problems in conceptual rainfall-runoff (CRR) models.

Under this framework, first, this application addresses the issue of automatic parameter estimation in conceptual rainfall-runoff (CRR) models applied to the SMAP model (Lopes et al. 1981), which is a lumped rainfall-runoff model widely used in Brazil by the electric sector to simulate streamflows based on forecasted rainfall for estimating energy generation in hydroelectric power plants. Due to threshold structures commonly occurring in CRR models, the associated mathematical optimization problems have the significant characteristic of being strongly non-differentiable (Sorooshian and Gupta 1985). In order to face this enormous task, the resolution method proposed adopts a smoothing strategy using a special C^∞ differentiable class function. The final estimation solution is obtained by solving a sequence of differentiable subproblems which gradually approach the original conceptual problem. The use of this technique, called Hyperbolic Smoothing, makes possible the application of the most powerful minimization algorithms, and also allows for the main difficulties presented by the original CRR problem to be overcome.

Second, this application focuses on the soil moisture model behavior in contrast to a soil moisture dataset collected during the SMEX03 (Sano and Assad 2004; Sano et al. 2004), which was a 10-days experiment conducted in Barreiras region in Rio Grande watershed, state of Bahia in Brazil. This analysis was performed after the hyperbolic smoothing method has been applied to the SMAP model based on the dataset available, including streamflow and rainfall records jointly with evapotranspiration calculated through the seasonal water balance (Dias and Kan 1999).

14.5.1 Characteristics of the Conceptual Rainfall-Runoff Models

Automatic calibration of a conceptual rainfall-runoff model involves the solution of a non-linear optimization problem. The characteristics of the conceptual rainfall-runoff models that complicate automatic calibration are the complexity of the response surface and the discontinuities resulting from the model structure, which affect the expected better performance of derivative-based techniques (Johnston and Pilgrim 1976; Gupta and Sorooshian 1985; Hendrickson et al. 1988; Duan et al. 1992; Rotunno Filho 1995; Di Bello 2005; Araujo 2006; Kavetski and Kuczera 2007; Getirana 2009; Xavier 2012).

Johnston and Pilgrim (1976) list the main features of catchment models and their calibration methodology which preclude accurate parameter estimates: discontinuity in first derivatives, interdependence between the parameters of the model, flat objective function region, erroneous definition of the direction of the gradient, existence of multiple local minima, influence of scale of parameters and lack of physical meaning of the lumped parameters due to aggregations of small scale processes. More recently, Duan et al. (1992) presented five major characteristics of conceptual rainfall-runoff models that generate difficulties in the calibration stage: regions of attraction, minor local optima, roughness, sensitivity and shape. All features were already considered by Johnston and Pilgrim (1976).

However, it should be pointed out that regions of attraction and minor local optima are the most important features and they are related to the local minima challenge. Basically, Duan et al. (1992) explained that the structure of multiple optima exists on at least two scales. At the large scale, there are broad regions of attraction into which the sequence of points generated by an optimization algorithm may converge. At the small scale, each major region of attraction contains a myriad of minor local optima.

14.5.2 Origin of the Discontinuities

As quoted in Hendrickson et al. (1988) and Rotunno Filho (1995), the central difficulty in the calibration of conceptual rainfall-runoff models lies precisely in the discontinuities of the derivatives. The origin of these discontinuities must be analyzed. Gupta and Sorooshian (1985) defined the possible modes of operation of a typical rainfall-runoff model and developed their arguments based on the threshold structures commonly employed in this kind of models.

This typical structure of threshold values, that appears in rainfall-runoff models, leads to the multiplicity of possible ways of model operation, represented in the program code by IF structures. These different modes of operation cause the discontinuities in the derivatives of the model functions that represent a theoretical restriction to the applicability of first order and second order derivative-based techniques. According to a scheme adopted by Gupta and Sorooshian (1985), the reservoir having maximum capacity defined as a threshold M presents a lower output, S_t, that is a function of a recession coefficient K and its level x_t, an overflow output, R_t, due to the reservoir spill, and an input u_t, due to the rain, or superior reservoirs. In this simple scheme, the two parameters to be adjusted are K and M. The total flow z_t, although continuous, presents discontinuity in the first-order derivative and indetermination in the second-order derivative at the point $x_t = M$. As the hydrological model becomes more complex, the number of possible modes of operation increases and, at each time step, there are more distinct possibilities of paths in the model's general flowchart. Each mode has a specific system of equations relating the model variables. The basic equations of the model describe the reservoir level as well as the output flows at each time step:

$$x_t = x_{t-1} + u_t \tag{14.1}$$

$$R_t = \max(0, x_t - M) \tag{14.2}$$

$$S_t = K(x_t - R_t) \tag{14.3}$$

$$z_t = S_t + R_t \tag{14.4}$$

14.5.3 Smoothing Technique Used

Analyzing Eqs. 14.1, 14.2, 14.3 and 14.4 of output flow z_t, the obvious conclusion is that the response surface of such simplified model contains a derivative discontinuity which corresponds to the presence of the upper outlet R_t (x_t, M).

In order to overcome this difficulty, a natural idea is to use a smoothing approach. Kitanidas and Bras (1980) suggested two possible alternative approximations to a threshold behavior represented by the equation (2), in a real-time estimation and forecasting model of river flows. However, such suggestions were not used in calibration procedures of rainfall-runoff models. As Hendrickson et al. (1988) state, the proposal made by Kitanidas and Bras (1980) of replacing discontinuities with smooth S-shaped jumps would introduce perturbations in the derivatives and not completely solve the problem of non-smoothness.

On the other hand, we examine a continuously differential function that properly approximates the function $R_t(x_t, M)$ in an extremely simple and more convenient way. To this end, the following function was adopted (Xavier 2001):

$$\phi\ (x_t, M, d) = \frac{1}{2} \cdot \left[x_t - M + \left[(x_t - M)^2 + 4 \cdot d^2 \right]^{\frac{1}{2}} \right] \tag{14.5}$$

The function $\phi(x_t, M, d)$ presents the following properties:

(a) $\phi(x_t, M, d)$ is asymptotically tangent to the straight lines $r_1(x_t, M) = 0$ and $r_2(x_t, M) = x_t - M$;
(b) $\lim_{d \to 0} \phi(x_t, M, d) = 0$ if $x_t \le M$;
$\lim_{d \to 0} \phi(x_t, M, d) = x_t - M$ if $x_t > M$;

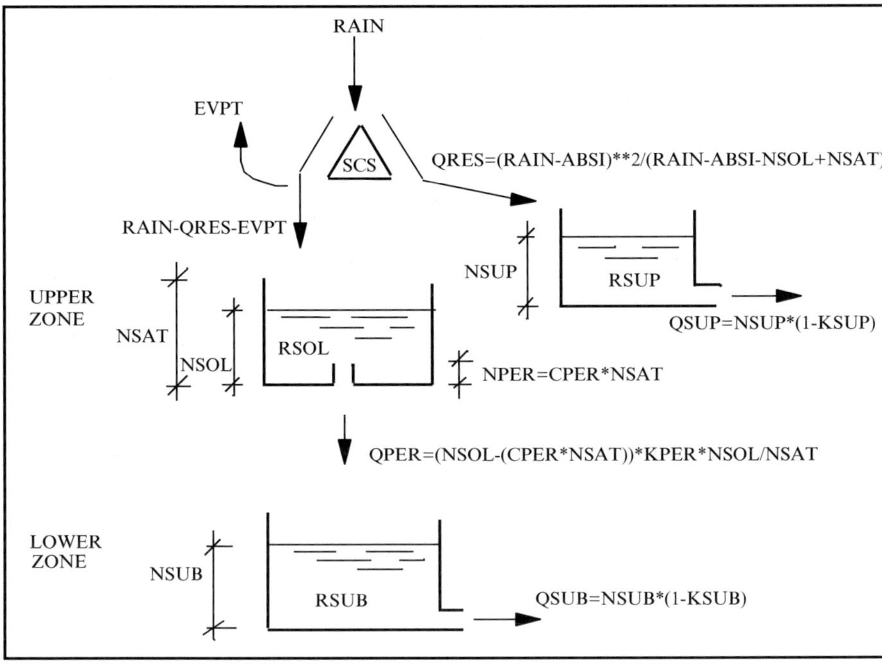

Fig. 14.1 Scheme of the SMAP rainfall-runoff model

(c) $\phi(x_t, M, d)$ is continuous and continuously differentiable in the variables x_t and M;
(d) $\phi(x_t, M, d)$ is convex in x_t and M (increasing in x_t and decreasing in M).

14.5.4 Description of the Application

The smoothing technique specified was implemented in the SMAP model (Lopes et al. 1981) (Fig. 14.1). The model simulates the land phase of the hydrological cycle by means of three linear reservoirs that represent respectively the surface runoff, the soil surface zone and the groundwater flow. The flow calculated by the model at each instant is the result of the sum of the parts contributed by the surface and underground reservoirs. The variables are constantly updated based on precipitation (RAIN) and potential evapotranspiration (EVPT) data and considering the mass conservation principle. The objective function (FO) considers the minimization of the sum of squares of differences between simulated (QGER) and observed streamflows (QOBS). The model, in its more simplified version, has six parameters: initial losses caused by vegetation retention and soil depressions (ABSI); surface recession coefficient (KSUP); soil saturation level (NSAT); field capacity (NPER); top soil reservoir recession coefficient (KPER); and groundwater recession coefficient (KSUB).

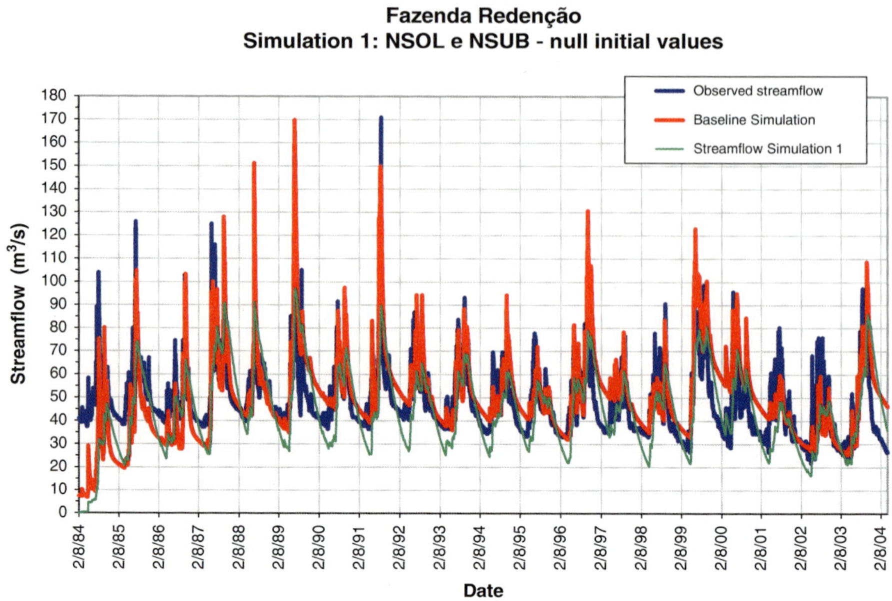

Fig. 14.2 Observed streamflow, baseline simulation and streamflow simulation 1 (m^3/s) – Fazenda Redenção streamflow gauge

The calibration period was chosen from 01/08/1984 to 31/07/1990, while the validation period was defined from 01/08/1990 to 30/09/2004. The initial set of parameters was established based on the basin seasonal water balance (Dias and Kan 1999) and on the physical characteristics of the Fazenda Redenção watershed, whose drainage area is about 5,000 km^2.

As we run the model with this initial set and calibrate it with the hyperbolic smoothing technique, the baseline simulation is established. A set of other computational simulations were then performed with changes made mainly to the initial conditions. Figure 14.2 illustrates one of these simulation runs (simulation 1), in which the initial soil moisture and the base flow were assumed to be zero. Figure 14.3 shows the volumetric soil moisture (%) temporal variability for different sets of simulation.

The chosen sites for soil moisture sampling in SMEX03 field campaign are located between the Branco River watershed and Ondas River watershed, both incoming tributaries to the Grande River, which is a tributary to São Francisco River, as depicted in Fig. 14.4. The division between sets A and B is due to the fact that soil moisture ground-truth information showed different daily pattern behavior during the analysis conducted for the 8-days experiment.

Figures 14.5 and 14.6 show the comparison among simulated soil moisture and soil moisture dataset collected at some fields located in Fazenda Redenção watershed during the SMEX03 experiment, which took place in December 2003.

In general, it is observed that the SMAP simulated volumetric soil moisture follow the behavior observed for the set of sampling points B, that is, the soil moisture tends to rise between December 2nd and December 3rd, 2003, entering

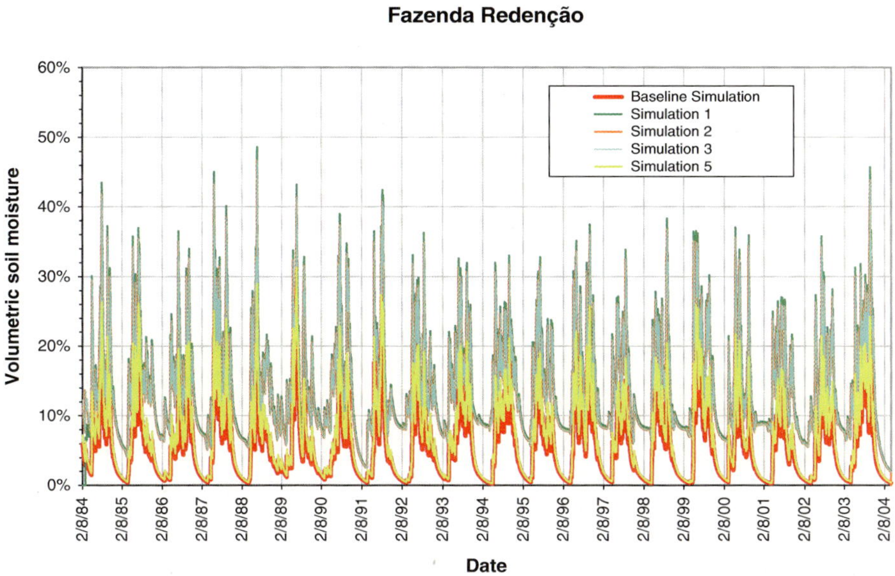

Fig. 14.3 Volumetric soil moisture (%) behavior for different simulations– streamflow gauge Fazenda Redenção – from August 1984 to September 2004

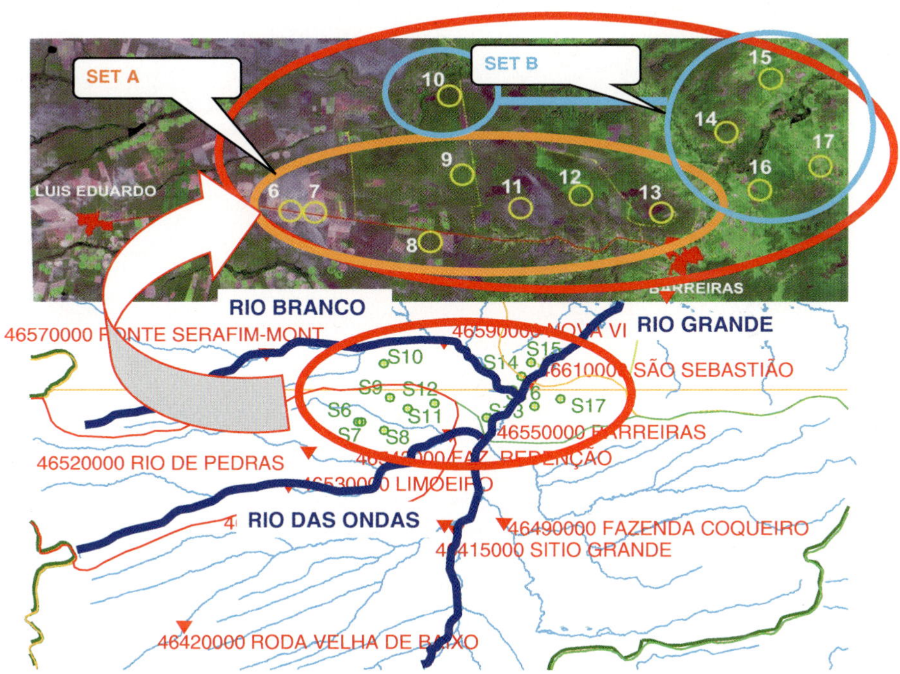

Fig. 14.4 Sites (*highlighted by circles*) sampled in the SMEX03 experiment in a Landsat-TM image (*upper part*) and over the drainage river network mapping (*lower part*) both areas in Fazenda Redenção watershed (*lower part*) – datasets A and B presented different soil moisture behavior in SMEX03 experiment (Source: Di Bello 2005 – Adapted from Sano and Assad 2004)

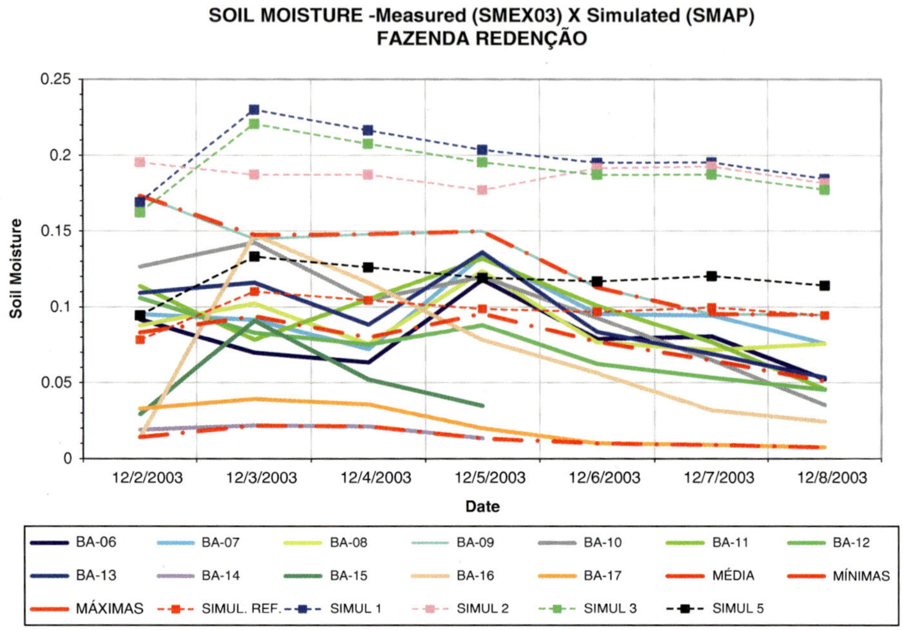

Fig. 14.5 Simulated soil moisture by the SMAP model at Fazenda Redenção (*dotted lines with markers*) and measured soil moisture at different sites (S6 to S17 – see Fig. 14.4) during the SMEX03 (*markers without lines*) – from 02/12/2003 to 08/12/2003

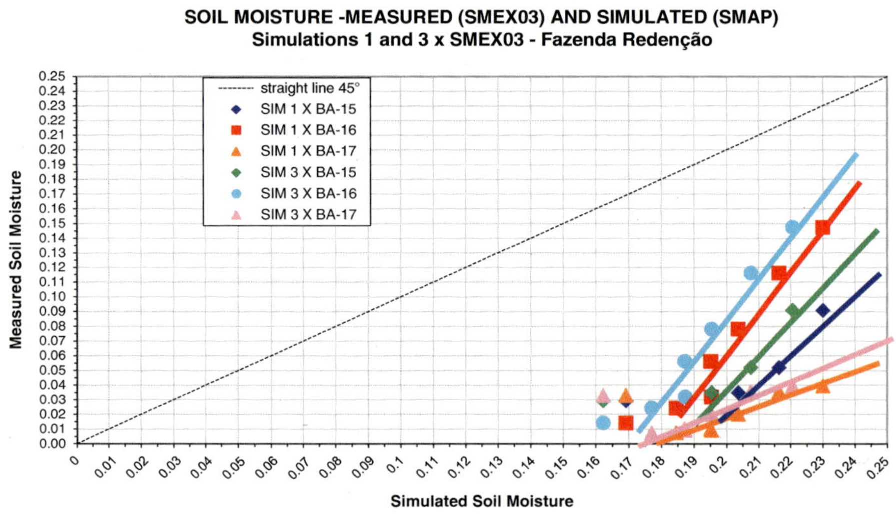

Fig. 14.6 Relationship between simulated and measured soil moisture for a set of fields-SMEX03

into decline in the following period until December 8th, 2003. The only exception was the result obtained on simulation 2, with a decrease until December 5th, 2003 and reversal of this trend on the following days. It stands out among the simulations the one identified as the baseline simulation. The behavior of the volumetric soil moisture generated in this baseline simulation could be considered very good, since it was situated between the actual maximum and average measured values of moisture, standing at around 10 % with respect to the average during the entire period. The results of the other simulations could also be considered good, since they could reproduce behavior obtained from the baseline simulation, though with higher values. Although the simulated values clearly overestimate basin average soil moisture, the lumped conceptual SMAP model could be able to simulate quite well the overall soil moisture behavior during the considered period. It is noteworthy to mention that a scarce rainfall database was available for modeling and that the evapotranspiration rate had to be estimated by the basin seasonal water balance approach (Dias and Kan 1999).

14.5.5 Concluding Remarks

In this application, a new methodology for automatic calibration of a rainfall-runoff model was adopted. By using the hyperbolic smoothing, the problem has been reformulated, in an approximation approach, as a completely differentiable constrained optimization problem. The final solution is obtained by solving a sequence of intermediate problems which gradually approach the original problem, so preserving the conceptual structure of the model.

A set of computational experiments was performed in order to evaluate the proposed approach, focusing on the simulated soil moisture behavior in comparison to an experimental dataset of measured soil moisture collected at some fields in Fazenda Redenção watershed, located in Barreiras, Bahia, Brazil. The results show that there is a reasonable relationship for the temporal behavior between simulated and measured soil moisture dataset, in spite of the use of a lumped model and the fact that there was measured soil moisture data available for only a few days. The authors have wishful expectations that the techniques presented herein may just as successfully be used in other conceptual rainfall-runoff models.

14.6 Monthly Hydrological Modeling and GRACE Data: Amazon Basin

The initial analysis framework adopted herein lies on the consideration of TWS obtained (derived from GRACE solution or integrated from model results) at the large scale. This approach is due to two main reasons:

(i) The coarse GRACE resolution: as stated previously, GRACE TWS data is usually derived at 3–4° resolution. Too coarse a resolution makes its use in hydrological modeling very difficult. The many studies presenting comparisons from GRACE versus model TWS consider model results (obtained at finer scales) averaged onto GRACE coarser spatial (and temporal) resolution. Therefore, the small scale basin TWS variability represented by models are severely smoothed, which in the worst situation leads to the complete loss of TWS spatial correlation patterns. As a practical consequence, GRACE TWS data can be matched with innumerous model TWS results averaged over the basin, increasing the indetermination character of the calibration process.

(ii) There could be a strong correlation between GRACE TWS (at 3–4° resolution) and basin streamflow: this fact is observed by some studies on GRACE data behavior along the Amazon watershed (e.g. Almeida Filho 2009; Xavier et al. 2010), which could represent that, at such a coarse scale, GRACE TWS variability in the Amazon basin would correspond mainly to the huge storage variations taking place in its rivers. The results presented by Han et al. (2009) concerning storage variations on floodplains and river bed at the Amazon main stem reinforced this assumption. As shown therein, the calculated runoff once routed leads to great discharge values on the Amazon River, thus representing the natural integration of all runoff generated over the drainage area. Being true this assumption, GRACE TWS series would represent in the Amazon basin a redundant source of information for a hydrological model calibration.

Following what has been exposed above, a true gain of hydrological information from current GRACE TWS series could only be fully experienced if these data presented a finer resolution (at least, at model resolution). Although a fine GRACE TWS resolution cannot be currently achieved, the modeling framework presented hereinafter aims at giving new insights on TWS use on hydrological modeling. Instead of focusing on the basin as a whole attention will be paid hereinafter on TWS variability on sub-basin scale. In order to achieve this goal, the following steps are envisaged:

(i) Use SMAP conceptual lumped hydrological model (Lopes et al. 1981) at the basin scale, capable of representing average basin TWS variability and streamflow behavior at basin mouth at monthly time step. Obviously, this is an extreme modeling simplification, especially when one realizes that the dimensions of the areas considered herein (~200,000 km^2) cannot be reasonably represented as homogeneous regions. However, in view of the exposed above and as the main modeling purpose herein is to estimate averages states at basin scale, this simplification is considered as a reasonable one.

(ii) Evaluate model calculated discharge and TWS series throughout a sensitivity analysis with the GLUE (*Generalized Likelihood Uncertainty Estimation*) methodology (Beven and Binley 1992). Thus, it is proposed a calibration procedure permitting the minimum error of modeled (i) monthly streamflow, and (ii) average TWS series with respect to the correspondent observed series. Ideally, a multiobjective calibration method such as the MOCOM-UA (Yapo et al. 1998) should be utilized to achieve this goal. Herein, it is used a simplified approach taking into account only streamflow data.

(iii) Case study: two catchments of Negro River basin defined by the gauge stations of Serrinha and Caracaraí, both encompassing areas about the dimension of GRACE TWS resolution.

14.6.1 Case Study: Negro River Basin

The Negro River basin was adopted for the application of the proposed framework, because of the following reasons:

(i) It has a reasonable gauged datasets from the 1980s on, especially for the catchments areas defined by the Serrinha (at Negro River) and Caracaraí (at Branco River, main Negro tributary) gauge stations;
(ii) Its has been considered as case study for many studies published last years (e.g. Frappart et al. 2006; Silva et al. 2010), thus a comprehensive knowledge on its hydrological main features has been accumulated;
(iii) It has been the object of an in depth hydrological modeling effort in the work of Getirana et al. (2009), which had been taken as a valuable initial step for the present work.

For Negro River basin, the modeling efforts have been concentrated on its two main sub-basins as defined by the gauge stations of Serrinha and Caracaraí. This choice has been made based on the following arguments:

(i) Most of gauge stations with available data on Negro River basin are located in these two basins. There are reasonable precipitation and streamflow data for both of them.
(ii) Their drainage areas are about GRACE resolution: Serrinha (~290,000 km^2) and Caracaraí (~120,000 km^2).
(iii) They are the basin utmost downstream gauge stations on Negro and Branco Rivers having discharge series. The utmost downstream gauge station, located in Manaus, has a long stage record, but not discharge series. This is due to the immense difficulty in establishing there a rating curve owing to the very strong backwaters effects from Amazon River (Meade et al. 1991) it is subjected to, as Manaus is located only a few kilometers upstream of the confluence Negro-Solimões.

The existing gauge stations in Colombian and Venezuelan parts of Serrinha sub-basin would be essential in estimating the runoff generation from important tributaries on Negro right bank upstream Serrinha station, as stated previously by Getirana et al. (2009). An alternative to overcome this limitation should be the combined use of gauge data with remotely sensed estimated precipitation data like the ones made available by GPCP and TRMM missions.

14.6.2 *Results*

A sensitivity analysis was performed applying the GLUE methodology, whereafter the parameter sets leading to the best results (evaluated by the Nash-Sutcliffe index) have been taken as the optimum ones. From these values, model TWS has been calculated, and then compared to GRACE TWS average series. In what follows these results will be presented and discussed.

Simulation with the Best Parameters Set

Even though an overall good agreement could be achieved between calculated and observed discharge series on 1992–2010 period for Serrinha, especially over the 2002–2010 period wherein a correlation of 0.90 ($R^2 = 0.8214$) has been found, some differences must be noticed (Fig. 14.7). First of all, model showed difficulties in matching the extreme values, especially the peaks. In fact, some large discrepancies were noticed, especially when two significant peaks occur in an interval of few months. In this situation, model usually matched the greatest one (see, e.g., 1993–1994 period). In average, the model tends to underestimate the discharge series.

On the other hand, much better results could be achieved for Caracaraí (Fig. 14.8); over the 2002–2010 period, a correlation of 0.96 ($R^2 = 0.9208$) was found between simulated and observed discharges. Although some significant errors had been observed in peaks, the recession limbs from the observed series could be very well matched. As it had been previously noted for Serrinha, the calculated series for Caracaraí are also underestimated.

Comparison with GRACE TWS Data

Although modeled discharge series could match reasonably well the observed series at Serrinha, the model TWS series on this basin shows a great discrepancy when compared to the GRACE one (Fig. 14.9).

As it can be expected, model TWS series shows a lower variability. Moreover, the well defined seasonality present in model TWS series cannot be observed in GRACE TWS series (e.g., see the 2004–2006 period) and model TWS series leads GRACE TWS for about 1 month. This fact can be due to the following reasons:

(i) Model lacks a reservoir representing water retention on floodplains (or on secondary channels), which would have contributed to water retention (and thus reducing the found phase-difference);
(ii) The anomalous GRACE TWS behavior observed on the basin utmost upstream region, which could distort the basin TWS as represented in GRACE data.

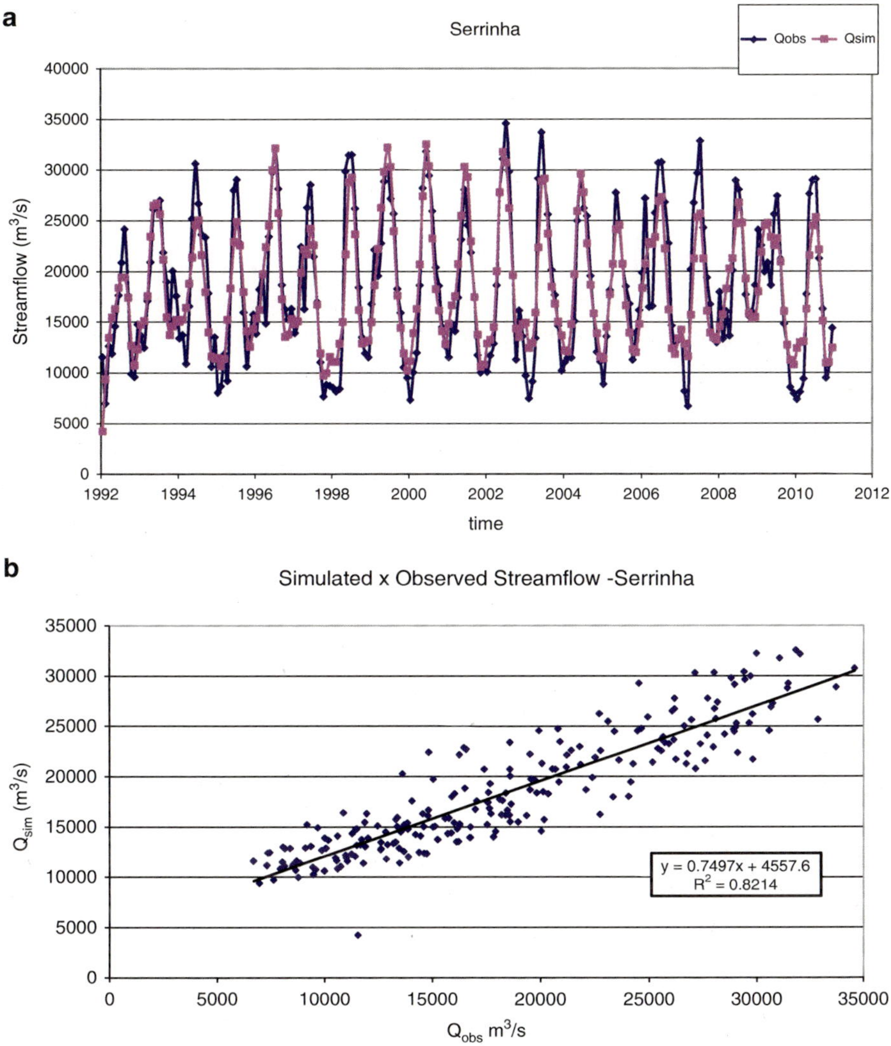

Fig. 14.7 Comparison between simulated (Q_{sim}) and observed (Q_{obs}) monthly streamflow: time series (*left panel*) and Q_{sim} x Q_{obs} dispersion (Serrinha)

Some discrepancies have been found in the relationship between precipitation and GRACE TWS series in Serrinha, what could be attributed to:

(i) An anomalous (when compared to the basin as a whole) behavior of the utmost upstream region of Serrinha basin, whose TWS variability is similar to the one verified on the Orinoco basin;
(ii) The basin's drainage area definition from coarse 1° TRIP flow direction data;

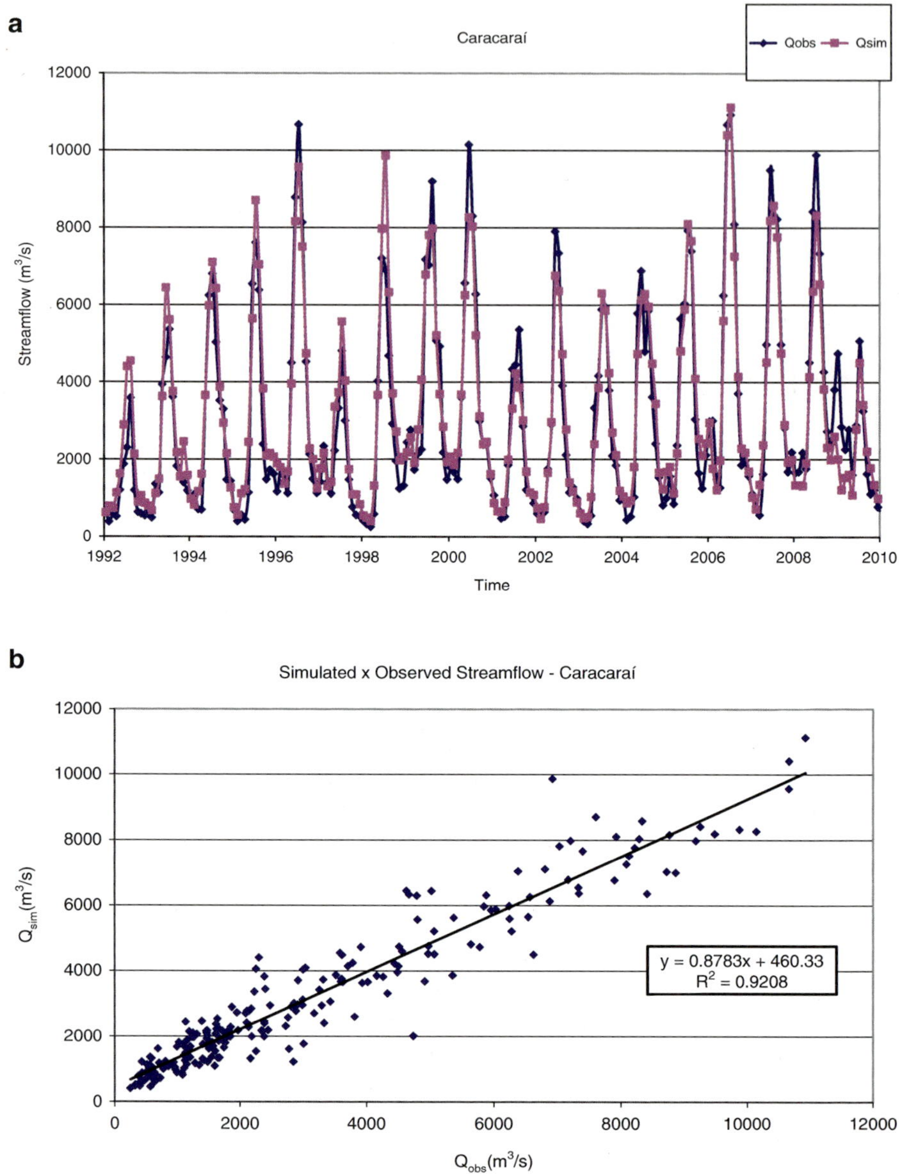

Fig. 14.8 Comparison between simulated (Q_{sim}) and observed (Q_{obs}) monthly streamflow: time series (*left panel*) and Q_{sim} x Q_{obs} dispersion (Caracaraí)

(iii) A possible influence of Casiquiare River catchment area contribution to the total runoff at Negro basin at Serrinha. The Casiquiare River consists of a permanent link (~300 km long) between Negro and Orinoco basins, whereby a small portion of the runoff generated on the utmost upstream part of Orinoco

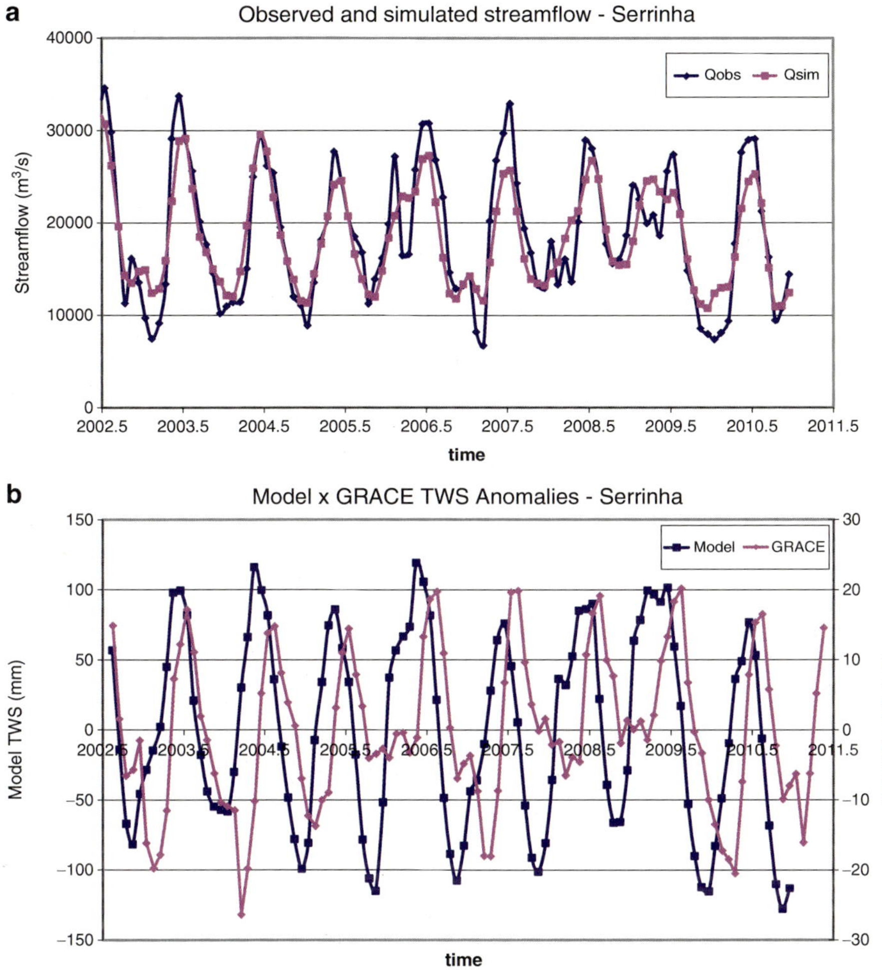

Fig. 14.9 Observed (Q_{obs}) and simulated (Q_{sim}) streamflow (*left panel*) and modeled basin TWS and GRACE TWS anomalies (*right panel*) for the 2002–2010 period – Serrinha

basin contributes to Negro River basin. Moreover, along its course the Casiquiare River receives the contribution of some tributaries, over a total drainage area of about 40,000 km^2. Following the data presented at GRDC database, its mean annual discharge (peaking at July as the Branco River one) is of about 2,200 m^3/s. Were its discharge series available, its contribution could be accounted for in the Serrinha sub-basin water balance.

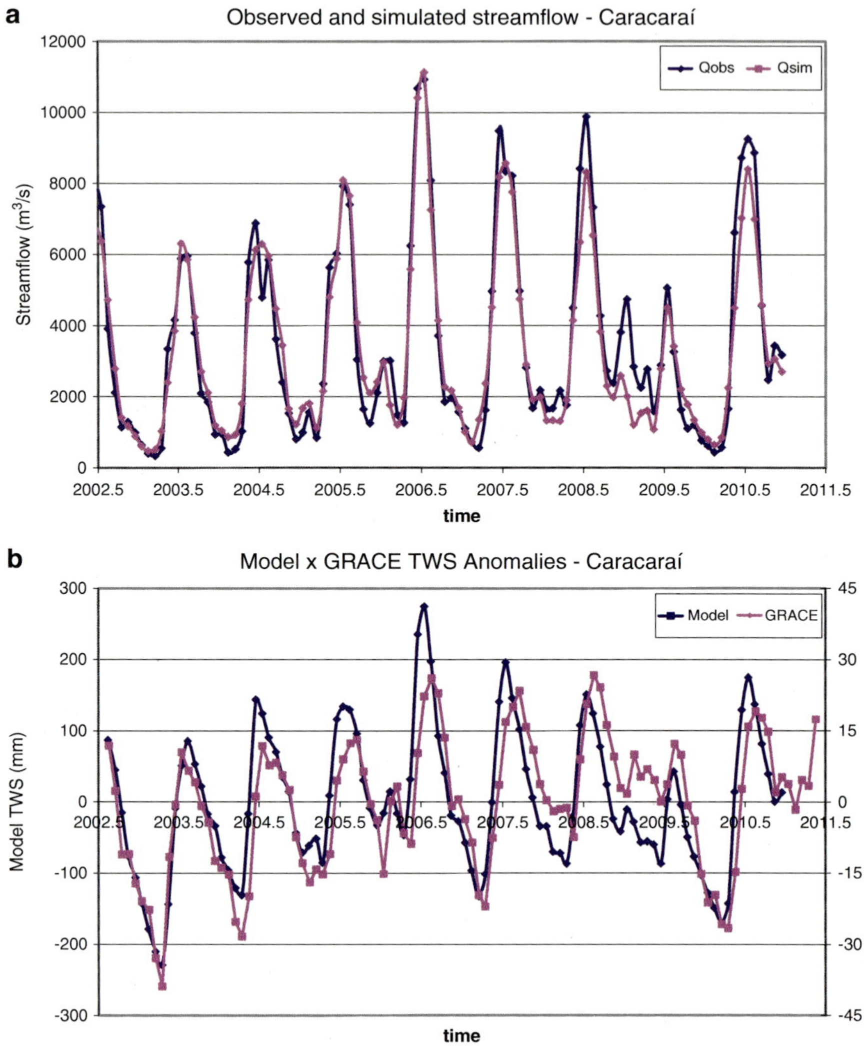

Fig. 14.10 Observed (Q_{obs}) and simulated (Q_{sim}) streamflow (*left panel*) and modeled basin TWS and GRACE TWS anomalies (*right panel*) for the 2002–2010 period – Caracaraí

On the other hand, a very good agreement has been found when comparing model and GRACE TWS at Caracaraí, with correlations values of 0.82, 0.86 and 0.65 at lags 0–2, respectively. Model TWS lags GRACE TWS for about 1 month (Fig. 14.10).

Thus, the simple conceptual lumped model adopted herein has been able to reproducing very well the monthly discharge and the Total Water Storage (even if lagged by 1 month) for Caracaraí. It can be argued that once constrained to being in phase with GRACE TWS series, model would not be able to reproduce so well the discharge at Caracaraí.

Discussion on the Proposed Framework and the Limitations Suggested by the Presented Results

The results presented herein showed that, depending on basin characteristics, even a simple model could be used in order to estimate monthly basin average storage. However, some important discrepancies were found, what should lead to a further discussion. In what follows, the main issues arisen from the analysis of the presented results will be summarized.

As it should be expected, the greater the basin, the more likely a simple lumped model will not be able to simulate well its hydrological behavior. In other terms, it is more likely that their natural heterogeneity cannot be represented by average states. Even though a good agreement can be found between simulated and observed discharges, this not necessarily means that model states (like the storage in each model reservoir) will be well represented. On the other hand, large basin hydrological modeling should be preferentially done by a distributed model. However, data scarcity and parameter identifiability issues must be posed against this rather obvious statement. For large South American basins, the first difficulty could be rarely overcome. Processes representation in hydrological models constitutes itself an intrinsic difficulty as many parameters have to be estimated by calibration. This statement would remain true even if a hypothetical large database would be available.

Even though the limitations of the adopted model become evident when addressing hydrological modeling of Negro basin at Serrinha, its application could be viewed as a first step in modeling its hydrological cycle. Following the “downward modeling” (or “top-down”) approach as defined by Klemes (1983), once reasonably well represented the main hydrological processes at basin scale, one should try to enhance modeling by adding more processes in model structure. Naturally, this leads to an increase in model complexity that does only make sense when enough data are available for calibration and verification, thus validation of the improved modeling structure.

In the case of the Amazon basin, modeling is a much more complex task due to specific characteristics like the presence of large floodplain areas and the occurrence of backwaters effects (not the case for the basins considered herein, but of great importance when taking the whole Negro basin into account). So, how could floodplains dynamics and backwaters effects be properly represented avoiding model overparameterization? Balancing model complexity and data availability is a fundamental issue to be paid attention when modeling basins such as the Amazon one.

An alternative should be the elaboration of a simple distributed conceptual model aiming at representing the main processes at basin scale. For instance, the processes of interception and the runoff generation could be represented in a distributed way, and the other ones, in a lumped average way. Moreover, each sub-basin could be divided into a small number of homogeneous hydrologic response units (HRU) based on vegetation cover (assuming implicitly that the vegetation cover could be directly related to the subjacent soil characteristics). In doing so, a simple conceptual framework representing the main hydrological processes could be adopted to each sub-basin (defined by the existence of a streamflow series permitting model calibration), each one subdivided into two or three HRU's.

The correspondence between model and data resolution is another fundamental issue to be addressed for avoiding information loss and masking model non-identifiability. Averaging model results obtained in a fine scale onto GRACE TWS much coarser resolution do not improve model identifiability nor fully take advantage of data informativeness (at least available on their intrinsic resolution).

Concluding Remarks

For hydrological applications, a first important step is to verify GRACE ability in representing main hydrological characteristics. This had been done in several papers from 2006 on. In this chapter, an analysis of GRACE solutions on the Amazon basin has been performed, which showed that seasonal and interannual features were well captured by the data.

In order to address data scarcity over large regions, remotely sensed data is of primordial importance, even though these data must always be validated against observed data or data estimated from observations. Especially when combined with gauged data, they can much enhance the current hydrological modeling praxis. For doing so, besides new satellite missions planning and launching, investments on network densification and operation are of the utmost importance.

Concerning GRACE TWS data, their use on large basin modeling remains to be further evaluated, especially in addressing the fundamental question: when (if so) they represent an independent (or "orthogonal", following Winsemius 2009) new information source suitable for enhancing model performance? Beyond data many sources of error (like the leakage, which might be very important in rivers subjected to such huge volume variations as the ones experienced by Amazon River and its main tributaries), its fundamental coarse spatial resolution makes very difficult answering this question. However, some regional approximated ("MASCONS") GRACE TWS data could furnish series on finer resolutions, which would permit this question to be addressed.

Once GRACE data (or the ones from the next spatial gravity mission) become available at a finer resolution, how should these data be integrated in hydrological models? This is authors' opinion that the most important information that could be extracted from GRACE TWS series concerns the spatiotemporal patterns of TWS variability revealed from these data. This kind of information could help inferring the spatiotemporal patterns of subsurface (at monthly or finer time resolution) and/or groundwater (at seasonal/annual or coarser time resolution) flow variability.

Therefore, understanding these patterns should much increase GRACE series value as an additional data source for calibration. Along with the traditional streamflow series, the use of this new information in a calibration or data assimilation framework could allow large watershed models to better represent TWS dynamics at the model resolution. This remains as a big challenge for future researchers.

14.7 Agenda for Research on Hydrological Modeling and Soil Physics

14.7.1 State-of-the-Art Calibration-Validation Aspects Applied to Distributed Physically-Based Hydrological Models

As Jensen and Mantoglou (1992) state, distributed physically-based models are still in the development phase where the theoretical bases of the individual process description consistent with the selected scales are not fully resolved. There is little rigorous testing and validation of full catchment models published in the literature. However studies are being conducted to address these problems.

Bathurst and O'Connell (1992) point out that efficient calibration of distributed physically-based models will require field studies to form an integrated part of the calibration procedure, rather than appearing as an isolated activity. Lack of enough data is always a problem or reason for complaining in hydrological studies, since the time and costs involved are usually high. However, these barriers should be overcome to open the possibility to new progress in this area.

First, the incorporation of a field program means that a full range of calibration and validation data can be made available, based on the requirements of the model. Such data can include internal variables, such as soil moisture, water table levels, snowpack depths, etc., essential to providing a convincing calibration and validation exercise. Dependency on past data records, or the implementation of a traditional water balance study as an isolated activity, is likely to allow calibration against catchment outlet discharge only. As the outlet discharge represents an integration of effects within the catchment, it might be possible for an apparently satisfactory calibration to be produced on the basis of erroneous but compensating representations of internal processes. Such a calibration would not form a sound basis for subsequent predictions. With data on internal model variables, however, it would be possible to achieve a much more satisfactory and internally consistent calibration. Second, as initial data begin to become available, or if there are existing past data record, the model can be used to make a preliminary investigation of different hypotheses of basin behavior, providing a guide as to what further measurements should be made, and at what locations. Third, sensitivity testing can indicate to which parameters the simulation results are most sensitive, that is, the ones which cause the greatest uncertainty in output, and thus which should be measured in greatest detail and which ones can be examined more sparingly.

As Bathurst and O'Connell (1992) mention, new thinking will also be required over what constitutes a successful model validation. Traditionally, split-sample testing is applied in which an existing time series of measured output data is divided into two parts, one used for calibration and one used for validation of the calibrated model. This provides a guide to the ability of the model to represent the observed behavior, but it is incomplete in terms of testing the predictive capability of the model. In particular, through its dependency on existing output data, it is open to charges of curve-fitting. In addition, as the user has sight of the data, it may be questioned whether the calibration can be carried out without any bias towards the conditions represented by the validation data. To avoid that and to examine uncertainty in predicted responses as a function of the amount and quality of data, the following procedure is presented by Bathurst and O'Connell (1992):

(a) choose a set of test statistics, such as peak outflow discharge or percentage of time the ground adjacent to the channel is saturated; decide on pass/fail criteria, e.g. for success the observed hydrograph should fall within the bounds determined for the simulated hydrograph for 95 % of the time;
(b) using hydrological judgement, define bounds on the values of the model parameters, e.g. maximum and minimum values for soil hydraulic conductivity;
(c) run the model for the various parameter bound values (including desired combinations of values for different parameters) to produce a range of possible output responses; extract the bounds for the various test statistics (up to this point the modeler is not allowed to have sight of the observed output);
(d) with all runs completed, compare the predicted bounds with the measured output and assign pass/fail to each statistic, according to the relevant criteria;
(e) assess the overall pass/fail outcome for the test;
(f) assess the performance of the model and the modeler, i.e., whether the prediction bounds have been determined with a range narrow enough for meaningful conclusions to be derived from the test; a wide range, although increasing the chance of a successful pass outcome, means that the output is being presented with a high degree of uncertainty; the acceptability or otherwise of the results will depend on the impact of the uncertainty on the decision to be taken: if the impact is unacceptable, more resources will need to be devoted to collecting further data for parameter evaluation, improving model representation and, indeed, improving the a priori knowledge and experience brought to the problem by the modeler.

A new approach for calibration of hydrological models has also been presented by Beven and Binley (1992), which they call Generalized Likelihood Uncertainty Estimation (GLUE). Such an approach is different from most calibration procedures used in hydrology, in which some global optimum parameter set is sought, and any assessment of parameter and predictive uncertainty is made with respect to that global optimum. The GLUE procedure incorporates the following features:

(a) the possibility that different sets of parameter values may be equally likely as simulators of a catchment system within the limitations of a given specific model structure and errors in the definition of boundary conditions and observed variables;

(b) the expectation that the errors between observed and predicted responses may be biased, non-Gaussian, autocorrelated, and heterokedastic;
(c) the possibility of updating the likelihood weight associated with a particular set of parameter values as more observations become available by a simple application of Bayes equation;
(d) the possibility of evaluating the worth of different types of data by examination of the effect on the likelihood distribution;
(e) procedures for evaluating the sensitivity of the model to individual parameter values.

14.7.2 Criticism and Directions for Distributed Physically-Based Hydrological Models

As Grayson et al. (1992a, b) comment, there is certain arrogance associated with physically-based models regarding their superiority over lumped-parameter or empirical models, and this needs to be overcome if expectations for modeling are to be changed. The assumption is that the use of equations that are theoretically correct at a certain scale should enable a universal application, so that the resulting model is of far greater use than the model requiring field calibration of empirical parameters. The real issue is whether, in practice, there is any difference between such models except for the vastly increased time required to calibrate the numerous parameters associated with physically-based models. Having more parameters allows the user more flexibility to obtain a good fit to measure data, but whether the fitted parameters have a unique physical interpretation on a specific catchment is debatable, as is their advantage in terms of predictive capability.

The use of these models presupposes that the physical processes can be represented in a deterministic way, leading Beven (1989) to refer to distributed physically-based models as lumped, conceptual models. There are certain assumptions underlying the notion of physically-based models like: the processes being modeled can be lumped into homogeneous units at some scale; the algorithms and parameters used to define the subprocesses are appropriate to the scale of the homogeneous units, and the input data are available at a resolution appropriate for the model. Even the most rigorous of mathematical models describing catchment response is a crude representation of reality, and while the theoretical rigor of some models is impressive, it is also seductive and implies a degree of accuracy that may not exist. Also model applications in which results were poor are rarely reported, perpetuating the notion the models hold all the answers (Klemes 1983).

Despite all these valid criticisms raised in the literature, there are hopes for the development of a physically-based hydrological model. Recognizing its limitation is the first step to make progress in this area.

What, then, are the principles that should be applied, and what are the future directions of physically-based models in catchment hydrology? Hillel (1986) lists four guiding principles: (a) parsimony, i.e. the number of parameters must be minimized and the values must be obtained from data; (b) modesty, i.e., the scope and applicability of a model should not be overstated; (c) accuracy, i.e., the accuracy of prediction need not be better than the accuracy of measurement; and (d) testability, i.e., a model must be testable so the limits of its validity can be defined. Beven (1989) is more specific about what should be taken into account: theory of lumping of subgrid scale processes; closer correspondence in scale between model predictions and real world processes; and rigorous assessment of uncertainty in model predictions.

Through a better understanding of natural systems, models will improve, and their applications will become more general. However, this requires active efforts to examine and represent the basic processes in a more realistic framework and in a way more closely related to the scale at which the heterogeneity of the phenomena can be represented.

In summary, increasing spatial scale tends to introduce greater variability, or disorder, into the catchment system and its rainfall-runoff response. The disorder arises from the spatial variability of soil properties, vegetation, topography, and the spatial and temporal variability of climatic variables. These variabilities give rise to a pattern of complexity of runoff production and routing. For example, evidence suggests that many different mechanisms of runoff production may occur in the same catchment at the same or different times.

The tremendous increase in disorder or complexity at large spatial scales makes an analysis of the catchment response, on the face of it, an almost impossible task. However, there is evidence that indicates that the increasing disorder is superimposed on, or may even have given rise to an underlying order of regularity in macroscale responses.

Some of this order in macroscale responses is yet to be identified or exploited and constitutes the challenges that rainfall-runoff models have to overcome. Merging deterministic concepts and stochastic modeling to handle spatial variability and upscaling is the recommended way to be pursued. Parsimony in the number of parameters used with appropriate calibration, sensitivity and validation tools are required. Finally the various sources of uncertainties should be studied in an integrated approach in order that reliability on the results produced by a rainfall-runoff model as an accurate representation of the real system can be achieved.

14.7.3 Trends on Hydrological Modeling and Soil Physics with the Support of Remotely Sensed Technology

In this general background overview on hydrological modeling and on a proposed research agenda associated to soil physics, rainfall-runoff models have been analyzed with respect to their parameters and, mainly, with regard to their improvement through the use of soil moisture information and soil water storage. Further, ground-truth surface soil moisture was extensively collected and investigated for different field experiments and its relationship with remotely sensed images was studied under different situations such as for bare and pasture fields. Two different applications were presented to give a taste of the enormous task still to be done in the next years. New concepts and ideas for estimating soil moisture and soil water storage were presented within the field of remotely sensing.

Different approaches of sensitivity analysis for the Brazilian lumped model named SMAP, which was developed for two different time steps, daily and monthly, applied to sub-basins of the São Francisco basin and the Amazon basin, respectively, allowed us to better understand the behavior and role of their respective parameters. Some problems were detected in the way lumped rainfall-runoff models are constructed. Identifiability of parameter values is also a concern in these lumped models. Therefore, problems are expected to occur during the optimization of parameters in distributed models, due to the increase in complexity. Results obtained showed that the essential nature of some parameters in the lumped model can be detected and that other parameters may be left out from the optimization procedure. Expert system technology may contribute to improve this type of results as well. It was also shown through the lumped model that the discontinuities arising from threshold structures in the model may be overcome using smoothing techniques. In addition, the percolation equation in the lumped model presented parameters that have some interaction. This may lead to problems in the calibration process and suggests that other data sets such as soil moisture information should be provided to the model.

In addition, initial soil moisture was shown to be an important information. Further, parameters associated with different land covers will present different effects in the model. At the present stage of knowledge, soil permeability, which may vary over up to 5 or 6 orders of magnitude, is still a challenge to be pursued in the construction of physically based distributed models. Soil permeability, though an important soil parameter in the infiltration process, should still be calibrated at this moment for obtaining an effective parameter. On the other hand, in terms of remotely sensed data, we can take into account different land covers, which are related, to a certain extent, to the underlying soil. This is a fortuitous result, since visible, infrared, radar and gravimetric images and their corresponding datasets are able to presently produce better useful information about soil moisture and soil water storage than soil permeability, due to the physical relationship between the microwave response and soil moisture and also due to the modeling of the land surface temperature jointly with the energy balance at the continental surface.

There is still room to lots of improvements for better modeling of the soil moisture and soil water storage variation within scenes. Nevertheless, it was evident that a relationship does exist between brightness values of radar images and soil moisture for pasture fields for example. The results across scenes are much better than the ones obtained for within scenes. It seems that the recognition of different wetness scenarios can be detected using active and passive radar images, visible and infrared images and gravimetric images, which is an important result for soil scientists and hydrologists and their rainfall-runoff models. Also, the effect of soil texture on the soil moisture-backscatter relationship needs to be better discriminated for example in the microwave range of the electromagnetic spectrum. Much lower frequencies with respect to the X and C bands should be used for this purpose.

Spatial analysis such as geostatistical analysis deserves to be pursued for estimating the spatial distribution of soil moisture, since some field experiments and image analysis strongly support this approach. Within-field soil moisture variability is as high as between-fields soil moisture variability. Pitfalls of using ANOVA (*Analysis of Variance*) and F distributions to test the significance of trends are hidden in the dazzle of attractive trend surfaces and hypersurface maps having impressive correlation coefficients assigned to them. If the original data are not representative or if they have a large, local, natural variation, the results of any trend-surface analysis using this data are pure speculation. These factors corroborate and strength the suggested approach in the agenda to explore, for instance, a geostatistical approach based on soft data (remotely sensed) coupled with hard data (collected in the field) to derive a soil moisture map. The fact that these methodologies make use of the spatial structure information, extractable from a remotely sensed image, for estimation purposes, seems to guide us to a better approach than just using average values as required by regression equations. Therefore suggested improvements built upon these models and comparisons to geostatistical techniques are important and desirable. Remotely sensed active radar images at adequate frequencies seems to be an appropriate choice to fulfill the requirements of becoming the soft data set related to surface soil moisture, due to the extensive coverage at the watershed scale and high resolution. Regression equations, though valuable tools for exploratory analysis and for uncovering relationships among variables in a data set, seem to the authors not suitable to derive a soil moisture map at the watershed scale. Of course, a regression equation might be the best choice when one has few data or insufficient information. Other than that, the results of a regression equation in the specific case of deriving a soil moisture map are quite limited and might be misleading. However, due to the enormous complexity of such a task of deriving soil moisture maps, regression models are presently part of the state of the art in this field.

Field level and watershed level soil moisture and soil water storage mapping therefore still needs to be addressed. Groundwater studies and irrigation studies, which currently operate at the field level, might benefit directly from results of such work. From the perspective of rainfall-runoff models, which operate at the watershed scale, the complexities involved in providing accurate and reliable soil moisture and soil water storage maps or upscaling the input information to the grid level used by the hydrological model will require considerable effort in future research. It is suggested to explore soil moisture and soil water storage spatial variability using visible, infrared and radar imagery using land cover as a subdomain in the watershed.

In pursuing the objective of deriving a reliable soil moisture and soil water storage map with information that operationally is viable to collect, a new approach is being introduced in the area of remote sensing applied to hydrology which includes:

- sensitivity analysis conducted for the parameters of typical lumped and distributed rainfall-runoff models;
- sensitivity analysis of lumped and distributed hydrological models for different initial soil moisture and soil water storage maps focusing on the spatial variability that is recognized to be present at the watershed scale;
- the investigation of vegetation indices from optical images and radar imagery for evaluating biomass, roughness characteristics, vegetation cover and land use and land cover for soil moisture and soil water storage within a year and along different years;
- a focus not only on soil moisture and soil water storage distribution based on field averages, but also focus on the within-field soil moisture distribution;
- the application of different spatiotemporal models to obtain soil moisture and soil water storage maps at the field level and at the watershed scale jointly with electromagnetic in situ measurements using TDR (time domain reflectometry) and GPR (ground penetrating radar) measurements.

It is hoped that this review might bring new insights to the development of physically-based rainfall-runoff models and soil physics jointly with current advances in remotely sensed techniques.

Acknowledgments The authors would like to thank the Civil Engineering Program of Instituto Alberto Luiz Coimbra de Pos-Graduação e Pesquisa de Engenharia (COPPE) – Universidade Federal do Rio de Janeiro (UFRJ) through the support of the Laboratory of Water Resources and Environmental Issues (LABH2O) with respect to data and infrastructure provided by these institutions for this research. The authors would like to express their sincere appreciation to the financial support for the work, which came through CAPES (Fundação Coordenação de Aperfeiçoamento de Nível Superior) – CAPES/COFECUB No. 516/05 (2005–2012) and Project IME-PEC/COPPE – CAPES – Aux-PE-PRO-Defense 1783/2008 (2008–2012), CNPq (Conselho Nacional de Ciência e Tecnologia) – Project MCT/FINEP/CT-HIDRO/EIBEX-1 (2005–2011), which addressed representative watersheds, Project PROSUL (Programa Sul-Americano de Apoio às Atividades de Cooperação em Ciência e Tecnologia) – Process 490684/2007-6 (2007–2012), which deals with remote sensing techniques applied to hydrological monitoring and climate change, and Project CNPq Edital Universal No. 14/2013 – Process 485136/2013-9, which is focused on rainfall-runoff modeling and on the corresponding issue of water balance and soil moisture with respect to extreme events, and Secretaria de Educação Superior (SESu) – Ministério da Educação (MEC) – CAPES – AUX-PE-PET-1228/2009 (PET CIVIL UFRJ), educational agencies of the Brazilian government. The authors would like to acknowledge the support provided by Fundação de Amparo à Pesquisa do Estado do Rio de Janeiro (FAPERJ) – Project PEC/COPPE – FAPERJ 014/2010 (2010–2012), Project FAPERJ – Process E-26/103.116/2011 (2012–2014) and Project FAPERJ – Pensa Rio – Edital 19/2011 (2012–2014) – E26/110.753/ 2012. The authors would like also to recognize the support of CPRM, INPE, ANA, EMBRAPA, CEPEL, ONS and INMET, for the continuous support of the research in hydrology in Brazil. In addition, the authours would like to recognize the support of *Laboratoire d'Études en Gé ophysique et Océanographie Spatiales* (LEGOS) (Université Paul Sabatier – UPS-Toulouse III, *Centre Nationale d'Études Spatiales* (CNES), ESA (*European Space Agency*), *National Aeronautics and Space Administration* (NASA), e ORE-HYBAM (*Observatoire de Recherche en Environment – Contrôles géodynamique, hidrologique et biogéochimique de l'érosion/alteration et dês* *transferts de matière dans le bassin de l'Amazone*). A special thank you goes to Dr. Edson Eijy Sano (EMBRAPA), to Dr. Eduardo Assad (EMBRAPA) and to Dr. Thomas J. Jackson (USDA), who stimulated the use of the soil moisture dataset collected during SMEX03 experiment.

References

Ahmad MD, Bastiaanssen WGM (2003) Retrieving soil moisture storage in the unsaturated zone using satellite imagery and bi-annual phreatic surface fluctuations. J Irrig Drain Syst 17:141–161

Almeida Filho FGV (2009) Variação temporal do campo gravitacional detectada pelo satélite GRACE: aplicação na bacia Amazônica, São Paulo. Ph.D. thesis, EP/USP, São Paulo, SP, Brasil

Allen RG, Tasumi M, Morse A, Trezza R (2005) A Landsat-based energy balance and evapotranspiration model in western US water rights regulation and planning. J Irrig Drain Syst 19:251–268

Allen RG, Tasumi M, Trezza R (2007) Satellite-based energy balance for mapping evapotranspiration with internalized calibration (METRIC) model. J Irrig Drain Eng 133(4):380–394

Araujo AAM (2006) Um novo esquema de parametrização hidrológica da superfície terrestre com redistribuição lateral da água no solo (A new hydrologic land surface parametrization scheme with lateral soil water redistribution). Ph.D. thesis, Civil Engineering Department, COPPE/UFRJ, Rio de Janeiro, RJ, Brasil, 236 p

Bastiaanssen WGM (1995) Regionalization of surface flux densities and moisture indicators in composite terrain: a remote sensing approach under clear skies in Mediterranean climates. Ph.D. thesis, CIP Data Koninklijke Bibliotheek, Den Haag, The Netherlands, 273 p

Bastiaanssen WGM (2000) Sebal-based sensible and latent heat flux in the irrigated Gediz basin, Turkey. J Hydrol 229:87–100

Bastiaanssen WGM, Cheema MJM, Immerzeel WW, Miltenburg IJ, Pelgrum H (2012) Surface energy balance and actual evapotranspiration of the transboundary Indus basin estimated from satellite measurements and the ETLook model. Water Resour Res 48(11):W11512

Bathurst JC, O'Connell PE (1992) Future of distributed modelling: the Système Hydrologique Europeen. Hydrol Process 6:265–277

Beven KJ (1989) Changing ideas in hydrology: the case of physically-based models. J Hydrol 105:157–172

Beven KJ, Binley A (1992) The future of distributed models: model calibration and uncertainty prediction. Hydol Process 6:279–298

Canedo PM (1979) The reliability of conceptual catchment model calibration. Ph.D. dissertation, University of Lancaster, U.K.

Cazenave A, Chen J (2010) Time-variable gravity from space and present-day mass redistribution in the Earth system. Earth Planet Sci Lett 298(3–4):263–274

Di Bello RC (2005) Análise do comportamento da umidade do solo no modelo chuva-vazão SMAP II – versão com suavização hiperbólica – Estudo de caso: região de Barreiras na bacia do rio Grande-BA (Analysis of the soil moisture behavior in the rainfall-runoff SMAPII model – hyperbolic smoothing version – Case study: Barreiras region in Rio Grande watershed – Bahia, Brazil). M.Sc. thesis, Civil Engineering Department, COPPE/UFRJ, Rio de Janeiro, RJ, Brasil, 225 p

Dias NL, Kan A (1999) A hydrometeorological model for basin-wide seasonal evapotranspiration. Water Resour Res 35(11):3409–3418

Diskin MH, Simon E (1977) A procedure for the selection of objective functions for hydrologic simulation models. J Hydrol 20:129–149

Dooge JCI (1986) Looking for hydrologic laws. Water Resour Res 22(9):46S–58S

Duan Q, Sorooshian S, Gupta V (1992) Effective and efficient global optimization for conceptual rainfall-runoff models. Water Resour Res 28:1015–1031

Dubois PC, Van Zyl J, Engman T (1995) Measuring soil moisture with imaging radars. IEEE Trans Geosci Remote Sens 33(4):915–926

Fortin JP, Villeneuve JP, Guilbot A, Seguin B (1986) Development of a modular hydrological forecasting model based on remotely sensed data for interactive utilization on a microcomputer. In: Johnson AI (ed) Proceedings cocoa beach workshop 1985: hydrologic applications of space technology, IAHS publication no. 160, Int. Assoc. of Hydrologic Sci., Wallingford, England, pp 307–319

Frappart F, Calmant S, Cauhopé M, Seyler F, Cazenave A (2006) Preliminary results of ENVISAT RA-2 derived water levels validation over the Amazon basin. Remote Sens Environ 114:252–264

Getirana ACV (2009) Contribuições da altimetria espacial à modelagem hidrológica de grandes bacias na Amazônia (Spatial altimetry contributions to the hydrological modeling of large basins in Amazon). Ph.D. thesis, Civil Engineering Department, COPPE/UFRJ, Rio de Janeiro, RJ, Brasil, 273 p

Getirana ACV, Bonnet MP, Calmant S, Roux E, Rotunno Filho OC, Mansur WJ (2009) Hydrological monitoring of poorly gauged basins based on rainfall-runoff modeling and spatial altimetry. J Hydrol 379:205–219

Grayson RB, Moore ID, McMahon TA (1992a) Physically-based hydrologic modelling, 1, A terrain-based model for investigative purposes. Water Resour Res 28:2639–2658

Grayson RB, Moore ID, McMahon TA (1992b) Physically based hydrologic modelling, 2, Is the concept realistic? Water Resour Res 28:2659–2666

Gupta VK, Sorooshian S (1983) Uniqueness and observability of conceptual rainfall-runoff model parameters: the percolation process examined. Water Resour Res 19:269–276

Gupta VK, Sorooshian S (1985) The automatic calibration of conceptual catchment models using derivative-based optimization algorithms. Water Resour Res 21:473–485

Han SC, Kim H, Yeo LY (2009) Dynamics of surface water storage in the Amazon inferred from measurements of inter-satellite distance change. Geophys Res Lett 36(9):L09403

Harvey KD, Solomon SI (1984) Satellite remotely-sensed land-use data for hydrologic modelling. Can J Remote Sens 10(1):68–91

Hemakumara HM, Chandrapala L, Moene A (2003) Evapotranspiration fluxes over mixed vegetation areas measured from large aperture scintillometer. Agric Water Manage 58:109–122

Hendrickson JD, Sorooshian S, Brazil LE (1988) Comparison of Newton-type and direct search algorithms for calibration of conceptual rainfall-runoff models. Water Resour Res 24:691–700

Hillel D (1986) Modelling in soil physics: a critical review. In: Future developments in soil science research. Soil Science Society of America, Madison, Wisconsin, pp 35–42

Jackson TJ, Ragan RM, Fitch WN (1977) Test of Landsat-based urban hydrologic modelling. J Water Resour Plann Manage Div ASCE 103(WR1):141–158

Jensen KH, Mantoglou A (1992) Future of distributed modelling. Hydrol Process 6:255–264

Johnston PR, Pilgrim DH (1976) Parameter optimization for watershed models. Water Resour Res 12:477–486

Kavetski D, Kuczera G (2007) Model smoothing strategies to remove microscale discontinuities and spurious secondary optima in objective functions in hydrological calibration. Water Resour Res 43, W03411

Kitanidas PK, Bras RL (1980) Real-time forecasting with a conceptual hydrologic model, 1, Analysis of uncertainty. Water Resour Res 16:1025–1033

Klemes V (1983) Conceptualization and scale in hydrology. J Hydrol 65:1–23

Kouwen N (1988) WATFLOOD: a micro-computer based flood forecasting system based on real-time weather radar. Can Water Resour J 13(1):62–77

Kuczera G (1983) Improved parameter inference in catchment models, 1, Evaluating parameter uncertainty. Water Resour Res 19(5):1151–1162

Leconte R, Brissette F, Galarneau M, Rousselle J (2004) Mapping near-surface soil moisture with RADARSAT- a synthetic aperture radar data. Water Resour Res 40:1029–1038

Lopes JEG, Braga BPF, Conejo JGL (1981) SMAP, a simplified hydrologic model. International symposium on rainfall-runoff modelling, Mississippi State University, Mississippi

Lopes HL, Accioly LJO, Silva FHBB, Sobral MCM, Araújo Filho JC, Candeias ALB (2011) Espacialização da umidade do solo por meio da temperatura da superfície e índice de vegetação. Revista Brasileira de Engenharia Agrícola e Ambiental 15(9):973–980

Lucena AJ, Rotunno Filho OC, França JRA, Peres LF, Xavier LNR (2013) Urban climate and clues of heat island events in the metropolitan area of Rio de Janeiro. Theor Appl Climatol 111(3–4):497–511

Lumb AM, McCammon RB, Kiltle JL Jr (1994) Users manual for an expert system (HSPEXP) for calibration of the hydrological simulation program – Fortran, U.S. Geological Survey, Water Resources Investigations Report 94-4168

Magette WL, Shanholtz VO, Carr JC (1976) Estimating selected parameters for the Kentucky watershed model from watershed characteristics. Water Resour Res 12(3):472–476

McCabe MF, Wood EF, Wójcik R, Pan M, Sheffield J, Gao H, Sua H (2008) Hydrological consistency using multi-sensor remote sensing data for water and energy cycle studies. Remote Sens Environ 112:430–444

Meade RH, Rayol JM, Conceição SC, Natividade JRG (1991) Backwater effects in the Amazon River basin of Brazil. Environ Geol Water Sci 18(2):105–114

Moore RJ, Clarke RT (1981) A distribution function approach to rainfall-runoff modelling. Water Resour Res 17:1367–1382

Njoku EG, Jackson TJ, Lakshimi V, Chan TK, Nghiem SV (2003) Soil moisture retrieval from AMSR-E. IEEE Trans Geosci Remote Sens 41(2):215–229

Ragan RM, Jackson TJ (1980) Runoff synthesis using Landsat and SCS model. J Hydraul Div ASCE 106(HY5):667–678

Rango A, Feldman A, George TS III, Ragan RM (1983) Effective use of Landsat data in hydrologic models. Water Resour Bull 19(2):165–174

Rotunno Filho OC (1995) Soil moisture mapping using remote sensing and geostatistics applied to rainfall-runoff models. Ph.D. thesis, Department of Civil Engineering, University of Waterloo, Canada, 396 p

Rotunno Filho OC, Soulis ED, Kouwen N, Abdeh-Kolahchi A, Pultz TJ, Crevier Y (1996) Soil moisture in pasture fields using ERS-1 SAR data: preliminary results. Can J Remote Sens 22(1):95–107

Sano EE, Assad ED (2004) Projeto de Pesquisa SMEX03-Brasil. Relatório de Campo – Barreiras Bahia, Dezembro de 2003. Embrapa Cerrados, Brasil

Sano EE, Assad ED, Jackson TJ, Crow W, Hsu A (2004) Overview of the aqua/AMSR-E 2003 soil moisture experiment in Brazil (SMEX03 Brazil). International geoscience and remote sensing symposium, IGARSS'2004, IEEE, pp 329–331

Scott C, Bastiaanssen W, Ahmad M (2003) Mapping root zone soil moisture using remotely sensed optical imagery. J Irrig Drain Eng 129(5):326–335

Silva JS, Calmant S, Seyler F, Rotunno Filho OC, Cochonneau G, Mansur WJ (2010) Water levels in the Amazon basin derived from the ERS-2 and ENVISAT radar altimetry missions. Remote Sens Environ 114:2160–2181

Soil Conservation Service (1972) National engineering handbook – section 4: hydrology. SCS-USDA, Washington, DC

Sorooshian S, Arfi F (1982) Response surface parameter sensitivity analysis methods for postcalibration studies. Water Resour Res 18:1531–1538

Sorooshian S, Dracup JA (1980) Stochastic parameter estimation procedures for hydrologic rainfall-runoff models: correlated and heterocedastic error cases. Water Resour Res 16:430–442

Sorooshian S, Gupta VK (1985) The analysis of structural identifiability: theory and application to conceptual rainfall-runoff models. Water Resour Res 21(4):487–495

Sorooshian S, Gupta VK, Fulton JC (1983) Evaluation of maximum likelihood parameter estimation techniques for conceptual rainfall-runoff models: influence of calibration data variability and length on model credibility. Water Resour Res 10:251–259

Tao T, Kouwen N (1989) Remote sensing and fully distributed modelling for flood forecasting. J Water Resour Plan Manage ASCE 115:809–823

Tapley BD, Bettadpur S, Watkins W, Reigber C (2004) The gravity recovery and climate experiment: mission overview and early results. Geophys Res Lett 31(9):L09607. doi:10.1029/2004GL019920

Tasumi M (2003) Progress in operational estimation of regional evapotranspiration using satellite imagery. Ph.D. thesis. University of Idaho, Moscow, Idaho, EUA, 357 p

Tasumi M, Allen RG, Trezza R, Wright JL (2005a) Satellite-based energy balance to assess within-population variance of crop coefficient curves. J Irrig Drain Eng 131(1):94–109

Tasumi M, Trezza R, Allen RG, Wright JL (2005b) Operational aspects of satellite-based energy balance models for irrigated crops in the semi-arid U. S. J Irrig Drain Syst 19:355–376

Trezza R (2002) Evapotranspiration using a satellite-based surface energy balance with standardized ground control. Ph.D. thesis, Biological and Agriculturas Engineering, Utah State University, Logan, Utah, 255 p

Troutman BM (1985a) Errors and parameter estimation in precipitation-runoff modelling, 1, theory. Water Resour Res 21:1195–1213

Troutman BM (1985b) Errors and parameter estimation in precipitation-runoff modelling, 2, a case study. Water Resour Res 21:1214–1222

U. S. Army Corps of Engineers (1976) Urban storm water runoff model (STORM) – Computer program 723-S8-L2520. Hydrologic Engineering Center, Davis

U. S. Army Corps of Engineers (1981) HEC-1 flood hydrograph package. Hydrologic Engineering Center, Davis

Vereecken H, Huisman JA, Bogena H, Vanderborght J, Vrugt JA, Hopmans JW (2008) On the value of soil moisture measurements in vadose zone hydrology: a review. Water Resour Res 44:1–21

Viana LQ, Gonçalves RC, Rotunno Filho OC (2013) Avaliação espaço-temporal do NDVI com a precipitação e com a evapotranspiração na bacia do rio Preto RJ/MG, XX Simpósio Brasileiro de Recursos Hídricos, ABRH, Bento Gonçalves/RS, 8 pp

Webb RP, Lermak R, Feldman A (1980) Determination of land use from satellite imagery for input into hydrologic models. Fourteenth international symposium on remote sensing of the environment, Anne Arbour

Winsemius HC (2009) Satellite data as complementary information for hydrological modelling. Ph.D. thesis, DUT, Delft, Holanda

Xavier AE (2001) Hyperbolic penalty: a new method for nonlinear programming with inequalities. Int Trans Oper Res 8:659–671

Xavier LNR (2012) Modelagem hidrológica com o aporte de dados da missão espacial GRACE: aplicação a bacias brasileiras (Hydrological modeling with GRACE data: application to Brazilian watersheds. Ph.D. thesis, Civil Engineering Department, COPPE/UFRJ, Rio de Janeiro, RJ, Brasil, 250 p

Xavier AE, de Oliveira AAF (2005) Optimal covering of plane domains by circles via hyperbolic smoothing. J Global Optim 31:493–504

Xavier LNR, Becker M, Cazenave A, Longuevergne L, Llovel W, Rotunno Filho OC (2010) Interannual variability in water storage over 2003-2008 in the Amazon Basin from GRACE space gravimetry in situ river level and precipitation data. Remote Sens Environ 114:1629–1637

Yapo PO, Gupta HV, Sorooshian S (1998) Multi-objective global optimization for hydrological models. J Hydrol 204:83–97

Part V
Creating Databases and Models Applied to Soil Physics

Chapter 15
The Use of Numerical Flow and Transport Models in Environmental Analyses

Martinus Th. van Genuchten, Carolina Naveira-Cotta, Todd H. Skaggs, Amir Raoof, and Elizabeth M. Pontedeiro

Abstract This chapter provides an overview of alternative approaches for modeling water flow and contaminant transport problems in soils and groundwater. Special focus is on flow and transport processes in the variably saturated vadose zone between the soil surface and the groundwater table. The governing flow and transport equations are discussed for both equilibrium and nonequilibrium flow conditions, followed by three examples. The first example shows how one-dimensional root-zone modeling can be used to estimate short- and long-term recharge rates, including contaminant transport through the vadose zone. A second example illustrates a two-dimensional application involving drip irrigation, while the third example deals with two-dimensional nonequilibrium transport of a

M.Th. van Genuchten (✉)
Department of Mechanical Engineering, POLI&COPPE, Universidade Federal do Rio de Janeiro, P.O. Box 68503, Rio de Janeiro, RJ 21945-970, Brazil

Department of Earth Sciences, Utrecht University, Budapestlaan 4, 3584CD Utrecht, The Netherlands
e-mail: rvangenuchten@hotmail.com

C. Naveira-Cotta
Department of Mechanical Engineering, POLI&COPPE, Federal University of Rio de Janeiro, P.O. Box 68503, Rio de Janeiro, RJ 21945-970, Brazil
e-mail: carolina@mecanica.coppe.ufrj.br

T.H. Skaggs
U.S. Salinity Laboratory, USDA-ARS, 450 W. Big Springs Rd., Riverside, CA 92507, USA
e-mail: todd.skaggs@ars.usda.gov

A. Raoof
Department of Earth Sciences, Utrecht University, Budapestlaan 4, 3584CD Utrecht, The Netherlands
e-mail: a.raoof@uu.nl

E.M. Pontedeiro
Department of Nuclear Engineering, POLI&COPPE, Federal University of Rio de Janeiro, P.O. Box 68503, Rio de Janeiro, RJ 21945-970, Brazil
e-mail: bettinadulley@hotmail.com

W.G. Teixeira et al. (eds.), *Application of Soil Physics in Environmental Analyses: Measuring, Modelling and Data Integration*, Progress in Soil Science, DOI 10.1007/978-3-319-06013-2_15,

pesticide in a tile-drained field soil. Also discussed are alternative pore-scale modeling approaches that may provide a better understanding of the basic physical and geochemical processes affecting fluid flow and contaminant transport in saturated and variably saturated media.

Keywords Numerical models • Vadose zone • Unsaturated flow • Contaminant transport • HYDRUS • Recharge • Drip irrigation • Pesticide transport • Preferential flow • Pore network models

15.1 Introduction

Soil and groundwater contamination by a broad range of agricultural and industrial pollutants is an ever-increasing problem facing this planet. Major sources of both point and non-point source contamination are the use of fertilizers and pesticides in agricultural operations, leaking underground storage tanks, chemical spills, municipal landfills and mine tailings. Specific contaminants include a range of organic and inorganic chemicals such as chlorinated hydrocarbons, phenols, heavy metals, radionuclides, pesticides, nitrates and ammonia, salts, pharmaceuticals and hormones, and many others. Once released into the environment, these contaminants are subject to a large number of often simultaneous physical, chemical, and biological processes, including advective-dispersive transport, sorption-desorption, precipitation-dissolution, volatilization, and biodegradation. Because of these complexities, mathematical models have become critical components of any effort to understand and predict site-specific subsurface water flow and contaminant transport processes. For example, models can be helpful tools for designing, testing and implementing soil, water and crop management practices in agriculture to minimize soil and water contamination by salts, pesticides and toxic trace elements. Models are equally needed for designing or remediating industrial waste disposal sites and landfills, or for long-term stewardship of nuclear waste repositories.

A large number of specialized analytical and numerical models now exist to simulate the various processes at various levels of approximation and for different applications. Modeling approaches range from relatively simple analytical and semi-analytical solutions, to much more complex numerical codes that permit consideration of a large number of simultaneous nonlinear processes. While analytical and semi-analytical solutions undoubtedly remain popular for many applications, especially for solute transport problems (e.g., Vanderborght et al. 2005; van Genuchten et al. 2012; Naveira-Cotta et al. 2013), the development of more versatile and numerically stable solution techniques and the ever-increasing power of personal computers are now facilitating the much wider use of numerical models. The use of numerical models is now also very much facilitated by the availability of specific software packages in both the public and commercial domains, including the development of sophisticated graphical user interfaces that dramatically simplify their use (Šimůnek and Bradford 2008; Mallants et al. 2011). Examples of widely used codes for flow and transport in variably saturated or multiphase systems are MACRO (Jarvis 1994), UNSATCHEM

(Šimůnek et al. 1996), FEHM (Zyvoloski et al. 1997), HYDROBIOGEOCHEM (Yeh et al. 1998), VS2DI (Healy 2008), MODFLOW-SURFACT (Panday and Huyakorn 2008), STOMP (White et al. 2008), SWAP (van Dam et al. 2008), and the HYDRUS (Šimůnek et al. 2008) and TOUGH (Finsterle et al. 2008) family of codes.

In this chapter we give an overview of various approaches for modeling fluid flow and contaminant transport in soils and groundwater. Our focus is especially on transport processes in the vadose zone. We first give a brief overview of the classical flow and transport equations generally used for modeling equilibrium contaminant transport processes in variably saturated media. This includes two applications. One of these considers root-zone modeling to estimate recharge rates in a semi-arid region of Brazil, but also includes downward transport of a contaminant following a chemical spill at the surface. The second example considers a two-dimensional application involving drip irrigation. Next we provide a brief discussion of possible nonequilibrium transport formulations often needed for flow processes in macroporous (structured) soils or unsaturated fractured rock. This includes an example dealing with the two-dimensional nonequilibrium transport of a pesticide in a tile-drained field soil. For our calculations we used the most recent or earlier versions of HYDRUS-1D (Šimůnek et al. 2013) for the one-dimensional problems, and HYDRUS 2/3D (Šimůnek et al. 2012) for the two-dimensional applications. In a separate section we also discuss briefly various pore-scale modeling approaches that may lead a better understanding of the basic physical and biogeochemical processes affecting fluid flow and contaminant transport in saturated and variably saturated media.

15.2 Classical Modeling Approaches for Variably Saturated Media

15.2.1 Governing Flow and Transport Equations

Classical descriptions of water and solute movement in the vadose zone are typically based on the Richards equation (Richards 1931) for flow in unsaturated porous media and the advection-dispersion equations for solute transport. Here we present the general equations for variably-saturated multidimensional systems. Water flow in such systems is most often described using the three-dimensional form of the Richards equation as follows

$$\frac{\partial \theta(h)}{\partial t} = \frac{\partial}{\partial x_i}\left[K\left(K_{ij}^A \frac{\partial h}{\partial x_j} + K_{ij}^A\right)\right] - S_w \qquad (15.1)$$

in which θ is the volumetric water content, h is the soil water pressure head, x_i ($i = 1,2,3$) are the spatial coordinates, t is time, S_w is a general source/sink term (often used to account for water uptake by plant roots), ${K_{ij}}^A$ are components of a dimensionless anisotropy tensor, and K is the unsaturated hydraulic conductivity function given by

$$K(h;x,y,z) = K_s(x,y,z)K_r(h;x,y,z) \tag{15.2}$$

where K_r is the relative hydraulic conductivity ($0 \leq K_r \leq 1$), and K_s the saturated hydraulic conductivity. The anisotropy vector $K_{ij}{}^A$ may be used to account for anisotropic media. For an isotropic medium, the diagonal entries of $K_{ij}{}^A$ are equal to one and the off-diagonal entries zero. If Eq. 15.1 is used for flow in a vertical cross-section, $x_1 = x$ is the horizontal coordinate, and $x_2 = z$ is the vertical coordinate of the medium, the latter being directed positive upward.

Contaminant transport can be described similarly using the general advection-dispersion equation. Ignoring any partitioning into to the gas phase, the general equation is given by

$$\frac{\partial(\theta c)}{\partial t} + \frac{\partial(\rho s)}{\partial t} = \frac{\partial}{\partial x_i}\left(\theta D_{ij}\frac{\partial c}{\partial x_j}\right) - \frac{\partial q_i c}{\partial x_i} - S_s \tag{15.3}$$

where c and s are solute concentrations associated with the liquid and solid phases, respectively, S_s is a general source-sink term, and q_i is the volumetric fluid flux given by Darcy-Buckingham's law as (Narasimhan 2005)

$$q_i = -K\left(K_{ij}^A\frac{\partial h}{\partial x_I} + K_{ij}^A\right) \tag{15.4}$$

while D_{ij} represents the dispersion coefficient tensor, generally assumed to be of the form (e.g., Bear 1972)

$$\theta D_{ij} = \theta D_o \tau \delta_{ij} + (D_L - D_T)\frac{q_i q_j}{|q|} + D_T|q|\delta_{ij} \tag{15.5}$$

in which D_o is the ionic or molecular diffusion coefficient, τ is a water content dependent tortuosity factor (e.g., Millington and Quirk 1961), δ_{ij} is the Kronecker delta function ($\delta_{ij} = 1$ if $i = j$, and $\delta_{ij} = 0$, if $i \neq j$), and D_L and D_T are the longitudinal and transverse dispersivities, respectively.

The above equations are for water flow and contaminant transport. Many applications may further include heat transport. The governing equation is then of the form (e.g., Šimůnek et al. 2012)

$$C_p(\theta)\frac{\partial T}{\partial t} = \frac{\partial}{\partial x_i}\left(\lambda_{ij}\frac{\partial T}{\partial x_j}\right) - C_w q_i \frac{\partial T}{\partial x_{i_s}} - S_T \tag{15.6}$$

where T is temperature, C_p and C_w are volumetric heat capacities of the porous medium and the liquid phase, respectively, S_T is a general source-sink term, and λ_{ij} is the apparent thermal conductivity of the soil given by

$$\lambda_{ij}(\theta) = \lambda_o(\theta)\delta_{ij} + (\lambda_L - \lambda_T)\frac{q_i q_j}{|q|} + \lambda_T C_w |q| \delta_{ij} \tag{15.7}$$

in which λ_o is a thermal conductivity accounting for the tortuosity of the porous medium, and λ_L and λ_T are the longitudinal and transverse thermal dispersivities, respectively. We refer to the HYDRUS 2/3D technical manual (Šimůnek et al. 2012) for a more detailed discussion of the above equations for water, solute and heat movement.

Applications of the above equations require definition of problem-specific initial and boundary conditions. A large number of boundary conditions are available for this purpose, including transient atmospheric conditions as explained by Šimůnek et al. (2012). Several of these conditions are discussed later in this chapter as part of the applications. Also needed is information about the unsaturated hydraulic functions. A number of specific models are available for this purpose (e.g., Brooks and Corey 1964; van Genuchten 1980; Durner 1994; Kosugi 1996; and many others), as summarized well in a study by Leij et al. (1997).

15.2.2 One-Dimensional Application to Recharge

Here we use the classical models to illustrate water flow and solute transport processes in a 8-m deep one-dimensional soil profile, with special attention to root water uptake and estimating recharge using time-dependent atmospheric boundary conditions. The governing flow Eq. 15.1 for this problem reduces then to

$$\frac{\partial \theta(h)}{\partial t} = \frac{\partial}{\partial z}\left(K(h)\frac{\partial h}{\partial z} + K(h)\right) - S_w \tag{15.8}$$

The soil hydraulic (constitutive) relationships needed in applications of Eqs. 15.1 and 15.8 are described here using the functions (van Genuchten 1980)

$$\theta(h) = \theta_r + \frac{\theta_s - \theta_r}{\left[1 + |\alpha h|^n\right]^m} \qquad (m = 1 - 1/n) \tag{15.9}$$

$$K(h) = K_s S_e^l \left[1 - \left(1 - S_e^{1/m}\right)^m\right]^2 \tag{15.10}$$

where θ_r and θ_s denote the residual and saturated water contents, respectively, α and n are empirical shape factors, K_s is the saturated hydraulic conductivity, l is a pore-connectivity parameter, and S_e is effective saturation given by

$$S_e(h) = \frac{\theta - \theta_r}{\theta_s - \theta_r} \tag{15.11}$$

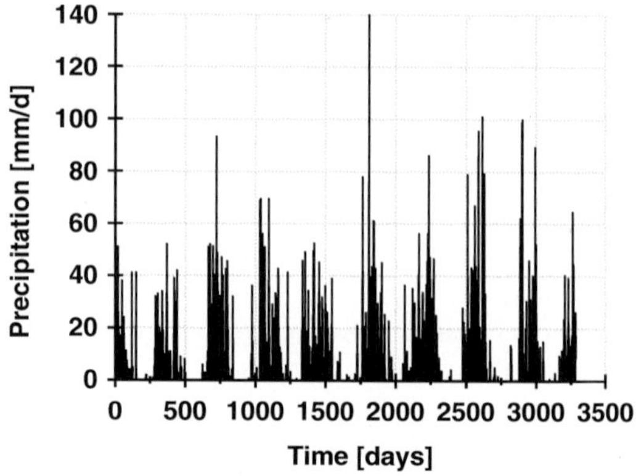

Fig. 15.1 Daily rainfall rates used for the recharge example

For our example, Fig. 15.1 shows the rainfall pattern we used, typical of a semi-arid region of Minas Gerais State in the Cerrado area of Brazil. Much of the rainfall occurs in the Brazilian summer months (January-March), while the winters are relatively dry. The example was selected to show many of the features involved in water flow through the vadose zone, including water uptake by plant roots and evaluating deep drainage and recharge. Root zone modeling has become a popular approach for estimating recharge (Jimenez-Martinez et al. 2009; Lu et al. 2011; Assefa and Woodbury 2013), among many other approaches (Gee and Hillel 1988; Scanlon et al. 2002). For the recharge calculations using HYDRUS-1D (Šimůnek et al. 2012) we assumed a one-dimensional vertical profile of 8 m. Atmospheric boundary conditions were assigned to the soil surface with the possibility of runoff. Daily values of precipitation and potential evapotranspiration were used, in combination with the water stress response model of Feddes et al. (1978) to account for root water uptake by the grass cover.

Daily potential evapotranspiration rates were calculated using the approach of Hargreaves (Hargreaves 1975; Jensen et al. 1990) which is considered one of the best when limited weather data are available. The method uses daily minimum and maximum temperature data, the latitude and altitude of the site, the leaf area index of the vegetative cover (2.0 in our example), and the rooting depth (70 cm). HYDRUS-1D default parameter values for the water stress response function of Feddes et al. (1978) were used in this hypothetical example. We further assumed a linearly decreasing root distribution from the soil surface (maximum) to a depth of 70 cm. Soil hydraulic parameters were estimated from available soil texture data of the site using the pedotransfer functions of Schaap et al. (2001) as implemented in HYDRUS-1D.

Several alternatives were considered for the initial and lower boundary conditions. One set of calculations simply assumed a constant initial condition of -150 cm versus depth, which is close to the long-term average pressure head distribution found with the calculations, along with a free-draining soil profile (represented by the condition that $\partial h/\partial z$ at the lower boundary is zero) which neglects the presence of a water table.

Fig. 15.2 Calculated root water uptake rates for the recharge example

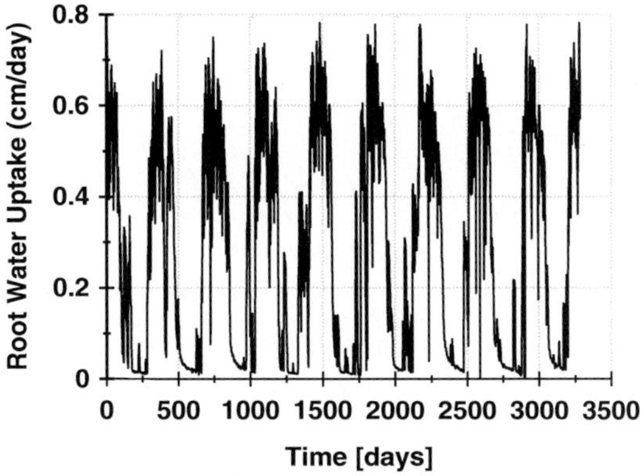

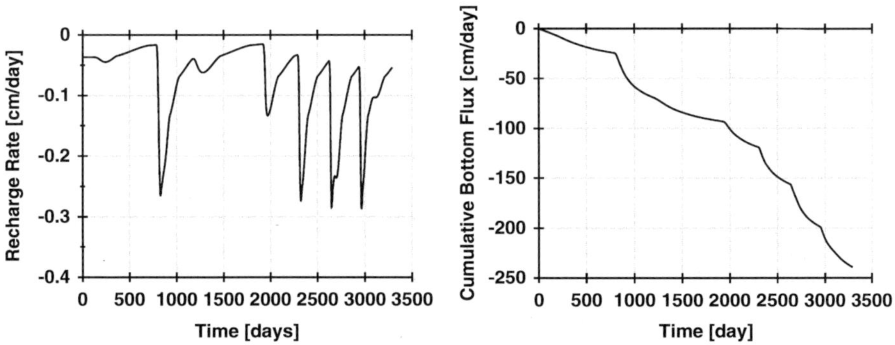

Fig. 15.3 Calculated instantaneous (*left*) and cumulative (*right*) recharge rates

Figure 15.2 shows the calculated root water uptake during the 9 years of simulation, while the calculated recharge rate for the free-draining profile is plotted in Fig. 15.3. Figure 15.3 further shows the calculated cumulative recharge rate (the cumulative bottom flux from the 8-m deep profile). The recharge rate in Fig. 15.3 and the root water uptake rate in Fig. 15.2 show close correlation with the precipitation data in Fig. 15.1, with the recharge rate being highest shortly after the maximum precipitation rates occur in the winter months. The average recharge rate during the 9 years was approximately 270 mm/year (best calculated as the average slope versus time of the cumulative bottom flux shown in Fig. 15.3).

The results in Fig. 15.3 were obtained assuming that the recharge rate does not depend on transient changes in the phreatic surface, including the upward/downward movement of the capillary fringe due to regional flow effects or time-dependent weather conditions. One alternative boundary condition would be the use of a transient condition specifying the water table consistent with local measurements, or perhaps even considering the presence of a constant water table. The latter case was considered by fixing a permanent water table at the bottom boundary. The calculated recharge rates in that case remained very close to those in Fig. 15.2, except that the peaks occurred slightly earlier.

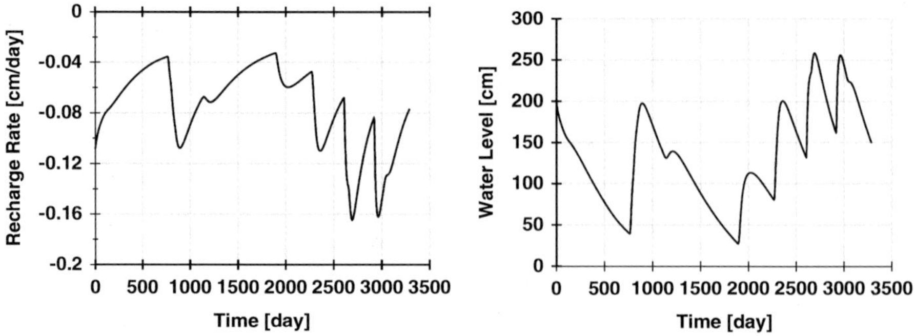

Fig. 15.4 Calculated recharge rates (*left*) and position of the water table (above the *bottom* boundary of the flow domain at a depth of 800 cm) for the recharge example

We further experimented with a special deep drainage lower boundary condition available within HYDRUS-1D to account in an approximate way for regional flow effects on the recharge rate. This boundary is given by (Hopmans and Stricker 1989):

$$q_L(t) = A \exp\left[-B\left|h_L - h_{gw}\right|\right] \tag{15.12}$$

where q_L is the imposed lower boundary flux, h_L is the transient pressure head at the lower boundary (to be calculated), A and B are adjustable parameters, and h_{gw} is some long-term equilibrium water table position relative to the lower boundary. The three parameters (A, B and h_{gw}) are essentially empirical. Their values in this example were guided by water table data of the Rio Claro Aquifer as observed in monitoring wells drilled on the Rio Claro campus of São Paulo State University in Rio Claro, SP, Brazil (Neto et al. 2014).

The initial condition for the deep drainage condition was taken to be −150 cm within the upper part of the profile, the same as before, but now with a pressure head distribution in the lower part of the profile that is in equilibrium with the water table at 600 cm depth. The initial pressure head at the lower boundary of the 8-m deep profile hence was 200 cm. Figure 15.4 shows the calculated recharge rates for the deep drainage boundary condition. Results were obtained using values for A, B and h_{gw} of −2.5 cm/day, −0.007 cm^{-1} and 600 cm, respectively. As compared to the plot in Fig. 15.2, the recharge rate is now less variable in time since the net rate also includes the upward (during the rainy winter seasons) and downward (the summer seasons) movement of the capillary fringe. The upward and downward movement of the water table as such moderated the local fluid fluxes close to the water table and within the capillary fringe. However, the downward fluxes in the vadose zone above the capillary fringe were affected only minimally. Also, the overall cumulative (recharge) fluxes through the lower boundary during the 9-year simulation for the two boundary conditions were relatively close: 239 versus 253 during 9 years (26.6 versus 28.1 cm/year) for the free drainage and deep drainage conditions, respectively.

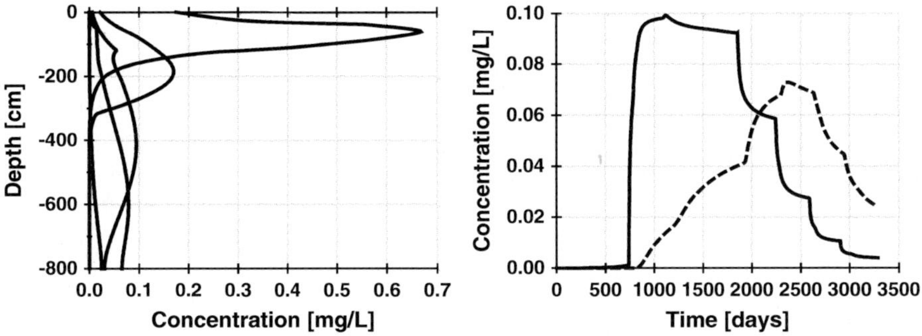

Fig. 15.5 Calculated concentration distributions versus depth 1, 2, 4, 6 and 9 years after a 1 cm chemical spill (*left*) and concentrations versus time at two observation nodes (*right*) located at depths of 4 m (*solid line*) and 8 m (*dashed line*) for the free drainage scenario

Figure 15.4 also shows the transients of the calculated water table position (in terms of the simulated pressure heads above the lower boundary at 800 cm below the soil surface). Results indicate an oscillating water table, very much in response to the local precipitation, potential evapotranspiration and root water uptake conditions.

Finally, we also used the recharge example to estimate the transport of a contaminant at the site, assuming that on day 1 a 1-cm spill occurred of a nonreactive pollutant having a concentration of 20 mg/L. Simulations assumed applicability of the equilibrium transport model given by Eq. 15.2, which for the current one-dimensional scenario reduces to the relatively simple

$$\frac{\partial(\theta c)}{\partial t} = \frac{\partial}{\partial z}\left(\theta D \frac{\partial c}{\partial z}\right) - \frac{\partial qc}{\partial z} \tag{15.13}$$

where q is the Darcy-Buckingham fluid flux given by

$$q = -K(h)\frac{\partial h}{\partial z} - K(h) \tag{15.14}$$

and D is the dispersion coefficient given by

$$D = D_o \tau + D_L \frac{|q|}{\theta} \tag{15.15}$$

in which D_o is the diffusion coefficient of the contaminant (1 cm^2/day in our example), and D_L the longitudinal dispersivity (assumed to be 40 cm). Equation 15.13 holds for the relatively simple case of no contaminant sorption onto the solid phase of the soil (an inert tracer), and no partitioning of the contaminant into the air phase.

Figure 15.5 shows calculated concentration distributions versus depth, 1, 2, 4, 6 and 9 years after the 1-cm chemical spill (left), and concentrations versus time at two observation nodes located at depths of 4 and 8 m deep (right) for the free drainage example. Notice that the contaminant front reached the bottom of the profile after about 1,000 days, and that the maximum concentration of the solute pulse decreased from 20 mg/L (the concentration of the 1 cm spill) to about 0.07 mg/L at 8 m depth, mostly because of dilution with initially present water and infiltrating rain water. Results for the deep drainage bottom boundary were essentially identical in the upper part of the profile (e.g., the distribution versus time at 4 m in Fig. 15.5 was exactly the same), and deviated only minimally in the lower part of the profile because of some additional mixing with transient changes of the capillary fringe and the groundwater table.

The above example was selected to show a number of features that can be captured relatively easily in numerical solutions of the governing flow and transport equations, in this case infiltration, root water uptake, deep drainage and recharge, and contaminant transport. All calculations were carried out using HYDRUS-1D, which can be downloaded freely from www.pc-progress.com. Other features included in HYDRUS-1D, not further discussed here, are such processes as linear and nonlinear sorption, physical and chemical nonequilibrium transport, multicomponent transport, and virus and colloid transport, including consideration of both direct (forward) and inverse (parameter estimation) problems. Several examples for multicomponent transport are provided by Šimůnek et al. (2014, this issue).

15.2.3 Two-Dimensional Application to Subsurface Drip Irrigation

This example considers a typical agricultural problem in which irrigation water is applied to a two-dimensional soil profile using subsurface drip irrigation (Fig. 15.6). Drip irrigation has become a popular method for improving water use efficiency in agricultural operations, especially in arid and semi-arid areas (Skaggs et al. 2004; Lazarovitch et al. 2005; Hanson et al. 2008; Kandelous et al. 2011; Rodriguez-Sinobas et al. 2012). Advantages are improved management of water, soil salinity, fertilizers and pesticides, including through fertigation practices in which agricultural chemicals are supplied directly with the applied irrigation water to the root zone of crops.

Figure 15.6 shows the finite element grid used for one set of simulations documented in more detail by Skaggs et al. (2004). The simulated right side of the profile consisted of a 60 cm wide, 70 cm deep cross-section, with a dripper represented by a half-circle along the left side of the domain 6 cm below the soil surface. Water was applied at a constant rate of 6.37 cm/h through the drip tubing boundary, equivalent to a constant water flux of 4 L/h per m drip tubing. The left, right and soil surface boundaries were assumed to be no-flow boundaries, thereby assuming that evaporation and drainage were negligible during the considered

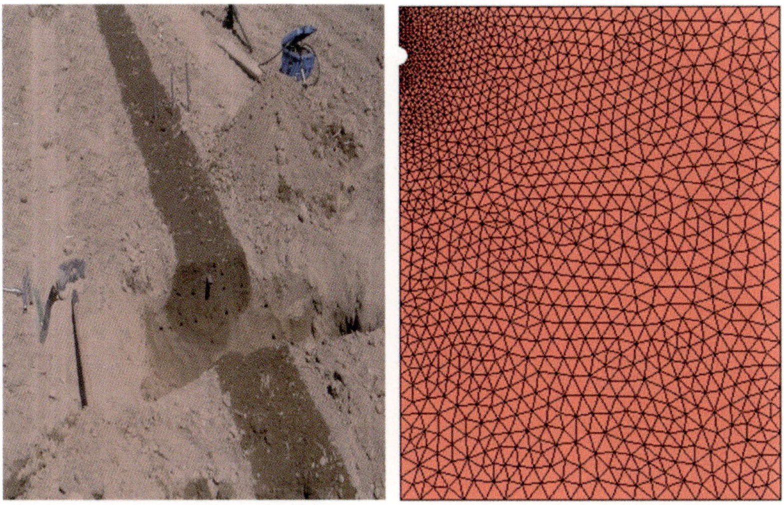

Fig. 15.6 Picture of the simulated drip irrigation experiment (*left*) in the San Joaquin Valley of California as documented by Skaggs et al. (2004), and the invoked finite element grid used for the simulations (*right*). Because of symmetry, only half of the cross-section was considered

irrigation and water redistribution times. A free drainage condition was applied to the lower boundary. Simulations were carried out using hydraulic parameters in Eqs. 15.9 and 15.10 estimated using pedotransfer functions developed by Schaap et al. (2001) leading to the following values: $\theta_r = 0.021$, $\theta_s = 0.34$, $\alpha = 0.023\ \text{cm}^{-1}$, $n = 1.4$, $K_s = 1.6$ cm/h, and $l = -0.92$.

Figure 15.7 shows the measured and simulated water content contour plots for one of the experiments by Skaggs et al. (2004). The irrigation rate for this example was 40 L per m of tubing. The figure shows excellent agreement between the measured and simulated water contents, with root-mean-square errors (RMSEs) being 0.027 and 0.013 $\text{cm}^3\ \text{cm}^{-3}$ at times of 10.75 h (45 min after the 10-h irrigation), and 31 h following an extended period of redistribution. The relatively close fit of simulated and measured values was due in part to having a relatively precise prediction of the hydraulic properties using soil texture data as well as measured values of the water contents at 330 and 15,000 cm as required by the most general option in Rosetta (Schaap et al. 2001). Equally important, the simulation involved a scenario where the irrigation rate is known, thus involving flux-controlled infiltration in which the simulated and calculated total infiltration rates must be the same. More deviations are expected for profile-controlled irrigation scenarios, such as during ponding of a soil surface or, as could be possible with the present example, when a relatively high flux is given such that saturated conditions near the dripper may develop. This may lead to a back pressure that could significantly reduce the drip discharge rate (e.g., Lazarovitch et al. 2005).

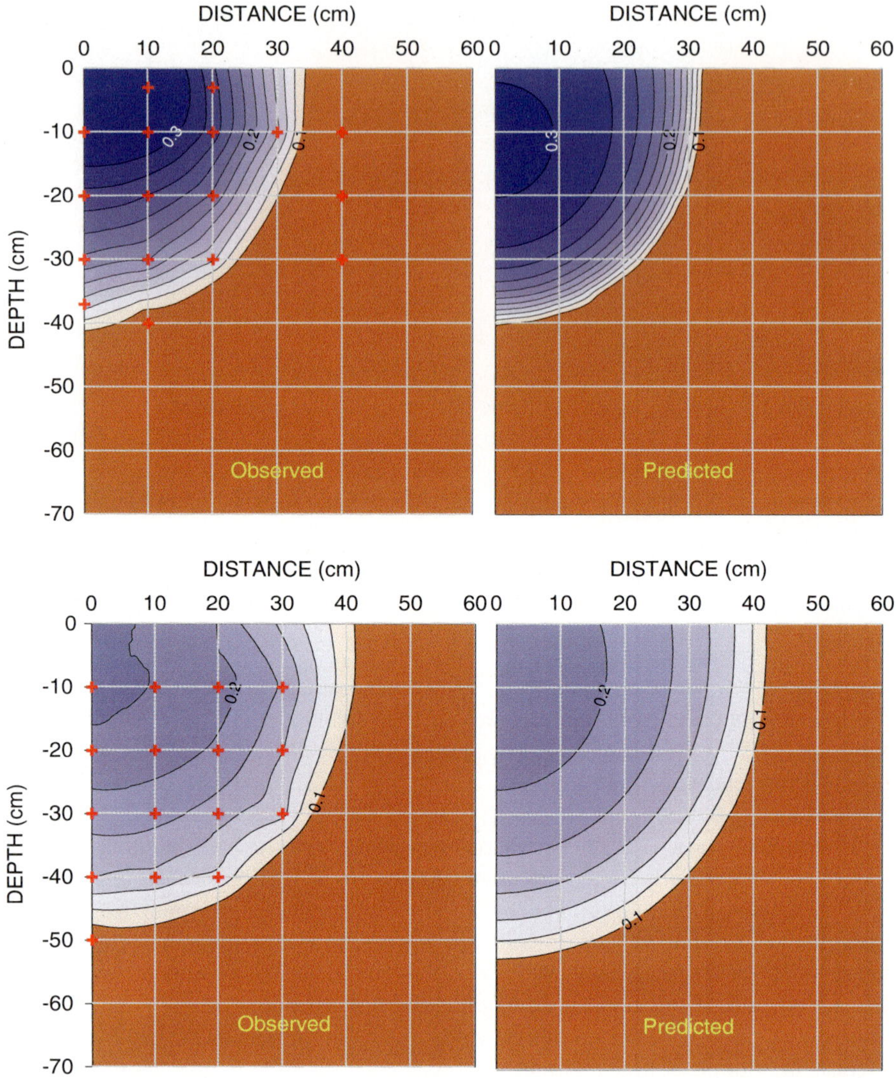

Fig. 15.7 Measured and calculated volumetric water content contours for a 10 h long, 40 L/m subsurface drip irrigation experiment. Plots are for times $t = 10.75$ h (45 min after the 10 h irrigation terminated; *top* plots), and $t = 31$ h (after 21 h of redistribution; *bottom* plots)

The above drip irrigation example showed very good agreement between measured and predicted water content distributions. As indicated by Skaggs et al. (2004), the results provide support for using numerical models such as HYDRUS 2/3D as a tool for investigating and designing drip irrigation management practices. This may include estimating optimal depths and spacing of the drip irrigation line as a function of soil texture, emitter discharge, initial water content, irrigation frequency and crop rooting patterns (e.g. Skaggs et al. 2010).

15.3 Nonequilibrium Model Formulations

The various model descriptions thus far involved very classical descriptions of water flow and solute transport assuming equilibrium conditions. Much evidence exists that water and solutes in many or most field soils do not move according to the uniform flow pattern typically predicted with the Richards equations and the advection-dispersion equations (Hendrickx and Flury 2001; Jarvis 2007; Gerke et al. 2010). This is due to the presence of macropores, fractures or other structural voids or biological channels through which water and contaminants may move preferentially, thereby bypassing parts of the matrix pore-space. Unstable flow in homogeneous or heterogeneous soil, hydrophobicity and the presence of sloping soil textural layers often also contribute to this apparent larger-scale non-equilibrium situation. The resulting nonequilibrium or preferential flow processes seriously hamper accurate predictions of contaminant transport in soils and fractured rocks (Šimůnek et al. 2003; Gerke 2006; Gerke et al. 2010). Contrary to uniform flow, preferential flow results in irregular wetting of the soil profile as a direct consequence of water moving faster in certain parts of the soil profile than in others. Water and dissolved contaminants for these reasons can move quickly to far greater depths, and much faster, than would be predicted with the Richards equation describing uniform flow.

15.3.1 Governing Equations

Nonequilibrium or preferential flow in macroporous soils and fractured rocks can be described using a variety of dual-porosity and dual-permeability models (Liu 1998; Šimůnek et al. 2003; Gerke 2006; Larsbo et al. 2005; Köhne et al. 2009; Vogel et al. 2010). Figure 15.8 shows a schematic of possible equilibrium and

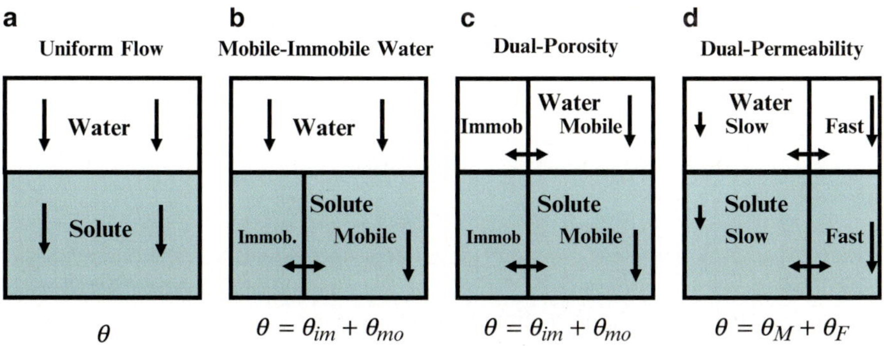

Fig. 15.8 Conceptual models for equilibrium and preferential flow of water and solutes. In the figure, θ is the water content, θ_{mo} and θ_{im} in (b) and (c) are water contents of the mobile and immobile flow regions, respectively, θ_M and θ_F in (d) are water contents of the matrix and macropore (fracture) regions, respectively, and c are concentrations of corresponding regions, with subscripts having the same meaning as for the water contents (After Šimůnek and van Genuchten 2008)

physical nonequilibrium models for water flow and solute transport (Šimůnek and van Genuchten 2008). The nonequilibrium models typically assume that the porous medium consists of two interacting pore regions, one associated with the inter-aggregate, macropore, or fracture system, and one comprising the micropores (or intra-aggregate pores) inside soil aggregates or the rock matrix. Dual-porosity models assume that water in the matrix is stagnant, while the more complex dual-permeability models allow transient water flow to occur also within the soil or rock matrix.

In the dual-permeability approach, Richards equations are applied to both the macropore or fracture (subscript f) and matrix regions (subscript m) as represented by the schematic in Fig. 15.8. For one-dimensional systems, the flow equations for the fracture and matrix regions are given by

$$\frac{\partial \theta_f(h_f)}{\partial t} = \frac{\partial}{\partial z}\left[K_f(h_f)\,\frac{\partial h_f}{\partial z} - K(h_f)\right] - \frac{\Gamma_w}{w} \tag{15.16a}$$

$$\frac{\partial \theta_m(h_m)}{\partial t} = \frac{\partial}{\partial z}\left[K_m(h_m)\,\frac{\partial h_m}{\partial z} - K_m(h_m)\right] + \frac{\Gamma_w}{1-w} \tag{15.16b}$$

respectively, where w is the ratio of the volumes of the fracture domain and the total soil system, and Γ_w is the transfer term for fluid exchange between the two regions. The mobile and immobile water contents θ_{mo} and θ_{im} in Fig. 15.8 are related to the local fracture and matric water contents θ_f and θ_m in Eqs. 15.16a and 15.16b through $\theta_{mo} = w\theta_f$ and $\theta_{im} = (1\text{-}w)\theta_m$, respectively.

Different dual-permeability formulations arise depending upon how water and/or solute movement in the micropore region is modeled, and how water and solutes in the fracture and matrix regions are allowed to interact through the exchange term Γ_w (Šimůnek et al. 2003; Larsbo et al. 2005). One approach is to use a first-order driving force in the pressure head as follows (Gerke and van Genuchten 1993):

$$\Gamma_w = \alpha_w(h_f - h_m) \tag{15.17}$$

in which α_w is a first-order mass transfer coefficient of the form:

$$\alpha_w = \frac{\beta\gamma_w K_a(h)}{a^2} \tag{15.18}$$

where β is a dimensionless geometry-dependent coefficient, a is the characteristic length of the matrix structure (e.g., half of the fracture spacing or the radius of spherical aggregates), γ_w is a dimensionless scaling coefficient, and K_a is the effective hydraulic conductivity of the fracture-matrix interface, which may not be the same as the matrix conductivity because of fracture coatings. Various expressions for $K_a(h)$ have been used in the literature (Zimmerman et al. 1993; Gerke and van Genuchten 1996; Köhne et al. 2004).

Analogous to Eqs. 15.16a and 15.16b, the dual-permeability formulation for solute transport is based on advection-dispersion type equations for transport in both the fracture and matrix regions as follows (Gerke and van Genuchten 1993):

$$\frac{\partial \theta_f R_f c_f}{\partial t} = \frac{\partial}{\partial z}\left(\theta_f D_f \frac{\partial c_f}{\partial z}\right) - \frac{\partial q_f c_f}{\partial z} - \phi_f - \frac{\Gamma_s}{w} \tag{15.19a}$$

$$\frac{\partial \theta_m R_m c_m}{\partial t} = \frac{\partial}{\partial z}\left(\theta_m D_m \frac{\partial c_m}{\partial z}\right) - \frac{\partial q_m c_m}{\partial z} - \phi_m + \frac{\Gamma_s}{1-w} \tag{15.19b}$$

where as before the subscripts f and m refer to the fracture and matrix pore regions, respectively, and Γ_s is the mass transfer term for solute exchange between the two regions. The latter includes both diffusional exchange as well as advective mass transport between the fracture and matrix regions.

The above dual-permeability formulation presents a potentially very powerful model for simulating preferential flow and transport processes in the subsurface as shown by several recent examples in the literature (Pot al. 2005; Kodesova et al. 2005; Köhne et al. 2006; Vogel et al. 2010; among others). While relatively complicated mathematically, the complexity is merely a reflection of the complicated nature in which preferential flow occurs, and the many physical and chemical processes and parameters involved. Unfortunately, application of dual-permeability models such as Eqs. 15.16a and 15.16b typically requires two water retention functions, one for the matrix and one for the fracture pore system, and two or three hydraulic conductivity functions: $K_f(h_f)$ for the fracture network, $K_m(h_m)$ for the matrix, and possibly a separate conductivity function $K_a(h_a)$ for the fracture/matrix interface as embedded in the exchange term, Γ_w (Gerke and van Genuchten 1993). Such information is not readily available for most practical applications. For these reasons a number of simplifications are often invoked, such as the use of dual-porosity formulations, exemplified by the schematic of Fig. 15.8c, which still allow exchange between the fracture and matrix regions, but assume that no longitudual flow and transport occurs within the matrix domain.

To avoid over-parameterization of the governing equations, a further simplification is possible by assuming instantaneous hydraulic equilibration between the fracture and matrix region during flow such that $h_f = h_m$ $(=h)$. In that case the coupling term Γ_w can be eliminated from Eqs. 15.16a and 15.16b to recover Eq. 15.1, but now with composite hydraulic properties of the form

$$\theta(h) = w\theta_f(h) + (1-w)\theta_m(h) \tag{15.20a}$$

$$K(h) = wK_f(h) + (1-w)K_m(h) \tag{15.20b}$$

While still leading to uniform flow, models using such composite hydraulic properties do allow for enhanced flow during conditions near saturation, and as such may provide more realistic simulations of field data than the standard Richards equation with unimodal hydraulic properties (Peters and Klavetter 1988; Mohanty et al. 1997; Zurmühl and Durner 1996; Schaap and van Genuchten 2006). For example, the hydraulic functions of Vogel et al. (2000) are given by

$$\theta(h) = \begin{cases} \theta_r + \dfrac{\theta_m - \theta_r}{\left(1 + |\alpha h|^n\right)^m} & h < h_s \\ \theta_s & h \geq h_s \end{cases} \tag{15.21}$$

$$K(h) = \begin{cases} K_k \left(\dfrac{\theta - \theta_r}{\theta_k - \theta_r}\right)^l \left[\dfrac{1 - F(\theta)}{1 - F(\theta_k)}\right]^2 & h \leq h_k \\ K_k + \dfrac{(h - h_k)(K_s - K_k)}{h_s - h_k} & h_k < h < h_s \\ K_s & h \geq h_s \end{cases} \tag{15.22}$$

where

$$F(\theta) = \left[1 - \left(\frac{\theta - \theta_r}{\theta_m - \theta_r}\right)^{1/m}\right]^m \tag{15.23}$$

Equation 15.21 introduces a very small but non-zero air entry value in $\theta(h)$ near saturation by replacing θ_s in Eq. 15.5 with by an extrapolated parameter θ_m slightly larger than θ_s. While this modification has little or no effect on the retention curve, the effect on the shape and value of the hydraulic conductivity function can be considerable for fine-textured soils (Vogel et al. 2000; Schaap and van Genuchten 2006). Equation 15.22 assumes that the predicted hydraulic conductivity function is matched to a measured value of the conductivity, $K_k = K(\theta_k)$, at some water content, $\theta_k \leq \theta_s$ and $K_k \leq K_s$. The conductivity function is then assumed to increase linearly between h_k and saturation to account for macropore flow.

When the Richards equation Eq. 15.1 is used in conjunction with composite (dual-porosity) hydraulic functions such as Eqs. 15.20a and 15.20b or 15.21 and 15.22, the solute transport model reduces to a relatively standard dual-porosity formulation which assumes that the liquid phase can be partitioned into mobile, $\theta_{mo} = w\theta_f$, and immobile, $\theta_{im} = (1\text{-}w)\theta_f$, regions, with advective-dispersive transport being restricted to the mobile region as follows (van Genuchten and Wagenet 1989):

$$\frac{\partial \theta_{mo} R_{mo} c_{mo}}{\partial t} = \frac{\partial}{\partial z}\left(\theta D \frac{\partial c_m}{\partial z}\right) - \frac{\partial q c_m}{\partial x} - \alpha_s (c_m - c_{im}) - \phi_{mo} \tag{15.24a}$$

$$\frac{\partial \theta_{im} R_{im} c_{im}}{\partial t} = \alpha_s (c_{mo} - c_{im}) - \phi_{im} \tag{15.24b}$$

in which

$$R_{mo} = 1 + \frac{f \rho K_d}{\theta_{mo}} \qquad R_{mo} = 1 + \frac{(1 - f) \rho K_d}{\theta_{im}} \tag{15.25}$$

where the subscripts *mo* and *im* refer to the mobile (fracture) and immobile (matrix) regions, respectively (Fig. 15.8), *f* is the dimensionless fraction of sorption sites in

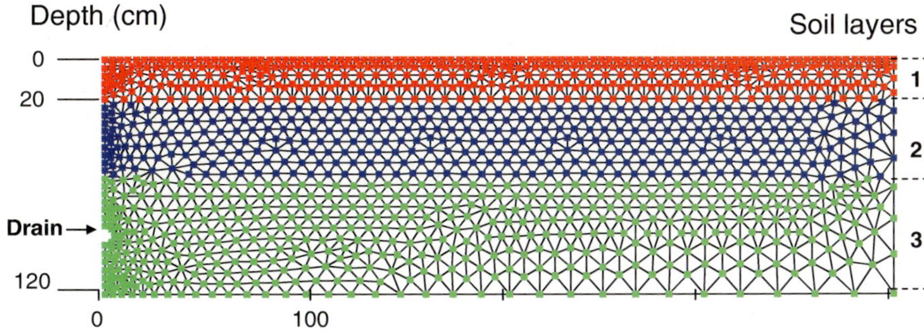

Fig. 15.9 Unstructured finite element grid used for the layered transport domain representing half the drain spacing of the modeled Bouzule-2 silty clay site. The grid consisted of 2,298 triangular elements and 1,237 nodes

contact with mobile water, α_s is the solute mass transfer coefficient between the two regions, and K_d is the distribution coefficient for linear sorption. Please note that the mass transfer terms in Eqs. 15.24a and 15.24b do not contain the mobile water content, θ_{mo}, such as was the case with Eqs. 15.19a and 15.19b. This was done to keep the same notation as used previously by van Genuchten and Wierenga (1976) and van Genuchten and Wagenet (1989).

15.3.2 Application to Field-Scale Pesticide Transport

We now briefly show one application of the physical nonequilibrium model consisting of the standard Richards equation for water flow in conjunction with composite hydraulic functions (in this example given by Eqs. 15.21 and 15.22, and the dual-porosity model for solute transport Eqs. 15.24a and 15.24b. The example concerns the subsurface transport of the herbicide bentazone in a tile-drained agricultural field in northeastern France (Boivin et al. 2006). The field site (La - Bouzule-2) consisted of relatively fine-textured silty clay. The site was equipped with a subsurface tile drain system (0.05 m diameter), with tiles at an average depth of 0.9 m and a drain spacing of 8 m. Precipitation rates and other meteorological variables were recorded at a weather station close to the site. Bentazone was sprayed on the field on March 11, 2002. We used HYDRUS to simulate two-dimensional variably-saturated flow and pesticide transport. The governing equations are exactly the same Eqs. 15.24a and 15.24b as before, except for their extension to two dimensions (Boivin et al. 2006). For the study we assumed isotropic media, included provisions for root water uptake, but neglected pesticide degradation since laboratory experiments indicated little or no degradation over time periods pertinent to the field experiments. We refer to Boivin et al. (2006) for a detailed discussion of the experiments and the data used for the simulations.

Figure 15.9 shows a cross-section of the simulated domain with the unstructured finite element grid used in the simulations. The grid was stratified in accordance

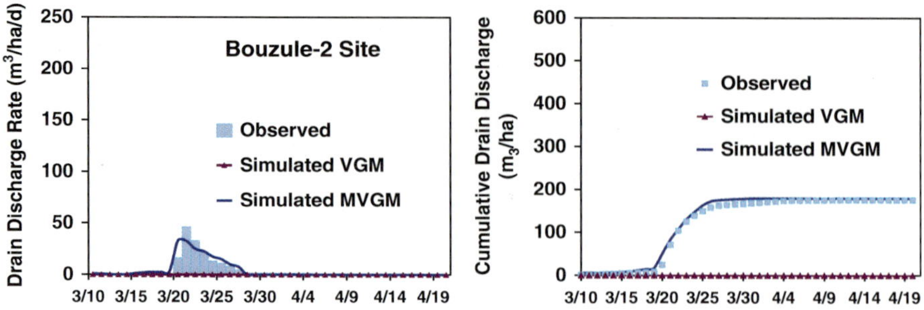

Fig. 15.10 Observed and simulated instantaneous and cumulative tile drain discharge rates

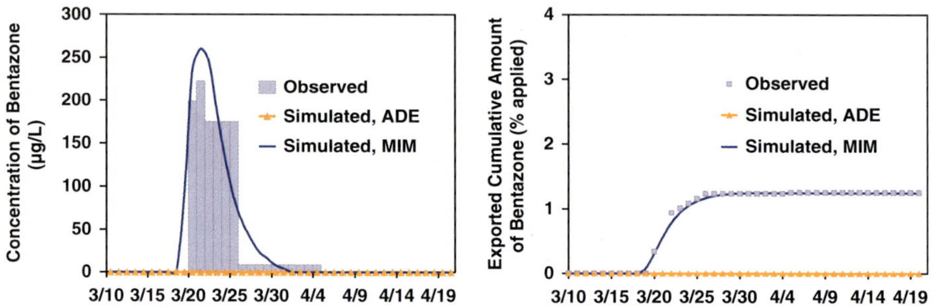

Fig. 15.11 Observed and simulated bentazone concentrations in the drainage water (*left*) and cumulative amount of bentazone exported (*right*) for the Bouzule-2 silty clay site. Solute transport simulations were carried out with both the advection-dispersion equation (ADE) and the mobile-immobile non-equilibrium transport model (MIM)

with observed soil profile layering. Boundary conditions included no-flow boundaries on the sides, atmospheric conditions at the soil surface (daily precipitation and evaporation rates), a free drainage condition bottom boundary, and a seepage face along the tile drain (which implies that the drain functions only when the surrounding soil is saturated). Preliminary simulations started on January 10, 2002, about 2 months before the pesticide application, with an equilibrium pressure head profile so as to obtain realistic soil water contents in the profile at the time of the pesticide application.

Observed and measured tile drainage discharge rates and concentrations, as well as cumulative amounts, are shown in Figs. 15.10 and 15.11. Observed concentration data (Fig. 15.11) indicated very high pesticide concentrations in the drainage water (a maximum concentration of 223 µg L^{-1}), substantially higher than the limit of 0.1 µg L^{-1} set by the European Union for pesticide concentrations in groundwater (FLMW 1995), and also well above 3 µg L^{-1} used by the U.S. Environmental Protection Agency for the maximum contaminant level for bentazone (USEPA 2004). Evidence of preferential flow at this site was clearly demonstrated by several other experiments conducted with a bromide tracer and the pesticide metolachlor in drainage water, especially during the relatively wet spring and summer seasons (Novak et al. 2003).

Two-dimensional HYDRUS simulations using the traditional van Genuchten-Mualem (VGM) hydraulic functions given by Eqs. 15.9 and 15.10 were found to severely underestimate the observed drain discharge rates (Fig. 15.10). K_s values obtained from laboratory experiments did not account for the influence of cracks, and thus substantially underestimated actual K_s values in the field, especially in the spring and summer months when drying cracks were known to develop in especially the soil surface horizon. To account for these cracks, the modified (MVGM) soil hydraulic functions of Vogel et al. (2000) were used to mimic preferential flow of water and solute.

Using Eqs. 15.21 and 15.22 with K_s values up to about 20 times the measured K_k values for some of the layers did lead to much more accurate simulations of the drain discharge rate during the spring of 2002. It is very unlikely that the selected combination of K_k values in the simulations represented a single unique description of the macropore flow processes at the site. Only a few hydraulic parameters in Eqs. 15.21 and 15.22 were obtained from direct measurements (notably θ_s, θ_s and K_k). Others (α and n) were estimated from soil texture using the Rosetta pedotransfer functions of Schaap et al. (2001), while K_s and h_k were calibrated against the observed drainage data.

Figure 15.11 shows that the concentrations and amounts of leached pesticide were severely underestimated using the equilibrium (ADE) model. An excellent fit could be obtained with the physical nonequilibrium mobile-immobile (MIM) model upon calibration of the mass transfer coefficient, α_s, and equating the immobile water content, θ_{im}, to the residual water content, θ_r for all three layers. A single α_s value of 0.027 d^{-1} for all three layers produced excellent agreement with the measured concentrations as (Fig. 15.11) as well as the total amount of bentazone exported with the drainage water. The fitted value of 0.027 d^{-1} for a_s is well within the range of values used by Maraqa (2001) for intermediate-scale and field studies, but lower than the laboratory studies analyzed by Maraqa (2001).

The pesticide study shows that a limited amount of input data could be used to successfully simulate drain discharge rates and chemical concentrations using an equilibrium flow model with composite hydraulic conductivity function, and the mobile-immobile physical nonequilibrium model. The modified soil hydraulic functions in this example provided sufficient flexibility to permit reasonable simulations at the site.

15.4 Pore-Scale Modeling

The various modeling applications thus far were based on macroscopic descriptions of the processes involved, notably the Richards equation for variably saturated flow and the advection-dispersion equation for contaminant transport. These equations provide an approximate, large-scale description of the processes involved. In reality, pore-scale physical and geochemical processes govern the fundamental behavior of water and solutes in porous media such as soils and groundwater. The complexities of these systems, our inability to obtain direct pore-scale observations, and difficulties in upscaling the processes have made it difficult to study porous media at the scale at which the processes occur, and then to extrapolate results to the larger (continuum or Darcy) scale using various upscaling and averaging techniques. For these reasons studies of most multiphase properties (such as the unsaturated soil hydraulic properties and solute dispersion properties) have been based mostly on empirical or quasi-empirical investigations that are inherently limited in detail and applicability.

In an attempt to improve our understanding of multiphase (air-water or air-oil-water) systems, and to develop alternative means of predicting fluid movement, many have turned to pore-scale modeling. As reviewed by Meakin and Tartakovsky (2009), a number of pore-scale modeling approaches are available. Techniques such as traditional grid-based computational fluid dynamics simulation and particle-based Lattice Boltzmann (LB) methods can be used to simulate directly fluid flow and biogeochemical processes within individual pores having realistic complex geometries. In combination with imaging technologies such as X-ray computed tomography (Wildenschild et al. 2002; O'Donnell et al. 2007; Blunt et al. 2013), these modeling techniques can be powerful tools for studying flow and transport processes. However, the models are relatively expensive in terms of computational storage and run-time requirements, particularly for multiphase systems. For these reasons, only limited work has been done modeling real multiphase porous media.

Alternatively, pore network models (PNMs) use idealized representations of complex pore geometries to permit computationally more efficient calculations of water flow and transport processes. PNMs offer a systematic approach for developing improved parametric relationships for the unsaturated soil hydraulic properties, and for simulating fluid flow and contaminant transport processes in homogeneous and heterogeneous media, including multiphase systems. PNMs permit one to focus on pore-scale properties or processes such as residual saturation, interfacial areas, mass transfer rates across interfaces, nonequilibrium transport, the fate of nonaqueous phase liquids (NAPLs), and evaluating alternative methods of simulating single and multiphase flow and transport (Celia et al. 1995; Raoof and Hassanizadeh 2010, 2012). Thus, these models have considerable potential to improve our understanding of the underlying processes, with the end result being improved predictive modeling capabilities and parametric relationships at the larger scale.

Since the seminal work of Fatt (1956), considerable literature has focused on network modeling. It is beyond the scope of the present chapter to present many of the details of pore network modeling techniques. Instead, we consider only briefly a few examples that are indicative of the ways in which pore scale modeling may improve our understanding of multiphase flow and transport processes. More comprehensive reviews are provided by Celia et al. (1995), Meakin and Tartakovsky (2009), and Blunt et al. (2013).

Pore network models (PNMs) can be constructed in various ways, but a typical arrangement consists of relatively large pore bodies located at the nodes of a

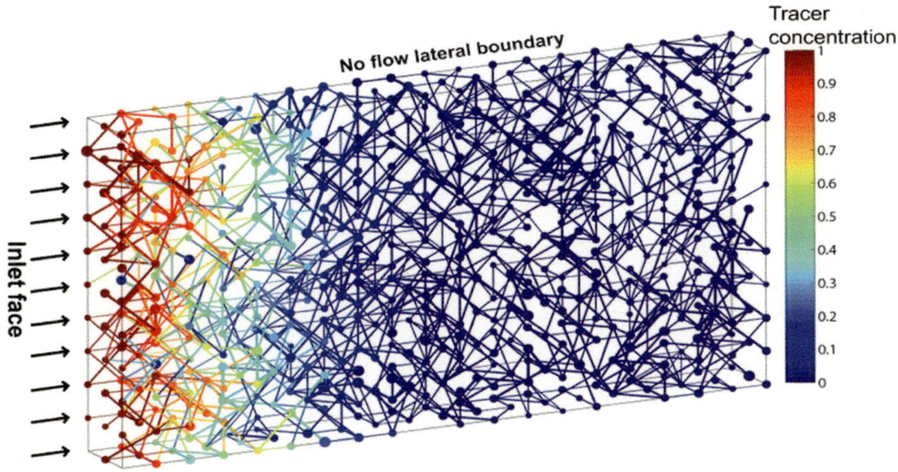

Fig. 15.12 An example of a pore network domain where a tracer is being injected across the left inlet boundary. The *colors* show the concentration while the *lines* show the pore spaces of the soil

network, with connections between nodes/pores being made by smaller capillary tubes that are referred to as pore throats (Raoof and Hassanizadeh 2010). Due to the use of idealized geometries, it is possible to specify laws or rules for fluid passage in pores and throats, and thereby compute flows in the network for given boundary and initial conditions.

Raoof et al. (2013) recently developed the PoreFlow code for simulating pore scale flow and transport processes. This code may be used to simulate fluid flow and multi-component reactive transport under saturated and variably saturated conditions. The simulated porous medium is created by defining capillary tubes of different sizes (which define the "geometry" of the porous medium) and connecting these in various ways to each other (the "topology" of the porous medium). The fluid flow and solute transport processes of interest are then simulated at the pore scale, with the relevant physics implemented on a pore to pore basis. At this scale, flow and transport are simulated by explicitly modeling the phase interfaces and mass exchange at surfaces. Average values and properties can be obtained by integration over the entire network domain, which provides then also the upscaling relationships. We refer to Raoof and Hassanizadeh (2012) and Raoof et al. (2013) for a detailed discussion of the construction of PNMs, and the mathematical equations used for this purpose.

Figure 15.12 shows an example of a pore network in which solute is being injected across the left side at a given rate. Once PNM calculations are carried out using pore networks of the type shown in this figure, results can be compared with those obtained using a model representing macroscale behavior (such as those based on the Darcy and Darcy-Buckingham laws). This comparison permits one then to study various relationships between the two scales.

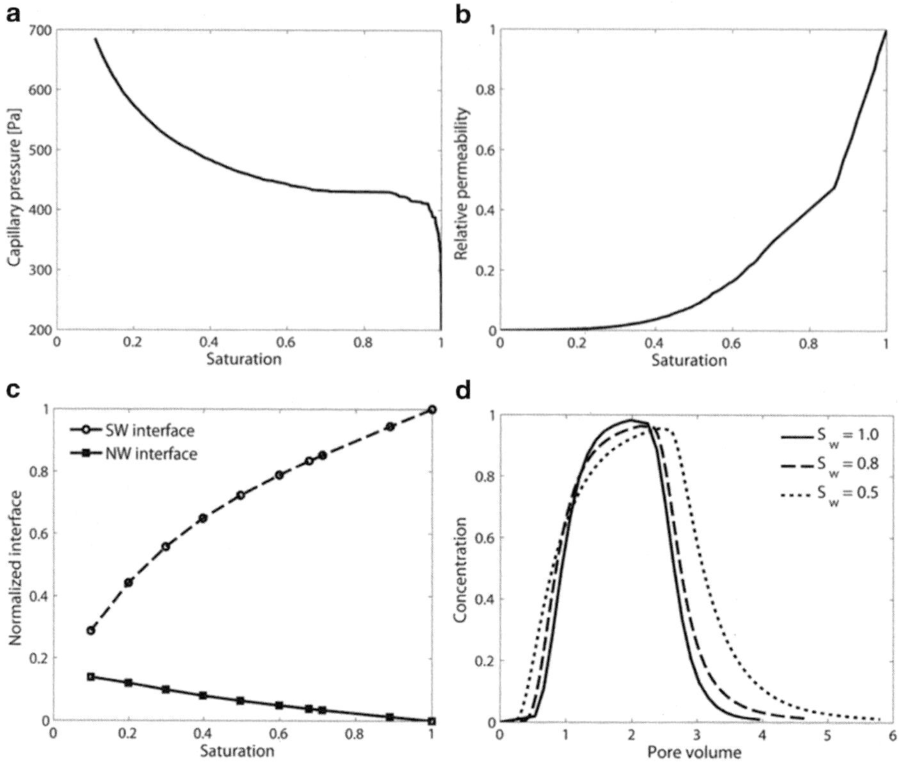

Fig. 15.13 Pore network modeling results obtained for (**a**) the capillary pressure-saturation curve, (**b**) the relative permeability-saturation curve, (**c**) the normalized total interfacial area of solid-wetting (SW) and nonwetting-wetting (NW) phases as a function of wetting-phase saturation, S_w (the areas are normalized to the total area under saturated conditions, which only belongs to the solid-wetting surfaces), and (**d**) breakthrough curves at three relative saturations

Figure 15.13 shows an application in which the PoreFlow pore network model of Raoof et al. (2013) was used to simulate drainage processes, leading to information about the unsaturated soil hydraulic properties. Results are presented in terms of capillary pressure-saturation and relative permeability-saturation curves, as well as changes in the interfacial areas as a function of relative saturation, S_w (= θ/θ_s).

Also shown in Fig. 15.13 are PNM calculated solute breakthrough curves obtained for three different values of relative saturation. The curves show higher solute dispersion at lower saturations. These types of results can be used to study the effect of water content and other parameters on solute dispersion. Although dispersion is known to be strongly dependent upon both flow velocity and water content (Maciejewski 1993), not much information exists on the exact nature and functional form of the water content dependency (Bear and Cheng 2008). Many studies have reported higher values of the dispersion coefficient at smaller water contents (Kirda et al. 1973; De Smedt and Wierenga 1984; Maraqa et al. 1997). This effect has been

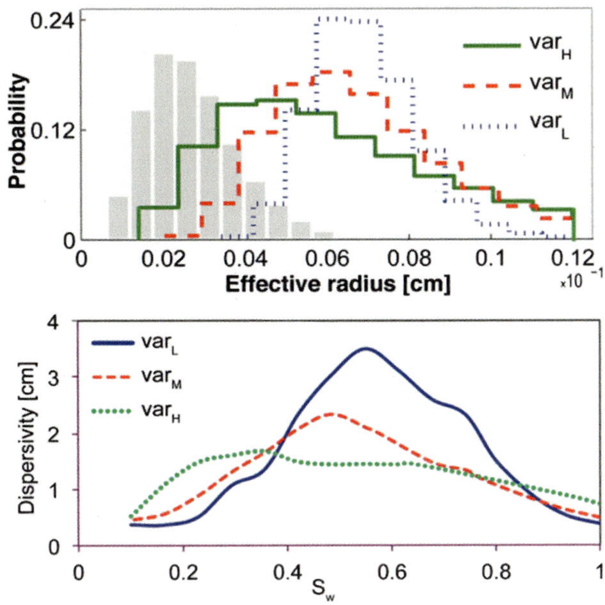

Fig. 15.14 Distributions of pore-body sizes together with a distribution of pore throat sizes shown as columns (*top*). The pore-body size distributions have different variances: high ($\mathrm{var_H}$), medium ($\mathrm{var_M}$), and low ($\mathrm{var_L}$), representing different soils. The relationship between the dispersivity (D_L) and relative saturation (S_w) for each pore network is shown in the *bottom* plot

attributed in part to the presence of immobile water (De Smedt and Wierenga 1979, 1984; van Genuchten and Dalton 1986), with the fraction of immobile water likely depending upon the pore structure as well as saturation.

Pore-scale network models are excellent tools for studying these type of immobile water and fluid velocity effects on the dispersion coefficient. Let us restate for this purpose the often-used relationship between the dispersion coefficient and the fluid velocity Eq. 15.15 in a more general form:

$$D(\theta, v) = D_o\tau(\theta) + D_L(\theta)|v|^{\varepsilon} \tag{15.26}$$

where D_o is the diffusion coefficient, D_L is the longitudinal dispersivity, v is the average pore-water velocity ($=q/\theta$) and ε an empirical constant presumably very close but not necessarily equal to 1.0 under variably saturated conditions (e.g., Kirda et al. 1973; De Smedt et al. 1986). When the water content or relative saturation decreases, travel times and microscopic travel distances tend to become longer, leading to a broader velocity distribution and possibly more dispersion. Consequently, the variability in the microscopic velocity, and its directions at the pore scale, can be (some or much) larger than in saturated porous media, thus increasing tailing in the observed breakthrough curves.

Raoof and Hassanizadeh (2013) used pore network modeling to calculate exactly this saturation dependency of the solute dispersion coefficient for different pore sizes. Results are presented in Fig. 15.14. They indeed show that the dispersivity is strongly dependent upon relative saturation. The relationship was found to be not monotonic in that a maximum dispersivity occurred at some intermediate saturation value referred to as critical saturation by Raoof and Hassanizadeh (2013). This behavior has been observed also experimentally, most notably by Toride et al. (2003).

Unsaturated dispersion is affected by imperfect solute mixing within pores due to the presence of a nonwetting phase (generally air) and changes in the connectivity among pores. Having a collection of saturated pores may create a relatively fast flow domain overall, while film flow within drained pores together with trapped phases (immobile water) create a relatively slow flow process. Under such a condition, advection-dispersion equation may not accurately simulate transport of solutes at the macro scale and other formulations, such as the mobile-immobile (dual-porosity) model given by Eqs. 15.24a and 15.24b, may become more appropriate for modeling unsaturated solute transport processes.

15.5 Concluding Remarks

In this chapter we reviewed a broad range of classical as well as dual-porosity and dual-permeability formulations for modeling equilibrium and nonequilibrium (or preferential) fluid flow and contaminant transport. We focused especially on transport processes in the vadose zone between the soil surface and the groundwater table. A large number of ready-to-use numerical models are now available in both the public and commercials domains. While equilibrium flow and transport models undoubtedly remain applicable to many situations, including macroporous soils and fractured rock when saturated conditions can be avoided, most practical applications may require the use of more parameter-intensive dual-porosity and dual-permeability formulations. As illustrated by the pesticide transport problem, one effective modeling approach requiring far fewer hydraulic parameters is the use of composite (dual-porosity type) functions for the unsaturated hydraulic conductivity to account for the separate effects of macropores and micropores, and then to combine this approach with a mobile-immobile water type nonequilibrium formulation for solute transport. Still, the example shows that preferential flow can have a major effect on the simulation results. The physical nonequilibrium dual-porosity formulation used in this study requires an estimate of the mass transfer coefficient (α_s) governing diffusive exchange between the fracture and matrix regions of structured media. While some guidance is provided by literature data, more research is needed to determine the exact scale dependency of this parameter, especially for larger scale field applications. Pore network modeling may well be very useful for studying this scale-dependency, as well as for other lingering problems such as deciphering the effects of fluid velocity and water content on the solute dispersivity.

References

Assefa KA, Woodbury AD (2013) Transient, spatially varied groundwater recharge modeling. Water Resour Res 49:1–14, doi:10.1002/wrcr.20332

Bear J (1972) Dynamics of fluids in porous media. Elsevier Sci, New York

Bear J, Cheng AHD (2008) Modeling groundwater flow and contaminant transport. Springer, New York

Blunt MJ, Bijeljic B, Dong H, Gharbi O, Iglauer S, Mostaghimi P, Paluszny A, Pentland C (2013) Pore-scale imaging and modelling. Adv Water Resour 51:197–216

Boivin A, Šimůnek J, Schiavon M, van Genuchten MT (2006) Comparison of pesticide transport processes in three tile-drained field soils using HYDRUS-2D. Vadose Zone J 5(3):838–849

Brooks RH, Corey AT (1964) Hydraulic properties of porous media, Hydrol. Paper no 3, Colorado State University, Fort Collins, CO

Celia MA, Reeves PC, Ferrand LA (1995) Recent advances in pore scale models for multiphase flow in porous media. Rev Geophys 33(S2):1049–1057

De Smedt F, Wierenga PJ (1979) Mass transfer in porous media with immobile water. J Hydrol 41 (1):59–67

De Smedt F, Wierenga PJ (1984) Solute transfer through columns of glass beads. Water Resour Res 20(2):225–232

De Smedt F, Wauters F, Sevilla J (1986) Study of tracer movement through unsaturated sand. J Hydrol 85(1):169–181

Durner W (1994) Hydraulic conductivity estimation for soils with heterogeneous pore structure. Water Resour Res 32(9):211–183

Fatt I (1956) The network model of porous media I. Capillary pressure characteristics. Trans AIME 207:144–159

Feddes RA, Kowalik PJ, Zaradny H (1978) Simulation of field water use and crop yield. Wiley, New York

Finsterle S, Doughty C, Kowalsky MB, Moridis GJ, Pan L, Xu T, Zhang Y, Pruess K (2008) Advanced vadose zone simulations using TOUGH. Vadose Zone J 7:601–609

FLMW (1995) Leaching models and EU registration. Guidance document 4952/VI/95, Focus Leaching Modelling Workgroup (FLMW), Commission of the European Communities, Directorate-General for Agriculture VI B II-I, Brussels

Gee GW, Hillel D (1988) Groundwater recharge in arid regions: review and critique of estimation methods. Hydrol Process 2(3):255–266

Gerke HH (2006) Preferential flow descriptions for structured soils. J Plant Nutr Soil Sci 169:382–400

Gerke HH, van Genuchten MT (1993) A dual-porosity model for simulating the preferential movement of water and solutes in structured porous media. Water Resour Res 29:305–319

Gerke HH, van Genuchten MT (1996) Macroscopic representation of structural geometry for simulating water and solute movement in dual-porosity media. Adv Water Resour 19:343–357

Gerke HH, German P, Nieber J (2010) Preferential flow: from the pore to the catchment scale. Vadose Zone J 9(2):207–212

Hanson BR, Šimůnek J, Hopmans JW (2008) Leaching with subsurface drip irrigation under saline, shallow ground water conditions. Vadose Zone J 8:810–818

Hargreaves GH (1975) Moisture availability and crop production. Trans Am Soc Agric Eng 18 (5):980–984

Healy RW (2008) Simulating water, solute, and heat transport in the subsurface with the VS2DI software package. Vadose Zone J. Special Issue "Vadose Zone Modelling" 7:632–639

Hendrickx JMH, Flury M (2001) Uniform and preferential flow, mechanisms in the vadose zone. In: Conceptual models of flow and transport in the fractured vadose zone. National Research Council. National Academy, Washington, DC, pp 149–187

Hopmans JW, Stricker JNM (1989) Stochastic analysis of soil water regime in a watershed. J Hydrol 105:57–84

Jarvis NJ (1994) The MACRO model (Version 3.1), technical description and sample simulations. Reports and Dissertations 19. Department of Soil Science, Swedish University of Agricultural Sciences, Uppsala, Sweden, 51 pp

Jarvis NJ (2007) A review of non-equilibrium water flow and solute transport in soil macropores: principles, controlling factors and consequences for water quality. Eu J Soil Sci 58(3):523–546. doi:10.1111/j.1365-2389.2007.00915.x

Jensen ME, Burman RD, Allen RG (1990) Evapotranspiration and irrigation water requirement, ASCE manuals and reports on engineering practice, vol 70. ASCE, New York

Jimenez-Martinez J, Skaggs TH, van Genuchten MT, Candela L (2009) A root zone modelling approach to estimating groundwater recharge from irrigated areas. J Hydrol 367:138–149

Kandelous MM, Šimůnek J, van Genuchten MT, Malek K (2011) Soil water content distributions between two emitters of a subsurface drip irrigation system. Soil Sci Soc Am J 75(2):488–497

Kirda C, Nielsen DR, Biggar JW (1973) Simultaneous transport of chloride and water during infiltration. Soil Sci Soc Am J 37(3):339–345

Kodesova R, Kozak J, Šimůnek J, Vacek O (2005) Field and numerical study of chlorotoluron transport in the soil profile: comparison of single and dual-permeability model. Plant Soil Environ 51(7):310–315

Köhne JM, Mohanty B, Šimůnek J, Gerke HH (2004) Numerical evaluation of a second-order water transfer term for variably saturated dual-permeability models. Water Resour Res 40. doi:10.1029/2004WR00385

Köhne S, Lennartz B, Köhne JM, Šimůnek J (2006) Bromide transport at a tile-drained field site: experiment, one- and two-dimensional equilibrium and non-equilibrium numerical modeling. J Hydrol 321(1–4):390–408

Köhne JM, Köhne S, Šimůnek J (2009) A review of model applications for structured soils: a) Water flow and tracer transport. J Contam Hydrol. Special Issue "Flow Domains" 104(1–4):4–35. doi:10.1016/j.jconhyd.2008.10.002

Kosugi K (1996) Lognormal distribution model for unsaturated soil hydraulic properties. Water Resour Res 32(9):2697–2703

Larsbo M, Roulier M, Stenemo F, Kasteel R, Jarvis N (2005) An improved dual-permeability model of water flow and solute transport in the vadose zone. Vadose Zone J 4(2):398–406. doi:10.2136/vzj2004.0137

Lazarovitch NJ, Šimůnek J, Shani U (2005) System-dependent boundary condition for water flow from subsurface source. Soil Sci Soc Am j 69:46–50

Leij FJ, Russel WB, Lesch SM (1997) Closed-form expressions for water retention and conductivity data. Ground Water 35(5):848–858

Liu H-H (1998) An active fracture model for unsaturated flow and transport in fractured rocks. Water Resour Res 34:2633–2646

Lu X, Jin M, van Genuchten MT, Wang B (2011) Ground water recharge at five representative sites in the Hebei Plain of China: case study. Ground Water 49(2):286–294

Maciejewski S (1993) Numerical and experimental study of solute transport in unsaturated soils. J Contam Hydrol 14(3):193–206

Mallants D, van Genuchten MT, Simunek J, Jacques D, Seetharam S (2011) Leaching of contaminants to groundwater. In: Swartjes FA (ed) Dealing with contaminated sites; from theory to practical application, Chapter 18. Springer, Dordrecht, pp 787–850

Maraqa MA (2001) Prediction of mass-transfer coefficient for solute transport in porous media. J Contam Hydrol 53:153–171

Maraqa MA, Wallace RB, Voice TC (1997) Effects of degree of water saturation on dispersivity and immobile water in sandy soil columns. J Contam Hydrol 25(3):199–218

Meakin P, Tartakovsky AM (2009) Modeling and simulation of pore-scale multiphase fluid flow and reactive transport in fractured and porous media. Rev Geophys 47(3), RG3002. doi:10.1029/2008RG000263

Millington RJ, Quirk JM (1961) Permeability of porous solids. Trans Faraday Soc 57:1200–1207

Mohanty BP, Bowman RS, Hendrickx JMH, van Genuchten MT (1997) New piecewise-continuous hydraulic functions for modeling preferential flow in an intermittent flood-irrigated field. Water Resour Res 33(9):2049–2063

Narasimhan TN (2005) Buckingham, 1907; an appreciation. Vadose Zone J 4:434–441

Naveira-Cotta CP, Pontedeiro EM, Cotta RM, Su J, van Genuchten MT (2013) Environmental impact assessment of liquid waste ponds in uranium milling installations. Waste Biomass Valoriz 4:197–211. doi:10.1007/s12649-012-9156-0

Neto DC, Chang HK, van Genuchten MTh (2014) Groundwater level and rainfall: A mathematical view of water level variations of the Rio Claro Aquifer, Brazil. Hydrogeol J (submitted)

Novak SM, Banton O, Schiavon M (2003) Modeling metolachlor exports in subsurface drainage water from two structured soils under maize (Eastern France). J Hydrol 270(3–4):295–308

O'Donnell AG, Young IM, Rushton SP, Shirley MD, Crawford JW (2007) Visualization, modelling and prediction in soil microbiology. Nat Rev Microbiol 5(9):689–699

Panday S, Huyakorn PS (2008) MODFLOW SURFACT: a state-of-the-art use of vadose zone flow and transport equations and numerical techniques for environmental evaluations. Vadose Zone J 7:610–631

Peters RR, Klavetter EA (1988) A continuum model for water movement in an unsaturated fractured rock mass. Water Resour Res 24:416–430

Pot V, Šimůnek J, Benoit P, Coquet Y, Yra A, Martínez-Cordón M-J (2005) Impact of rainfall intensity on the transport of two herbicides in undisturbed grassed filter strip soil cores. J Contam Hydrol 8:63–88

Raoof A, Hassanizadeh SM (2010) A new method for generating pore-network models of porous media. Transp Porous Media 81(3):391–407

Raoof A, Hassanizadeh SM (2012) A new formulation for pore-network modeling of two-phase flow. Water Resour Res 48, W01514. doi:10.1029/2010WR010180

Raoof A, Hassanizadeh SM (2013) Saturation-dependent solute dispersivity in porous media: pore-scale processes. Water Resour Res 49:1943–1951. doi:10.1002/wrcr.20152

Raoof A, Nick HM, Hassanizadeh SM, Spiers CJ (2013) PoreFlow: a complex pore-network model for simulation of reactive transport in variably saturated porous media. Comput Geosci 61:160–174

Richards LA (1931) Capillary conduction of fluid through porous mediums. Physics 1:318–333

Rodríguez-Sinobas L, Gil M, Sanchez R, Benitez J (2012) Evaluation of drip and subsurface drip in a uniform loamy soil. Soil Sci 177(2):147–152

Scanlon BR, Healy RW, Cook PG (2002) Choosing appropriate techniques for quantifying groundwater recharge. Hydrogeol J 10(1):18–39

Schaap MG, van Genuchten MT (2006) A modified Mualem-van Genuchten formulation for improved description of the hydraulic conductivity near saturation. Vadose Zone J 5:27–34

Schaap MG, Leij FJ, van Genuchten MT (2001) Rosetta: a computer program for estimating soil hydraulic parameters with hierarchical pedotransfer functions. J Hydrol 251:163–176

Šimůnek J, Bradford S (2008) Vadose zone modeling: introduction and importance. Vadose Zone J. Special Issue "Vadose Zone Modelling" 7(2):581–586

Šimůnek J, van Genuchten MTh (2008) Modeling nonequilibrium flow and transport with HYDRUS. Vadose Zone J. Special Issue "Vadose Zone Modeling" 7:782–797

Šimůnek J, Suarez DL, Šejna M (1996) The UNSATCHEM software package for simulating one-dimensional variably saturated water flow, heat transport, carbon dioxide production and transport, and multicomponent solute transport with major ion equilibrium and kinetic chemistry. Version 2.0. Research Report No. 141, U.S. Salinity Laboratory, USDA, ARS, Riverside, CA, 186 pp

Šimůnek J, Jarvis NJ, van Genuchten MT, Gärdenäs A (2003) Nonequilibrium and preferential flow and transport in the vadose zone: review and case study. J Hydrol 272:14–35

Šimůnek J, van Genuchten MT, Sejna M (2008) Development and applications of the HYDRUS and STANMOD software packages and related codes. Vadose Zone J. Special Issue "Vadose Zone Modelling" 7(2):587–600

Šimůnek J, van Genuchten MTh, Šejna M (2012) The HYDRUS Software package for simulating two- and three-dimensional movement of water, heat, and multiple solutes in variably-saturated media, Technical manual, Version 2.0, PC Progress, Prague, Czech Republic, 258 pp

Šimůnek, Šejna M, Saito H, Sakai M, van Genuchten MTh (2013) The HYDRUS-1D software package for simulating the one-dimensional movement of water, heat, and multiple solutes in variably-saturated media. Version 4.16, HYDRUS Software Series 3, Department of Environmental Sciences, University of California, Riverside, CA, 340 pp

Šimůnek J, Jacques D, Ramos TB, Leterme B (2014) The use of multicomponent solute transport models in environmental analyses. In: Texeira WG, Ceddia MB, Ottono MV, Donnagema GK (eds) Application of soil physics in environmental analyses, Springer (this publication)

Skaggs TH, Trout TJ, Šimůnek J, Shouse PJ (2004) Comparison of HYDRUS-2D simulations of drip irrigation with experimental observations. J Irrig Drain Eng 130:304–310

Skaggs TH, Trout TJ, Rothfuss Y (2010) Drip irrigation water distribution patterns: effects of emitter rate, pulsing, and antecedent water. Soil Sci Soc Am J 74:1886–189. doi:10.2136/sssaj2009.0341

Toride N, Inoue M, Leij FJ (2003) Hydrodynamic dispersion in an unsaturated dune sand. Soil Sci Soc Am J 67(3):703–712

USEPA (2004) Drinking water standards and health advisories. Office of Water, U.S. Environmental Protection Agency (USEPA), Washington, DC, 20 pp

van Dam JC, Groenendijk P, Hendriks RFA, Kroes JG (2008) Advances of modeling water flow in variably saturated soils with SWAP. Vadose Zone J 7:640–635

van Genuchten MT (1980) A closed-form equation for predicting the hydraulic conductivity of unsaturated soils. Soil Sci Soc Am J 44:892–898

van Genuchten MT, Dalton FN (1986) Models for simulating salt movement in aggregated field soils. Geoderma 38:165–183

van Genuchten MT, Wagenet RJ (1989) Two-site/two-region models for pesticide transport and degradation: theoretical development and analytical solutions. Soil Sci Soc Am J 53:1303–1310

van Genuchten MT, Wierenga PJ (1976) Mass transfer studies in sorbing porous media: I. Analytical solutions. Soil Sci Soc Am J 40(4):473–480

van Genuchten MT, Šimůnek J, Leij FJ, Toride N, Šejna M (2012) STANMOD: model use, calibration and validation. Trans ASABE 55(4):1353–1366

Vanderborght J, Kasteel R, Herbst M, Javaux M, Thiéry D, Vanclooster M, Mouvet C, Vereecken H (2005) A set of analytical benchmarks to test numerical models of flow and transport in soils. Vadose Zone J 4(1):206–221. doi:10.2136/vzj2005.0206

Vogel T, van Genuchten MT, Cislerova M (2000) Effect of the shape of the soil hydraulic functions near saturation on variably-saturated flow predictions. Adv Water Resour 24 (2):133–144

Vogel T, Brezina J, Dohnal M, Dusek J (2010) Physical and numerical coupling in dual-continuum modeling of preferential flow. Vadose Zone J 9(2):260–267. doi:10.2136/vzj2009.0091

White MD, Oostrom M, Rockhold ML, Rosing M (2008) Scalable modeling of carbon tetrachloride migration at the Hanford site using the STOMP simulator. Vadose Zone J 7:654–666

Wildenschild D, Vaz CMP, Rivers ML, Rikard D, Christensen BSB (2002) Using X-ray computed tomography in hydrology: systems, resolutions, and limitations. J Hydrol 267(3):285–297

Yeh GT, Salvage KM, Gwo JP, Zachara JM, Szecsody JE (1998) HYDROBIOGEOCHEM: a coupled model of hydrological transport and mixed biochemical kinetic/equilibrium reactions in saturated-unsaturated media. Rep. ORNL/TM-13668, Oak Ridge National Laboratory, Oak Ridge, TN

Zimmerman RW, Chen G, Hadgu T, Bodvarsson GS (1993) A numerical dual-porosity model with semi-analytical treatment of fracture/matrix flow. Water Resour Res 29:2127–2137

Zurmühl T, Durner W (1996) Modeling transient water and solute transport in a biporous soil. Water Resour Res 32:819–829

Zyvoloski GA, Robinson BA, Dash ZV, Trease LL (1997) Summary of the models and methods for the FEHM application – a finite element heat- and mass-transfer code. Los Alamos National Laboratory Rept. LA-13307-MS, Los Alamos, NM

Chapter 16
The Use of Multicomponent Solute Transport Models in Environmental Analyses

Jiří Šimůnek, Diederik Jacques, Tiago B. Ramos, and Bertrand Leterme

Abstract This chapter provides a brief overview of multicomponent solute transport models, which simulate the subsurface transport of multiple ions that may mutually interact, can create various complex species, can compete with each other for sorption sites, and/or can precipitate or dissolve. These models are broadly divided into two major groups, those with specific chemistry and general models, and typical examples of these models are given. More detail is provided for the UnsatChem and HPx modules of HYDRUS. The applicability of the UnsatChem module to simulate the multicomponent transport of major ions is demonstrated using the data from a field experiment involving irrigation with waters of different quality, carried out in Portugal. The second example illustrates the use of HP1 to simulate the fate of mercury in a contaminated soil. Finally, the third example demonstrates the versatility of HP2 to simulate the release and migration of uranium from a simplified uranium mill tailings pile toward a river. Using these three examples, it is demonstrated that in spite of the considerable demand on input data, the multicomponent solute transport models can be effective and versatile tools for evaluating complex agricultural and environmental problems.

J. Šimůnek (✉)
Department of Environmental Sciences, University of California, Riverside, CA 92521, USA
e-mail: Jiri.Simunek@ucr.edu

D. Jacques • B. Leterme
Institute for Environment, Health and Safety, Belgian Nuclear Research Centre (SCK•CEN), Boeretang 200, Mol 2400, Belgium
e-mail: djacques@sckcen.be

T.B. Ramos
CEER-Biosystems Engineering, Institute of Agronomy, Technical University of Lisbon, Tapada da Ajuda, 1349-017 Lisbon, Portugal
e-mail: tiago_ramos@netcabo.pt

W.G. Teixeira et al. (eds.), *Application of Soil Physics in Environmental Analyses: Measuring, Modelling and Data Integration*, Progress in Soil Science,
DOI 10.1007/978-3-319-06013-2_16,

Keywords Variably-saturated • Water flow • Soil • Solute transport • Multicomponent models • Hydrus • HP1 • UnsatChem • Uranium transport • Salinity • SAR • Tailings pile

16.1 Introduction

In Chap. 15 of this volume, van Genuchten et al. provided a detailed description of various approaches to modeling water flow and solute transport in soils and in unsaturated porous media in general. The focus of that chapter was on modeling the transport of only one chemical species while assuming that the behavior of this solute is independent of that of other species that may be present in the soil solution. The interaction of this species with the environment (soil) is usually described assuming linear reversible sorption (the K_d approach) and zero- or first-order reactions (production and/or decay). The advantage of this approach is that a large number of analytical solutions exist for one-, two-, and three-dimensional solute transport problems with different types of boundary conditions and steady-state water flow conditions (e.g., van Genuchten 1981; Yeh 1981; Toride et al. 1993; Leij and Bradford 1994; Šimůnek et al. 1999). These analytical models are useful for a variety of applications, such as for "providing initial or approximate analyses of alternative pollution scenarios, conducting sensitivity analyses to investigate the effects of various parameters or processes on contaminant transport, extrapolating results over large times and spatial scales where numerical solutions become impractical, serving as screening models, estimating transport parameters from laboratory or well-defined field experiments, providing benchmark solutions for more complex transport processes that cannot be solved analytically, and for validating more comprehensive numerical solutions of the governing transport equations" (van Genuchten et al. 2012).

A large number of numerical models (e.g., Šimůnek 2005; Šimůnek et al. 2008; van Genuchten et al. 2014) exists for more complex solute transport situations, such as those involving one- or multi-dimensional transient water flow, heterogeneous systems, or nonequilibrium solute transport with nonlinear reactions, for which analytical solutions are generally not available and/or cannot be derived. However, even such models represent a great simplification of real solute transport problems, since these models usually consider the transport of a single solute species, while the soil solution is always a mixture of many ions. These ions may mutually interact, can create various complex species, can compete with each other for sorption sites, and/or can precipitate or dissolve, depending on actual conditions in the soil profile (e.g., Šimůnek and Valocchi 2002; Šimůnek and van Genuchten 2006; Seaman et al. 2012). The transport codes accounting for these more complex reactions, usually called the hydrogeobiochemical transport models, offer the possibility to include processes such as aqueous complexation, cation and anion exchange, surface complexation, and/or equilibrium or kinetic precipitation/dissolution of pure solid phases or solid-solutions. In addition, they may also consider complex parallel and/or sequential geochemical and/or microbiological kinetic networks, including soil carbon cycles, which may involve more advanced kinetic rate equations than simple zero-order or first-order reactions (e.g., Regnier et al. 2005; Thullner et al. 2005; Centler et al. 2010; Langergraber and Šimůnek 2012; Jacques et al. 2013). There are many important agricultural and environmental problems that require models capable of analyzing transport and reactions of multiple chemical species involved in such mutual interactions. In this chapter, we briefly review multicomponent solute transport models, give few typical examples (e.g., BioGeoChem, MIN3P, CrunchFlow, UnsatChem, HP1/2/3), and show a few examples of their use in agricultural (i.e., modeling of soil salinity and the fate of mercury in contaminated soil using HP1) and environmental (i.e., leaching of uranium from U tailings pile) applications.

16.2 HydroGeoBioChemical Transport Models

Reviews of the development over the last two decades of hydrogeochemical transport models involving reactive multiple components are given by Mangold and Chin-Fu Tsang (1991), Lichtner (1996), Steefel and MacQuarrie (1996), Šimůnek and Valocchi (2002), Bell and Binning (2004), Steefel et al. (2005), Seaman et al. (2012), and Šimůnek et al. (2013). Most coupled reactive transport codes were developed for steady-state flow conditions for one, two, or three dimensions, while the flow field (water contents and water fluxes) has to be specified at the input or calculated outside of the reactive transport code using other flow and transport numerical models and imported (Parkhurst and Appelo 1999; Steefel 2009; Parkhurst et al. 2004). Several multicomponent transport models have been published recently that also consider variably-saturated flow (Zyvoloski et al. 1997; Mayer et al. 2002; Jacques et al. 2006, 2008a; Šimůnek et al. 2013).

16.2.1 Models with Specific Chemistry

Geochemical models can be broadly divided into two major groups: those with specific chemistry and general models (Šimůnek and Valocchi 2002). Models with specific chemistry are usually constrained to very specific applications since they are restricted to certain prescribed chemical systems. They are, however, much easier to use since the chemical system is already predefined by model developers and can be more computationally efficient than general models because the numerics can be optimized for this particular chemical system. Typical examples of models with specified chemistry are those simulating the transport of major ions, such as LEACHM (Hutson and Wagenet 1992) and UnsatChem (Šimůnek and Suarez 1994, 1997), and various reclamation models (Šimůnek and Valocchi 2002).

These models typically consider the transport of major ions and their mutual reactions such as complexation, cation exchange, and precipitation/dissolution. The UnsatChem model, as an add-on module to HYDRUS-1D and HYDRUS (2D/3D) (Šimůnek et al. 2008), and its application to a field study will be discussed below.

Another group of models with specified chemistry are those simulating carbon and nitrogen cycles. Some examples are CENTURY (Parton et al. 1987), LEACHM (Hutson and Wagenet 1992), RZWQM (Ahuja and Hebson 1992), COUP (Jansson and Karlberg 2001), and the Wetland module of HYDRUS (2D/3D) (Langergraber and Šimůnek 2005, 2012). These models typically distribute organic matter, carbon, and organic and mineral nitrogen over multiple computational pools, while allowing organic matter to be decomposed by multiple microbial biomass populations. They can account for most of the major reaction pathways, such as inter-pool transfer of carbon and nitrogen, nitrification (ammonium to nitrate-N), denitrification (leading to the production of N_2 and N_2O), volatilization losses of ammonia (NH_3), and microbial biomass growth and death (e.g., Yuan et al. 2011).

16.2.2 General Models

Models with generalized chemistry, on the other hand, provide users with much more flexibility in designing particular chemical systems, thus permitting a much broader range of applications. Users can then either select species and reactions from large geochemical databases or are able to define their own species with particular chemical properties and reactions. However, general models require much better understanding of the chemical system involved from their users, since they themselves are responsible for correctly defining it.

As discussed above, most codes with general geochemistry are limited to solute transport and biogeochemical reactions, while water flow paths must be calculated outside of the reactive transport code. Typical examples are PHREEQC (Parkhurst and Appelo 1999), CRUNCH (Steefel 2000), and PHAST (Parkhurst et al. 2004). Only a few models allow the velocity field to be internally calculated (e.g., MIN3P, Mayer et al. 2002). Several codes for transient, variably-saturated flow have also been coupled to general biogeochemistry models. These include 3DHYDROGEOCHEM (Yeh and Cheng 1999), TOUGHREACT (Xu et al. 2012), MIN3P (Mayer et al. 2002), HYTEC (Van der Lee et al. 2003), OpenGeoSys-Gem (Kosakowski and Watanabe 2014), and HP1/2/3 (Jacques and Šimůnek 2005; Šimůnek et al. 2013). HP1/2/3, an add-on module to HYDRUS-1D and HYDRUS (2D/3D), and its applications for site remediation and risk assessment of mercury-contaminated sites and for evaluating leaching of uranium from a uranium mine tailings pile will be discussed below.

16.3 Applications of HydroGeoBioChemical Transport Models

16.3.1 Literature Applications

There are many important agricultural and environmental applications for general hydrogeobiochemical models, and many of these appeared in the literature during the last two decades. For example, water leaching from various tailing piles (such as from uranium mills) often contains many trace metals including molybdenum, selenium, arsenic, and chromium (Brookins 1984). Using the HYDROGEOCHEM model, Yeh and Tripathi (1991) simulated the release of trace metals and acidity from an acidic uranium mill tailings pile. Narasimhan et al. (1986) used the DYNAMIX model to study groundwater contamination from an inactive uranium mill tailings pile. Similar studies have been carried out by Walter et al. (1994), Lichtner (1996), and Gerke et al. (1998).

Another significant application for hydrogeobiochemical models is to design and evaluate facilities for the safe disposal and long term isolation of hazardous and radioactive waste, especially high-level nuclear waste. The migration of neptunium between the repository and the ground water table after a hypothetical repository breach at the high-level nuclear waste repository at Yucca Mountain, Nevada, was studied using a multicomponent solute transport model by Viswanathan et al. (1998). The fate of metal-organic mixed wastes was studied by Rittmann and VanBriesen (1996) and VanBriesen (1998). Possible consequences of long term interactions of engineered barriers (clay plugs, concrete components) with the host rock of a disposal facility or various perturbations as a consequence of construction (e.g. oxidation fronts) or waste content are also frequently studied using geochemical transport models (Martens et al. 2010, 2011; Kosakowski and Watanabe 2014). The analysis of redox zone formation in organic-contaminated aquifers was carried out by Abrams et al. (1998) and Essaid et al. (1995), while reactive permeable barriers for aquifer remediation were evaluated by Fryar and Schwartz (1994). Deep geological CO_2 storage (e.g., Gherardi et al. 2012; Walsh et al. 2013) represents another environmental application of reactive transport models. In a study of Jacques et al. (2008a), the transport of major cations and anions and U associated with inorganic P-fertilization were computed for transient flow conditions. This study clearly demonstrated that the distribution coefficient K_d, which is often used to simulate the transport of reactive solutes, may vary significantly during the year as a consequence of transient flow conditions. Thus, the "K_d approach" may represent an inadequate over-simplification of a complex environmental system and may produce erroneous predictions.

Generic multicomponent transport models are sufficiently versatile to allow consideration of soil biological processes such as those involved in the carbon cycle (Jacques et al. 2013), nitrogen cycle (Maggi et al. 2008; Gu et al. 2009), and microbial dynamics (e.g., Thullner et al. 2005; Lim et al. 2007; Centler et al. 2010). For example, Thaysen et al. (2014a, b) implemented the CO_2 module of UnsatChem into a generic biogeochemical model HP1 and extended it to include the effects of mixed cementitous materials in the soil profile on enhancing inorganic C-leaching to groundwater. Zhang et al. (2013) evaluated the bacterial transport using the HYDRUS-PHREEQC framework and transport parameters depending on solution chemistry.

Although the models with specific chemistry are limited to simulating only the transport and reactions of a limited number of ions (e.g., major ions), there are many important agricultural and environmental problems, such as salinization and sodification of irrigated soils, reclamation of saline or sodic soils, sustainability of various irrigation practices, or the release of saline brine water during mining operations, that can be evaluated using these types of models (Šimůnek and Suarez 1997; Gonçalves et al. 2006; Corwin et al. 2007; Ramos et al. 2011; Rasouli et al. 2013).

Several interesting applications of both the specific and general chemistry models have recently appeared in a special issue of Vadose Zone Journal entitled "Reactive Transport Modeling" (Seaman et al. 2012). The model with specific chemistry was used in a study of Reading et al. (2012), who used the UNSATCHEM code to evaluate the use of gypsum to ameliorate a nonsaline sodic clay soil in North Queensland, Australia. General reactive transport models have been used in the recent studies of Bea et al. (2012), Chang et al. (2012), and Lichtner and Hammond (2012). Bea et al. (2012) used the MIN3P-D multicomponent reactive transport code to describe processes occurring within a mine tailing impoundment and their impact on CO_2 sequestration. Chang et al. (2012) used PHREEQC, in combination with laboratory column experiments, to evaluate the efficacy of Fe(II)-containing solutions as an in-situ means for reducing toxic hexavalent chromium (Cr(VI)) to trivalent chromium (Cr(III)), which subsequently coprecipitates with Fe(III). Lichtner and Hammond (2012) used PFLOTRAN to evaluate geochemical processes controlling the aqueous phase distribution and solid phase speciation of a hexavalent uranium [U(VI)] plume below the Hanford 300 Area bordering the Columbia River in Hanford, WA.

16.3.2 Modeling of Soil Salinity Using the UnsatChem Module of HYDRUS

The UnsatChem geochemical module (Šimůnek and Suarez 1994) simulates the transport of major ions (i.e., Ca^{2+}, Mg^{2+}, Na^{+}, K^{2+}, SO_4^{2-}, CO_3^{2-}, and Cl^{-}) and their equilibrium and kinetic geochemical interactions, such as complexation, cation exchange, and precipitation-dissolution (e.g., of calcite, gypsum and/or dolomite). This module has been implemented into both the one- and two-dimensional computational versions of HYDRUS (Šimůnek et al. 2008). Possible applications of this module include studies of the salinization/reclamation of agricultural soils (Šimůnek and Suarez 1997), sustainability of various irrigation systems (Gonçalves et al. 2006; Ramos et al. 2011; Rasouli et al. 2013), the disposal of brine waters from mining operations, and many other applications (e.g., Schoups et al. 2006; Skaggs et al. 2006; Corwin et al. 2007). Gonçalves et al. (2006) and

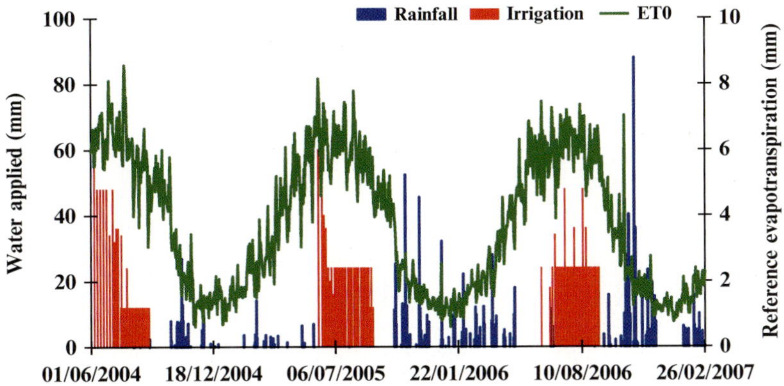

Fig. 16.1 Daily values of precipitation, irrigation and reference evapotranspiration from June 2004 until February 2007

Ramos et al. (2011) recently demonstrated the applicability of UnsatChem to the simulation of the multicomponent transport of major ions both in soil lysimeters and in the field, respectively, irrigated with waters of different quality. They used the UnsatChem module of HYDRUS-1D to describe field measurements of the water content, overall salinity (expressed as Electric Conductivity, EC), the concentration of individual soluble cations, the Sodium Adsorption Ratio (SAR), and the Exchangeable Sodium Percentage (ESP). The experimental data and their analysis using the UnsatChem module are described in detail by Gonçalves et al. (2006) and Ramos et al. (2011) and only selected results from the latter study will be given here.

Problem Description

The field experiment was conducted from June 2004 to February 2007 in the Alentejo region in Southern Portugal on a medium texture soil. The climate in this region is mostly dry, sub-humid to semi-arid, with hot dry summers, and mild winters with irregular rainfall. The experiment involved irrigating maize with either locally available water or with synthetic (with added NaCl) saline water. The experimental field was divided into three groups, A, B, and C (only results for plots A and C are reported below), each having a surface area of 6.75 m^2 (2.25 m wide × 3 m long; 0.75 m between maize lines), with the Na^+ gradient decreasing from A (saline waters with EC of 7.8 dS m^{-1} during the first two seasons and 15.6 dS m^{-1} during the third season) to C (local water, $EC \leq 1.2$ dS m^{-1}). A drip irrigation system with an overall constant cumulative discharge of 18 L/h/m (24 mm/h) was used to deliver water, salts (NaCl), and fertilizer (N) to the crop. The quality of irrigation water was thus the key variable between plots. The reference evapotranspiration rates (ET_0) were calculated using the Penman-Monteith method. Crop evapotranspiration rates (ET_c) were then calculated using the product of ET_0 and K_c, where K_c is a crop coefficient accounting for both soil evaporation and crop transpiration (Fig. 16.1).

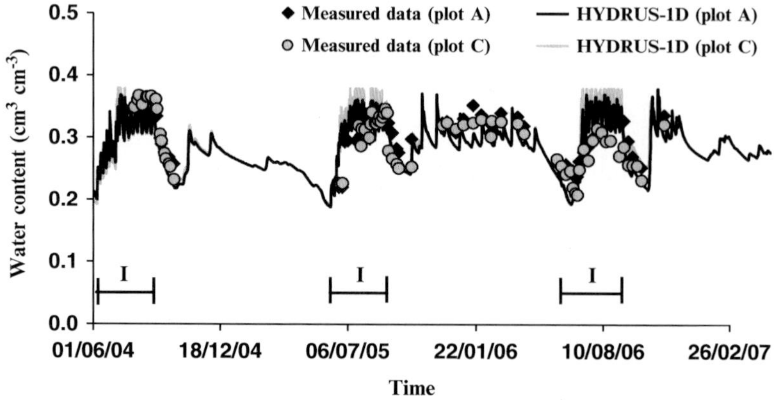

Fig. 16.2 Measured and simulated water contents at a depth of 40 cm. Irrigation seasons are indicated by the letter *I*

TDR probes and ceramic cups were installed at 20, 40, and 60 cm depths to measure soil water contents and collect soil solutions twice a week during irrigation seasons, generally 24 h after an irrigation event, and twice a month during the rest of the year. The soil solution was monitored for the concentrations of soluble Na^+, Ca^{2+}, Mg^{2+}, EC_{sw} (Electrical Conductivity of Soil Water), and *SAR* (Sodium Adsorption Ratio). Soil hydraulic properties for three soil layers were measured using suction tables, the pressure plate apparatus, and the evaporation method on undisturbed soil samples, and fitted using the van Genuchten model. The Gapon selectivity coefficients for four cations (Ca, Mg, K, and Na) were calculated from the initial measured soluble and exchangeable cations concentrations.

Results

Figure 16.2 shows measured and simulated water contents at a depth of 40 cm during the three seasons. During irrigation periods (indicated by a letter I in Fig. 16.2), the amount of water applied in the experimental field was considerably higher than the usual amount of water used to irrigate maize in this region in order to increase water contents above soil field capacity so that soil solution samples could be collected using the installed ceramic cups. As a result, soil water content rapidly increased in the beginning of each irrigation season, and then varied between soil saturation and soil field capacity (Ramos et al. 2011). Between the end of the irrigation seasons (September) and the beginning of rainy seasons (October or November), soil water contents gradually decreased, allowing maize to mature and be harvested. During rainy seasons, soil water contents reflected the occurrence of individual rainfall events (Fig. 16.2).

Figure 16.3 shows that soil salinity increased considerably in the plot irrigated with saline waters. Measured EC_{sw} in plot A reached values higher than 6.0 dS m^{-1}

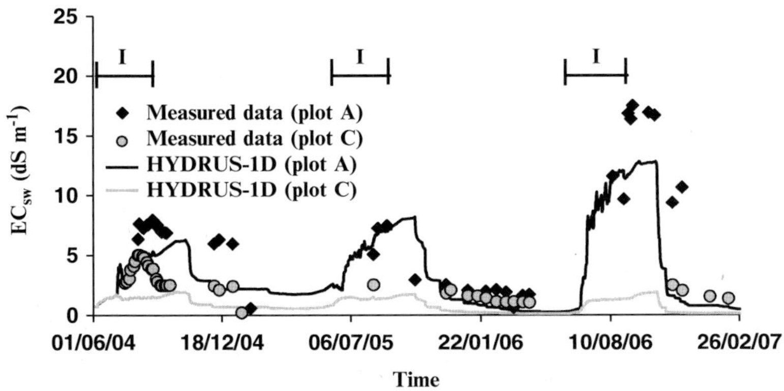

Fig. 16.3 Measured and simulated electrical conductivity of soil water (EC_{sw}) at a depth of 40 cm. Irrigation seasons are indicated by the letter *I*

during the first two irrigation seasons, and higher than 16.0 dS m^{-1} during the third year, as applied water became more saline. In the plot irrigated with the locally available water (plot C), soil salinity remained below 5.0 dS m^{-1} throughout all irrigation seasons. Soil salinity decreased in all plots during rainfall periods, due to soil leaching. Only in the rainy season of 2004–2005, rainfall was not sufficient to completely remove salts from the root zone in plot A, where EC_{sw} values between 2.0 and 6.3 dS m^{-1} were observed in the soil profile (Ramos et al. 2011).

Figure 16.4 shows measured and simulated concentrations of soluble Na^+, Ca^{2+}, and Mg^{2+} for plots A and C at a depth of 40 cm. As the only cation being added to synthetic saline irrigation waters, the general behavior of sodium was similar to EC_{sw}. The highest concentrations were reached in plot A during the irrigation seasons. The highest measured Na^+ concentrations were about 50 $mmol_{(c)}L^{-1}$ during the first 2 years and 130 $mmol_{(c)}L^{-1}$ during the third year. During the rainy seasons, Na^+ concentrations decreased considerably due to soil leaching, similar to EC_{sw}.

The general dynamics of calcium and magnesium concentrations were similar to those of sodium and EC_{sw}. The highest measured Ca^{2+} and Mg^{2+} concentrations were about 20 and 10 $mmol_{(c)}L^{-1}$, respectively. During rainfall events, these values also decreased due to soil leaching. As Ca^{2+} and Mg^{2+} were not added to the synthetic saline irrigation waters applied in plot A, concentrations of these cations in saline waters were the same as in the fresh waters in plot C. Since Na^+ was applied in large concentrations to plot A, both Na^+ concentrations in the soil solution and in the solid phase (not shown) increased, leading to soil sodification. While the exchangeable Na^+ concentration in the solid phase increased, the other cations, namely Ca^{2+} and Mg^{2+}, were inevitably released to the soil solution. For this reason, concentrations of Ca^{2+} and Mg^{2+} were higher in plot A than in plot C where low concentrations of Na^+ were present in the irrigation waters. Since compared to simpler models based on adsorption isotherms, the UnsatChem module is able to consider cation exchange, it was able to simulate the release of Ca^{2+}

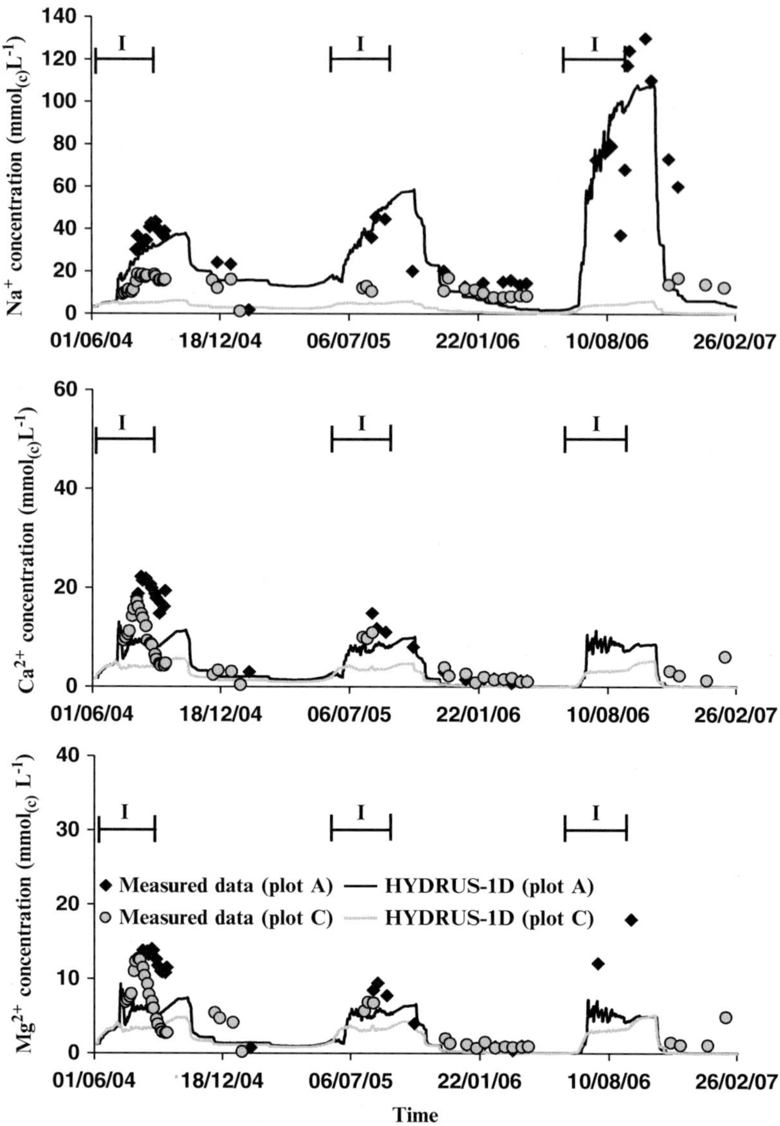

Fig. 16.4 Measured and simulated Na^+ (*top*), Ca^{2+} (*middle*), and Mg^{2+} (*bottom*) concentrations at a depth of 40 cm

and Mg^{2+} from the solid phase as soil sodification increased. The UnsatChem module thus proved to more adequately represent reality than linear models with adsorption isotherms.

Figure 16.5 presents the results for measured and simulated *SAR* values. *SAR* is an integral variable that characterizes salt-affected soils and provides information on comparative concentrations of Na^+, Ca^{2+}, and Mg^{2+} in soil solutions. This variable takes into consideration that the adverse effects of sodium are moderated

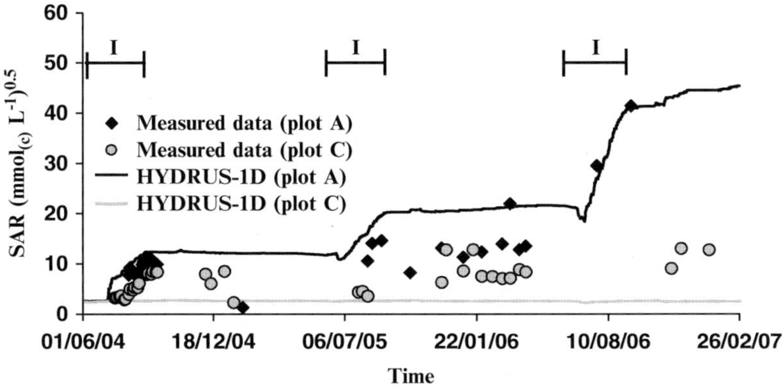

Fig. 16.5 Measured and simulated SAR at a depth of 40 cm

by the presence of calcium and magnesium ions. This makes *SAR* an important variable to consider when managing saline irrigation waters. In plot A, *SAR* increased rapidly in the surface layers after irrigation events, and then also gradually at deeper depths. The application of saline irrigation waters (*SAR* ranged from 38.8 to 80.2 $(\text{mmol}_{(c)}\text{L}^{-1})^{0.5}$) led to very high *SAR* values in the soil solution, reaching about 54 $(\text{mmol}_{(c)}\text{L}^{-1})^{0.5}$ at the end of the experiments. In plot C, irrigated with fresh waters, *SAR* values did not vary significantly, having roughly the same values in the beginning and at the end of the experiments.

Various statistical measures evaluating the goodness of correspondence between the model (the UnsatChem module of HYDRUS-1D) and measurements are given by Ramos et al. (2011) and will not be given here. The correspondence between measurements and model results is very good, considering that most model inputs were independently measured in the laboratory and used in simulations without any further adjustments and/or calibration. The correspondence between measurements and model results obviously could have been better, had the input parameters been calibrated (Ramos et al. 2011). An extensive discussion on the possible causes of deviations between measured data and simulation can also be found in Ramos et al. (2011, 2012).

16.3.3 Modeling the Fate of Mercury in Contaminated Soils Using HP1

The HP1 module, which is used in this example, couples the PHREEQC geochemical code (Parkhurst and Appelo 1999) with HYDRUS-1D (Šimůnek et al. 2008). The one-dimensional version (HP1) was first released in 2005 (Jacques and Šimůnek 2005; Jacques et al. 2008a, b), and used successfully in many applications. HP1, which is an acronym for HYDRUS-PHREEQC-1D, is a relatively comprehensive simulation module that can be used to simulate (1) transient water flow,

(2) the transport of multiple components, (3) mixed equilibrium/kinetic biogeochemical reactions, and (4) heat transport in one-dimensional variably-saturated porous media. HP2 is a corresponding two-dimensional version of HP1, which couples PHREEQC with HYDRUS (2D/3D). Both HP1 and HP2 modules can simulate a broad range of low-temperature biogeochemical reactions in water, the vadose zone and/or ground water systems, including interactions with minerals, gases, exchangers and sorption surfaces based on thermodynamic equilibrium, and kinetic or mixed equilibrium-kinetic reactions.

The example described below illustrates the possibility to simulate the fate of mercury (Hg) in anthropogenic contaminated soil systems (Leterme and Jacques 2013; Leterme et al. 2014).

Problem Definition

Mercury (Hg) poses threats to human health and the environment, notably due to its persistence in the environment and its ability to bioaccumulate in ecosystems (Liu et al. 2012). Mercury pollution can result from either direct contamination (e.g., spills, landfills, mine tailings) or indirect pathways, such as atmospheric deposition (Guédron et al. 2013). Mercury mining, gold and silver mining, manufacturing (chlor-alkali plants, manometer spill), wood preservation and cemeteries (through the release of Hg from dental amalgams) have been identified as the main sources of soil contamination (UNEP 2002). In addition, Hg geochemistry is quite complex (Fig. 16.6) and thus Hg contamination can occur in different forms: as aqueous inorganic and organic Hg^{II} species, in elemental form (Hg^0) in a (residual) non-aqueous liquid phase (NAPL), or as a solid phase (e.g., cinnabar). Hg transport is highly influenced by its strong binding to organic matter, especially to the thiol groups, which decreases (sorption to soil organic matter) or increases (complexation with dissolved organic matter) the mobility of mercury. Reduction of Hg^{II} to Hg^0 enhances the Hg transport through the gaseous phase, which is an important pathway in mercury-related risk assessment studies. Note that under reducing conditions, organic forms of mercury (monomethylmercury and dimethylmercury) may occur and may have a strong tendency for bioaccumulation in wetland systems.

Leterme et al. (2014) implemented the conceptual model presented in Fig. 16.6 into HP1. This model addresses Hg-contaminated soils under oxidizing conditions (organic forms of Hg are excluded). They investigated seven scenarios with initial conditions representing different types of contamination (cinnabar, residual NAPL, and aqueous Hg^{II}, and any combination of these three contamination sources) polluting the first 10 cm of the soil (Fig. 16.7). Only the first three scenarios are presented below. They also evaluated the sensitivity of four different indicators, illustrated in Fig. 16.7, to different parameters and processes.

Daily climatic data (50 years) at Dessel, Belgium, were used to define surface atmospheric boundary conditions and free drainage was considered as the bottom boundary condition (representing deep groundwater and oxidising conditions).

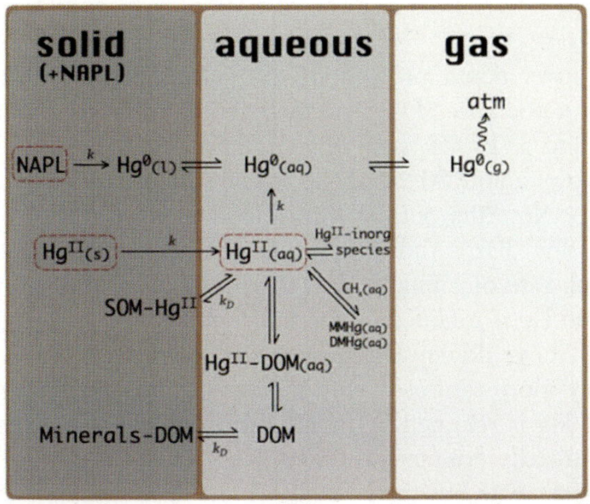

- - - -	Hg contamination source	MMHg	Monomethylmercury (CH_3Hg^+)
k	Kinetic reaction	DMHg	Dimethylmercury ($(CH_3)_2Hg$)
k_D	Sorption coefficient		

Fig. 16.6 Conceptual model of Hg speciation and reactions in soil systems (*DOM* dissolved organic matter, *SOM* soil organic matter, *MMHg* monomethylmercury, *DMHg* dimethylmercury)

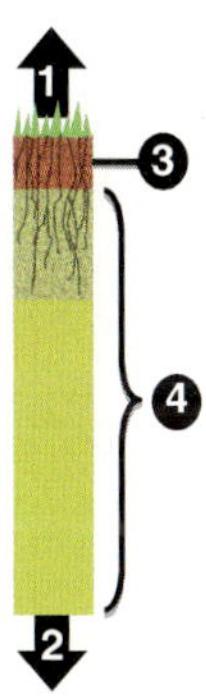

#	Indicator
1	Percentage of Hg volatilized to atmosphere
2	Percentage of Hg leached out of the soil profile
3	Percentage of Hg present in the originally-polluted soil horizon (top 10 cm)
4	Percentage of Hg present in the soil profile (10-100 cm) below the originally-polluted soil horizon

Fig. 16.7 Problem definition and evaluated indicators. Initial Hg contamination is in the first 10 cm of the soil (*red shaded*). The top 30 cm of the soil contains soil organic matter (*grey shaded*), the remaining part of the soil profile is inert

Incoming water has a typical rain water composition and contains dissolved organic matter (DOM). Soil texture is sandy (this a priori corresponds to coarse-textured soil types, likely to be encountered at polluted sites such as mine tailings) with a grass cover (30 cm rooting depth). Soil organic matter (SOM) and the initial Hg contamination are uniform in the top 30 and 10 cm of the soil profile, respectively. There is a background atmospheric Hg(g) concentration of 1 ng m^{-3}. Soil and dissolved organic matter are both represented by four different reactive sites (three representing oxygen sites of fulvic and humic acids and one representing thiols).

Results

Figure 16.8 shows depth profiles of the initial contamination source and Hg concentrations sorbed to SOM for three contamination scenarios and selected times (initial, and 5, 10, 25, and 50 years). Since the dissolution rate of cinnabar is relatively slow, about 50 % of the initial amount still remains after 50 years (Fig. 16.8, top). Note the non-uniform dissolution profile of cinnabar due to the dependence of the dissolution kinetics on DOM concentrations. Because of the slow dissolution rate of cinnabar and the presence of the strongly binding DOM thiols, almost no Hg is retained on the SOM in the first few centimetres of the soil. As DOM thiols move downward through the contaminated zone (top 10 cm), they are increasingly more saturated with mercury due to continued cinnabar dissolution, and more Hg is retained by SOM, resulting in a peak retention of Hg at the bottom of the initially polluted soil layer (Fig. 16.8, top).

In the scenario with residual NAPL (Fig. 16.8, middle), all elemental Hg is dissolved after about 10 years. Because of this gradual release, mercury is mostly retained within the top 10 cm during the first 25 years. Since the release of Hg from NAPL is fast enough to saturate the thiol groups of DOM, a significant portion of mercury is retained on SOM. After 25 years, the gradual leaching of mercury to the lower soil horizons is obvious.

In the scenario with initial aqueous contamination (Fig. 16.8, bottom), a large portion of Hg leaches very quickly (mainly in the form of inorganic species such as $Hg(OH)_2$, $HgCl_2$, HgOCl) from the top 10 cm and sorbs significantly to the SOM in the underlying 20 cm. Note that the "Initial" concentration profile is drawn after the first time step, which explains the "Initial" profile of Hg sorbed to SOM. Also from the start, all Hg is available for leaching via complexation on the thiols of DOM (although binding is much stronger on the thiols of DOM than of SOM, their capacity, i.e., the number of thiol sites, is very limited), resulting in a gradual leaching of Hg from the top 10 cm.

Figure 16.9 shows Hg^{II} fluxes from the top 1 m to underlying layers for the first 25 years of the simulations. A quick initial release of mobile Hg in the scenario with a $HgCl_2$(aq) contamination (a green line in Fig. 16.9) leads to earlier Hg leaching than in scenarios with cinnabar and NAPL contaminations (red and blue lines, respectively). In these two scenarios, Hg leaching occurs only after a few years, but remains at only about half of the Hg flux in the scenario with the $HgCl_2$(aq) contamination. This is because in the latter case Hg was rapidly redistributed in the top 30 cm (Fig. 16.8, bottom right) and less re-sorption occurs when it complexes with DOM thiols. Hg fluxes for scenarios with cinnabar and Hg NAPL are quite similar because the limiting factor for Hg mobility is not the Hg release through dissolution nor the sorption capacity of the soil solid phase, but rather the availability of DOM thiol groups. Figure 16.9 also shows that the temporal dynamics (not its absolute values) of leaching is driven by the climate.

The sensitivity analysis indicated that the most sensitive parameters are related to SOM and DOM. DOM concentration strongly influence the mobility and

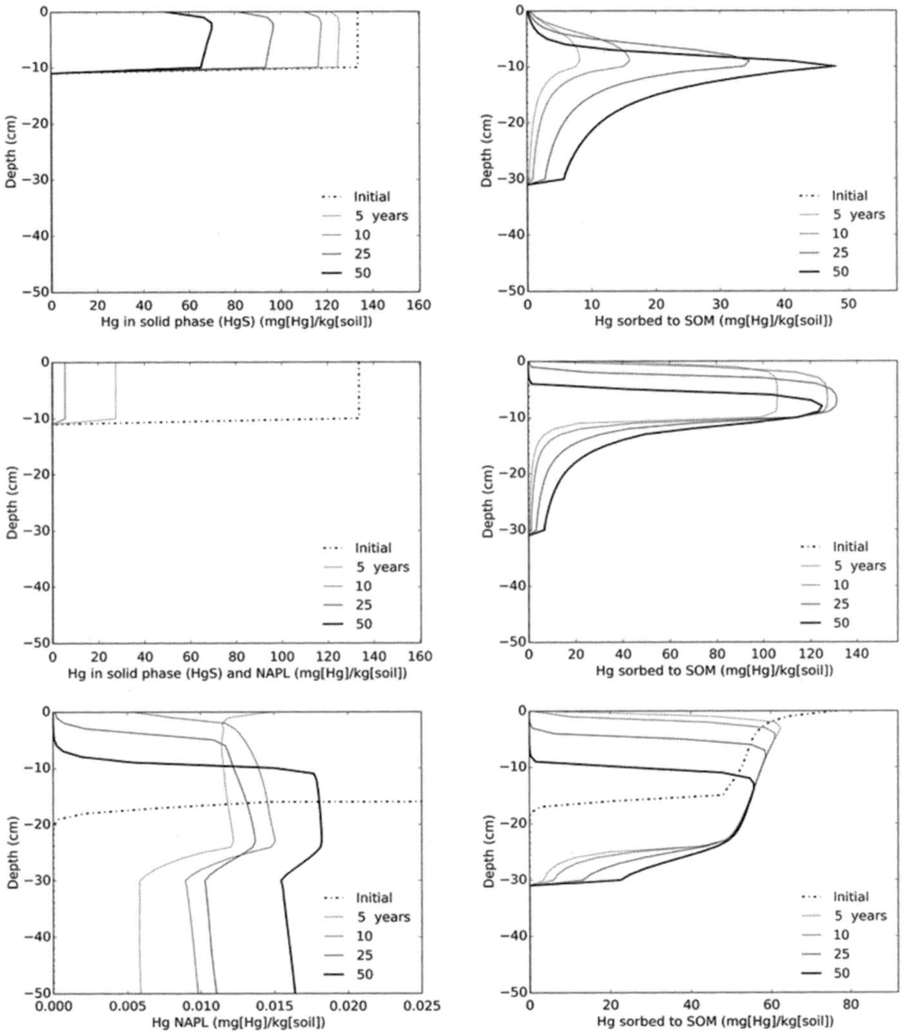

Fig. 16.8 Depth profiles at selected times for different forms of Hg for different types of initial contamination: cinnabar (*top*), residual NAPL (*middle*), and aqueous Hg species (*bottom*). The *left column* shows Hg concentrations of the original contamination, the *right columns* shows Hg sorbed to SOM. Note different scales of the x-axis

leaching of mercury due to the high affinity of Hg to DOM thiol groups. On the other hand, the exchange capacities and the selectivity coefficients of the fulvic and humic acids of SOM greatly influence the retention of mercury in the top soil. Note also that the selectivity coefficient of the thiol groups in SOM has only a minor effect because in anthropogenic mercury-contaminated sites, the thiol groups in SOM are initially saturated with mercury. The dissolution rate of the initial

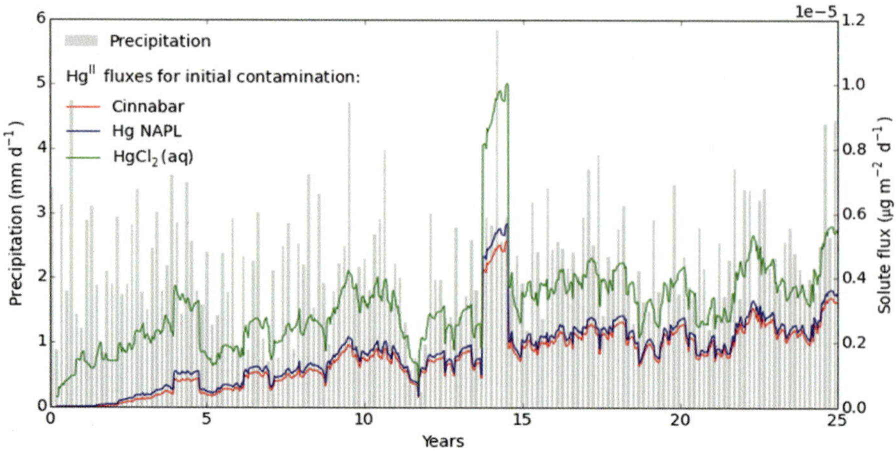

Fig. 16.9 Precipitation (*left axis*) and Hg^{II} fluxes (monthly moving averages) at the bottom of the soil profile for three initial contamination scenarios (*right axis*)

contamination is also a sensitive parameter as illustrated above in scenarios with cinnabar and residual NAPL.

16.3.4 Modeling of Uranium Leaching from a U Tailings Pile Using the HP2 Module of HYDRUS (2D/3D)

The HP2 is a two-dimensional equivalent of the one-dimensional HP1 program (Šimůnek et al. 2008). We will demonstrate below the versatility of HP2 using an example involving the release and migration of uranium from a simplified uranium mill tailings pile toward a river. This example includes the processes of water flow, solute transport, precipitation/dissolution of the solid phase, cation exchange, complexation, and many other reactions. This problem was inspired by, and is a modification of (to make it more realistic), a problem reported by Yeh and Tripathi (1991). Detailed information about this project, as well as instructions on how to implement the project using HP2, can be found in Šimůnek et al. (2012) and on the HYDRUS website.

Problem Description

The schematic of the transport domain is shown in Fig. 16.10. The mill tailings pile is located adjacent to a surface that slopes down to a river. The horizontal bottom of the region is impermeable. The vertical left-edge has the Dirichlet boundary

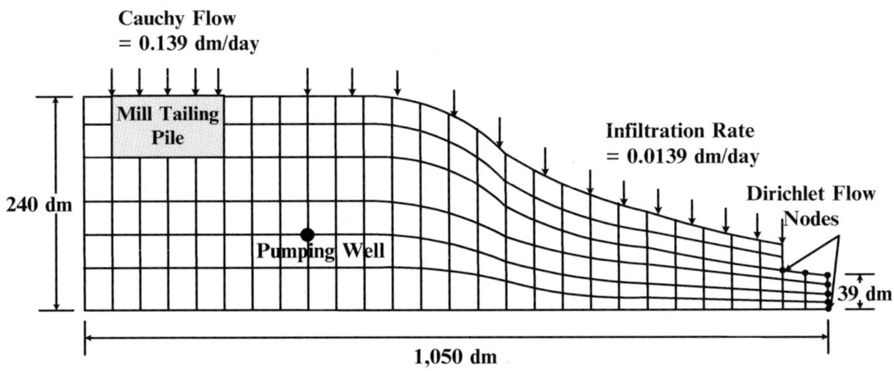

Fig. 16.10 Problem description for the uranium tailings problem (Adopted from Yeh and Tripathi 1991)

condition with a groundwater table 12 m above the bottom of the transport domain. The top boundary (except for the mill tailings pile and the river) has a flux boundary condition with a net rainfall rate of 0.139 cm/day. The horizontal region on the top of the mill tailings pile is a flux boundary with an infiltration rate of 1.39 cm/day. The nodes on the vertical line on the right side of the transport domain and the nodes on the river bottom have the Dirichlet boundary condition, reflecting the position of water in the river (4.5 m above the bottom of the transport domain). A hypothetical pumping well with a withdrawal rate of 271.58 cm^2/day is located at $(x, z) = (400, 100)$. The region is discretized using a structured finite element mesh with 1,564 elements and 852 nodes.

The medium has the hydraulic properties of a loam (the residual water content, $\theta_r = 0.078$, the saturated water content, $\theta_s = 0.430$, and the van Genuchten (1980) parameters $\alpha = 0.036$ cm^{-1}, and $n = 1.56$; Carsel and Parrish 1988) with the saturated hydraulic conductivity of $K_s = 3.78$ m/day. A longitudinal dispersivity of 2.5 m and a transverse dispersivity of 0.25 m are assumed.

With respect to solute transport, uranium was modeled as a complex multicomponent system involving multiple interacting components. Table 16.1 lists chemical reactions and their thermodynamic equilibrium constants considered in this example. For reactive hydrogeochemical transport, the problem consists of eight components: Total H, Total O, Ca, C, uranium, sulfate, phosphate, and Fe. A total of 35 aqueous species, 18 exchange species, 9 surface species and 2 minerals are defined for the problem; redox reactions were not considered. Precipitation/dissolution and sorption are the main chemical processes considered in this example. Gypsum (0.0037 mol/kg) and calcite (0.00047 mol/kg) are assumed to be present in the waste zone and the rest of the profile, respectively.

Sorption of U was described using a multi-site cation exchange model (see Jacques et al. 2008a). This sorption model also buffers the acid pH due to proton exchange. In the current model, protons will replace Ca in the exchange complex. Parameters of the multi-site cation exchange model are also given in Table 16.1 (adapted from Jacques

Table 16.1 Reaction network for the uranium tailings pile leaching problem

Reaction	No.	Log K[a]
Aqueous complexation reactions		
$H_2O \Leftrightarrow H^+ + OH^-$	(R1)	−14.00
$Ca^{2+} + CO_3^{2-} \Leftrightarrow$ CaCO3 (aq)	(R2)	3.22
$Ca^{2+} + H^+ + CO_3^{2-} \Leftrightarrow CaHCO_3^+$	(R3)	11.43
$Ca^{2+} + SO_4^{2-} \Leftrightarrow CaSO_4$ (aq)	(R4)	2.31
$Ca^{2+} + 2H^+ + PO_4^{3-} \Leftrightarrow CaH^2PO_4^+$	(R5)	20.96
$Ca^{2+} + PO_4^{3-} \Leftrightarrow CaPO_{4-}$	(R6)	6.46
$Ca^{2+} + H^+ + PO_4^{3-} \Leftrightarrow CaHPO_{4\,(aq)}$	(R7)	15.08
$Ca^{2+} + H_2O \Leftrightarrow H^+ + CaOH^+$	(R8)	−12.58
$Fe^{2+} + SO_4^{2-} \Leftrightarrow FeSO_{4\,(aq)}$	(R9)	2.20
$Fe^{2+} + H_2O \Leftrightarrow H^+ + FeOH^+$	(R10)	−9.50
$Fe^{2+} + 2H_2O \Leftrightarrow 2H^+ + Fe(OH)_{2\,(aq)}$	(R11)	−20.57
$Fe^{2+} + 3H_2O \Leftrightarrow 3H^+ + Fe(OH)_3^-$	(R12)	−31.00
$Fe^{2+} + 4H_2O \Leftrightarrow 4H^+ + Fe(OH)_4^{2-}$	(R13)	−43.00
$UO_2^{2+} + H_2O \Leftrightarrow H^+ + (UO_2)(OH)^+$	(R14)	−5.30
$2UO_2^{2+} + 2H_2O \Leftrightarrow 2H^+ + (UO2)_2(OH)_2^{2+}$	(R15)	−5.68
$3UO_2^{2+} + 4H_2O \Leftrightarrow 4H^+ + (UO2)_3(OH)_4^{2+}$	(R16)	−11.88
$3UO_2^{2+} + 5H_2O \Leftrightarrow 5H^+ + (UO2)_3(OH)_5^+$	(R17)	−15.82
$4UO_2^{2+} + 7H_2O \Leftrightarrow 7H^+ + (UO2)_4(OH)_7^+$	(R18)	−21.90
$3UO_2^{2+} + 7H_2O \Leftrightarrow 7H^+ + (UO_2)_3(OH)_7^-$	(R19)	−28.34
$UO_2^{2+} + CO_3^{2-} \Leftrightarrow (UO_2)(CO_3)_{(aq)}$	(R20)	9.65
$UO_2^{2+} + 2CO_3^{2-} \Leftrightarrow (UO_2)(CO_3)_2^{2-}$	(R21)	17.08
$UO_2^{2+} + 3CO_3^{2-} \Leftrightarrow (UO_2)(CO_3)_3^{4-}$	(R22)	21.70
$2UO_2^{2+} + CO_3^{2-} + 3H_2O \Leftrightarrow 3H^+ + (UO_2)_2(CO_3)(OH)_3^-$	(R23)	−1.18
$UO_2^{2+} + SO_4^{2-} \Leftrightarrow (UO_2)(SO_4)_{(aq)}$	(R24)	2.95
$UO_2^{2+} + 2SO_4^{2-} \Leftrightarrow (UO_2)(SO_4)_2^{2-}$	(R25)	4.00
$2H^+ + UO_2^{2+} + PO_4^{3-} \Leftrightarrow H_2(UO_2)(PO_4)^+$	(R26)	23.20
$3H^+ + UO_2^{2+} + PO_4^{3-} \Leftrightarrow H_3(UO_2)(PO_4)^{2+}$	(R27)	22.90
$Ca^{2+} + 4H^+ + UO_2^{2+} + 2PO_4^{3-} \Leftrightarrow CaH_4(UO_2)(PO_4)_2^{2+}$	(R28)	45.24
$Ca^{2+} + 5H^+ + UO_2^{2+} + 2PO_4^{3-} \Leftrightarrow CaH_5(UO_2)(PO_4)_2^{3+}$	(R29)	46.00
$H^+ + CO_3^{2-} \Leftrightarrow HCO_3^-$	(R30)	10.32
$2H^+ + CO_3^{2-} \Leftrightarrow H_2CO_{3(aq)}$	(R31)	16.67
$H^+ + SO_4^{2-} \Leftrightarrow HSO_4^-$	(R32)	1.99
$H^+ + PO_4^{3-} \Leftrightarrow HPO_4^{2-}$	(R33)	12.35
$2H^+ + PO_4^{3-} \Leftrightarrow H_2PO_4^-$	(R34)	19.55
$3H^+ + PO_4^{3-} \Leftrightarrow H_3PO4$	(R35)	21.74
Precipitation-dissolution reactions		
$Ca^{2+} + SO_4^{2-} \Leftrightarrow CaSO_{4(s)}$ (gypsum)	(R36)	4.62
$Ca^{2+} + CO_3^{2-} \Leftrightarrow CaCO_{3(s)}$ (calcite)	(R37)	8.48
Multi-site cation exchange model[b]		
$Ca^{2+} + 2\ Y_i^- = CaY_{i2}$	(R37)	−0.2
$UO_2^{2+} + 2\ Y_i^- = UO2Y_{i2}$	(R38)	−0.2
$H^+ + Y_a = HY_a$	(R39)	1.65
$H^+ + Y_b = HY_b$	(R40)	3.30
$H^+ + Y_c = HY_c$	(R41)	4.95
$H^+ + Y_d = HY_d$	(R42)	6.85
$H^+ + Y_e = HY_e$	(R43)	9.60
$H^+ + Y_f = HY_f$	(R44)	12.35

(continued)

Table 16.1 (continued)

Reaction	No.	Log K[a]
Specific sorption – surface complexation		
$Hfo\_OH + H^+ = Hfo\_OH_2^+$	(R45)	7.29
$Hfo\_OH = Hfo\_O^- + H^+$	(R46)	−8.93
$Hfo\_OH + Ca^{2+} = Hfo\_OCa^+ + H^+$	(R47)	−5.85
$Hfo\_OH + UO_2^{2+} = Hfo\_OUO_2^+ + H^+$	(R48)	2.8
$Hfo\_OH + PO_4^{3+} + 3H^+ = Hfo\_H_2PO_4 + H_2O$	(R49)	31.29
$Hfo\_OH + PO_4^{3+} + 2H^+ = Hfo\_HPO_4^- + H_2O$	(R50)	25.39
$Hfo\_OH + PO_4^{3+} + H^+ = Hfo\_PO_4^{2-} + H_2O$	(R51)	17.72
$Hfo\_OH + SO_4^{2-} + H^+ = Hfo\_SO_4^- + H_2O$	(R52)	7.78
$Hfo\_OH + SO_4^{2-} = Hfo\_OHSO_4^{-2}$	(R53)	0.79

[a]K is a thermodynamic equilibrium constant
[b]Y_i represents one of the six different exchange sites (Y_a, Y_b, Y_c, Y_d, Y_e, Y_f). Exchange coefficients are the same for each exchange site for Ca^{2+} and UO_2^{2+}, but are different for H^+

Table 16.2 Initial and boundary compositions of recharge water and pore water in the tailings and regions outside of the tailings for the uranium tailings problem (mol/l)

Species	Inside the tailings	Outside the tailings
Ca^{2+}	1.14×10^{-2}[a]	1.42×10^{-2}[a]
CO_3^{2-}	1.0×10^{-2}	1.5×10^{-3}
UO_2^{2+}	5.0×10^{-4}	1.0×10^{-7}
PO_4^{3-}	1.0×10^{-6}	7.5×10^{-7}[b]
SO_4^{2-}	5.33×10^{-2}[c]	1.35×10^{-2}[d]
Fe^{2+}	3.5×10^{-2}	1.0×10^{-7}
pH	2.3	7.14[c]

[a]Calculated in equilibrium with calcite
[b]Calculated slightly under-saturated with hydroxyapatite (the saturation index of −0.1)
[c]Used as charge balance
[d]Calculated slightly under-saturated with gypsum (the saturation index of −0.1)

et al. 2008a). A relatively low (8.1×10^{-3} mol/kg) sorption capacity and a bulk density of 1.5 g/cm^3 were considered. Additionally, specific sorption of U and other cations and anions on Fe-oxides were described using a non-electrostatic surface complexation model with a capacity of 8.1×10^{-4} mol/kg. Parameters of this non-electrostatic surface complexation model are also given in Table 16.1.

Two different water compositions were considered: (i) one solution composition was considered for pore water and recharge water in the tailings, and (ii) the second solution composition was used for pore water and recharge water outside of the tailings pile and for recharge water at the left hand side of the domain. Both water types are given in Table 16.2. Cauchy boundary conditions were considered on all boundaries where boundary conditions are specified. A total of 1,000 days was simulated.

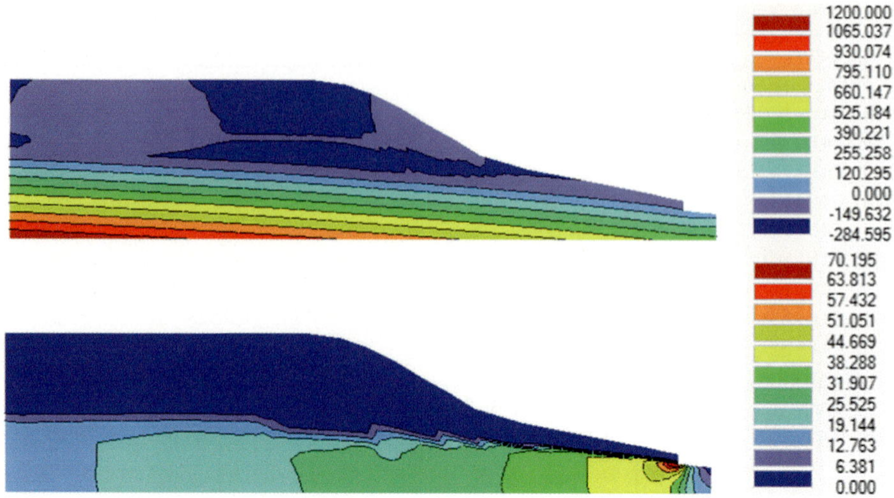

Fig.16.11 The steady-state pressure head (cm) (*top*) and flux (cm/d) (*bottom*) profiles (at 1,000 days) for the tailings pile leaching example

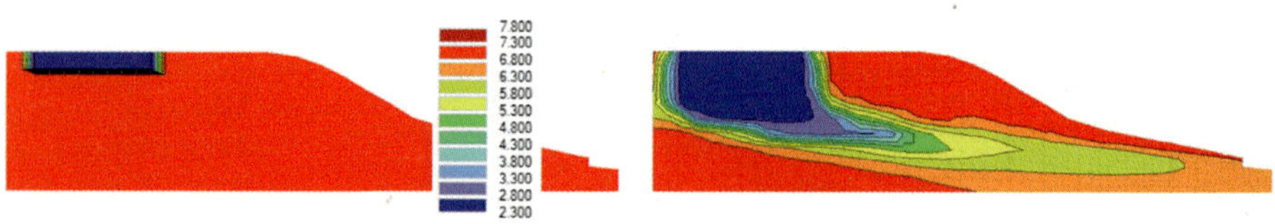

Fig. 16.12 Spatial distribution of pH initially (*left*) and at 1,000 days (*right*) for the tailings pile leaching example

Results

The steady-state pressure heads and velocity fields are depicted in Fig. 16.11. Figure 16.11 shows that water fluxes in the groundwater (about 20-50 cm/day) were significantly larger than in the vadose zone (below 1.5 cm/day).

Figure 16.12 shows contour plots of the initial (left) and final (right) distributions of pH. Acidity was initially confined to the tailings pile and was larger than 7 outside of the tailings pile. The proton exchange on the cation exchanger buffered the acidity by replacing calcium with protons on the exchanger. The spatial extent of the pH-altered region extended all the way to the river.

Figure 16.13 depicts contour plots of precipitated carbonate (Calcite, $CaCO_{3(s)}$) and sulfate (Gypsum, $CaSO_{4(s)}$) at 1,000 days. $CaCO_{3(s)}$, which was initially present everywhere outside of the tailings, is dissolved completely in the zone where pH is lower than about 5.5 (see Fig. 16.12). The influence of the acidity on calcite precipitation is present almost in the entire domain, except for the upper right and bottom left parts of the domain. In these zones, the geochemistry was governed by incoming rain water (upper part) or ground water (bottom left). Dissolution of

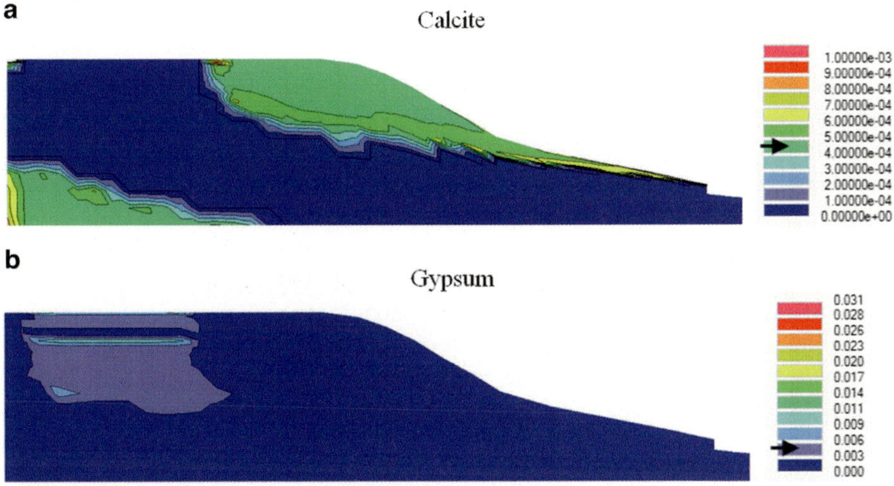

Fig. 16.13 Spatial distribution of calcite (**a**) and gypsum (**b**) at 1,000 days for the tailings pile leaching example. *Arrows* indicate the color representing the initial conditions for calcite in the soil region or for gypsum in the tailings pile

Fig. 16.14 Spatial distribution of aqueous U concentration (mol/l) at 1,000 days

Fig. 16.15 Spatial distribution of sorbed U concentration (mol/dm^3) at 1,000 days

calcite and the abundance of sulphate in tailings water led to gypsum precipitation. Gypsum, which was initially present only inside of the tailings, precipitated in the domain where groundwater encountered the leachate from the tailings.

Figures 16.14 and 16.15 show dissolved and sorbed uranium concentration profiles, respectively, as calculated using HP2 (i.e., using complex biogeochemical calculations). U migration within the ground water system is considerably retarded. U sorption seems to coincide quite well with the downstream edge of the aqueous U plume.

16.4 Conclusions

In spite of the considerable demand on input data, the multicomponent solute transport models such as HYDRUS, UnsatChem, or HP1/2, can be effective and versatile tools that are very useful for evaluating complex agricultural and environmental problems, such as irrigation management in regions with scarce water resources, where suitable waters are not always available for irrigation or for evaluating the leaching of acidic waters from mine tailings. We have demonstrated the use of the UnsatChem and HP2 modules of the HYDRUS software for these two types of applications, respectively. These models, after proper calibration and validation, should be considered useful tools for establishing sound irrigation and environmental policies to mitigate various problems such as soil salinization/ sodification or point and non-point source pollution from mining operations. Additional examples of the use of the complex hydrobiogeochemical transport models can be found at (and downloaded from) the HYDRUS website.

References

Abrams RH, Loague K, Kent DB (1998) Development and testing of a compartmentalized reaction network model for redox zones in contaminated aquifers. Water Resour Res 34:1531–1541

Ahuja LR, Hebson C (1992) Root zone water quality model. GPSR Tech. Rep. No. 2. USDA, ARS, Fort Collins

Bea SA, Wilson SA, Mayer KU, Dipple GM, Power IM, Gamazo P (2012) Reactive transport modeling of natural carbon sequestration in ultramafic mine tailings. Vadose Zone J 11(2). doi:10.2136/vzj2011.0053

Bell LSJ, Binning PJ (2004) A split operator approach to reactive transport with the forward particle tracking Eulerian-Lagrangian localized adjoint method. Adv Water Resour 27:323–334

Brookins DG (1984) Geochemical aspects of radioactive waste disposal. Springer, New York

Carsel RF, Parrish RS (1988) Developing joint probability distributions of soil water retention characteristics. Water Resour Res 24:755–769

Centler F, Shao H, De Biase C, Park C-H, Regnier P, Kolditz O, Thullner M (2010) GeoSysBRNS – a flexible multidimensional reactive transport model for simulating biogeochemical subsurface processes. Comput Geosci 36:397–405

Chang H, Singer JH, Seaman JC (2012) In situ chromium(VI) reduction using iron(II) solutions: modeling dynamic geochemical gradients. Vadose Zone J 11(2). doi:10.2136/vzj2011.0172

Corwin DL, Rhoades JD, Šimůnek J (2007) Leaching requirement for soil salinity control: steady-state vs. transient-state models. Agric Water Manage 90(3):165–180

Essaid HI, Bekins BA, Godsy EM, Warren E, Baedecker MJ, Cozzarelli IM (1995) Simulation of aerobic and anaerobic biodegradation processes at a crude oil spill site. Water Resour Res 31(12):3309–3327

Fryar AE, Schwartz FW (1994) Modeling the removal of metals from groundwater by a reactive barrier: experimental results. Water Resour Res 30(12):3455–3469

Gerke HH, Molson JW, Frind EO (1998) Modelling the effect of chemical heterogeneity on acidification and solute leaching in overburden mine spoils. J Hydrol 209:166–185

Gherardi F, Audigane P, Gaucher EC (2012) Predicting long-term geochemical alteration of wellbore cement in a generic geological CO_2 confinement site: tackling a difficult reactive transport modeling challenge. J Hydrol 420–421:340–359

Gonçalves MC, Šimůnek J, Ramos TB, Martins JC, Neves MJ, Pires FP (2006) Multicomponent solute transport in soil lysimeters irrigated with waters of different quality. Water Resour Res 42:W08401, 17 pp. doi:10.1029/2006WR004802

Gu CH, Maggi F, Riley WJ, Hornberger GM, Xu T, Oldenburg CM, Spycher N, Miller NL, Venterea RT, Steefel C (2009) Aqueous and gaseous nitrogen losses induced by fertilizer application. J Geophys Res 114, G01006

Guédron S, Grangeon S, Jouravel G, Charlet L, Sarret G (2013) Atmospheric mercury incorporation in soils of an area impacted by a chlor-alkali plant (Grenoble, France): contribution of canopy uptake. Sci Total Environ 445–446:356–364

Hutson JL, Wagenet RJ (1992) LEACHM: leaching estimation and chemistry model. Research series no. 92-3, Cornell University, Ithaca

Jacques D, Šimůnek J (2005) User manual of the multicomponent variably-saturated flow and transport model HP1, description, verification and examples, version 1.0, SCK•CEN-BLG-998, Waste and disposal. SCK•CEN, Mol, 79 pp

Jacques D, Šimůnek J, Mallants D, van Genuchten MT (2006) Operator-splitting errors in coupled reactive transport codes for transient variably saturated flow and contaminant transport in layered soil profiles. J Contam Hydrol 88:197–218

Jacques D, Šimůnek J, Mallants D, van Genuchten MT (2008a) Modeling coupled hydrological and chemical processes: long-term uranium transport following mineral phosphorus fertilization. Vadose Zone J 7(2):698–711. doi:10.2136/VZJ2007.0084

Jacques D, Šimůnek J, Mallants D, van Genuchten MT (2008b) Modelling coupled water flow, solute transport and geochemical reactions affecting heavy metal migration in a podzol soil. Geoderma 145:449–461

Jacques D, Šimůnek J, Mallants D, van Genuchten MTh (2013) The HPx reactive transport models: summary of recent developments and applications. In: Šimůnek J, van Genuchten MTh, Kodešová R (eds) Proceedings of the 4th international conference HYDRUS software applications to subsurface flow and contaminant transport problems. Invited paper, 21–22 Mar 2013. Department of Soil Science and Geology, Czech University of Life Sciences, Prague. ISBN: 978-80-213-2380-3, pp 7–16

Jansson P-E, Karlberg L (2001) Coupled heat and mass transfer model for soil-plant-atmosphere systems. Royal Institute of Technology, Department of Civil and Environmental Engineering, Stockholm, 325 pp

Kosakowski G, Watanabe N (2014) OpenGeoSys-Gem: a numerical tool for calculating geochemical and porosity changes in saturated and partially variable saturated media. Phys Chem Earth (in press). doi:http://dx.doi.org/10.1016/j.pce.2013.11.008

Langergraber G, Šimůnek J (2005) Modeling variably-saturated water flow and multi-component reactive transport in constructed wetlands. Vadose Zone J 4(4):924–938

Langergraber G, Šimůnek J (2012) Reactive transport modeling of subsurface flow constructed wetlands. Vadose Zone J 11(2). doi:10.2136/vzj2011.0104, 14 pp

Leij FJ, Bradford SA (1994) 3DADE: a computer program for evaluating three-dimensional equilibrium solute transport in porous media. Research report no. 134. U. S. Salinity Laboratory, USDA, ARS, Riverside

Leterme B, Jacques D (2013) Modeling Hg reactive transport in soil system using HP1. In: Šimůnek J, van Genuchten MTh, Kodešová R (eds) Proceedings of the 4th international conference HYDRUS software applications to subsurface flow and contaminant transport problems. Invited paper, 21–22 Mar 2013. Department of Soil Science and Geology, Czech University of Life Sciences, Prague. ISBN: 978-80-213-2380-3, pp 225–234

Leterme B, Blanc P, Jacques D (2014) A reactive transport model for mercury fate in soils – application to different anthropogenic pollution sources. Soil Pollut (in review)

Lichtner PC (1996) Continuum formulation of multicomponent-multiphase reactive transport. In: Lichtner PC, Steefel CI, Oelkers EH (eds) Reactive transport in porous media, reviews in mineralogy, Chapter 1, vol 34. Mineralogical Society of America, Washington, DC, pp 1–81

Lichtner PC, Hammond GE (2012) Using high performance computing to understand roles of labile and nonlabile uranium(VI) on Hanford 300 area plume longevity. Vadose Zone J 11(2). doi:10.2136/vzj2011.0097

Lim M-S, Yeo IW, Clement TP, Roh Y, Lee K-K (2007) Mathematical model for predicting microbial reduction and transport of arsenic in groundwater systems. Water Res 41:2079–2088

Liu G, Cai Y, O'Driscoll N, Feng X, Jiang G (2012) Overview of mercury in the environment. In: Liu G, Cai Y, O'Driscoll N (eds) Environmental chemistry and toxicology of mercury. Chantilly, VA, pp 1–12

Maggi F, Gu C, Riley WJ, Venterea R, Hornberger GM, Venterea RT, Xu T, Spycher N, Steefel C, Miller NL, Oldenburg CM (2008) A mechanistic treatment of the dominant soil nitrogen cycling processes: model development, testing, and application. J Geophys Res Biogeosci 113, G02016. doi:10.1029/2007JG000578

Mangold DC, Chin-Fu Tsang (1991) A summary of subsurface hydrological and hydrochemical models. Rev Geophys 29(1):51–79

Martens E, Jacques D, Van Gerven T, Wang L, Mallants D (2010) Geochemical modelling of leaching of Ca, Mg, Al, and Pb from cementitious waste forms. Cement Concr Res 40(8):1298–1305

Martens E, Jacques D, Wang L, De Cannière P, Moors H, Van Gompel M, Mariën A, Valcke E (2011) Modelling of cation concentrations in the outflow of $NaNO_3$ percolation experiments through boom clay cores. Phys Chem Earth 36:1693–1699

Mayer KU, Frind EO, Blowes DW (2002) Multicomponent reactive transport modeling in variably saturated porous media using a generalized formulation for kinetically controlled reactions. Water Resour Res 38:1174. doi:10.1029/2001WR000862

Narasimhan TN, White AF, Tokunaga T (1986) Groundwater contamination from an inactive uranium mill tailings pile, 2. Application of a dynamic mixing model. Water Resour Res 22(13):1820–1834

Parkhurst DL, Appelo CAJ (1999) User's guide to PHREEQ C (version 2) – a computer program for speciation, batch-reaction, one-dimensional transport and inverse geochemical calculations, water-resources investigations. Report 99–4259, Denver, 312 pp

Parkhurst DL, Kipp KL, Engesgaard P, Charlton SC (2004) PHAST – a program for simulating ground-water flow, solute transport and multicomponent geochemical reactions. U.S. Geol Surv Tech Methods 6-A8, 154 pp

Parton WJ, Schimel DS, Cole CV, Ojima DS (1987) Analysis of factors controlling soil organic matter levels in Great Plains grasslands. Soil Sci Soc Am J 51:1173–1179

Ramos TB, Šimůnek J, Gonçalves MC, Martins JC, Prazeres A, Castanheira NL, Pereira LS (2011) Field evaluation of a multicomponent solute transport model in soils irrigated with saline waters. J Hydrol 407(1–4):129–144

Ramos TB, Šimůnek J, Gonçalves MC, Martins JC, Prazeres A, Pereira LS (2012) Two-dimensional modeling of water and nitrogen fate from sweet sorghum irrigated with fresh and blended saline waters. Agric Water Manage 111:87–104

Rasouli F, Pouya AK, Šimůnek J (2013) Modeling the effects of saline water use in wheat-cultivated lands using the UNSATCHEM model. Irrig Sci 31(5):1009–1024. doi:10.1007/s00271-012-0383-8

Reading LP, Baumgartl T, Bristow KL, Lockington DA (2012) Applying HYDRUS to flow in a sodic clay soil with solution composition–dependent hydraulic conductivity. Vadose Zone J 11 (2). doi:10.2136/vzj2011.0137

Regnier P, Dale AW, Pallud C, van Lith Y, Bonneville S, Hyacinthe C, Thullner M, Laverman AM, Van Cappellen P (2005) Incorporating geomicrobial processes in reactive transport models of subsurface environments. In: Nützmann G, Viotti P, Aagaard P (eds) Reactive transport in soil and groundwater. Springer, Berlin Heidelberg, pp 109–125

Rittmann BE, VanBriesen JM (1996) Microbiological processes in reactive modeling. In: Lichtner PC, Steefel CI, Oelkers EH (eds) Reactive transport in porous media, reviews in mineralogy, Chapter 7, vol 34. Mineralogical Society of America, Chantilly, VA, pp 311–334

Schoups G, Hopmans JW, Tanji KK (2006) Evaluation of model complexity and space–time resolution on the prediction of long-term soil salinity dynamics, western San Joaquin Valley, California. Hydrol Process 20:2647–2668

Seaman JC, Chang H, Goldberg S, Šimůnek J (2012) Reactive transport modeling. Vadose Zone J 11(2). doi:10.2136/vzj2012.0066, 7 pp

Šimůnek J (2005) Chapter 78: models of water flow and solute transport in the unsaturated zone. In: Anderson MG, McDonnell JJ (eds) Encyclopedia of hydrological sciences. Wiley, Chichester, pp 1171–1180

Šimůnek J, Suarez DL (1994) Two-dimensional transport model for variably saturated porous media with major ion chemistry. Water Resour Res 30(4):1115–1133

Šimůnek J, Suarez DL (1997) Sodic soil reclamation using multicomponent transport modeling. ASCE J Irrig Drain Eng 123(5):367–376

Šimůnek J, Valocchi AJ (2002) Geochemical transport. In: Dane JH, Topp GC (eds) Methods of soil analysis, part 4, physical methods, Chapter 6.9, 3rd edn. SSSA, Madison, pp 1511–1536

Šimůnek J, van Genuchten MT (2006) Contaminant transport in the unsaturated zone: theory and modeling, Chapter 22. In: Delleur J (ed) The handbook of groundwater engineering, 2nd edn. CRC Press, Boca Raton, pp 22.1–22.46

Šimůnek J, van Genuchten MTh, Šejna M, Toride N, Leij FJ (1999) The STANMOD computer software for evaluating solute transport in porous media using analytical solutions of convection-dispersion equation. Versions 1.0 and 2.0, IGWMC – TPS – 71. International Ground Water Modeling Center, Colorado School of Mines, Golden, 32 pp

Šimůnek J, van Genuchten MT, Sejna M (2008) Development and applications of the HYDRUS and STANMOD software packages, and related codes. Vadose Zone J 7(2):587–600. doi:10.2136/VZJ2007.0077

Šimůnek J, Jacques D, Šejna M, van Genuchten MTh (2012) The HP2 program for HYDRUS (2D/3D): a coupled code for simulating two-dimensional variably-saturated water flow, heat transport, and biogeochemistry in porous media, version 1.0, PC Progress, Prague, 76 pp

Šimůnek J, Jacques D, Langergraber G, Bradford SA, Šejna M, van Genuchten MTh (2013) Numerical modeling of contaminant transport with HYDRUS and its specialized modules, invited paper for the special issue "water management in changing environment", Mohan Kumar MS (ed) J Indian Inst Sci 93(2) 265–284. ISSN: 0970-4140 Coden-JIISAD

Skaggs TH, Shouse PJ, Poss JA (2006) Irrigation of forage crops with saline drainage waters: 2. Modeling drainage and root water uptake. Vadose Zone J 5:824–837

Steefel CI (2000) New directions in hydrogeochemical transport modeling: incorporating multiple kinetic and equilibrium pathways. In: Bentley LR, Sykes JF, Brebbia CA, Gray WG, Pinder GF (eds) Computational methods in water resources XIII. A.A. Balkema, Rotterdam, pp 331–338

Steefel CI (2009) CrunchFlow: software for modeling multicomponent reactive flow and transport, user's manual, 12 Oct 2009

Steefel CI, MacQuarrie KTB (1996) Approaches to modeling of reactive transport in porous media. In: Lichtner PC, Steefel CI, Oelkers EH (eds) Reactive transport in porous media. Reviews in mineralogy, Chapter 2, vol 34. Mineralogical Society of America, Chantilly, VA, pp 83–129

Steefel CI, DePaolo DJ, Lichtner PC (2005) Reactive transport modeling: an essential tool and a new research approach for the earth sciences. Earth Planet Sci Lett 240:539–558

Thaysen EM, Jacques D, Jessen S, Andersen CE, Laloy E, Ambus P, Postma D, Jakobsen I (2014a) Controls on carbon dioxide fluxes across the unsaturated zone of cropped and unplanted soil mesocosms. Biogeosciences (in review)

Thaysen EM, Jessen S, Postma D, Jakobsen R, Jacques D, Ambus P, Laloy E, Jakobsen I (2014b) Plant-mediated transfer of CO2 to aquifers as influenced by crushed concrete waste and lime amendments. Environ Sci Technol (in review)

Thullner M, Van Cappellen P, Regnier P (2005) Modeling the impact of microbial activity on redox dynamics in porous media. Geochim Cosmochim Acta 69:5005–5019

Toride N, Leij FJ, van Genuchten MT (1993) A comprehensive set of analytical solutions for nonequilibrium solute transport with first-order decay and zero-order production. Water Resour Res 29:2167–2182

United Nations Environment Programme (UNEP) (2002) Global mercury assessment. United Nations Environment Programme (UNEP), Geneva

VanBriesen J (1998) Modeling coupled biogeochemical processes in mixed waste systems, Ph.D. thesis, Civil Engineering. Northwestern University

Van der Lee J, De Windt L, Lagneau V, Goblet P (2003) Module-oriented modeling of reactive transport with HYTEC. Comput Geosci 29:265–275

van Genuchten MTh (1980) A closed-form equation for predicting the hydraulic conductivity of unsaturated soils. Soil Sci Soc Am J 44:892-898

van Genuchten MTh (1981) Analytical solutions for chemical transport with simultaneous adsorption, zero-order production and first-order decay. J Hydrol 49:213-233

van Genuchten MT, Šimůnek J, Leij FL, Toride N, Sejna M (2012) STANMOD: model use, calibration and validation, special issue standard/engineering procedures for model calibration and validation. Trans ASABE 5(4):1353–1366

van Genuchten MTh et al (2014) The use of numerical flow and transport models in environmental analyses. In: Teixeira WG, Ceddia MB, Ottoni MV, Donnagema GK (eds) Application of soil physics in environmental analyses: measuring, modelling and data integration. Springer, Cham

Viswanathan HS, Robinsons BA, Valocchi AJ, Triay IR (1998) A reactive transport model of neptunium migration from potential repository at Yucca Mountain. J Hydrol 209:251–280

Walsh SDC, Frane WLD, Mason HE, Carroll SA (2013) Permeability of wellbore-cement fractures following degradation by carbonated brine. Rock Mech Rock Eng 46:455–464

Walter AL, Frind EO, Blowes DW, Ptacek CJ, Molson JW (1994) Modeling of multicomponent reactive transport in groundwater 1. Model development and evaluation. Water Resour Res 30 (11):3137–3148

Xu T, Spycher N, Sonnenthal E, Zheng L, Pruess K (2012) TOUGHREACT user's guide: a simulation program for non-isothermal multiphase reactive geochemical transport in variably saturated geologic media, version 2.0. Report LBNL-DRAFT. Lawrence Berkeley National Laboratory, Berkeley

Yeh GT (1981) Analytical transient one-, two-, and three-dimensional simulation of waste transport in the aquifer system, ORNL-5602. Oak Ridge National Laboratory, Oak Ridge

Yeh G-T, Cheng H-P (1999) 3DHYDROGEOCHEM: a 3-dimensional model of density-dependent subsurface flow and thermal multispecies-multicomponent HYDROGEOCHEMical transport. EPA/600/R-98/159, 150 pp

Yeh GT, Tripathi VS (1991) A model for simulating transport of reactive multispecies components: model development and demonstration. Water Resour Res 27(12):3075–3094

Yuan F, Meixner T, Fenn ME, Šimůnek J (2011) Impact of transient soil water simulation on estimated nitrogen leaching and emission at high- and low-deposition forest sites in Southern California. J Geophys Res 116:G03040. doi:10.1029/2011JG001644, 15 pp

Zhang H, Nordin NA, Olsen MS (2013) Evaluating the effects of variable water chemistry on bacterial transport during infiltration. J Contam Hydrol 150:54–64

Zyvoloski GA, Robinson BA, Dash ZV, Trease LL (1997) Summary of the models and methods for the FEHM application – a finite element heat- and mass-transfer code. Los Alamos National Laboratory Rept. LA-13307-MS, Los Alamos

Chapter 17
Hydraulic Properties and Non-equilibrium Water Flow in Soils

Wolfgang Durner, Efstathios Diamantopoulos,
Sascha C. Iden, and Benedikt Scharnagl

Abstract Accurate knowledge of hydraulic properties for unsaturated soils is critical in the estimation of soil water fluxes by simulation models that are based on the Richards equation. The purpose of this chapter is to review the characterization of unsaturated soil hydraulic properties for their applicability in models simulating unsaturated water transport. We start with a short review of the fundamentals that lead to the definition of the hydraulic functions in the framework of continuum hydromechanics. Next, we address problems of common parameterizations of hydraulic functions in the critical regions near and at saturation, towards dryness, and on hysteresis. We find that traditional approaches have deficiencies, but recent progress has been significant in particular with respect to including film and corner flow components in the hydraulic conductivity function. The chapter closes with a discussion of the phenomenon of dynamic non-equilibrium in soil water flow, which shows to our opinion toward the need for a paradigm change in the modeling of soil water transport.

Keywords Soil hydrology • Unsaturated zone • Vadose zone • Hydraulic properties • Hydraulic functions • Hydraulic conductivity • Retention characteristic • Unimodal • Bimodal • Parametric models • Hysteresis • Dynamic effects • Spatial variability • Effective properties • Upscaling

17.1 Modeling Soil Water Flow with the Richards Equation

Soil moisture refers to the water present in the uppermost part of a field soil and affects almost any ecological, hydrological, geotechnical, and meteorological process (Romano 2014). Simulation of soil water dynamics under variably-saturated

W. Durner (✉) • E. Diamantopoulos • S.C. Iden • B. Scharnagl
Institute for Geoecology, Soil Science and Soil Physics, Technische Universität Braunschweig, Langer Kamp 19c, 38106 Braunschweig, Germany
e-mail: w.durner@tu-bs.de

W.G. Teixeira et al. (eds.), *Application of Soil Physics in Environmental Analyses: Measuring, Modelling and Data Integration*, Progress in Soil Science,
DOI 10.1007/978-3-319-06013-2_17,

conditions is therefore of crucial importance for a variety of disciplines such as hydrology, ecology, environmental sciences, soil science, and agriculture. Modeling water transport in soils is most frequently based on the conceptual model of Richards (1931). In this model, water flow in a porous medium is described in the framework of continuum mechanics (Cushman 1984) by a partial differential equation that combines a volume conservation equation for water in porous media with an equation for pressure-gradient driven water flux. The volume conservation equation relates the temporal changes of water content at a point to the divergence of the macroscopic water flux. In one-dimensional (1D) form this reads:

$$\frac{\partial \theta}{\partial t} = -\frac{\partial q}{\partial z} \tag{17.1}$$

where θ is volumetric water content (L^3L^{-3}), q is volumetric water flux density ($L^3L^{-2}\,T^{-1}$), t is time (T) and z is the spatial coordinate (L). θ and q are defined as averages within a representative elementary volume (REV) of the porous medium (Bear 1972). Equation 17.1 is valid for a rigid pore space, i.e., no system-dependent or temporal changes of the spatial configuration of the pore space and the physico-chemical interactions at water-solid interfaces and for water as an incompressible fluid. The rigidity assumption limits the applicability of Eq. 17.1 for swelling or shrinking soils, or for soils that change their structure due to natural forcings (seasonal weather, frost, vegetation) or anthropogenic interventions (soil tillage). However, if temporal changes are slow as compared to the water movement, they can be considered by a temporal adjustment of the effective hydraulic properties. Considering total masses and fluxes, θ and q must include liquid and gas phase components. The vapor component in θ can be safely neglected. In standard modeling, also the water flux q is often reduced to liquid water flux, and vapor flux is ignored. This might be problematic in drying situations, as discussed in Sect. 2.3.

Water flux in saturated and unsaturated soils is driven by spatial gradients of the water potential. In formulating the Richards equation, the mobility of air as complementary part of water in the pore space is considered to be unrestricted at any system state. Thus, the two-phase flow problem is reduced to single flow which can be expressed by the Darcy-Buckingham law (Darcy 1856; Buckingham 1907). In 1D notation, it is given by

$$q = -K(\theta)\frac{\partial H}{\partial z} \tag{17.2}$$

where H [L] is the head of the total water potential and K is the hydraulic conductivity function (LT^{-1}), which depends strongly non-linearly on water content θ or pressure head h (L). In general terms, the conductivity must be expressed as a tensor with direction-dependent components (see van Genuchten, Chap. 15), but here we stick to a 1D notation, which allows to reduce it to a scalar function. Note that throughout this paper we will address the water potential as energy per weight and express it as "pressure head", which is negative for unsaturated conditions and decreases when a soil becomes drier. In most practical applications of

Eq. 17.2, gradients in the thermal, osmotic, and pneumatic components of the water potential are considered negligible. Under these conditions, H reduces to the hydraulic potential $H = h + h_z$, where h [L] represents the matric potential and h_z [L] the gravitational potential. Schmid et al. (2014) note that chemical gradients could be important drivers of multi-phase flow, in particular, in porous media with micro-porosity and should be included in flow models, which is currently not the case. A further requirement for the validity of Eq. 17.2 is that flow in the porous medium takes place as viscous laminar flow. This assumption might be violated in very small pores (e.g. in pure clays), where water flux may cease despite finite potential gradients, or in macropores, i.e. voids in the size range of mm, where turbulent flow might occur.

Combination of Eqs. 17.1 and 17.2 and replacing $\partial\theta/\partial t = (\partial\theta/\partial h)\cdot(\partial h/\partial t)$ leads to the pressure-head based form of the Richards equation, RE:

$$C(h)\frac{\partial h}{\partial t} = \frac{\partial}{\partial z}\left[K(h)\left(\frac{\partial h}{\partial z} + 1\right)\right] \tag{17.3}$$

where *C(h)* is the specific water capacity (L^{-1}). Despite severe criticisms (e.g., Hunt et al. 2013), it is the widely accepted recognition that Eq. 17.3 is currently the most suitable model for describing water flow in the unsaturated zone at the macroscopic scale (Vanclooster et al. 2004).

17.2 Soil Hydraulic Properties

Solution of the RE for a given domain relies on specified initial and boundary conditions and a proper specification of the material properties. These are characterized by two non-linear constitutive relationships, commonly called "hydraulic functions" (Durner and Flühler 2005). The *hydraulic conductivity function* describes the dependency of the hydraulic conductivity K on the water content or pressure head and the *water capacity function* $C(h) = \partial\theta/\partial h$ describes the functional relation between changes of the soil water content and the matric potential h. The water capacity function is a direct mirror of the effective pore-size distribution of a soil (Durner 1994). A key paradigm for the validity of the Richards model in the form of Eq. 17.3 is a quick local equilibration between matric potential, water content, and hydraulic conductivity. In a strict sense, this requirement is equivalent to the assumption of a time-invariant spatial distribution of the liquid and gas phase at the pore scale at any given liquid saturation, independent of the actual flow dynamics and wetting history. From basic hydrodynamic pore-scale considerations, it is obvious that this can never be fully the case (e.g., Hassanizadeh et al. 2002). However, evidence from numerous laboratory experiments has proven the practical applicability of the RE approach to describe water transport in soils on spatial scales from mm to m.

17.2.1 Standard Expressions for the Retention Curve

Solving the RE for soil water flow under transient conditions requires numerical methods because of the strong non-linear dependence of the soil hydraulic properties (SHP) on h. Accurate knowledge of the SHP is crucial, because water flow depends very sensitively on their particular shape (Schelle et al. 2013). For use in simulation models, SHP are preferably formulated by empirical analytical expressions, i.e., the *retention function* $\theta(h)$ and the *conductivity function* $K(h)$. Numerous functions have been proposed (see reviews of Leij et al. 1997; Kosugi et al. 2002; and Assouline and Or 2013), most popular being the expressions of Brooks and Corey (1964), van Genuchten (1980), and Kosugi (1996). A fundamental assumption in using these functions is the independence of the hydraulic properties from the dynamics of the water flux. In Sect. 17.3 of this paper, we will show that this paradigm must be questioned.

To calculate the retention functions, the water content is typically scaled between a maximum value, i.e., the saturated water content θ_s, and a minimum value, often expressed as "residual" or "irreducible" water content θ_r. The range between θ_r and θ_s is expressed as effective saturation, $S_e = (\theta - \theta_r)/(\theta_s - \theta_r)$. If θ_r is replaced by zero, effective saturation reduces to saturation $S = \theta/\theta_s$, which is often used in geotechnical and civil engineering (Fredlund and Xing 1994).

Most standard retention functions express $S_e(h)$ by simple sigmoidal shape functions with two or three parameters that are related to the characteristic pore size and the width of the pore-size distribution. If the pore system is conceptualized as a bundle of capillaries (not necessarily straight, circular, and continuous), these sigmoidal functions reflect smooth unimodal equivalent pore-size distributions. Exemplarily, in the commonly used shape function proposed by van Genuchten (1980)

$$S_e = \left[1 + (\alpha|h|)^n\right]^{-m} \tag{17.4}$$

the characteristic pore-size of the porous medium is related to the inverse of the shape coefficient α [L^{-1}], and the width of the pore-size distribution is related to the shape coefficients n [−] and m [−]. With the availability of numerical codes to solve the RE on computers, numerous expressions have been (and still are) proposed as general shape functions (Brooks and Corey 1964; King 1965; Brutsaert 1966; Visser 1968; Laliberte 1969; Farrell and Larson 1972; Rogowski 1972; Gillham et al. 1976; Vauclin et al. 1979; Simmons et al. 1979; d'Hollander 1979; van Genuchten 1980; Tani 1982; McKee and Bumb 1984, 1987; Driessen and Konijn 1992; Globus 1987; Hutson and Cass 1987; Russo 1988; Ross and Smettem 1993; Zhang and van Genuchten 1994; Kosugi 1994, 1996; Rossi and Nimmo 1994; Pachepsky et al. 1995; Assouline et al. 1998; Groenevelt and Grant 2004; Khlosi et al. 2006).

According to Laplace's law, the suction at which a water-filled pore starts to drain is related inversely to the equivalent radius of the pore necks. Most of the above-listed functions approximate the equivalent pore-size distribution by more or less a log-normal distributed density function. If the specific water capacity, $d\theta/dh$ is plotted versus h on a linear scale, pores with large radii of different orders of magnitude are squeezed together, which masks the important range close to saturation. Durner (1994) therefore proposed to visualize the pore-size density by plotting $d\theta/ln(|h|)$ versus $log\ |h|$, which is equivalent to plotting it versus the $log\ r$. The standard retention functions then in most cases lead to smooth almost symmetrical bell-shaped functions of the pore-size density. Exceptions are functions with a distinct air entry which have a discontinuity when $S_e(h)$ reaches its maximum value $S_e = 1$ (e.g., Brooks and Corey 1964). While empirical evidence has shown that the assumption of an *approximately* log-normal distributed pore-size distribution appears justified for artificially homogenized soil materials, there is no reason to assume that the distribution for natural soils *exactly* follows any of the proposed shapes or that one of the models might be more "physically-based" than another. Theoretical pore-size distribution models, e.g. based on fractal representation of the pore space (e.g., Tyler and Wheatcraft 1989), also give inspiring insights, but cannot describe real soils with a sufficient degree of accuracy. The question therefore is, how accurate should a description of a real pore-size distribution be.

The problem of the adequacy of the hydraulic function is not evident, if the number of supporting retention data to be fitted is small, as is commonly the case by traditional methods to determine the retention curve (Dane and Hopmans 2002; Durner and Lipsius 2005). However, new automated measuring methods, e.g. the HYPROP™ implementation of the evaporation method (Schindler et al. 2010a, b), yield water retention data in almost arbitrary high resolution and across a relatively wide pressure head range. When fitting these data, it becomes apparent that the simple sigmoidal functions with two or three parameters can describe measured data of some soils well, but grossly fail for others. Furthermore, comparisons have shown that some models are on average better suited than others, but there is never a single "best" model. According to our experience, the likelihood for general misfits is particularly given for soils with a wide range of particle sizes, such as morainic soils. The corresponding loams often show a gradual release of water over a very wide pressure head range, which cannot be represented by sigmoidal curves. It appears remarkable in that context that the shape function of van Genuchten (1980) is applied in an overwhelming majority of all cases with two shape parameters, by constraining the parameter $m = 1 - 1/n$, despite the proven superiority of the three-parameter model with variable m (e.g., van Genuchten and Nielsen 1985).

To circumvent the problem, and to be able to describe measured retention data with sufficient flexibility, models with higher numbers of degrees of freedom must be used, such as bi- or multimodal expressions (e.g., Othmer et al. 1991; Durner 1992, 1994; Ross and Smettem 1993; and recently Romano et al. 2011). Due to their flexibility, these expressions improve the fitting accuracy and can help to avoid systematic misfits. This is most evident for structured soils with bimodal pore-size distributions that obviously cannot be described by unimodal models. However, also in cases where no obvious bimodality is apparent, models with more degrees of freedom clearly better approach measured shapes of retention curves. According to our experience, this is of particular benefit for loamy soils (Mallants et al. 1997). The use of a free-formed function (Iden and Durner 2007) can be seen as an ultimate flexible expression for that purpose. In that approach, the retention curve is described by a Hermitian spline with an arbitrary number of degrees of freedom, and the only constraint is that the water content must decrease strongly monotonically from its maximum value at saturation to zero at oven dryness, i.e., around $h = -10^{-6,8}$ (Schneider and Goss 2012). As Durner (1994) pointed out, the question of how close measured data should be represented by a fitted function is not problematic in the use of the retention curve to express the soil water capacity, but becomes virulent if the function is used to estimate the conductivity function.

The concept of describing the effective saturation with a shape function implies that the water contents at very high suctions reach a distinct residual water content, θ_r. This has been subject to discussions. Corey and Brooks (1999) interpreted the residual water content as the water held by adsorptive forces. More commonly, it is interpreted as a mere fitting parameter (van Genuchten and Nielsen 1985). In the very dry range, the concept of a residual water content is inappropriate (e.g. Nimmo 1991), because it contradicts the observation that measured water contents approach zero. We will come back to this issue in Sect. 17.2.3.

17.2.2 Standard Expressions for the Conductivity Curve

For use in simulation models, the conductivity function must be described over the full moisture range, too. There exist some early empirical expressions for the *K* function (e.g., Gardner 1958) that can be fitted to measured data. However, since the unsaturated conductivity is hard to measure and since there is no single method that yields data over a wide moisture range (Durner and Lipsius 2005) it is common practice to estimate the unsaturated conductivity based on parameters that characterize other basic soil properties. The most widely applied concept is to derive the conductivity from the retention curve, by representing a soil by an equivalent bundle of capillaries. Flow in the water-filled capillaries is then modelled assuming viscous flow according to the Poiseuille equation. Based on the seminal works of Purcell (1949), Childs and Collis-George (1950), Fatt and Dykstra (1951), and Burdine (1953), a generalized form of the resulting conductivity is (Mualem 1986; Hoffmann-Riem et al. 1999; Peters 2013)

$$K_r(S_e) = \frac{K(S_e)}{K_s} = S_e^{\tau}\left[\left(\int_0^{S_e} R^{\kappa}(x)\,dx\right)^{\beta} \Bigg/ \left(\int_0^{1} R^{\kappa}(x)\,dx\right)^{\beta}\right] \tag{17.5}$$

where K_r indicates the so-called *relative* conductivity function, which reaches a value of 1 at full saturation, K_s is the saturated conductivity, R is the equivalent pore radius, which is inversely proportional to the pressure head h, and x is the dummy variable of integration. With $\tau = 2$, $\kappa = 2$, and $\beta = 1$, Eq. 17.5 represents the model of Burdine (1953), with $\tau=0.5$, $\kappa=1$, and $\beta=2$, it becomes the widely used model of Mualem (1976). The parameter τ accounts for tortuosity and connectivity in the original interpretation, hence, in a physical sense must be positive. Any functional expression for $R(S_e)$ (resp. $h(S_e)$) can be inserted in Eq. 17.5. In general, this will require numerical integration methods (see Mualem 1986 for a comprehensive overview). However, for some retention curve formulations, closed form expressions can be derived, the most widely distributed being the unimodal van Genuchten-Mualem model (van Genuchten 1980). Combining the retention model of van Genuchten (Eq. 17.4 with the constraint $m=1-1/n$) with the conductivity model of Mualem (1976) leads to (van Genuchten 1980)

$$K_r(S_e) = \frac{K(S_e)}{K_s} = S_e^\tau\left(1 - \left[1 - S_e^{1/m}\right]^m\right)^2 \qquad (17.6)$$

Equation 17.6 is the conductivity part of the van Genuchten-Mualem (VGM) model. Similar compact algebraic expressions exist for a variety of other combinations of retention curve and conductivity prediction models (Russo 1988; Kosugi 1996; Assouline 2001, 2004), including bimodal functions (Priesack and Durner 2006; Romano et al. 2011). Equation 17.5 shows that only the pore-size distribution parameter m is used to predict the shape of the conductivity function and that the predicted shape function is scaled by a matching factor and corrected by an empirical tortuosity-connectivity correction term. Commonly, the saturated conductivity K_s is used as matching factor, despite known problems of the prediction of the shape of the K function near saturation. Furthermore, τ must be regarded as additional empirical tortuosity/connectivity parameter. If unsaturated K measurements are available, τ can be estimated by fitting Eq. 17.5 to data. If no data are available, τ is set constant, in most cases to $\tau=0.5$ as proposed in Mualem's original paper (Mualem 1976). Both approaches are problematic. By fitting a generalized form of the analytical capillary bundle prediction model to data of the UNSODA data base (Nemes et al. 2001), already Schaap and Leij (2000) in accordance with Hoffmann-Riem et al. (1999) came to the conclusion that the conductivity function should not be interpreted as physically based. This has been confirmed again and again since, showing the need for amendments for the matching factors and poor agreement between predicted and measured conductivity functions. In a majority of all cases fitting measured data gave negative values of τ (Kosugi 1999; Schaap and Leij 2000; Shinomiya et al. 2001; Børgesen et al. 2006). Hunt et al. (2013) conclude that the need to correct for tortuosity and pore connectivity is simply compensating fundamental deficiencies of the underlying capillary bundle model concept. It is remarkable to note that Childs and Collis-George clearly recognized the areas of weakness in their model and were careful to point these out in their paper, but apparently, this has not hindered the wide application of the capillary bundle models.

Despite these conceptual problems, numerous investigations have shown that simulations with hydraulic functions of this type can match experimental data very well. Successful matching of soil water dynamics has been shown in particular for water flow regimes that (i) change in only one direction, i.e., monotonously drying or wetting processes, and (ii) remain in moist conditions in which capillary flow dominates the water regime. Primarily such conditions are applied in laboratory experiments which are performed to determine hydraulic properties (Hopmans et al. 2002). Nonetheless, problems are evident simulating water movement near saturation, under dry conditions, under variable imbibition and drying conditions, and under non-equilibrium water flow. In the following, we will shortly address some of the challenges.

17.2.3 Hydraulic Properties Near and at Saturation

Modeling water flow dynamics with the RE approach and with traditional parameterizations of soil hydraulic properties is most successful in the wet to medium moist range, where variably-saturated capillary flow dominates. This moisture region represents a typical state of water movement in many soils under humid climatic conditions. However, problems are likely to occur when modeling changes between full water saturation and unsaturated conditions, in particular for fine-textured soils. These problems are often caused by (i) the asymptotic shape of the K function at water saturation, and (ii) the shape of the K function in structured soils.

There are arguments that the asymptotic slope $\mathrm{d}K/\mathrm{d}h$ at water saturation should be zero or have at least a finite value to avoid extreme non-linearity of the conductivity function (van Genuchten and Nielsen 1985; Vogel et al. 2001). For the VGM model given by Eq. 17.5, this condition is only fulfilled if the parameter $n \geq 2$. In practice, fitting experimental data with Eq. 17.4 results in most cases in $n < 2$, with only well-sorted sandy soils usually yielding values of $n \geq 2$. If n is only slightly smaller than 2, we experience no problems in numerical simulations. However, if n approaches its theoretical lower bound of unity, which is often the case for fine-textured soils, the hydraulic conductivity function drops abrupt and near-instantaneous. The highly non-linear decrease in $K_r(h)$ is caused by the integrand R in Eq. 17.5 which goes to infinity (representing arbitrarily large pores), and hence leads to a singularity in the VGM conductivity equation. This drop of K (h) must be regarded as an artifact of the retention model, and applies similar to any smooth retention function with only two parameters. In these functions, the shape of the retention function towards the small pores fixes also the shape of the function towards large pores. A wide pore-size distribution will accordingly lead to a gradual increase of the water contents even at extremely small suctions, causing the observed behavior. The problem is illustrated in Fig. 17.1 where it becomes evident that values of n lower than 1.4 lead to a drastic decrease of $K(h)$ close to water saturation.

Not only does this negatively affect the stability of numerical simulations, but it also influences their results (Vogel et al. 2001; Ippisch et al. 2006). To resolve these difficulties, modifications of the VGM model have been proposed. The most popular remedy is the introduction of an "artificial" air-entry value above which

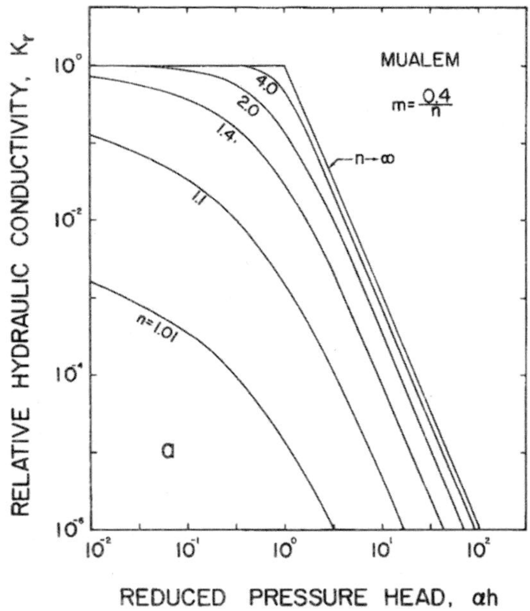

Fig. 17.1 Shape of the predicted relative conductivity function of the van Genuchten-Mualem model near saturation, for different values of n and the constraint $m = 0.4/n$. Reproduced from van Genuchten and Nielsen (1985)

the effective saturation and the relative conductivity are both equal to unity (Vogel et al. 2001). The same idea in a more elegant mathematical form can be found in Ippisch et al. (2006). The conceptual disadvantages of these corrections are that the water capacity function and the slope of the conductivity function at the artificial air-entry may become discontinuous. An alternative idea was proposed by Durner (1991) who suggested to leave the retention function unchanged while restricting the pore radius to a maximum permissible value, e.g. $R \leq 3$ mm, in Eq. 17.5. This correction does not suffer from the problems discussed above and seems to be the simplest and most elegant way to solve the problem.

Since structured soils often have multimodal pore-size distributions, their hydraulic properties cannot be parameterized by models which are based on the assumption of an unimodal pore-size distribution. Bimodal or even multimodal expressions have been used successfully in numerical vadose zone simulation studies (e.g., Mallants et al. 1997; Coppola et al. 2009). In general, the description of water content changes near water saturation is easily achieved with flexible water retention equations. However, it must be acknowledged that soils with extreme jumps in water content very close to saturation, reflecting quickly draining water held in macropores, are often parameterized in a way that causes numerical problems in the simulation (Dettmann et al. 2014). These problems are often not caused by the specific water capacity function, but again by the shape of the conductivity function near and at saturation. Similarly to fine-textured soils with an unimodal pore-size distribution, the drop of the conductivity near saturation can become unrealistically large. Remarkably, this drop can again be caused by the low value of the parameter n of the textural pore system, and cannot be compensated by a large value of n in the structural pore system. This lead Durner (1994) to the conclusion that conductivity predictions by capillary bundle models, where R in Eq. 17.5 is not limited to a maximum value, are intrinsically unstable, and that the saturated and unsaturated conductivity values should both be explicitly measured and included into the objective function which is minimized during parameter estimation.

For extremely large pores, the validity of the Darcy flow equation and thus the applicability of the RE must be questioned. Unsaturated flow in macropores is likely taking place on the macropore walls, and might become turbulent under certain conditions. Furthermore, the hydraulic equilibration within an REV, encompassing soil matrix and macropores, will not be instantaneous, thus a fundamental requirement for the validity of the RE is violated. For these reasons, flow of water in soils containing macropores is often simulated with dual permeability models, where water flow in the macropore domain is represented by a variety of approaches, which span a range from Darcy flow to gravitational driven flow in water films. Šimůnek et al. (2003), Gerke (2006) and Jarvis (2007) give comprehensive reviews on this topic.

17.2.4 Hydraulic Properties in the Medium to Dry Moisture Range

A major drawback of van Genuchten and Kosugi type models is that they do not define the soil-water retention curve beyond the residual water content. From a physical point of view, it makes no sense to define a dry endpoint of the WRC at some infinitely negative water potential, and indeed the usual models that make use of the concept of residual water content are not well suited to describe observations of water dynamics in drying soils. It was early recognized that water contents in drying soils approach zero (Schofield 1935). Zero water content is operationally defined by oven drying of soil at 105 °C until constant weight is reached. This state corresponds to a suction of $6.8 \cdot 10^6$ to 10^7 cm, depending on ambient relative humidity (Schneider and Goss 2012). Measurements of water contents at low potentials always show a gradual decrease of water contents towards zero at that suction. This clearly contradicts the idea of an irreducible water content, and in practice, the so-called residual water content is in most cases a mere result of the fitting procedure, which is largely affected by the range of available data.

The concept of a residual water content has been defended by some authors as a "*empirical constitutive parameter in geohydraulic flow models (*i.e. *for modeling the flow of mobile coherent liquid soil water) where the hydraulic gradient (expressed as grad h) is the pressure gradient in the mobile coherent liquid soil water phase*" (Luckner et al. 1991). For an overview about the different definitions of θ_r see Cornelis et al. (2005). Assuming local equilibration of the water potential in all phases, we should however not distinguish between the potential of the liquid phase and the total potential. Furthermore, as Hunt (2004) points out, allowing for an adjustable parameter θ_r in capillary-bundle based conductivity models is an internal contradiction. Nimmo (1991) argues that even if no-liquid-flow conditions exist, it would be no surprise if θ_r determined by curve fitting of retention data would differ substantially from it. Hunt et al. (2013) list the residual water content as being composed of water at various locations: capillary water in corners and bridges of grain contacts (Orr et al. 1975), thin films (Tuller et al. 1999; Tokunaga 2009; Wang et al. 2013), water in surface bumps and crevasses (Toledo et al 1990), and water that simply has not reached its equilibration drainage at drainage state (Hunt and Skinner 2005).

To be able to describe water retention measurements across the full moisture range, modified functions have been proposed (Ross et al. 1991; Rossi and Nimmo 1994; Fredlund and Xing 1994; Fayer and Simmons 1995; Morel-Seytoux and Nimmo 1999; Webb 2000; Groenevelt and Grant 2004; Khlosi et al 2006; Silva and Grifoll 2007; Lebeau and Konrad 2010; Zhang 2011; Peters 2013). In part, these models construct the total water content as a superposition of a capillary and a non-capillary fraction, where the non-capillary water shows a linear behavior on a semilog plot (Campbell and Shiozawa 1992). However, as Peters (2013) shows, these models might fail for fine-textured soils with wide pore-size distributions, where the capillary component of the model does not reach a value of zero at high suctions. Peters (2013) therefore proposes a function where also a capillary and a non-capillary part are added, and where the capillary part is, if necessary, scaled to reach zero at oven dryness by multiplying it with a correction function. In our view, however, the need for this correction function remains unsatisfactory because it points at a major deficiency in distinguishing between capillary and non-capillary water for fine textured soils.

If water transport in drying soils, e.g. during an evaporation experiment, is simulated with standard van Genuchten and Kosugi-type models, it is our experience that the drying dynamics cannot be reproduced. An important point here is that by parameter adjustment it is often possible to match the flux across the soil surface, as long as no additional information (water content depth distribution, temporal evolution of matric potential in the sample) is simultaneously considered (Fujimaki and Inoue 2003). Similarly, steady state depth profiles of water contents and pressure heads cannot be reproduced with the standard functions, because liquid conductivity in the medium to dry range is underestimated. Goss and Madliger (2007) found a similar phenomenon under in situ conditions. There is clear evidence that the underestimation of the liquid conductivity is caused by ignoring film and corner flow components by standard capillary bundle models. Tuller and Or (2001) proposed an alternative approach for the derivation of hydraulic conductivity functions for homogeneous porous media. Their model is based on liquid configurations in both angular and slit-shaped spaces, and accounts for capillary, corner and film flows. The study gave very useful insights into the principal mechanisms, but is to date rarely applied, because it is mathematically complex and must be used in conjunction with a water retention model that often fails to describe experimental data in the intermediate saturation range. The reason for this might be the limited flexibility of the Gamma probability distribution function that is used to characterize the pore-size distribution in their model. The results of Tuller and Or (2001) motivated the development of further extensions of capillary bundle models for film flow (Peters and Durner 2008; Lebeau and Konrad 2010; Zhang 2011; Peters 2013). Based on theoretical and empirical evidence, and in order to be able to describe the outcomes of evaporation experiments, Peters and Durner (2008) introduced a very simple and fully empirical model for liquid conductivity in the complete moisture range. In their model, capillary conductivity is described by a capillary bundle model (e.g., Mualem 1976) and a film conductivity component is added, which is expressed by a simple empirical power function of saturation. Their function mimics the general form of the Tuller and Or (2001) model, and its parameters are easy to estimate by fitting to data. However, it is conceptually inconsistent, because their film conductivity component was coupled with an inappropriate retention model.

In the physically based conductivity model of Lebeau and Konrad (2010) water in films is further partitioned into thick and thin film conductivity, accounting for different viscosities in the vicinity of solid surfaces. While being conceptually very appealing, their model is mathematical complex and therefore difficult to implement in simulation models. Furthermore, heterogeneities of mineral and organic matter surfaces as well as the usually unknown composition of the fluid make it hard to get the right physicochemical constants. An additional conceptual drawback of this modeling approach is that it attributes all non-capillary flow phenomena to film flow on planar surfaces. This neglects the rather complex topology of water films in drying soils as well as the existence of other non-capillary flow processes such as water flow in corners and crevices, which ultimately restricts the flexibility and predictive power of this model.

Peters (2013) proposes a conductivity model that is conceptually similar to the model of Lebeau and Konrad (2010), but it is much simpler to implement. The liquid conductivity model was furthermore coupled with an existing vapor conductivity model to describe conductivity in the complete moisture range.

In the very dry range, liquid water flow might completely cease, but water can still be conducted by vapor flow (e.g., Philip and de Vries 1957; Saito et al. 2006). Under isothermal conditions, the vapor flow follows the gradient in the total water potential. Assuming that the gravitational component of the gradient in the total water potential can be neglected, the hydraulic conductivity due to isothermal vapor flow can be simply added to the liquid conductivity to yield a total conductivity function (e.g., Saito et al. 2006). This approximation has the practical advantage that isothermal vapor flow can easily be simulated using the standard Richards equation.

Figure 17.2 illustrates the concept of extended retention and conductivity models that can be applied to the complete moisture range. These new models were successfully applied to literature data, which all reach the dry to very dry range and could not be well described with the classic capillary models. Furthermore, a simulation study with steady-state evaporation from a groundwater to the soil

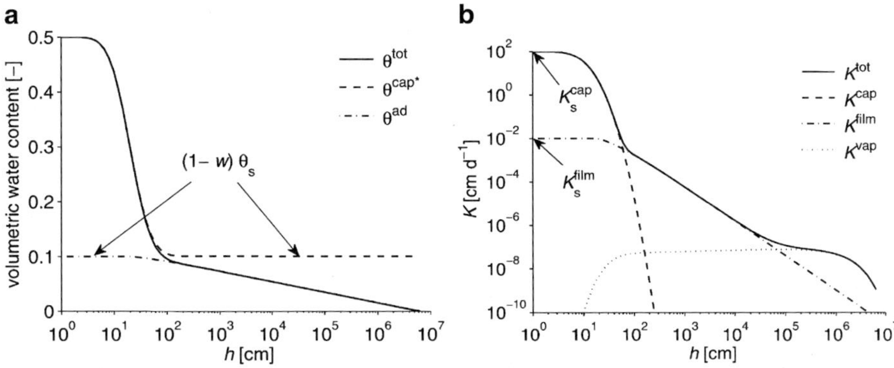

Fig. 17.2 Illustration of the contributions of capillary (cap) and non-capillary (ad) water content components to the retention curve (*left*) and corresponding hydraulic conductivity model, composed by addition of capillary, film and vapor phase components (*right*). Reproduced from Peters (2013)

surface showed that neglecting either film or vapor conductivity or both can lead to significant underestimation of water transport at low water contents (Peters 2013).

17.2.5 Hysteresis of Hydraulic Properties

Water flow modeling under changing wetting and drainage conditions in the framework of the RE needs to consider that hydraulic properties are hysteretic in nature, and that potentially dynamic non-equilibrium occurs, in particular if fast changes of system changes take place, e.g., by infiltration. The term hysteresis goes back to the work of J.E. Ewing (1881) in the context of research on ferromagnetism. For him, hysteresis was the temporally retarded reaction of a system state on the forcing. For water movement in porous materials, the term has been used in the past to describe both retarded equilibration of the state variables as well as permanent hysteresis, which can be seen as a special case of the former with infinite equilibration time (e.g. Funk 2012). Rate dependency of effective hydraulic properties is extensively discussed in the next main section of this paper. In this section, we focus shortly on some key features of "static" hysteresis.

The dependencies of θ and K on the pressure head h are hysteretic. This implies that the relationships, as shown in Fig. 17.3, are valid only for a certain history of wetting and drying, and for one particular saturation or desaturation path. Most measurement experiments rely on a uniquely defined initial hydraulic state, from which the system state is changed continually in one direction. Drainage experiments in the laboratory, such as multistep outflow experiments start usually from a maximum achievable saturation (Durner and Iden 2011). Re-wetting the soil from complete dryness will yield different relationships. Further, drying and wetting cycles from intermediate states between full saturation and oven dryness will again

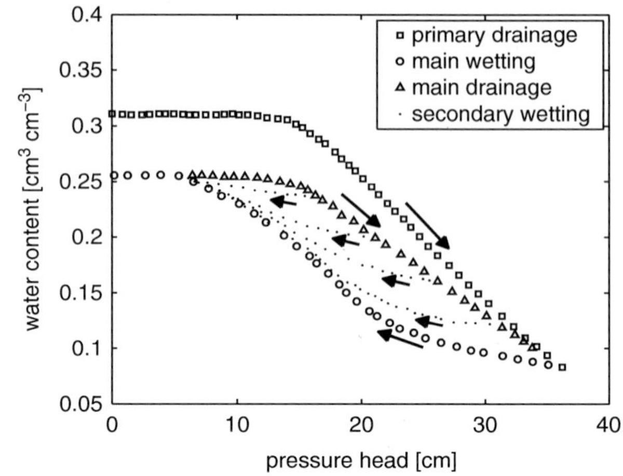

Fig. 17.3 Hysteresis of the moisture characteristic. Data from Poulovassilis (1970)

give different relationships. Hysteresis of the water retention characteristic is shown in Fig. 17.3. Drainage of a fully saturated medium yields the *primary drying curve*, imbibition from oven dryness yields the *primary wetting curve*, drainage from satiation yields the *main drying curve*, imbibition from the residual water content yields the *main wetting curve*, and any changes between imbibition and drainage at intermediate pressure values yield *secondary scanning curves*. The term satiation was proposed from Hillel (1980) for the typical maximum saturation in the field in order to distinguish it from the "saturated water content". It is the maximum water saturation of soils in the unsaturated zone under natural conditions and it is always significantly lower than the porosity, due to air entrapment, with typical saturation values of 80–90 % of pore space (Klute and Dirksen 1986). Full saturation of the pore space can be obtained only by specific experimental manipulations (slow wetting from the bottom, flushing with carbon dioxide, flushing with de-aired water, application of vacuum).

Hysteresis is typically assumed to be caused by the inkbottle effect, but other processes such as contact angle hysteresis, shrink-swell effects, or entrapped air play a significant role (Hillel 1980). Contact angle hysteresis is affected by numerous factors, like surface roughness, chemical heterogeneity, fluid dynamics, particle shape, and gas adsorption, and all these relations depend on the composition of the soil solution (Dury et al. 1998; Henry et al. 2001). Therefore, hysteresis effects on soil water redistribution may result from a single process or interactions between different processes, and "*a treatise on the role of individual factors controlling hysteresis is rather speculative*" (Kutilek and Nielsen 1994, p. 73). Hunt et al. (2013) argue that hysteresis in wetting and drying of a porous medium is fundamentally due to the distinction between "allowability" and "accessibility" of water or air to pore spaces. They list three conditions to be met for a given pore to drain: (i) it must be large enough that its water can no longer be held by capillary forces against the applied pressure, (ii) air must be able to enter it, and (iii) water must be able to leave it. The first condition makes the pore allowable, while the other two make it accessible. Analysis of the process by percolation theory indicates that efforts to arrive at a more valid description of hysteresis with models based on capillary bundle theories might be futile, because these are based on a fundamentally wrong conception, and that information about the statistical pore-size distribution alone is not sufficient for a characterization of the pore space. Also, the actual extent of hysteresis might depend on the system size.

The hysteresis of $\theta(h)$ leads to hysteresis of $K(h)$, since conductivity is controlled by the water-filled pore space. The hysteresis of the $K(\theta)$ relation of soils is generally regarded as negligible (Mualem 1986; Kool and Parker 1987; Assouline and Or 2013). Considerable effort has been put into the analysis and description of hysteretic soil hydraulic properties. This has led to numerous models for describing hysteresis in $\theta(h)$. Both, empirical models (Scott et al. 1983; Hogarth et al. 1988) and physically based models (Poulovassilis and Childs 1971; Parlange 1976; Mualem 1974, 1984) have been proposed. When comparing different models of hysteresis using 10 measured scanning curves, Viaene et al. (1994) concluded that the best models were the conceptual models needing two branches for calibration. Recent performance comparisons have been published by Pham et al. (2005), and Mualem and Beriozkin (2009).

Hysteresis has been recognized as an important component of soil water flow since the early work of Haines (1930). Jaynes (1992) showed that soil water redistribution is retarded by hysteresis, and its magnitude depends on the rate at which the soil water content varies. McNamara (2014) argues that agriculture would not be possible without this mechanism for trapping water close to the surface. Ignoring hysteresis in the K-θ-h relations may be partly based on an unjustifiable assumption that its effects are negligible, partly on the absence of a well-validated hysteretic model that is easy to calibrate (Lenhard et al. 1991), and partly on the difficulties of performing measurement experiments that are suitable to identify parameters of hysteresis models. Neglecting hysteresis leads to significant errors in the water redistribution under transient boundary conditions (Dane and Wierenga 1975; Gillham et al. 1979; Hoa et al. 1977; Lehmann et al. 1999; Si and Kachanoski 2000). Russo et al. (1989) and Mitchell and Mayer (1998) investigated the influence of hysteresis on solute transport and found that the magnitude of the deviations between hysteretic and non-hysteretic simulations was not a simple function of single parameters, but rather depended on the combined values of many or all of the hydraulic parameters. Hysteresis is also seen as one of the major obstacles for comparing different measurement methods for the $K(\theta)$ (Stolte et al. 1994; Basile et al. 2003).

Hysteresis has been incorporated into simulations of unsaturated flow since Rubin (1965). However, incorporating a physically-based hysteresis model in numerical codes is demanding, since bookkeeping of the wetting and drying history at any point in the soil must be provided. Therefore, simple approximations such as the Scott et al. (1983) model are preferred. This, and possibly the difficulty to determine the appropriate parameters, lead to the fact that to date most simulations even under transient boundary conditions are still performed with non-hysteretic properties.

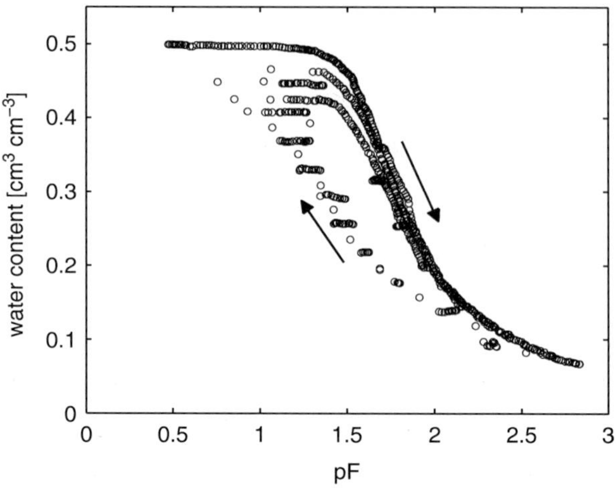

Fig. 17.4 Hysteresis and DNE concept: Shown are trajectories of water content and pressure head with time, obtained during wetting and drying system changes. The figure shows a static hysteresis loop and dynamic fringes, measured by three evaporation cycles and stepwise re-wetting cycles. Evaporation phases were interrupted by placing lids on the samples, and then continued. The interruption is visible as small relaxation of the matric potential at constant water contents. Wetting was performed by applying stepwise defined amounts of water to the otherwise sealed sample. This led in each case to a strong and quick increase in the matric potential, followed by a subsequent relaxation to the hysteretic equilibrium suction. Data are obtained with the measurement device HYPROP© by UMS GmbH, Munich, with 250 cm^3 samples of homogeneously packed sand. Initial capillary saturation was 0.5 cm^3/cm^3

It might be speculated that this, among other problems, in turn contributes to the often-found result that laboratory properties cannot be successfully used to simulate field water behaviour.

Characterization of static hysteresis by inverse modeling of inflow/outflow processes appears to be an attractive option to test and validate hysteresis concepts. However, our key point here is that even for a given soil sample, this approach cannot be successful, as long as we have not a clear conception of dynamic non-equilibrium effects. This is in accordance with a recent model of Hilfer (2006a, b) who developed a macroscopic theory of capillarity in porous media where different hydrodynamic properties of percolating (= connected) and non-percolating (= not connected) fluid components are conceptualized. This concept allows to simulate hysteresis and dynamic effects without the need to refer to a historic system state. Our conceptual picture is therefore, that trajectories of matric potential and water contents depend on the rate of changes at system boundaries, and converge to a quasi-equilibrium state which is not necessarily identical to a true thermodynamic equilibrium, but sufficiently stable to be regarded as static hysteresis in soil water modeling (Fig. 17.4).

17.3 Dynamic Nonequilibrium (DNE) Water Flow

17.3.1 Introduction and Terminology

Chapters 15 and 16 of this book and the first part of this chapter emphasize that in order to simulate practical problems in the vadose zone, accurate knowledge of the water retention curve and the hydraulic conductivity curve is mandatory. However, observations from the 1960s clearly show that the water retention curve is different if estimated under dynamic or static conditions (Topp et al. 1967). This phenomenon has been termed "dynamic nonequilibrium" (DNE) water flow, or nonequilibrium in the capillary pressure–saturation relationship. To distinguish it from the traditional hysteresis concept, we attribute it to a non-uniqueness of the water content-capillary pressure relation even in the case of a monotonous drainage or imbibition process (Hassanizadeh et al. 2002). We hypothesize that the importance of the phenomenon is yet not recognized, because the majority of the existing methods, which measure the water retention curves, are static methods (Durner and Lipsius 2005). The question is, whether these static retention curves are valid for modeling purposes in the field, where water flow takes place under transient conditions. It seems that although both phenomena, static hysteresis and dynamic effects, are qualitatively identified for a long time, they are not taken into account or they are often unjustifiably omitted from the majority of modeling studies. In the remainder of this chapter we define the phenomenon of dynamic nonequilibrium, present observations which support it and also discuss the reasons which have been proposed in the years for generating these results. A more thorough discussion of the observations, reasons and modeling approaches for DNE is given in Diamantopoulos and Durner (2012).

The term "nonequilibrium" is related in soil physics to different phenomena. Most apparent are phenomena which show that water transport in soil takes not place as expected from simulations in homogeneous porous media, but is routed in preferential flow paths (Fig. 17.5). These phenomena are caused by macropore flow (Jarvis 2007), strong and sharp heterogeneity of physical soil properties (Schlüter et al. 2012), or fingered or funneled water flow (van Dam et al. 1990). For chemical transport, similar patterns can be caused by specific chemical matrix-solute interaction phenomena. In this discussion we are going to focus on nonequilibrium in water flow.

One way to define "nonequilibrium water flow" is to attribute it in a very broad manner to any situation where water flow is not in equilibrium. With "equilibrium water flow", we understand a flow process which can be described by the Richards equation coupled with a hydraulic properties that are invariant with respect to the intensity of the water flux in the system. The specific relation between h, θ and K can depend on the history of the wetting or drying process (i.e., be hysteretic), but not on the rate by which the state changes. From this definition, it becomes obvious that the perception, whether state variables are in equilibrium or not, is intrinsically

Fig. 17.5 Nonequilibrium due to preferential flow, caused by macroscopic heterogeneity, visualized by a dye experiment. Grids are spaced 10 cm by 10 cm. Photo by courtesy of Dr. Christian Stamm, EAWAG, Switzerland

to the size definition of the REV, or in practical application to the spatial resolution of the simulation model. If the resolution is sufficiently high, a situation depicted in Fig. 17.5 could be *in principal* be resolved by a numerical model (Schlüter et al. 2012).

In this work, we define dynamic nonequilibrium water flow from a phenomenological point of view as the apparent non-uniqueness of the relationships between the (measured) state variables θ, h, and K, which is dependent on the rate of how the system state changes. In quasi-hydrostatic situations, DNE is indicated by drifting water contents despite apparent hydrostatic equilibrium, or by drifting matric potentials at constant water contents. In steady-state flux situations DNE is indicated by drifts in water contents and/or matric potential at constant fluxes. In monotonic drying or wetting situations, it becomes evident by different trajectories of the locally observed $\theta(h)$ relation that depend on the rate of drainage or wetting. It is mandatory in this definition that the state variables are measured within the same REV, or else in macroscopically homogeneous porous media in the same depth level, where one would expect from traditional modeling laterally homogeneous conditions. To observe this kind of non-equilibrium, experiments conducted in the laboratory for macroscopically homogeneous materials are required.

Preferential flow phenomena, as indicated in Fig. 17.5, are very important in particular for solute transport and are thus widely treated in the soil science literature. In this chapter, we exclude preferential flow and transport due to macroscopic heterogeneity, because it can be *in principal* be treated with Richards type models, if heterogeneities are in size above the usual REV scale of cm. The interested reader is referred to the reviews by Flühler et al. (1996), Šimůnek et al. (2003), Gerke (2006) and Jarvis (2007).

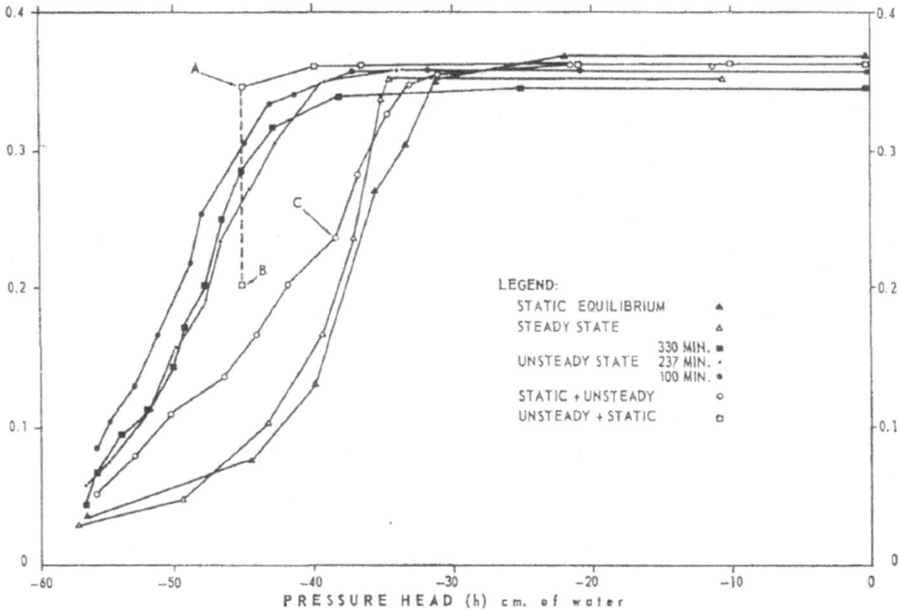

Fig. 17.6 Water retention curves estimated by Topp et al. (1967). The results show that in the case of drainage more water is retained by the soil matric in the case of a transient experiment than in the case of a steady-state or static experiments. Reproduction from Topp et al. (1967)

17.3.2 *Observations*

Observations of DNE are known since 1960s, and found attention since the pioneering work of Topp et al. (1967). That study initiated a first group of reported observations of DNE, where retention curves were estimated under transient conditions and compared with retention curves estimated under static and/or steady-state conditions (Wildenschild et al. 2001; Sakaki et al. 2010). DNE effects appear as differences between the transient and the steady state/static water retention curves. Topp et al. (1967) showed that although, in the case of drainage of a well sorted sand, the water retention curves estimated under static and steady-state conditions were similar, the retention curves estimated under transient conditions were substantially different (Fig. 17.6). Specifically, more water was retained by the soil matrix in the case of transient drainage conditions as in the previous cases.

The second group of observations contains experiments which have been designed for estimating the soil hydraulic properties in the laboratory. A classic experiment for estimating the soil hydraulic properties in the laboratory is the Multistep Outflow (MSO) experiment (Hopmans et al. 2002). In the case of drainage, different negative pressure heads are applied at the lower boundary of a saturated soil column in a stepwise manner. The volume of water that leaves the soil column is recorded along with the evolution of the pressure head inside the

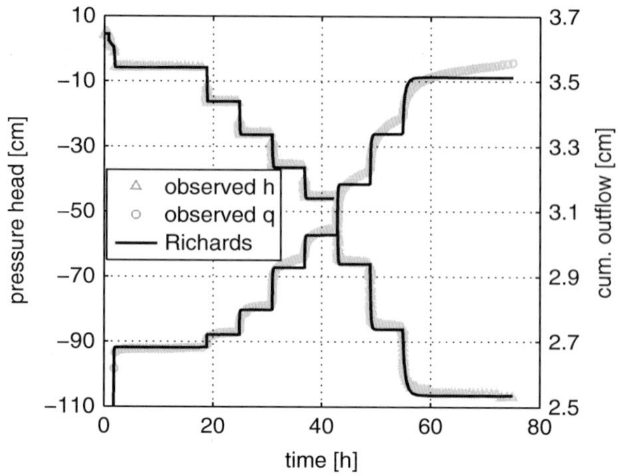

Fig. 17.7 Observed and simulated cumulative outflow and pressure head data for a loamy sand soil (Diamantopoulos et al. 2012). From 30 h on, the equilibrium kinetics of pressure head and outflow differed significantly. The fitted data were calculated using the Richards equation coupled with the van Genuchten-Mualem model. Richards equation predicts that the pressure head inside the soil column and the cumulative outflow data reach the equilibrium state simultaneously. This lies on the inherent assumption of the Richards equation that water content and pressure head are tightly coupled through the water retention curve

soil column, measured by one or more tensiometers. Figure 17.7 shows that after a pressure head change at the lower boundary, the tensiometer readings reach the new equilibrium level relatively quickly, whereas the outflow of water continues for hours or even days (Diamantopoulos et al. 2012). According to our experience, such DNE effects in MSO experiments do not occur for all soils, but are the rule rather than the exception.

The third group of observation tracks DNE effects in infiltration experiments under constant flux conditions. These experiments are similar to MSO experiments and were also designed for measuring the soil hydraulic properties in the laboratory. In the case of drainage, one may start again from a saturated condition and apply constant flux at the upper boundary while the pressure head at the lower boundary is set equal to the pressure head measured near to the upper boundary (Weller et al. 2011). In this way we are forcing the soil column to achieve unit gradient conditions which means that the pressure head, the water content and the water flux are the same throughout the soil column. Figure 17.8 shows experimental data for a steady state experiment presented by Weller et al. (2011). After diminishing the infiltration flux rate at the upper boundary, the pressure head recorded by a tensiometer in the soil column first reacts promptly, reaching a more negative value, but then shows a relaxation towards a less negative value.

All three experiments presented above cannot be described by the RE, if we assume that the water content and the pressure head within an REV are coupled through the water retention curve. If we would simulate the experiment type of the first group, the estimated water retention curve would be independent of the flow

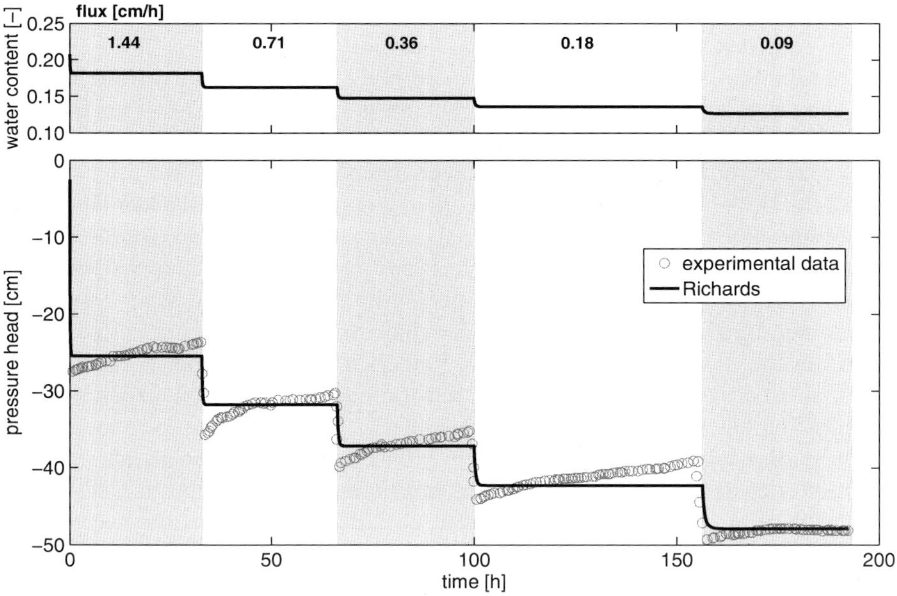

Fig. 17.8 Observed and simulated water content and pressure head data for a sandy soil (Weller et al. 2011). The experimental data were digitized from Weller et al. (2011). The fitted data were calculated using the Richards equation coupled with the van Genuchten-Mualem model. Richards equation predicts that after a flux change on the upper boundary the pressure head inside the soil column and the water content data reach the equilibrium state simultaneously. This lies on the inherent assumption of the Richards equation that water content and pressure head are tightly coupled through the water retention curve

conditions. Similar, for the second category, once the pressure head is at equilibrium, the water content should be in equilibrium and also the cumulative outflow of water should cease (Figure 17.7). In the case of the third experiment, once the unit gradient condition is reached, water content and pressure head inside the soil column should stay constant (Figure 17.8).

A point to note here is that a majority of studies which deal with the first group of observations conclude that the water retention curves estimated under steady-state and static conditions are similar. However, this is in contradiction with the experimental results of the third group of observations.

From a conceptual point of view, all the above experiments can be explained by the following process: In the case of drainage, an amount of water is left behind in the emptying pores and some portion of this water is conducted slowly toward the continuous water body. This process, which is in accordance to a concept published by Hilfer (2006a, b), can explain why the estimated water content under transient conditions is higher in the case of drainage in comparison with the static one. Furthermore, this can also explain the DNE observation in the case of the MSO experiments. After the application of a new pressure head at the lower boundary the main body of water reaches the equilibrium state relatively quickly, but there is a second part of the water which is left behind and is slowly approaching equilibrium. This is recorded as a delayed outflow period. Similarly, in the case of the constant flux infiltration experiments, only the main body of water maintains the applied flux, however the water which is left behind is contributing slowly to the flow and that's why we record an increase in the tensiometer reading. However we are not yet aware of the exact mechanisms which generate DNE effects.

All the DNE phenomena listed in the three groups were studied under laboratory conditions and for well sorted materials. Diamantopoulos and Durner (2012) reviewed 24 studies focused on DNE effects and they found that 20 of the 24 studies included sandy materials. Seventeen studies investigated only sandy materials. We do not think that this allows us to conclude that DNE effects are more pronounced in the case of sandy materials. Rather, it indicates sand as the favorite material used by soil physicists when performing controlled flow experiments in the lab. The few results of DNE studies with different materials are contradictory. Some studies suggest that DNE are more pronounced in sandy materials (Stauffer 1978; Wildenschild et al. 2001), whereas others found DNE effects in sandy loams (Simunek et al. 2001; Diamantopoulos et al. 2012). This shows the necessity of conducting more sophisticated experiments in order to improve our understanding on DNE effects.

17.3.3 Reasons Proposed for the Occurrence of DNE Effects

Reviews about the most hypothesized causes for emerging DNE effects can be found in Wildenschild et al. (2001), Hassanizadeh et al. (2002) and in Diamantopoulos and Durner (2012). In this section we will briefly list the proposed reasons of DNE in soils which can be divided into three main groups, i.e., (i) air-water reconfiguration, (ii) heterogeneity, and (iii) contact angle effects.

(i) Air-water interface reconfiguration
When air displaces water or water displaces air in a porous medium, the pore scale water-air configuration within the REV can by principal not change instantaneously, but requires a finite time to reach a new state (Barenblatt 1971; Sakaki et al. 2010). As an example, water which enters dry soil in an infiltration front proceeds on a pore scale in the large pores much quicker in the main direction of flow, but must then be sucked laterally into smaller pores. Dependent on the liquid's saturation and the porous medium properties, this local equilibration, which is not associated with a change of the macroscopic water content within the REV, will take place very quickly or takes a bit of time. Other factors which can contribute to temporal changes of air-water interface configuration are processes such as entrapment of air which subsequently slowly dissolves (Schultze et al. 1999), pore water blockage (Wildenschild et al. 2001), entrapment of water (Harris and Morrow 1964; Topp et al. 1967), or two-phase flow effects which occur particularly at the air entry pressure (Wildenschild et al. 2001).

(ii) Heterogeneity
Microscale heterogeneity, i.e. heterogeneity at the level of pore surfaces, can cause a finite reconfiguration time of the water at a pore scale, i.e, below the REV scale. Examples are clay coatings or lenses of fine sand which have different multiphase flow properties compared to the surrounding porous medium (Mirzaei and Das 2007). Heterogeneity on a scale above the REV is by principal observable, but might be below the measurement window of a sensor. If this is the case, e.g. by a heterogeneous distribution of different soil materials below the measurement scale, it can also provoke DNE in observations, even with relative small sample volumes as used in lab experiments.

(iii) Dynamic contact angle
In dynamic experiments, the contact angle between the solid-liquid-gas interfaces is dependent on the direction and velocity of the liquid-gas interface. Friedmann (1999) proposed that the concept of dynamic contact angle could be the reason for the observed flow rate dependence of the SHPs measured in dynamic experiments. A particular challenge is given in systems, where the contact angle depends not only on the speed of progression of a water front, but changes with time, even if no further addition of removal of water takes place. This can be caused, e.g., by a slow reconfiguration of the orientation of the polar groups of previously dry organic material that comes in contact with water (Doerr et al. 2000).

17.3.4 Discussion

In the previous paragraphs we defined DNE effects, presented observations of DNE and also listed some reasons provoking these kind of effects. The key question to be answered is: Are these kind of phenomena important for simulating water flow for practical problems? This question has not an easy and probably not a unique answer. It is true that under field conditions it's not very likely to meet high water flow velocities or boundary conditions as are applied for example in MSO experiments. But it is not uncommon to observe in field studies similar phenomena. Field tensiometers and TDR sensors, installed at almost the same position, show often reactions that indicate a different response time. Under infiltration conditions, tensiometers react quicker. We like to attribute this in fuzzy manner to somehow to the measurement characteristics of the sensors, to artefacts of the sensor placement, to spatial variability of hydraulic properties, to hysteresis, or to preferential flow paths, but in fact we are not able to identify the causes.

Although various reason have been proposed for generating DNE effects, it seems very difficult to isolate in a given situation the partial contribution from individual processes. We have shown in previous studies that DNE can be incorporated into RE simulations with relatively simplistic approaches, e.g., by replacing the direct coupling of the water content to the pressure head with a kinetic coupling, where water content equilibrates not instantaneously with the matric potential, but according to a first order kinetics (or any other kinetics) (Diamantopoulos and Durner 2012). Also, dual domain models of the type of Gerke and van Genuchten (1993) can be used to simulate the phenomenology of DNE, which is most applicable for structures soils, but less for the apparently homogeneous soils that have been used in the controlled lab experiments. However, such empirical modeling approaches must remain unsatisfying, if time and length scales of the systems affect the extent of DNE. Furthermore, some of the causes for DNE, such as air-water reconfiguration effects and dissolution of entrapped air, might occur simultaneously for one hydraulic process, for example under drainage conditions, but not under imbibition experiments. This even leads to hysteresis in the DNE. Mirzaei and Das (2013) conducted experiments in two-phase flow with observed DNE and they concluded that the local hydraulic equilibration at a given saturation was slower in the case of imbibition than in the case of drainage. The perception that DNE effects are more pronounced in the case of imbibition as in the case of drainage is also supported by unpublished data of our group.

A topic closely related to DNE is the study of hysteresis phenomena. "Static" Hysteresis, as described in the previous chapter, can significantly influence water flow in the unsaturated zone. Most of the models (if not all) which describe hysteresis have been compared against data of static experiments. However, if we take into account DNE then these models might not be adequate to describe hysteresis in dynamic experiments, since hysteresis includes reversals from wetting to drying and vice versa with certain rates. Keeping in mind that DNE maybe not be the same during imbibition and drainage, or furthermore may be different for every scanning curve, we can conclude that existing models may be inadequate to describe dynamic experiments which include reversals from wetting to drying and vice versa.

17.4 Conclusions

In this chapter, we gave a broad overview about parametric expressions of hydraulic properties that are used in the context of simulations of the Richards equation. We conclude that the standard hydraulic functions that are currently in use are not suitable as a general form of the water retention function, but might apply for some soils over a limited range of suction values. We focused on the critical regions of soil hydraulic properties that are located at the extremes of the moisture range, i.e., near saturation and toward dryness. Here, significant progress has been achieved during the last years, which enables users of numerical models to avoid misconceptions. Current work on the topic is intensive and will lead soon to a wide availability of more appropriate functions, which better represent the hydraulic conductivity including film flow in particular at medium to dry soil moisture conditions. These relationships can be easily incorporated into standard simulation codes, improving predictions of water fluxes under semi-arid and arid conditions.

The problem of hysteresis is more difficult: despite the availability of traditional concepts of hysteresis, hysteresis is rarely incorporated into models. Comprehensive experiments are required in order to calibrate hysteretic models. Inverse modeling of inflow/outflow experiments appears to be a premier choice to obtain the desired information, but we have shown that it interferes with the problem of dynamic nonequilibrium of water flow. We believe that static hysteresis and dynamic equilibrium need to be brought to a unifying and usable concept. It is doubtful whether this can be achieved by further modifications of pore-bundle concepts. Further improvements may come from percolation theory.

It is currently not clear to what degree dynamic effects affect water flow on time and spatial scales that are relevant for practical applications. Despite its early notion, the problem of dynamic non-equilibrium during water flow in soils had not been treated intensively in the past, a situation that currently appears to change. This might be also related to better exchange of knowledge between the communities of soil physicists, chemical, civil and petroleum engineers, which, due to modeling two phase problems, are confronted with similar problems as in soil hydrology. But our expectation for significant progress in this field within short time is nevertheless modest. Reasons for this are the high requirements with respect to suitable measurements, and equally the difficulties in tracking the theoretical problem. Without the availability of fast computers and fast measurement techniques, quantitative comparisons of observed transient flow process variables with simulation results obtained by models of increasing complexity, such as Richards equation, coupled two-phase flow models, or pore-network models, are not possible. Finally, the reality of soil confronts us with much more challenges than merely hydraulic properties, hysteresis and dynamic effects. We must have in mind (i) the reality of soil particles (their mineral composition and spatial distribution and their chemical, physical, organic and microbiological properties, including biofilms), (ii) the reality of soil water (its nonhomogeneous distribution of cations and anions within any single pore or multitude of pores that changes at each different soil water content regardless of the net flux of water), (iii) the reality of the gaseous phase within soils that continually changes its composition and distribution owing to the metabolism of micro- and macro-organisms regardless of the net flux of water, and (iv) the reality of water within plant tissues and membranes (Don Nielsen 2014, personal communication). Furthermore, water movement in natural soils is driven by thermal and solute gradients in addition to gravity and matric potential. Therefore, we conclude that further progress will need not only a better consideration of the specification of soil hydraulic properties including hysteresis and DNE, but also a situation-specific coupling of the compartments of the soil-water-atmosphere continuum.

References

Assouline S (2001) A model for soil relative hydraulic conductivity based on the water retention characteristic curve. Water Resour Res 37(2):265–271

Assouline S (2004) Correction to "A model of soil relative hydraulic conductivity based on water retention curve characteristics". Water Resour Res 40(2), W02901

Assouline S, Or D (2013) Conceptual and parametric representation of soil hydraulic properties: a review. Vadose Zone J 12(4)

Assouline S, Tessier D, Bruand A (1998) A conceptual model of the soil water retention curve. Water Resour Res 34(2):223–231

Barenblatt GI (1971) Filtration of two nonmixing fluids in a homogeneous porous medium, Soviet academy Izvestia. Mech Gas Fluids 5:857–864

Basile A, Ciollaro G, Coppola A (2003) Hysteresis in soil water characteristics as a key to interpreting comparisons of laboratory and field measured hydraulic properties. Water Resour Res 39:1355

Bear J (1972) Dynamics of fluids in porous media. American Elsevier, New York

Børgesen CD, Jacobsen OH, Hansen S, Schaap MG (2006) Soil hydraulic properties near saturation, an improved conductivity model. J Hydrol 324(1):40–50

Brooks RH, Corey AT (1964) Hydraulic properties of porous media, vol 3, Hydrology papers. Colorado State University, Fort Collins, pp 1–27

Brutsaert W (1966) Probability laws for pore-size distributions. Soil Sci 101(2):85–92

Buckingham E (1907) Studies on the movement of soil moisture. US Department of Agriculture, Washington, DC

Burdine NT (1953) Relative permeability calculations from pore size distribution data. Trans Am Inst Min Metall Pet Eng 198:71–77

Campbell GS, Shiozawa S (1992) Prediction of hydraulic properties of soils using particle-size distribution and bulk density data. In: van Genuchten MT, Leij FJ, Lund LJ (eds) Proceedings of the international workshop on indirect methods for estimating the hydraulic properties of unsaturated soils. University of California, Riverside, pp 317–328

Childs EC, Collis-George N (1950) The permeability of porous materials. Proc R Soc Lond A Math Phys Sci 201(1066):392–405

Coppola A, Comegna V, Basile A, Lamaddalena N, Severino G (2009) Darcian preferential water flow and solute transport through bimodal porous systems: experiments and modelling. J Contam Hydrol 104(1):74–83

Corey AT, Brooks RH (1999) The Brooks-Corey relationships. In: van Genuchten MT, Leij FJ, Wu L (eds) Proceedings of the international workshop on characterization and measurement of the hydraulic properties of unsaturated porous media. University of California, Riverside, pp 13–18

Cornelis WM, Khlosi M, Hartmann R, Van Meirvenne M, De Vos B (2005) Comparison of unimodal analytical expressions for the soil-water retention curve. Soil Sci Soc Am J 69 (6):1902–1911

Cushman JH (1984) On unifying the concepts of scale, instrumentation, and stochastics in the development of multiphase transport theory. Water Resour Res 20(11):1668–1676

d'Hollander EH (1979) Estimation of the pore size distribution from the moisture characteristic. Water Resour Res 15(1):107–112

Dane JH, Hopmans JW (2002) Water retention and storage. In: Dane JH, Topp GC (eds) Methods of soil analysis, Part 4, vol 5, Soil Science Society of America book series. SSSA, Madison, pp 739–757, 671–717

Dane JH, Wierenga PJ (1975) Effect of hysteresis o the prediction of infiltration, redistribution and drainage of water in a layered soil. J Hydrol 25:229–242

Darcy H (1856) Les fontains publiques de la ville de Dijon. Dalmont, Paris

Dettmann U, Bechtold M, Frahm E, Tiemeyer B (2014) On the applicability of unimodal and bimodal van Genuchten-Mualem based models to peat and other organic soils under drying conditions. J Hydrol, in press, doi: 10.1016/j.jhydrol.2014.04.047

Diamantopoulos E, Durner W (2012) Dynamic nonequilibrium of water flow in porous media: a review. Vadose Zone J 11(3). http://dx.doi.org/a10.2136/vzj2011.0197

Diamantopoulos E, Iden SC, Durner W (2012) Inverse modeling of dynamic non-equilibrium in water flow with an effective approach. Water Resour Res 48:W03503. doi:10.1029/2011WR010717

Doerr SH, Shakesby RA, Walsh RPD (2000) Soil water repellency: its causes, characteristics and hydro-geomorphological significance. Earth-Sci Rev 51:33–65

Driessen PM, Konijn NT (1992) Land-use systems analysis. Agricultural University, Wageningen

Durner W (1991) Vorhersage der hydraulischen Leitfähigkeit strukturierter Böden. PhD thesis. Universität Bayreuth, Bayreuth, pp 1–180

Durner W (1992) Predicting the unsaturated hydraulic conductivity using multi-porosity water retention curves. In: van Genuchten MT et al (eds) Proceedings of the international workshop on indirect methods for estimating the hydraulic properties of unsaturated soils. University of California, Riverside, pp 185–202

Durner W (1994) Hydraulic conductivity estimation for soils with heterogeneous pore structure. Water Resour Res 30(2):211–223. doi:10.1029/93WR02676

Durner W, Flühler H (2005) Soil hydraulic properties (Chapter 74). In: Anderson MG, McDonnell JJ (eds) Encyclopaedia of hydrological sciences. Wiley, USA, pp 1089–1102

Durner W, Iden SC (2011) Extended multistep outflow method for the accurate determination of soil hydraulic properties near water saturation. Water Resour Res 47(8):W08526

Durner W, Lipsius K (2005) Determining soil hydraulic properties (Chapter 75). In: Anderson MG, McDonnell JJ (eds) Encyclopaedia of hydrological sciences. Wiley, New York, pp 1121–1144

Dury O, Fischer U, Schulin R (1998) Dependence of hydraulic and pneumatic characteristics of soils on a dissolved organic compound. J Contam Hydrol 33:39–57

Ewing JA (1881) On the production of transient electric currents in iron and steel conductors by twisting them when magnetized or by magnetizing them when twisted. In: Proceedings of the Royal Society of London, November 17 1881–March 30 1882

Farrell DA, Larson WE (1972) Modeling the pore structure of porous media. Water Resour Res 8 (3):699–706

Fatt I, Dykstra H (1951) Relative permeability studies. Trans Am Inst Min Metall Pet Eng 192:249–256

Fayer MJ, Simmons CS (1995) Modified soil water retention functions for all matric suctions. Water Resour Res 31(5):1233–1238. doi:10.1029/95WR00173

Flühler H, Durner W, Flury M (1996) Lateral mixing processes – a key for understanding the solute transport regimes. Geoderma 70:165–183

Fredlund DG, Xing AQ (1994) Equations for the soil-water characteristic curve. Can Geotech J 31(4):521–532. doi:10.1139/t94-061

Friedmann S (1999) Dynamic contact angle explanation of flow rate-dependent saturation-pressure relationships during transient liquid flow in unsaturated porous media. J Adhes Sci Technol 13:1495–1518

Fujimaki H, Inoue M (2003) A transient evaporation method for determining soil hydraulic properties at low pressure. Vadose Zone J 2(3):400–408. doi:10.2113/2.3.400

Funk M (2012) Hysteresis der Feuchtespeicherung in porösen Materialien. PhD dissertation, Institut für Bauklimatik, TU Dresden, Dresden

Gardner WR (1958) Some steady-state solutions of the unsaturated moisture flow equation with application to evaporation from a water table. Soil Sci 85(4):228–232

Gerke HH (2006) Preferential flow descriptions for structured soils. J Plant Nutr Soil Sci 169(3):382–400

Gerke HH, van Genuchten M (1993) A dual porosity model for simulating the preferential movement of water and solutes in structured porous media. Water Resour Res 29(2):305–319

Gillham RW, Klute A, Heermann DF (1976) Hydraulic properties of a porous medium: measurement and empirical representation. Soil Sci Soc Am J 40(2):203–207

Gillham RW, Klute A, Heermann DF (1979) Measurement and numerical simulation of hysteretic flow in a heterogeneous porous medium. Soil Sci Soc Am J 43:1061–1067

Globus AM (1987) Soil-hydrophysical supporting of the agroecological mathematical models (in Russian). Hydrometeoizdat Press, Leningrad

Goss K-U, Madliger M (2007) Estimation of water transport based on in situ measurements of relative humidity and temperature in a dry Tanzanian soil. Water Resour Res 43:W05433. doi:10.1029/2006WR005197

Groenevelt PH, Grant CD (2004) A new model for the soil-water retention curve that solves the problem of residual water contents. Eur J Soil Sci 55(3):479–485. doi:10.1111/j.1365-2389.2004.00617.x

Haines WB (1930) Studies in the physical properties of soil: V. The hysteresis effect in capillary properties, and the modes of moisture distribution associated therewith. J Agric Sci 20:97–116

Harris CC, Morrow NR (1964) Pendular moisture in packings of equal spheres. Nature 203:706–708

Hassanizadeh SM, Celia MA, Dahle HK (2002) Dynamic effects in capillary pressure–saturation relationships and its impact on unsaturated flow. Vadose Zone J 1:38–57

Henry EJ, Smith JE, Warrick AW (2001) Surfactant effects on unsaturated flow in porous media with hysteresis: horizontal column experiments and numerical modeling. J Hydrol 245:73–88

Hilfer R (2006a) Macroscopic capillarity and hysteresis for flow in porous media. Phys Rev E 73(1):016307

Hilfer R (2006b) Capillary pressure, hysteresis and residual saturation in porous media. Phys Stat Mech Appl 359:119–128

Hillel D (1980) Fundamentals of soil physics. Academic Press, New York, 413

Hoa NT, Gaudu R, Thirriot C (1977) Influence of the hysteresis effect on transient flows in saturated–unsaturated porous media. Water Resour Res 13(6):992–996

Hoffmann-Riem H, van Genuchten MT, Flühler H (1999) General model for the hydraulic conductivity of unsaturated soils. In: van Genuchten MT, Leij FJ, Wu L (eds) Proceedings of the international workshop on characterization und measurement of the hydraulic properties of unsaturated porous media. University of California, Riverside, pp 31–42

Hogarth W, Hopmans JW, Parlange JY, Haverkamp R (1988) Application of a simple soil-water hysteresis model. J Hydrol 98:21–29

Hopmans JW, Šimůnek J, Romano N, Durner W (2002) Inverse methods. In: Dane JH, Topp GC (eds) Methods of soil analysis, part 4: physical methods, vol 5, 4th edn, Soil Science Society of America book series. American Society of Agronomy and Soil Science Society of America, Madison, pp 963–1008

Hunt AG (2004) Comparing van Genuchten and percolation theoretical formulations of the hydraulic properties of unsaturated media. Vadose Zone J 3(4):1483–1488

Hunt AG, Skinner TE (2005) Hydraulic conductivity limited equilibration. Vadose Zone J 4(1):145–150

Hunt AG, Ewing RP, Horton R (2013) What's wrong with soil physics? Soil Sci Soc Am J 77(6):1877–1887

Hutson JL, Cass A (1987) A retentivity function for use in soil–water simulation models. Soil Sci 38(1):105–113

Iden SC, Durner W (2007) Free form estimation of the unsaturated soil hydraulic properties by inverse modeling using global optimization. Water Resour Res 43(7):W07451. doi:2006WR005845

Ippisch O, Vogel HJ, Bastian P (2006) Validity limits for the van Genuchten–Mualem model and implications for parameter estimation and numerical simulation. Adv Water Resour 29(12):1780–1789

Jarvis NJ (2007) A review of non-equilibrium water flow and solute transport in soil macropores: principles, controlling factors and consequences for water quality. Eur J Soil Sci 58(3):523–546

Jaynes DB (1992) Estimating hysteresis in the soil water retention function. In: van Genuchten MT, Leij FJ, Lund LJ (eds) Proceedings of the international workshop on indirect methods for estimating the hydraulic properties of unsaturated soils. University of California, Riverside, pp 219–232

Khlosi M, Cornelis WM, Gabriels D, Sin G (2006) Simple modification to describe the soil water retention curve between saturation and oven dryness. Water Resour Res 42:W11501. doi:10.1029/2005WR004699

King LG (1965) Description of soil characteristics for partially saturated flow. Soil Sci Soc Am J 29(4):359–362

Klute A, Dirksen C (1986) Hydraulic conductivity and diffusivity: laboratory methods. In: Klute A (ed) Methods of soil analysis, part 1, vol 9, 2nd edn, Agronomy monographs. ASA and SSSA, Madison, pp 687–734

Kool JB, Parker JC (1987) Development and evaluation of closed-form expressions for hysteretic soil hydraulic properties. Water Resour Res 23(1):105–114

Kosugi K (1994) Three-parameter lognormal distribution model for soil water retention. Water Resour Res 30(4):891–901

Kosugi K (1996) Lognormal distribution model for unsaturated soil hydraulic properties. Water Resour Res 32(9):2697–2703. doi:10.1029/96WR01776

Kosugi K (1999) General model for unsaturated hydraulic conductivity for soils with lognormal pore-size distribution. Soil Sci Soc Am J 63(2):270–277

Kosugi K, Hopmans JW, Dane JH (2002) Parametric models. In: Dane JH, Topp GC (eds) Methods of soil analysis, Part 4, vol 5, Soil Science Society of America book series. SSSA, Madison, pp 739–757

Kutilek M, Nielsen D (1994) Soil hydrology. Catena Verlag, Cremlingen-Destedt, 370 pp

Laliberte GE (1969) A mathematical function for describing capillary pressure-desaturation data. Hydrol Sci J 14(2):131–149

Lebeau M, Konrad JM (2010) A new capillary and thin film flow model for predicting the hydraulic conductivity of unsaturated porous media. Water Resour Res 46(12): W12554. doi:10.1029/2010WR009092

Lehmann P, Stauffer F, Hinz C, Dury O, Flühler H (1999) Effect of hysteresis on water flow in a sand column with a fluctuating capillary fringe. J Contam Hydrol 33:81–100

Leij FJ, Russell WB, Lesch SM (1997) Closed form expressions for water retention and conductivity data. Ground Water 35:848–858. doi:10.1111/j.1745-6584.1997.tb00153.x

Lenhard RJ, Parker JC, Kaluarachchi JJ (1991) Comparing simulated and experimental hysteretic two-phase transient fluid flow phenomena. Water Resour Res 27:2113–2124

Luckner L, van Genuchten MT, Nielsen DR (1991) Reply to Nimmo's "comment on the treatment....". Water Resour Res 27(4):661–662

Mallants D, Tseng PH, Toride N, Tinunerman A, Feyen J (1997) Evaluation of multimodal hydraulic functions in characterizing a heterogeneous field soil. J Hydrol 195(1):172–199

McKee CR, Bumb AC (1984) The importance of unsaturated flow parameters in designing a monitoring system for hazardous wastes and environmental emergencies. In: Proceedings of Hazardous Materials Control Research Institute national conference, Houston, pp 50–58

McKee CR, Bumb AC (1987) Flow-testing coalbed methane production wells in the presence of water and gas. SPE Form Eval 2(4):599–608

McNamara H (2014) An estimate of energy dissipation due to soil-moisture hysteresis. Water Resour Res 50:725–735. doi:10.1002/2012WR012634

Mirzaei M, Das BS (2007) Dynamic effects in capillary pressure-saturation relationships for two-phase flow in 3D porous media: implication of micro-heterogeneities. Chem Eng Sci 62:1927–1947

Mirzaei M, Das BS (2013) Experimental investigation of hysteretic dynamic effect in capillary pressure-saturation relationship for two phase-flow in porous media. AIChE J 59(10):3958–3974

Mitchell RJ, Mayer AS (1998) The significance of hysteresis in modeling solute transport in unsaturated porous media. Soil Sci Soc Am J 62:1506–1512

Morel-Seytoux HJ, Nimmo JR (1999) Soil water retention and maximum capillary drive from saturation to oven dryness. Water Resour Res 35(7):2031–2041. doi:10.1029/1999WR900121

Mualem Y (1974) A conceptual model of hysteresis. Water Resour Res 10(3):514–520

Mualem Y (1976) A new model for predicting the hydraulic conductivity of unsaturated porous media. Water Resour Res 12(3):513–522. doi:10.1029/WR012i003p00513

Mualem Y (1984) A modified dependent-domain theory of hysteresis. Soil Sci 137(5):283–291

Mualem Y (1986) Hydraulic conductivity of unsaturated soils: prediction and formulas. In: Klute A (ed) Methods of soil analysis: part 1. Physical and mineralogical methods. American Society of Agronomy: Soil Science Society of America, Madison, pp 799–823

Mualem Y, Beriozkin A (2009) General scaling rules of the hysteretic water retention function based on Mualem's domain theory. Eur J Soil Sci 60:652–661

Nemes A, Schaap MG, Leij FJ, Wösten JHM (2001) Description of the unsaturated soil hydraulic database UNSODA version 2.0. J Hydrol 251(3–4):151–162. doi:10.1016/S0022-1694(01)00465-6

Nimmo JR (1991) Comment on the treatment of residual water content in "A consistent set of parametric models for the two-phase flow of miscible fluid in the subsurface" by L. Luckner et al., 1989. Water Resour Res 27:661–662. doi:10.1029/91WR00165

Orr FM, Scriven LE, Rivas AP (1975) Pendular rings between solids: meniscus properties and capillary force. J Fluid Mech 67(4):723–742

Othmer H, Diekkrüger B, Kutilek M (1991) Bimodal porosity and unsaturated hydraulic conductivity. Soil Sci 152(3):139–150. doi:10.1097/00010694-199109000-00001

Pachepsky YA, Shcherbakov RA, Korsunskaya LP (1995) Scaling of soil water retention using a fractal model. Soil Sci 159(2):99–104

Parlange J-Y (1976) Capillary hysteresis and relationship between drying and wetting curves. Water Resour Res 12:224–228

Peters A (2013) Simple consistent models for water retention and hydraulic conductivity in the complete moisture range. Water Resour Res 49:6765–6780. doi:10.1002/wrcr.20548

Peters A, Durner W (2008) A simple model for describing hydraulic conductivity in unsaturated porous media accounting for film and capillary flow. Water Resour Res 44:W11417. doi:10.1029/2008WR007136

Pham HQ, Fredlund DG, Barbour SL (2005) A study of hysteresis models for soil-water characteristic curves. Can Geotech J 42(6):1548–1568

Philip JR, de Vries DA (1957) Moisture movement in porous media under temperature gradients. Eos Trans AGU 38(2):222–232

Poulovassilis A (1970) Hysteresis of pore water in granular pore bodies. Soil Sci 109(1):5–12

Poulovassilis A, Childs EE (1971) The hysteresis of pore water: the non-independence of domains. Soil Sci 112:301–312

Priesack E, Durner W (2006) Closed-form expression for the multi-modal unsaturated conductivity function. Vadose Zone J 5(1):121–124

Purcell WR (1949) Capillary pressures-their measurement using mercury and the calculation of permeability therefrom. J Pet Technol 1(2):39–48

Richards LA (1931) Capillary conduction of liquids through porous media. Physics 1:318–333

Rogowski AS (1972) Estimation of the soil moisture characteristic and hydraulic conductivity: comparison of models. Soil Sci 114(6):423–429

Romano N (2014) Soil moisture at local scale: measurements and simulations. J Hydrol, in press. http://dx.doi.org/10.1016/j.jhydrol.2014.01.026. Available online 21 Jan 2014

Romano N, Nasta P, Severino G, Hopmans JW (2011) Using bimodal lognormal functions to describe soil hydraulic properties. Soil Sci Soc Am J 75(2):468–480

Ross PJ, Smettem KRJ (1993) Describing soil hydraulic properties with sums of simple functions. Soil Sci Soc Am J 57(1):26–29. doi:10.2136/sssaj1993.03615995005700010006x

Ross PJ, Williams J, Bristow KL (1991) Equation for extending water-retention curves to dryness. Soil Sci Soc Am J 55(4):923–927. doi:10.2136/sssaj1991.03615995005500040004x

Rossi C, Nimmo JR (1994) Modeling of soil water retention from saturation to oven dryness. Water Resour Res 30(3):701–708. doi:10.1029/93WR03238

Rubin J (1965) Numerical method for analyzing hysteresis-affected, post-infiltration redistribution of soil moisture. Soil Sci Soc Am Proc 31:13–27

Russo D (1988) Determining soil hydraulic properties by parameter estimation: on the selection of a model for the hydraulic properties. Water Resour Res 24(3):453–459

Russo D, Jury WA, Butters GL (1989) Numerical analysis of solute transport during transient irrigation: 1. The effect of hysteresis and profile heterogeneity. Water Resour Res 25(10):2109–2118

Saito H, Simunek J, Mohanty BP (2006) Numerical analysis of coupled water, vapor, and heat transport in the vadose zone. Vadose Zone J 5:784–800. doi:10.2136/vzj2006.0007

Sakaki T, O'Carroll DM, Illangasekare TH (2010) Direct quantification of dynamic effects in capillary pressure for drainage-wetting cycles. Vadose Zone J 9:424–437

Schaap MG, Leij FJ (2000) Improved prediction of unsaturated hydraulic conductivity with the Mualem-van Genuchten model. Soil Sci Soc Am J 64(3):843–851. doi:10.2136/sssaj2000.643843x

Schelle H, Durner W, Schlüter S, Vogel H-J, Vanderborght J (2013) Virtual soils: moisture measurements and their interpretation by inverse modeling. Vadose Zone J 12(3), doi:10.2136/vzj2012.0168

Schindler U, Durner W, Von Unold G, Müller L (2010a) Evaporation method for measuring unsaturated hydraulic properties of soils: extending the measurement range. Soil Sci Soc Am J 74(4):1071–1083

Schindler U, Durner W, von Unold G, Mueller L, Wieland R (2010b) The evaporation method: extending the measurement range of soil hydraulic properties using the air-entry pressure of the ceramic cup. J Plant Nutr Soil Sci 173(4):563–572

Schlüter S, Vogel H-J, Ippisch O, Bastian P, Roth K, Schelle H, Durner W, Kasteel R, Vanderborght J (2012) Virtual soils: assessment of the effects of soil structure on the hydraulic behavior of cultivated soils. Vadose Zone J 11(3), doi:10.2136/vzj2011.0174

Schmid KS, Gross J, Helmig R (2014) Chemical osmosis in two-phase flow and salinity-dependent capillary pressures in rocks with microporosity. Water Resour Res 50(2):763–789. doi:10.1002/2013WR013848

Schneider M, Goss K-U (2012) Prediction of the water sorption isotherm in air dry soils. Geoderma 170:64–69. doi:10.1016/j.geoderma.2011.10.008

Schofield RK (1935) The pF of the water in soil. In: Transactions of the 3rd international congress of soil science. T. Murby, Oxford, pp 38–48

Schultze B, Ippisch O, Huwe B, Durner W (1999) Dynamic nonequilibrium in unsaturated water flow. In: van Genuchten MT, Leij FJ, Wu L (eds) Proc. int. workshop on characterization und measurement of the hydraulic properties of unsaturated porous media. University of California, Riverside, pp 877–892

Scott PS, Farquhar GJ, Kouwen N (1983) Hysteretic effects on net infiltration, advances in infiltration. Am. Soc. Agric. Eng. Publ, St. Josephs, pp 11–83, 163–170

Shinomiya Y, Takahashi K, Kobiyama M, Kubota J (2001) Evaluation of the tortuosity parameter for forest soils to predict unsaturated hydraulic conductivity. J For Res 6(3):221–225

Si BC, Kachanoski RG (2000) Unified solution for infiltration and drainage with hysteresis: theory and field test. Soil Sci Soc Am J 64:30–36

Silva O, Grifoll J (2007) A soil-water retention function that includes the hyper-dry region through the BET adsorption isotherm. Water Resour Res 43:W11420. doi:10.1029/2006WR005325

Simmons CS, Nielsen DR, Biggar JW (1979) Scaling of field-measured soil-water properties. I. Methodology. II. Hydraulic conductivity and flux. Hilgardia Calif Agric Exp Stn 47:1–15

Simunek J, Wendroth O, Wypler N, van Genuchten MT (2001) Nonequilibrium water flow characterized from an upward infiltration experiment. Eur J Soil Sci 52:13–24

Šimůnek J, Jarvis NJ, Van Genuchten MT, Gärdenäs A (2003) Review and comparison of models for describing non-equilibrium and preferential flow and transport in the vadose zone. J Hydrol 272(1):14–35

Stauffer F (1978) Time dependence of the relations between capillary pressure water content and conductivity during drainage of porous media. In: IAHR symposium on scale effects in porous media, Thessaloniki, 29 Aug–1 Sept 1978. IAHR, Madrid

Stolte J, Freijer JI, Bouten W, Dirksen C, Halbertsma JM, Van Dam JC, Van den Berg JA, Veerman GJ, Wösten JHM (1994) Comparison of six methods to determine unsaturated soil hydraulic conductivity. Soil Sci Soc Am J 58:1596–1603

Tani M (1982) The properties of a water-table rise produced by a one-dimensional, vertical, unsaturated flow. J Jpn For Soc 64:409–418

Tokunaga TK (2009) Hydraulic properties of adsorbed water films in unsaturated porous media. Water Resour Res 45(6):W06415. doi:10.1029/2009WR007734

Toledo PG, Novy RA, Davis HT, Scriven LE (1990) Hydraulic conductivity of porous media at low water content. Soil Sci Soc Am J 54(3):673–679

Topp GC, Klute A, Peters DB (1967) Comparison of water content-pressure head data obtained by equilibrium, steady state, and unsteady-state methods. Soil Sci Soc Am Proc 31:312–314

Tuller M, Or D (2001) Hydraulic conductivity of variably saturated porous media: film and corner flow in angular pore space. Water Resour Res 37(5):1257–1276. doi:10.1029/2000WR900328

Tuller M, Or D, Dudley LM (1999) Adsorption and capillary condensation in porous media: liquid retention and interfacial configurations in angular pores. Water Resour Res 35(7):1949–1964

Tyler SW, Wheatcraft SW (1989) Application of fractal mathematics to soil water retention estimation. Soil Sci Soc Am J 53(4):987–996

van Dam JC, Hendrickx JMH, van Ommen HC, Bannink MH, van Genuchten MT, Dekker LW (1990) Water and solute movement in a coarse-textured water repellent field soil. J Hydrol 120:359–379

van Genuchten MT (1980) A closed-form equation for predicting the hydraulic conductivity of unsaturated soils. Soil Sci Soc Am J 44(5):892–898. doi:10.2136/sssaj1980.03615995004400050002x

Van Genuchten MT, Nielsen DR (1985) On describing and predicting the hydraulic properties of unsaturated soils. Ann Geophys 3(5):615–628

Vanclooster M, Boesten AJ, Tiktak N, Jarvis JG, Kroes R, Munoz-Carpena R, Clothier BE, Green SR (2004) On the use of unsaturated flow and transport models in nutrient and pesticide management. In: Feddes RA, de Rooij GH, van Dam JC (eds) Unsaturated-zone modeling, UR Frontis series. Kluwer Academic, Dordrecht, pp 331–362

Vauclin M, Haverkamp R, Vachaud G (1979) Résolution numérique d'une équation de diffusion non linéaire. Presses Univ de Grenoble, Grenoble, 183 pp

Viaene P, Vereecken H, Diels J, Feyen J (1994) A statistical analysis of six hysteresis models for the moisture retention characteristic. Soil Sci 157:345–355

Visser WC (1968) An empirical expression for the desorption curve. In: Rijtema PE, Wassink H (eds) Water in the unsaturated zone, Proceedings of the Wageningen Symposium, IASH/AIHS, Unesco, Paris, France, pp 329–335

Vogel T, van Genuchten MT, Cislerova M (2001) Effect of the shape of the soil hydraulic functions near saturation on variably-saturated flow predictions. Adv Water Resour 24:133–144

Wang Y, Ma J, Zhang Y, Zhao M, Edmunds WM (2013) A new theoretical model accounting for film flow in unsaturated porous media. Water Resour Res 49:5021–5028. doi:10.1002/wrcr.20390

Webb SW (2000) A simple extension of two-phase characteristic curves to include the dry region. Water Resour Res 36(6):1425–1430. doi:10.1029/2000WR900057

Weller U, Ippisch O, Köhne JM, Vogel HJ (2011) Direct measurement of unsaturated hydraulic conductivity in sandy soil including temporal dynamics, hydraulic non-equilibrium and hysteresis. Vadose Zone J 10:654–661

Wildenschild D, Hopmans JW, Simunek J (2001) Flow rate dependence of soil hydraulic characteristics. Soil Sci Soc Am J 65:35–48

Zhang ZF (2011) Soil water retention and relative permeability for conditions from oven-dry to full saturation. Vadose Zone J 10(4):1299–1308. doi:10.2136/vzj2011.0019

Zhang R, Van Genuchten MT (1994) New models for unsaturated soil hydraulic properties. Soil Sci 158(2):77–85

Chapter 18
Water Availability to Plants

Quirijn de Jong van Lier

Abstract Water is essential for plant growth, and modeling of water availability to plants, root water uptake and actual transpiration rates is important to support hydrological, meteorological and crop growth modeling. Soil texture, soil structure and plant rootability are the crucial factors in determining the amount of soil water available for plants. In a simple approximation, available and non-available water is often defined in terms of concepts like field capacity and permanent wilting point. In this chapter these concepts are discussed, together with more advanced approaches. Topics include the water potentials in the soil-plant-atmosphere system, the relation between transpiration and dry matter accumulation, zones of soil water availability to plants, the limiting water content for optimum plant transpiration and growth, the role of the matric flux potential, a composite soil hydraulic property, and aspects of root system geometry and how they affect crop water availability.

Keywords Soil water balance • Evapotranspiration • Root water uptake • Drought stress

18.1 Introduction

Water is essential for plant growth, but contrarily to nutrients that are absorbed by roots or leaves and incorporated in plant dry matter by chemical reactions, almost all water taken up by plant roots is lost to the atmosphere by transpiration from the leaves. Only a tiny portion, between 0.1 and 1 % of the absorbed water remains in the plant, as water or chemically transformed to other components. Water productivity, the volume of water needed for the production of a unit mass of crop dry

Q. de Jong van Lier (✉)
Center for Nuclear Energy in Agriculture, University of São Paulo (CENA/USP), São Paulo, Brazil
e-mail: qdjvlier@usp.br

W.G. Teixeira et al. (eds.), *Application of Soil Physics in Environmental Analyses: Measuring, Modelling and Data Integration*, Progress in Soil Science,
DOI 10.1007/978-3-319-06013-2_18,

matter, is reported to be of the order of 1–10 m^3 water per kg of grain production (Rockström and Barron 2007). These figures make clear the huge importance of water to plant growth and crop production.

Modeling of water availability to plants, root water uptake and its partitioning over depth and actual transpiration rates is important to support hydrological, meteorological and crop growth modeling. It is one of the major contributions of soil physics to irrigation management and engineering. Irrigation aims to supply the correct amount of water to plants to maximize yield. To achieve the correct timing and apply the right amount of water, soil physical knowledge or, more specifically, knowledge about the soil-water-root interactions, is of utmost importance.

Although most attention has been given to water availability to plants in the agronomic context, the importance of water availability to plants in an ecological approach is more and more recognized as it determines vegetation types and plant species strategies and adaptations to cope with stress due to drought or excess water (Eagleson 1982; Cui et al. 2009).

From a qualitative point of view, soil texture, soil structure and plant rootability are the crucial factors in determining the amount of soil water available for plants. Rootability can be constrained by physical (compaction, hardpans) or chemical (e.g. acidity, aluminum) soil characteristics. In a simple (soil physical) approximation, available and non-available water is often defined in terms of concepts like field capacity and permanent wilting point. In this chapter these concepts will be discussed, together with more advanced approaches.

18.2 Water Potentials is the Soil-Plant-Atmosphere Continuum

Water availability to plants reflects the ease with which plants can extract water from the soil. This water uptake is a dynamic process that depends on driving forces and conductances in the soil-plant pathway. Water and solutes move from positions with a higher to a lower total potential energy, which is from soil to plant to atmosphere in almost all situations.

The potential energy of water in the unsaturated soil is essentially determined by position or depth (gravitational component), by water content (matric component) and by solute concentration (osmotic component). The osmotic component is of relevance to soil water availability only in the presence of semi-permeable barriers like plant membranes or in the case of water transport in the vapor phase. The matric potential (or pressure head) varies, under field conditions, from zero at saturation to some hundreds of meters negative under very dry conditions, whereas the gravity component (compared to a zero reference at the soil surface) reduces to some decimeters or meters negative, corresponding to the rooted depth. From these orders of magnitude it becomes clear that variations in the matric component are those that usually prevail in total potential energy variation, an exception made for soils very close to saturation (when matric potential is very small) and those specific cases in which osmotic potential plays an important role.

In the atmosphere, the potential energy of water ψ (m) is mainly determined by the thermodynamic temperature T (K) and relative humidity Ω (Pa Pa^{-1}), according to

$$\psi = \frac{RT}{\rho_a g v_a} \ln(\Omega) \tag{18.1}$$

where R is the universal gas constant ($R = 8.314$ J mol^{-1} K^{-1}), ρ_a is the specific mass of liquid water ($\rho_a \approx 1{,}000$ kg m^{-3}), g is the gravitational acceleration at the earth surface ($g \approx 9.81$ m s^{-2}), and v_a is the molar volume of liquid water ($v_a \approx 1.8 \cdot 10^{-5}$ m^3 mol^{-1}). Substituting the values of R and approximate values for ρ_a, g and v_a in Eq. 18.1 yields

$$\psi \approx 47.1 \cdot T \ln(\Omega) \tag{18.2}$$

Except for atmospheric conditions very close to vapor saturation ($\Omega = 1$), vapor potential is much lower (more negative) than the soil water potential. To give an idea, at 50 % of relative humidity ($\Omega = 0.5$) and temperature around 300 K, the specific energy of atmospheric water vapor is of the order of -10^4 m. Even at a high relative humidity of 80 %, Eq. 18.2 yields values around −3,000 m, much more negative than the lower limit values of about −500 m that may occur in very dry soils. Consequently, it is a natural tendency for water to move from soil to atmosphere, directly (by evaporation) or through the plant (by transpiration). The partition of evapotranspiration in evaporation and transpiration depends on many factors, mainly on soil cover (leaf area index and crop residues) and on water availability in the rooted zone.

Inside a plant, the potential energy of water will equilibrate at values somewhere in between the value in the soil and in the atmosphere, decreasing from root to stem to leaf, as shown in Fig. 18.1. It is not the objective of this text to present details about plant water potentials which can be found in classical text books like Van Wijk (1963) and Slatyer (1967).

18.3 Transpiration and Dry Matter Accumulation

The gradients of potential energy of water in the soil-plant-atmosphere continuum give rise to water flow through the system. The intensity of this flow is determined by the gradients of potential themselves, but also by the hydraulic resistance of the pathway. The SPA-system can be considered similar to a network of resistors in an electrical circuit, taking water fluxes as analogues of electrical current, and gradients of potential energy corresponding to voltage difference (Van Den Honert 1948; Cowan 1965). Analogous to Ohm´s law, water flux and potentials in the system can then be described. The potentials tend to equilibrate at values which yield equal flow rates in all parts of the system. Under these steady state conditions, no excess or deficit of water will arise in any part of the system. As boundary

Fig. 18.1 Order of magnitude of water potentials in the soil-plant-atmosphere system

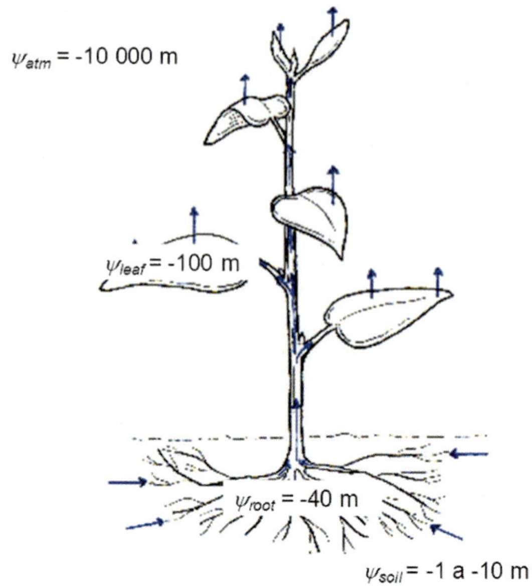

conditions in soil and atmosphere are constantly changing, this final equilibrium will, however, never be reached.

Without rainfall or irrigation, and when no water table is close to the surface, soils tend to dry out while plants extract water to transpire. Hydraulic conductivity is highly dependent on water content and decreases sharply when water content reduces. Consequently, the soil hydraulic resistance increases, and to maintain the intensity of water uptake imposed by the atmospheric vapor pressure deficit and leaf-to-atmosphere (stomatal) conductance, root and leaf potential will decrease, increasing the pressure gradient between soil and plant. There is a lower limit of plant (root and leaf) potential and close to this limit physiological mechanisms trigger the closure of stomata, increasing the leaf-to-atmosphere resistance and reducing water loss to the atmosphere by transpiration. Increasing stomatal resistance, however, reduces not only the water loss to the atmosphere, but also the uptake of CO_2 from the atmosphere. As CO_2 uptake is a limiting factor for plant growth, the reduction of transpiration by increasing the stomatal resistance has a collateral effect: the reduction of photosynthesis due to CO_2 depletion in the stomatal cavities. This mechanism is the basis of the following equation:

$$\frac{dY}{dt} = \frac{T_a}{T_p}\frac{dY_p}{dt} \tag{18.3}$$

in which dY/dt is the crop dry matter increase rate (kg d^{-1}), dY_p/dt is the potential crop dry matter increase rate (kg d^{-1}), T_a is the actual transpiration rate (mm d^{-1}) and T_p is the potential transpiration rate (mm d^{-1}). Note that when $T_a = T_p$, then $dY/dt = dY_p/dt$. In that case, rates are not limited by soil water availability and these

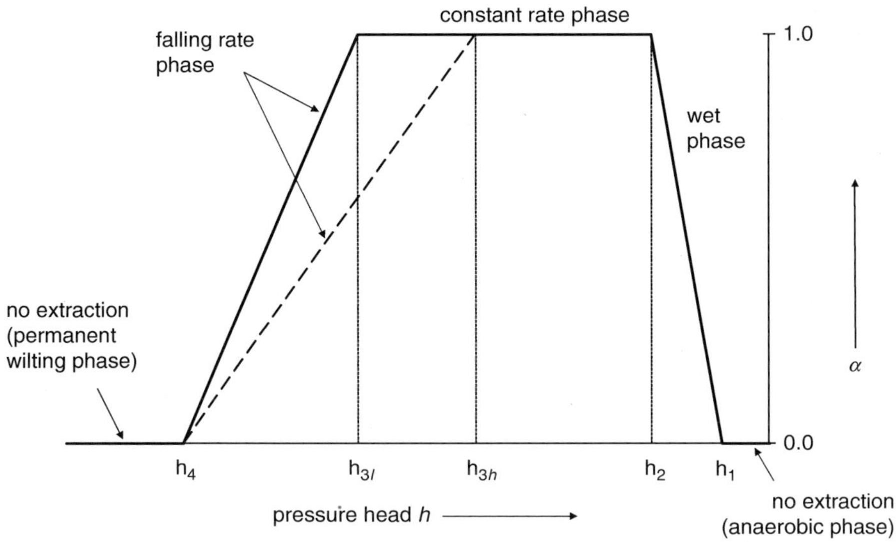

Fig. 18.2 Transpiration reduction factor α versus pressure head h (Adapted from Feddes et al. 1978)

conditions are referred to as hydraulically non-limiting. On the other hand, under limiting hydraulic conditions, stomata are partially or totally closed, T_a is smaller than T_p and $dY/dt < dY_p/dt$.

The ratio T_a/T_p, also called relative transpiration or transpiration *reduction factor* is very important for crop growth and ecological modeling. It is sometimes represented by the symbol α and its relation to soil hydraulic conditions (soil water content, pressure head or some other state variable) is called the (transpiration) *reduction function.* Prediction of this reduction function is important for any hydrological or yield prediction model. A frequently employed reduction function is the function proposed by Feddes et al. (1978), shown in Fig. 18.2. It is characterized by five ranges of pressure head. In the constant rate phase, delimited at the wet side by pressure head h_2 and at the dry side by h_3, the soil hydraulic conditions for plant growth are optimal and stomata are fully opened, making $T_a = T_p$ and $\alpha = 1$. At the right side of the constant rate phase (less negative pressure heads and a wetter soil) are the wet phase and the anaerobic phase, where plants become stressed due to a deficit aeration of the root zone. Left of the constant rate phase is the falling rate phase ($h < h_3$) and the permanent wilting phase ($h < h_4$). Following the empirical model of Feddes et al. (1978), the value of h_3 (limiting value between the constant and falling rate phase) depends on the atmospheric conditions. At low atmospheric demand (typically characterized by low temperature, low wind speed, high relative humidity), the onset of limiting hydraulic conditions is associated to a more negative pressure head (or: to a lower soil water content) than under high atmospheric demand conditions. The model uses two values of h_3 shown in Fig. 18.2: h_{3l} corresponds to low atmospheric demand, whereas h_{3h} corresponds to high demand. On its turn, h_4 corresponds to the onset of permanent wilting conditions.

18.4 Zones of Soil Water Availability

As a consequence of the relation between soil water content, soil hydraulic conductivity, pressure head and drainability, four zones of soil water content can be distinguished with respect to its availability (Fig. 18.3):

1. *Very wet soil* (with a water content higher than "field capacity"): Under these conditions, hydraulic conductivity is high and water flow to plant roots is easy, making water available to plants. On the other hand, the high hydraulic conductivity causes water to drain out of the rooting zone at a high rate, becoming unavailable in a relatively short time, in the order of days.
2. *Moist soil* (soil water content between "field capacity" and "limiting water content", the onset of the falling rate phase): Hydraulic conductivity is lower than in the very wet soil, making drainage to occur at a negligible rate, but it is still sufficient for root water uptake to occur at rates compatible to the potential transpiration rate. Under these conditions, water is easily available to plants while not subject to significant drainage.
3. *Dry soil* (water content between "limiting water content" and "permanent wilting point"): in this zone of water content, hydraulic conductivity is too low for the flow of water from bulk soil to root surface to occur at a rate compatible with transpiration demand. The plant closes its stomata to reduce transpiration and avoid wilting. In this range, soil water is available to plants, however, the rate at which it can be extracted from the soil is lower than the atmospheric demand, restricting photosynthesis and dry matter accumulation.
4. *Very dry soil* (water content below the "permanent wilting point"): The soil hydraulic conductivity is very low and high pressure head gradients are necessary to extract soil water. Soil water is unavailable to plants due to the low soil hydraulic conductivity and the plant cannot survive under these conditions.

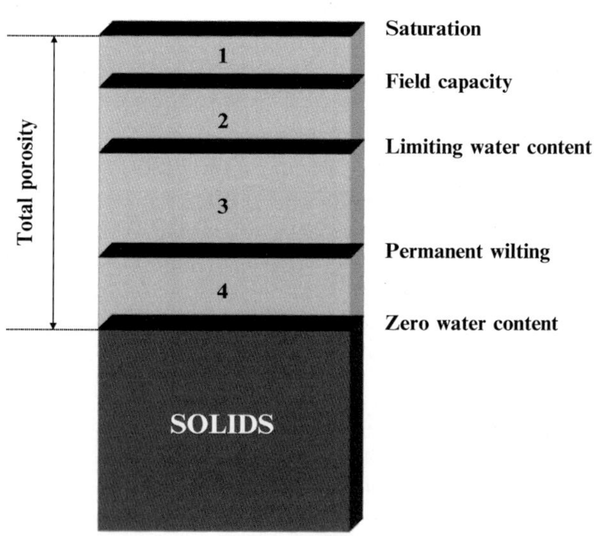

Fig. 18.3 Schematic representation of the soil as a water reservoir, showing the four zones of water content and their limits

It is common to express the total plant available soil water (θ_{taw}, $m^3\ m^{-3}$) as the difference between the water content at field capacity (θ_{fc}, $m^3\ m^{-3}$) and at the permanent wilting point (θ_{pwp}, $m^3\ m^{-3}$):

$$\theta_{taw} = \theta_{fc} - \theta_{pwp} \tag{18.4}$$

Multiplying θ_{taw} for the n rooted layers of a soil by their respective thickness (z, m) the soil available water storage in units of $m^3\ m^{-2} = m$ is obtained, also called the total plant available water storage (TAW, m):

$$TAW = \sum_{i=1}^{n} \theta_{taw,i} z_i \tag{18.5}$$

Analogously, the readily available soil water (θ_{raw}, $m^3\ m^{-3}$) is defined as the difference between θ_{fc} and θ_{lim}, and the readily available soil water storage (RAW, m) as the product of θ_{raw} and the respective thicknesses of the n rooted soil layers:

$$\theta_{raw} = \theta_{fc} - \theta_{\text{lim}} \tag{18.6}$$

$$RAW = \sum_{i=1}^{n} \theta_{raw,i} z_i \tag{18.7}$$

Although θ_{taw}, θ_{raw}, TAW and RAW are well defined, it is difficult to attribute an objective value due to the difficulty in defining or determining values for θ_{fc}, θ_{lim} and θ_{pwp}. In the following, these three hydraulic conditions will be discussed in more detail.

18.5 Field Capacity

Water availability is not only determined by the ease with which plants can extract it from the soil, but also by the time it resides in the rooted zone. After a rainfall or irrigation, in well-drained soils drainage will occur at high rates, causing part of the water to become rapidly unavailable for plants. For this reason, this draining fraction of soil water is normally considered to be unavailable to plants. The so-called "field capacity" refers to the hydraulic conditions (water content - θ_{fc} – or pressure head – h_{fc}) at which this drainage rate is considered to be negligible. It corresponds to the division of soil water in "rapidly draining" and "held by retention". In the words of Veihmeyer and Hendrickson (1931, 1949), who suggested methods for the determination of field capacity, it refers to the amount of soil moisture or water content held in a soil after the excess water has drained away and the rate of downward movement has substantially decreased. Possibly due to the lack of precision in this definition, field capacity is sometimes understood as the amount of water the soil retains against gravity. This is physically untrue, and it can be shown that in a semi-infinite deep soil without water table it is a matter of time for *all* the water to drain, and for the soil to become completely dry. However, under normal circumstances this final equilibrium is never reached.

A more objective definition is needed because of the fact that field capacity is an important feature for agronomy and agricultural management, related to the estimation of available water and the determination of the correct irrigation depth. Its value represents the maximum soil water content at which the water loss by drainage is small and for this reason it is also called the "upper limit of soil water availability". From the definition itself it becomes clear that it is impossible to determine the soil field capacity using an isolated sample from one of its layers, because the rate at which a soil layer loses water by drainage will depend on its insertion in the soil profile.

As a consequence, methods aiming to determine the water content corresponding to field capacity in soil samples, disturbed or undisturbed, are conceptually incorrect. Souza and Reichardt (1996) presented a revision of these methods. Most of them make an association between field capacity and pressure head. The ease of determination, in a soil physics laboratory, of the water content as function of pressure head, especially when compared to the laborious task to determine field capacity in situ, made many researchers to use the water content retained at suctions of 33, 10, 6 kPa or other values as equivalent to field capacity, ignoring the original definition and significance and reducing the validity of its use.

To come to a more complete definition of field capacity, additional criteria are necessary. A drainage rate considered to be negligible in one situation may be significant in another. Therefore, for an irrigation project, field capacity might be considered equivalent to the water content when drainage reduced to, for example, 1 mm/day. On the other hand, when studying pollution of aquifers, this same drainage rate may be considered an excess. The value of field capacity for these two cases would be different. Protocols to determine field capacity that are closer to the original definition apply a dependence of an acceptable drainage rate, as illustrated by Campbell and Norman (1998) and by Twarakavi et al. (2009). Thinking about field capacity without specifying what for and how its value was determined is of very limited use.

18.6 Limiting Water Content

A correct prediction of the hydraulic condition corresponding to the onset of the falling rate phase, the so-called limiting water content (θ_l, m^3 m^{-3}) or limiting pressure head (h_l, m), equal to h_3 from Fig. 18.2 is essential for hydrological predictions. Theoretical considerations from the classical paper of Gardner (1960), as well as experiments, have shown that root length density, soil hydraulic properties and hydraulic head gradients play an important role in determining these values. However, in hydrological or yield prediction models, these factors are generally

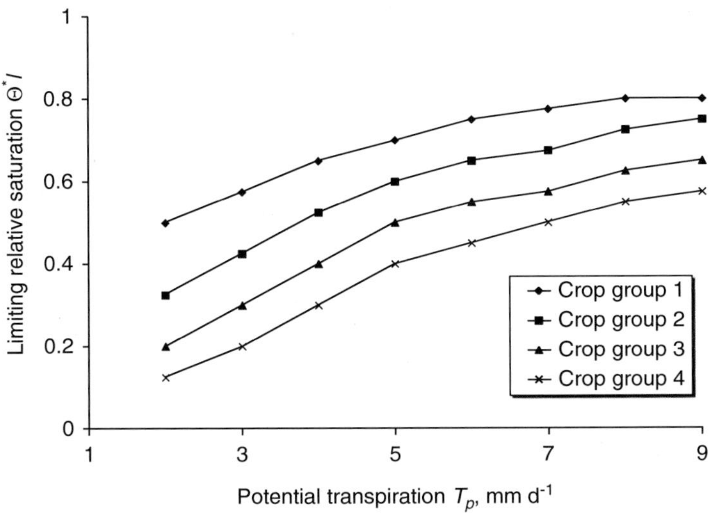

Fig. 18.4 Limiting relative saturation $\Theta^*_l = (\theta_l - \theta_{pwp})/(\theta_{fc} - \theta_{pwp})$ as a function of potential transpiration for four crop groups, according to FAO Irrigation and Drainage Paper 33 (Doorenbos and Kassam 1986)

not considered. Figure 18.4 shows the relation between limiting water content and potential transpiration, according to the FAO Irrigation and Drainage Paper 33 (Doorenbos and Kassam 1986). Four crop groups are defined and the limiting water content is considered a function of field capacity and permanent wilting point, but root system characteristics and soil hydraulic properties are not considered explicitly.

The prediction using the FAO method and values from Fig. 18.4 is frequently applied, but some of the associated problems are:

(1) The same plant species (hence: crop type) may, for several reasons, develop root systems with very different characteristics, both in depth and in root length density distribution. The FAO approach predicts the same limiting water content where common sense would indicate a crop with a sparser root system to experience drought stress at an earlier stage (at a higher water content) than a crop with a better developed root system;
(2) Comparing soil types and texture groups, large differences in hydraulic properties exist. Both retention (h-θ) and hydraulic conductivity (K-θ,h) affect the water availability in a soil and, therefore, the limiting soil water content and pressure head. These properties are only implicitly considered by the FAO method, as they are related to field capacity and permanent wilting point. The mechanism of soil water uptake by the roots is not implicitly included in the method;
(3) Different depths in a soil may present different hydraulic conditions. During a dry period, for example, surface layers tend to be drier than deeper layers. The inverse may occur when a rainfall event occurs at the end of a dry period. In these cases the FAO method is not suitable for the prediction of relative transpiration.

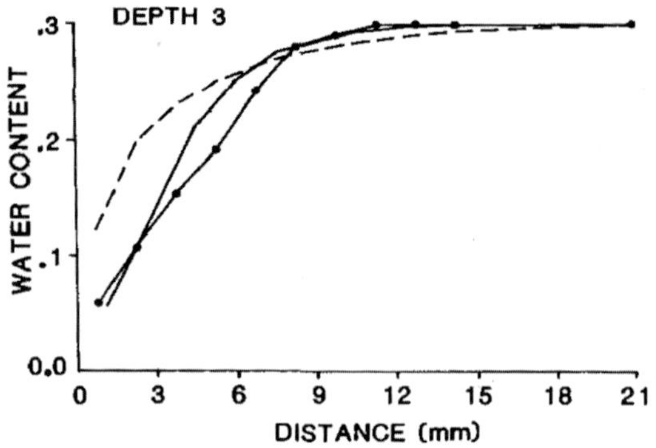

Fig. 18.5 Water content ($m^3\ m^{-3}$) versus distance (mm) for a water extracting radish (Raphanus sativus) root. The *solid line with dots* represents experimental data. The other two lines (*without dots*) show predictions of two models (For details see the original publication: reproduced from Hainsworth and Aylmore (1986))

To increase understanding about the physical interaction between soil and roots, and about how this interaction determines root water uptake, modeling can be a useful tool. Several models to predict root system water uptake exist. They describe water uptake from a single-root approach (the microscopic or mesoscopic approach) or from properties of the overall root system (the macroscopic approach). Analytical and numerical models and solutions, based on the Richards equation, are available for both approaches. Macroscopic models are easier to be applied as they require less and easier to obtain data; on the other hand, microscopic models make more process-based predictions.

Figure 18.5, reproduced from Hainsworth and Aylmore (1986), gives an illustration of these processes. They measured the water content in the close vicinity of a water extracting root using an X-ray CT scanning technique. The figure shows that closer to the root surface the water content is lower. This fact is explained by the water uptake process itself: a water extracting root creates a depletion zone around itself. This water depletion creates negative pressure heads and results in movement of water from bulk soil towards root. Distances being very small, in the order of millimeters, the role of the gravitational energy is negligible, and the ruling potential is the pressure head. Therefore, in most cases the pressure head gradient determines water movement to roots.

One of the consequences of this mechanism is that the pressure head gradient between soil and root surface increases with increasing water uptake rate. The uptake rate, on its turn, is correlated to the transpiration rate and to the total root surface area, which is a function of root length density. Then it can be understood that the times of the day representing the highest transpiration demand (low relative humidity and vapor potential, corresponding to high temperatures that occur, generally, during the afternoon) correspond to the lowest (most negative) pressure head at the root surface, whereas during the night, when transpiration almost ceases, the root surface pressure head tends to equalize with the soil water pressure head.

This phenomenon is illustrated in Fig. 18.6, reproduced from Slatyer (1967). The figure shows the water potential (or pressure head) versus time for soil water

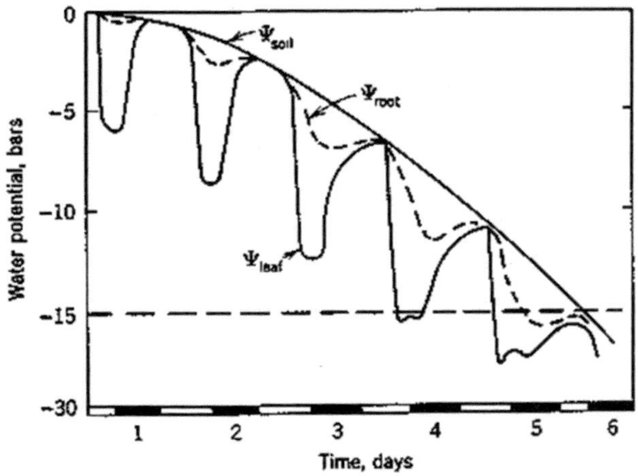

Fig. 18.6 Water potential versus time for soil water (ψ_{soil}), root water (ψ_{root}) and leaf water (ψ_{leaf}), from wet initial to dry final conditions (Reproduced from Slatyer 1967)

(ψ_{soil}), root water (ψ_{root}) and leaf water (ψ_{leaf}). The dark bars at the X-axis indicate the night period. From this figure it can be seen that:

(1) There is a continuous decrease of soil water pressure head (hence, of soil water content), while at the same time pressure heads in root and leaf have a daily cycle, decreasing during daytime and becoming equal to soil water pressure head at night;
(2) At any moment, $\psi_{leaf} \leq \psi_{root} \leq \psi_{soil}$, making water to flow always from soil to root and then to the leaf. From the leaf (stomata), the water evaporates (transpires) and moves to the atmosphere by diffusion;
(3) As the soil gets drier, the difference in pressure head from soil to root increases. This is due to the fact that hydraulic conductivity decreases with decreasing water content. To maintain the same water flow rate towards the root, the pressure head gradient has to increase.
(4) When the soil becomes very dry, with water potentials of the order of -15 atm (day 5 in Fig. 18.6), the difference between pressure heads in root and leaf is reduced, while there is no increase in pressure head difference between soil and root between days 4 and 5. This indicates that the transpiration rate decreased or, in other words, that stomata were closed to increase the canopy resistance to water flow. This is a symptom of crop water stress.

From Fig. 18.6 we see how several system parameters together determine the occurrence of drought stress. Atmospheric demand (potential transpiration rate), root length density (the area available for soil water uptake) and soil hydraulic properties, determining the root potential required for the uptake of the sufficient amount of water. The interaction of all these parameters makes it difficult to find a simple way of predicting the limiting water content or soil water pressure head.

To shed some light on this complex problem, a mathematical transform, the Kirchhoff transform or matric flux potential (M, $\mathrm{m}^2\ \mathrm{d}^{-1}$) can be useful. M is defined as the integral of hydraulic conductivity K (m d^{-1}) from a reference pressure head $\psi_{m,0}$ (m) to the pressure head of interest ψ_m (m):

$$M = \int_{\psi_{m,0}}^{\psi_m} K(\psi_m) d\psi_m \tag{18.8}$$

M is frequently used in studies that involve the flow of water in non-expanding and non-hysteretic soils. It is a parameter that is especially convenient to model flow phenomena in which pressure head is the driving force, for example, the uptake of soil water by plant roots. In these cases, as $\Delta\psi_t = \Delta\psi_m$, the Darcy-Buckingham equation can be written as

$$q = -K(\psi_m)\frac{d\psi_t}{dx} = -K(\psi_m)\frac{d\psi_m}{dx} \tag{18.9}$$

and the combination with Eqs. 18.8 and 18.9 gives

$$q = -\frac{dM}{dx} \tag{18.10}$$

Equation 18.10 shows that a gradient of M in space represents a water flux density, allowing the characterizing of soil water availability to plant roots. In this context, M was used by, for example, Cowan (1965), De Willigen and Van Noordwijk (1994), Heinen (2001), and De Jong Van Lier et al. (2006, 2008, 2013). Although M is a composed soil hydraulic property (Eq. 18.8), each value of M corresponds to a single water content and pressure head. Therefore, values of M can be correlated to values of θ or ψ_m for a specific soil.

Figure 18.7 contains an example of M-ψ relations for two soils from different texture classes. The functional relationship between M and ψ follows directly from the K-ψ relation, solving the integral from Eq. 18.8 analytically or numerically. $\Delta M/\Delta x$ equals water flux density (under the condition that there are no gradients of gravitational, osmotic, or other heads, Eq. 18.10), and the example in Fig. 18.7 shows that, for the case of a ΔM of, e.g., 50 cm^2 d^{-1} (the length of the vertical arrows in the figure) between two points at a distance of 10 cm, the flux density between these two points is 50 cm^2 d^{-1}/10 cm = 5 cm d^{-1}. This situation is to occur, in the loamy Clay soil, when the pressure head is −0.02 m at one point and −0.1 m at the other, equivalent to a difference of 0.08 m. If the same soil is drier, for example at a pressure head of −1 m at the first point, the pressure head at the second point should be −2.5 m (a difference of 1.5 m) to result in the same flux density. The increase in pressure head gradient results from the decrease in hydraulic conductivity.

Figure 18.7 allows also to compare water movement in both soils: whereas in the loamy clay soil two points at a distance of 10 cm at pressure heads of −0.02 and −0.1 m result in a flux density of 5 cm d^{-1} (the above example), in the loamy sand soil these values should be −0.02 m and −0.082 m, a smaller difference, indicating that, at this pressure head range, the loamy sand soil has a higher hydraulic conductivity than the loamy clay soil.

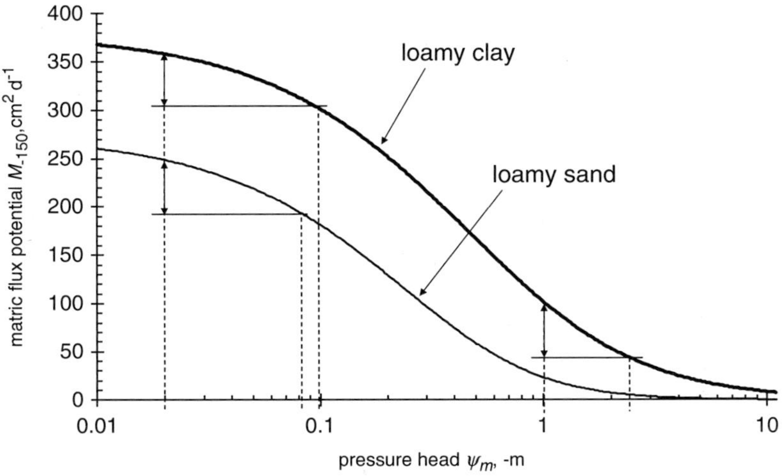

Fig. 18.7 Matric flux potential for $\psi_{m,0} = -150$ m (M_{-150}) as a function of pressure head (ψ_m) for two soils. *Dashed lines* and *arrows* are referred to in the text

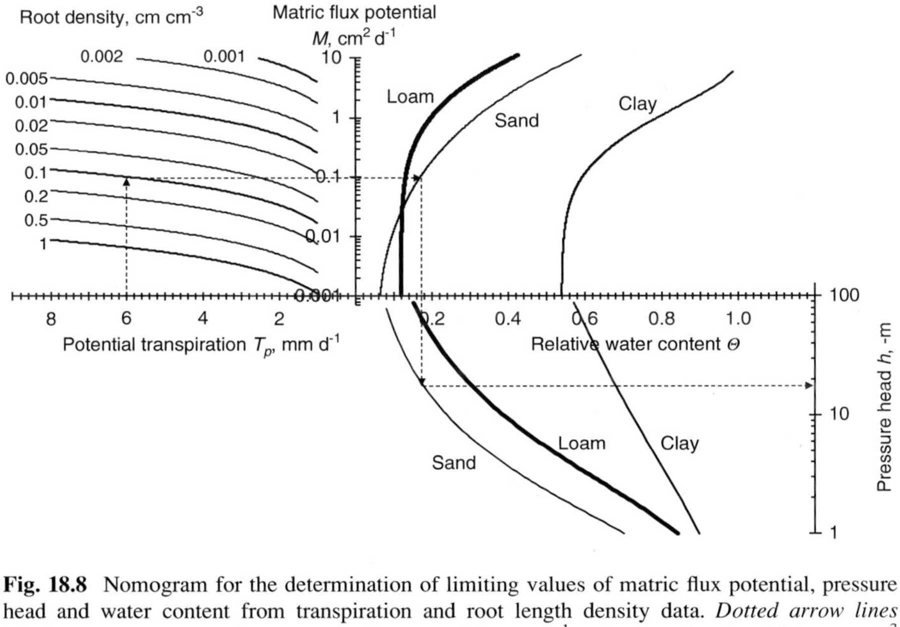

Fig. 18.8 Nomogram for the determination of limiting values of matric flux potential, pressure head and water content from transpiration and root length density data. *Dotted arrow lines* illustrate nomogram use for a transpiration rate of 6 mm d^{-1} and a root density of 0.1 cm cm^{-3} for a sandy soil (Reproduced from De Jong Van Lier et al. 2006)

De Jong Van Lier et al. (2006) developed a method for the determination of limiting values of matric flux potential as a function of root length density and potential transpiration rate. They illustrate their method as a nomogram which is reproduced in Fig. 18.8. The obtained limiting matric flux potential can be combined to soil hydraulic properties and transformed in a limiting pressure head or water content. In De Jong Van Lier et al. (2008), the approach was extended to layered soil systems, whereas in De Jong Van Lier et al. (2013) a model improvement was made by including plant internal hydraulic resistances.

18.7 Permanent Wilting Point

The definition of the permanent wilting point was proposed in publications by Briggs and Shantz (1912) and Hendrickson and Veihmeyer (1929), cited by Salter and Haworth (1961). The permanent wilting point was defined as the soil water content at which the leaves of a plant reach an irreversible wilting, irreversible even when the plant is placed in an environment at saturated vapor pressure (or: with a relative humidity of 100 % or with a zero vapor pressure deficit). The experimental determination of the permanent wilting point of a soil is complex and discussed in e.g. Veihmeyer and Hendrickson (1955) and Cutford et al. (1991).

It is normally considered that the soil water content at permanent wilting corresponds to the pressure head of $-15{,}000$ cm or -150 m (equivalent to -15 atm), but the definition of a fixed pressure head may lead to conceptual questions similar to those presented in the previous item. In the case of permanent wilting, however, the determination of an exact pressure head is not so important for two reasons. In the first place, the importance of the exact determination of the permanent wilting point is smaller because it is beyond the range of water contents that normally occur in soils, not being of practical use in irrigation management, for example. It is used mainly to characterize a soil in terms of available water (Eqs. 18.4 to 18.7), and a large associated error is already likely to occur at the upper limit (field capacity, see discussion in previous item). In the second place, in most soils the inclination of the water retention curve ($d\theta/d\psi_m$) is close to zero at high tensions, making large differences in pressure head to be associated to only very small differences in water content. To illustrate this, Savage et al. (1996) found, after careful experimentation with cotton plants in a lysimeter, the wilting point to correspond to a pressure head around -220 m, concluding that the corresponding water content differed less than 0.01 $m^3\ m^{-3}$ from the water content at -150 m. If we think about permanent wilting point in terms of water content, in order to determine the available soil water, it will make only a marginal difference which pressure head we chose, be it -100, -150, -220 m, or another value.

18.8 Plant Water Availability: A Matter of Access

From the foregoing it became clear that plant water availability can be understood as a matter of access of plant roots to soil water. In this item this access will be organized in different types (or constraints) that can be identified:

18.8.1 Root Length Density

Root length density, normally expressed in units of length of roots per volume of soil, determines the average travel distance of soil water to a plant root and is, therefore, intimately associated to plant water availability. In microscopic modeling of root water uptake, root length density is one of the main process-determining factors (Gardner 1960; Cowan, 1965; De Jong van Lier et al., 2006). Root length density is affected by depth, horizontal position relative to the stem, and soil physical, chemical and biological properties.

18.8.2 Rooting Depth

Total rooting depth defines the vertical extent of soil to which a root system has access. It is used to calculate available water in, e.g., Eqs. 18.5 and 18.7. It is determined by plant genetic factors, as well as by external factors like water and nutrient availability and soil biology. Rooting depth can be constrained by physical soil characteristics like compaction or hardpans (Horn et al. 2000; Grzesiak, 2009), or by chemical factors like acidity and aluminum content (Baligar et al., 1991; McDonald et al., 2013).

18.8.3 Root System Width

In the early stages of development of crops, horizontal distribution of root systems is far from homogeneous (Hund, 2010) and water in the less densely rooted parts of the soil (usually in between the planting lines) may remain unavailable to the crop. The root distribution tends to become more homogeneous as plants (and root systems) grow, but some horizontal heterogeneity is likely to remain. This is also the case for perennial crops and trees (Yanai et al., 2006; Morgan et al., 2007) with a larger distance between adjacent plants.

18.8.4 Within-Layer Heterogeneity

Whereas root length density is normally considered to represent a referred soil layer, heterogeneity within a layer may result in additional restrictions to crop water availability. Especially in well structured soils, roots tend to follow preexisting cracks (Wang et al., 1986; Jin et al., 2013). Therefore, macroscopically measured root length density may overestimate the real soil water availability in the respective soil layer as the water retained inside the structured elements has to travel a larger distance to the root than would be expected from the average distance between roots. The understanding of these interactions becomes more complex if we include the fact that water may infiltrate a soil preferentially, through cracks, leaving the zones between cracks dry. A modeling approach to this problem has been presented by Novak et al. (2000).

18.9 Conclusion

Soil water availability to plants is the result of dynamic processes, involving water flows from the root zone to below (drainage) as well as flow from the bulk soil towards roots (uptake). As a consequence, four zones of water availability to plants can be distinguished: excess water, readily available water, water with restricted accessibility and unavailable water. The limiting conditions between these zones, called respectively field capacity, limiting water content and permanent wilting point, are normally estimated by simplified methods that do not consider the underlying hydrological processes. To estimate field capacity, additional criteria about the tolerable drainage are mandatory. Estimates of the limiting hydraulic conditions and permanent wilting point can be performed involving the matric flux potential, a property that integrates the relation $K(\psi_m)$ over a range of pressure heads, together with potential transpiration and root length density. In the end, plant water availability is a matter of access of plant roots to soil water, determined by root length density distribution in the soil.

References

Baligar VC, Wright RJ, Ritchey KD, Ahlrichs JL, Woolum BK (1991) Soil and soil solution property effects on root growth of aluminium tolerant and intolerant wheat cultivars. Dev Plant Soil Sci 45:245–252

Briggs LJ, Shantz HL (1912) The wilting coefficient for different plants and its indirect determination. US Bur Plant Ind Bull 230

Campbell GS, Norman JM (1998) Introduction to environmental biophysics. Springer, New York, 286 pp

Cowan IR (1965) Transport of water in the soil–plant-atmosphere system. J Appl Ecol 2:221–239

Cui B, Tang N, Zhao X, Bai J (2009) A management-oriented valuation method to determine ecological water requirement for wetlands in the Yellow River Delta of China. J Nat Conserv 17:129–141

Cutford HW, Jefferson PG, Campbell CA (1991) Lower limit of available water for three plant species grown on a medium-textured soil in southwestern Saskatchewan. Can J Soil Sci 71:247–252

De Jong Van Lier Q, Metselaar K, Van Dam JC (2006) Root water extraction and limiting soil hydraulic conditions estimated by numerical simulation. Vadose Zone J 5:1264–1277

De Jong Van Lier Q, Van Dam JC, Metselaar K, De Jong R, Duijnisveld WHM (2008) Macroscopic root water uptake distribution using a matric flux potential approach. Vadose Zone J 7:1065–1078

De Jong Van Lier Q, Van Dam JC, Durigon A, Dos Santos MA, Metselaar K (2013) Modeling water potentials and flows in the soil–plant system comparing hydraulic resistances and transpiration reduction functions. Vadose Zone J 12. doi:10.2136/vzj2013.02.0039

De Willigen P, Van Noordwijk M (1994) Diffusion and mass flow to a root with constant nutrient demand or behaving as a zero-sink. Soil Sci 157:162–175

Doorenbos J, Kassam AH (1986) Yield response to water, FAO irrigation and drainage paper 33. Food and Agricultural Organization of the United Nations, Rome

Eagleson PS (1982) Ecological optimality in water-limited natural soil-vegetation systems: 1. Theory and hypothesis. Water Resour Res 18:325–340

Feddes RA, Kowalik PJ, Zaradny H (1978) Simulation of field water use and crop yield, Simulation monographs. Pudoc, Wageningen

Gardner WR (1960) Dynamic aspects of water availability to plants. Soil Sci 89:63–67

Grzesiak MT (2009) Impact of soil compaction on root architecture, leaf water status, gas exchange and growth of maize and triticale seedlings. Plant Root 3:10–16

Hainsworth JM, Aylmore LAG (1986) Water extraction by single plant roots. Soil Sci Soc Am J 50:841–848

Heinen M (2001) FUSSIM2: Brief description of the simulation model and application to fertigation scenarios. Agronomie 21:285–296

Hendrickson AH, Veihmeyer FJ (1929) Irrigation experiments with peaches in California. Calif Agric Exp Stn Bull 479

Horn R, Van den Akker JJH, Arvidsson J (2000) Subsoil compaction: distribution, processes and consequences, vol 32, Advances in geoecology. Catena Verlag, Reiskirchen

Hund A (2010) Genetic variation in the gravitropic response of maize roots to low temperatures. Plant Root 4:22–30

Jin K, Shen J, Ashton RW, Dodd IC, Parry MAJ, Whalley WR (2013) How do roots elongate in a structured soil? J Exp Bot. doi:10.1093/jxb/ert286

McDonald GK, Taylor JD, Verbyla A, Kuchel H (2013) Assessing the importance of subsoil constraints to yield of wheat and its implications for yield improvement. Crop Pasture Sci 63:1043–1065

Morgan KT, Obreza TA, Scholberg JMS (2007) Orange tree fibrous root length distribution in space and time. J Am Soc Hortic Sci 132:262–269

Novak V, Simunek J, Van Genuchten MT (2000) Infiltration of water into soil with cracks. J Irrig Drain Eng 126:41–47

Rockström J, Barron J (2007) Water productivity in rainfed systems: overview of challenges and analysis of opportunities in water scarcity prone savannahs. Irrig Sci 25:299–311

Salter PJ, Haworth F (1961) The available-water capacity of a sandy loam soil. J Soil Sci 12:326–334

Savage MJ, Ritchie JT, Bland WL, Dugas WA (1996) Lower limit of soil water availability. Agron J 88:644–651

Slatyer RO (1967) Plant-water relationships. Academic Press, London

Souza LD, Reichardt K (1996) Estimativas da capacidade de campo. Rev Bras Cienc Solo 20:183–189

Twarakavi NKC, Sakai M, Simunek J (2009) An objective analysis of the dynamic nature of field capacity. Water Resour Res 45:1–9

Van Den Honert TH (1948) Water transport in plants as a catenary process. Discuss Faraday Soc 3:146–153

Van Wijk WR (1963) Physics of plant environment. North-Holland Publishing, Amsterdam

Veihmeyer FJ, Hendrickson AH (1931) The moisture equivalent as a measure of the field capacity of soils. Soil Sci 32:181–193

Veihmeyer FJ, Hendrickson AH (1949) Methods of measuring field capacity and wilting percentages of soils. Soil Sci 68:75–94

Veihmeyer FJ, Hendrickson AH (1955) Does transpiration decrease as the soil moisture decreases? Trans Am Geophys Union 36:425–448

Wang J, Hesketh JD, Woolley JT (1986) Preexisting channels and soy-bean rooting patterns. Soil Sci 141:432–437

Yanai RD, Park BB, Hamburg SP (2006) The vertical and horizontal distribution of roots in northern hardwood stands of varying age. Can J Forest Res 36:450–459

Chapter 19
Experiences and Challenges in Developing European Soil Hydrological Databases

Allan Lilly, Attila Nemes, J. Henk M. Wösten, and Roland Hiederer

Abstract Development of the Hydraulic Properties of European Soils (HYPRES) database began in 1995 and was funded by the European Commission. The main aims of the project were to collate existing soil hydrological data held by Universities and Research Institutes into a single database and to use these data to derive a set of pedotransfer functions to predict soil hydrological properties for use in simulation models. Data were collated from 21 partners from 20 different Institutions throughout Europe encompassing 12 different countries. The database was structured around six main tables comprising two with either profile or horizon specific contextual data, three with 'raw' data and one with a suite of derived, standardized soil hydrological properties.

A recent review of the HYPRES database revealed some errors that were not identified during previous extensive checks and highlighted a number of decisions that, with hindsight, were not ideal. The database was subsequently revised such that an additional table with units was added, hierarchical approaches to describing land use and parent material were adopted, soil classification was standardized and the data in some fields were simplified.

HYPRES was a highly successful project and the derived pedotransfer functions have been widely used. However, some of the key lessons learnt from developing

A. Lilly (✉)
The James Hutton Institute, Craigiebuckler, Aberdeen AB15 8QH, Scotland, UK
e-mail: allan.lilly@hutton.ac.uk

A. Nemes
Division of Soil and Environment, Bioforsk, Fredrik A. Dahls vei 20, Ås N-1432, Norway

J.H.M. Wösten
Alterra – Wageningen University and Research Centre, Droevendaalsesteeg 4,
P.O. Box 47, 6700 AA, Wageningen, The Netherlands

R. Hiederer
European Commission, Joint Research Centre, Institute for Environment and Sustainability,
Via Enrico Fermi, 2749-21027 Ispra (VA), Italy

W.G. Teixeira et al. (eds.), *Application of Soil Physics in Environmental Analyses: Measuring, Modelling and Data Integration*, Progress in Soil Science,
DOI 10.1007/978-3-319-06013-2_19, © Springer International Publishing Switzerland 2014

the database were: ensure there is agreement from all contributors that the data can be used freely by other researchers and establish plans for the long term management and upkeep of the database.

Keywords HYPRES • Soil hydrology • Database

19.1 Introduction

The development of the HYPRES database began in 1995 when the European Commission funded a project under the Human Capital and Mobility (DG XII) program designed to develop networks of scientists and to allow scientists to work at other European Institutions for short periods of time.

The main driving force behind the development of the HYPRES database (Hydraulic Properties of European Soils) was to provide consistent, good quality soil hydrological data applicable to simulation modeling of soil water and solute transport at international, national and regional scales. The project team comprising a Project Coordinator (Henk Wösten); a Project Researcher (Allan Lilly) and a Research Student (Attila Nemes) set out to collate existing soil hydrological data from (mainly) western European countries, to standardize the data and then to develop a series of pedotransfer functions to predict soil hydrological properties for soils and parts of Europe where such hydrological data were lacking in order to provide data for simulation modelling at a range of scales (Wösten et al. 1998). During the three year lifespan of the project, it became a Working Group under the auspices of the European Soil Bureau.

19.2 Participants

Data were collated from 21 partners from 20 different Institutions throughout Europe encompassing 12 different countries (Fig. 19.1) including partners not directly funded by the European Commission. Additional data from non-participating countries that was already in the public domain were also incorporated into the database.

The initial aim of the project was to derive pedotransfer functions applicable to European soils and the database was only a secondary consideration. However, it soon became clear that the HYPRES database itself would be of considerable value but issues surrounding the Intellectual Property Rights of individual participating Institutions meant that certain restrictions were imposed on the subsequent use of the database at the end of the project.

The project funding was aimed at developing networks amongst scientist and so was limited primarily to allow participants to attend meetings and workshops and to allow the Project coordinator to visit various Institutions to discuss data transfer. The participants were not paid for contributing their data to the project.

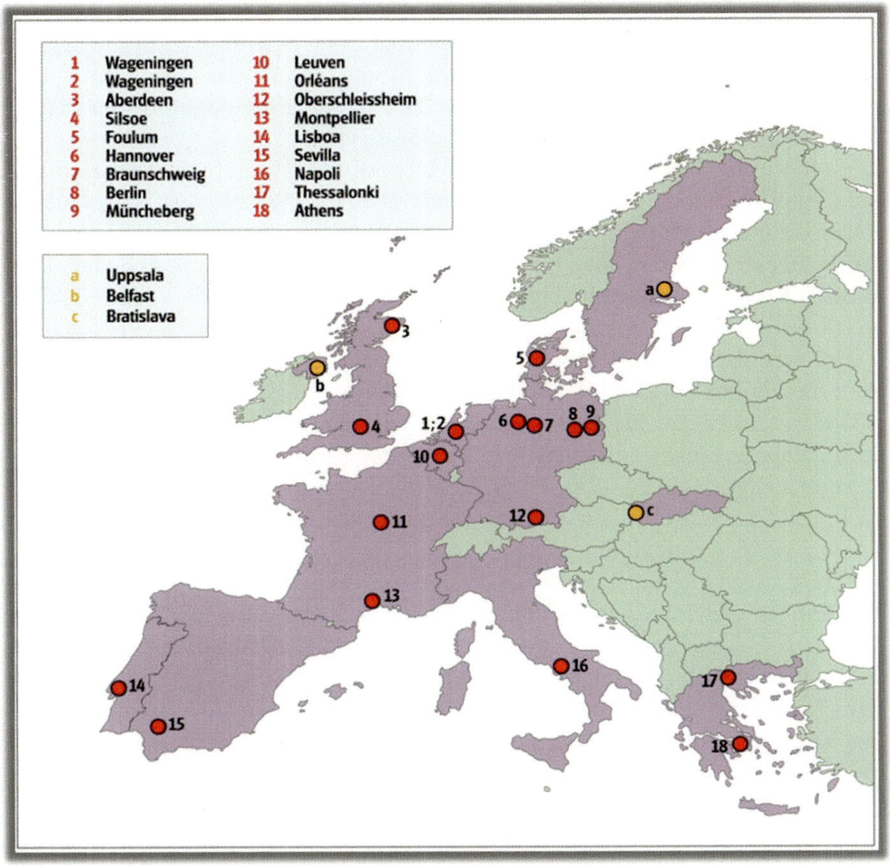

Fig. 19.1 Location of participating institutions (a–c were not directly funded)

19.3 Stages in the Development of HYPRES and Related Pedotransfer Functions

19.3.1 Overview

There were a number of key stages in the development of the HYPRES database and related pedotransfer functions led by different members of the project team. The first tasks were to develop a framework and identify an operational system for the database. Because of the likelihood that the data would be used for different purposes and due to the many different datasets that were being collated, it was decided that the most effective way to store and manipulate the data was in a relational database. Initially the Oracle Relational Database Management System™ (RDMS) was used and, although a master copy was retained within this Database

Management System, copies of the database were transferred to Microsoft Access™ for ease of handling and distribution. Long term storage of the database in Oracle™ RDMS by a single Institution means that the database was regularly migrated to new versions of the RDMS. This avoided the possibility of the data being stored in an old and obsolete operational system.

At the time HYPRES was being developed, the Unsaturated Soil Hydraulic Database (UNSODA, Leij et al. 1996) had just become available. This database contained much of the same types of data we had hoped to collate for European soils but had gathered data from all over the world. We took account of the format of UNSODA while developing HYPRES so that the two datasets would be compatible.

Once the database format was developed, the next stage was to collate the data from the various partners and contributors. Given the number of institutions and individuals involved, the data were delivered by various means (hard copy, csv files, database files, spreadsheets) resulting in considerable work to reformat or even manually enter the data. Each participant was responsible for checking the quality and accuracy of their data but checks for obvious errors were run when the data were stored in temporary tables before entry to the main database.

Again, given the disparate sources of the data, it was necessary to standardize the data prior to developing the pedotransfer functions. This involved parameterizing the water retention/pressure head (θ/h) and hydraulic conductivity/pressure head (K/h) curves and then back-calculating to derive a series of θ/h and K/h relationships for a fixed set of pressure heads. As different systems to determine particle size class distribution were used throughout Europe, the particle size distribution also had to be standardized before pedotransfer functions based on soil texture could be derived. The final stage in the project was to develop a suite of both class and continuous pedotransfer functions to predict the water retention and hydraulic conductivity of European soils (Wösten et al. 1999).

19.3.2 Database Structure

The data model for HYPRES is quite simple and comprises a table of basic, contextual and profile information (BASICDATA) that links to five other tables which contain information for the horizons within each of these profiles (Fig. 19.2). The tables designated as 'RAW' contain the particle size, water retention and hydraulic conductivity data while the SOIL_PROPS table has contextual information about each horizon and the HYDRAULIC_PROPS table has the standardized water retention and hydraulic conductivity data at 14 pressure heads. Each soil profile in the BASICDATA table was given a unique identifier (Primary Key) based primarily on its geographical location (latitude/longitude or national georeference). The five other tables that hold the horizon information and data use a combination of this unique primary key and the horizon designation to derive a unique identifier for each horizon in the database. Where there are replicate samples from

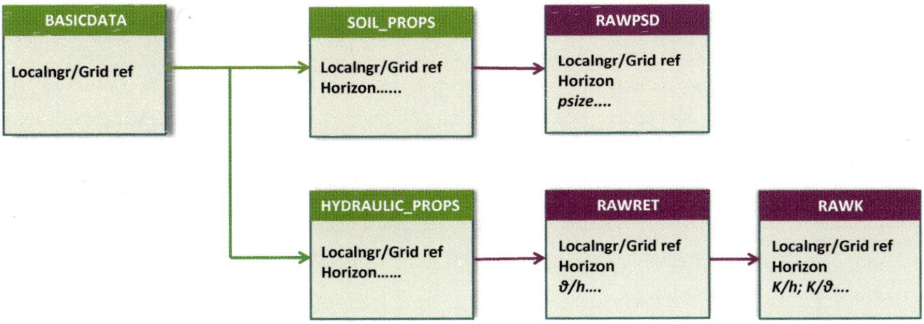

Fig. 19.2 Structure of the HYPRES database

a horizon and the data were not averaged, an additional alphabetic character was added to the horizon symbol to distinguish them. For example, a topsoil (Ap) sample from a Scottish site with a georeference 'NJ400500' would be labeled 'NJ400500 A p' and, if the sample was replicated within that horizon, it became 'NJ400500 A p a' with the spacing between characters being as important as the characters themselves.

Figure 19.3 shows an example of the data types stored in each table. The data arrived from the partners in a variety of ways resulting in a considerable effort to process the data before loading them into the main database. As far as possible, the data was also retained in its original format such as a proprietary spreadsheet software package, however, some of these software packages are no longer in use and are not compatible with modern spreadsheet or database software packages. Unfortunately this means that the original data held in these formats can no longer be checked in the event of any queries arising regarding these data.

Quality checks were run on the data to ensure that the primary keys and unique identifiers functioned properly and the data could be readily linked and extracted. A limited check on the values of the various properties was made to ensure they lay within acceptable limits and to ensure the data were in the correct units. Care was taken to establish whether soil organic carbon (SOC) values were indeed reported as organic carbon and not as soil organic matter since this is a key property in deriving pedotransfer functions.

In total, version 1 of HYPRES consisted of:

- 95 different Soil Types
- 1,777 Soil Profiles
- 5,521 Soil Horizons
- over 125,000 pairs of K(h)
- over 200,000 pairs of θ(h)

from Belgium, Denmark, France, Germany, Greece, Italy, Netherlands, Portugal, Spain, United Kingdom (Scotland, England, Wales and Northern Ireland) plus some additional data from Slovakia, Sweden, Russia, Poland, and Switzerland.

BASICDATA	SOIL_PROPS.	HYDR._PROPS.	RAWPSD	RAWRET	RAWK
Localngr/Grid ref	**Localngr/Grid ref**	**Localngr/Grid ref**	**Localngr/Grid ref**	**Localngr/Grid ref**	**Localngr/Grid ref**
	Horizon	**Horizon**	**Horizon**	**Horizon**	**Horizon**
local site name	upper sample depth	dMVG param θsat.	part. size range	flag lab./ field	flag laboratory / field
soil name (FAO)	lower sample depth	dMVG param. θres.	% of particles	pres. head (h)	indicator if k(h) / k(θ)
country of origin	prim. structure (FAO)	dMVG param. alpha		θ(h) value	value of (h) or (θ)
local geo-reference	sec. structure (FAO)	dMVG param. n			hydr. cond. at h/θvalue
local soil name	% of clay (FAO)	dMVG param. m			
local soil series	% of silt (FAO)	dMVG param. l			
highest gr.water depth	% of sand (FAO)	dMVG param. Ksat.			
lowest gr.water depth	Ksaturated	theta(1)			
site description	sat. water-content	theta(2)			
sampling date	bulk density	...			
annual rainfall	particle density	...theta(14)			
avg. temp. in January	porosity	conductivity(1)			
avg. temp. in July	organic material	conductivity(2)			
contact person	MVG param. θsat.	...			
contact address	MVG param. θres.	...conductivity(14)			
contact email	MVG param. alpha				
relevant publication	MVG param. n				
comments1	MVG param. m				
comments2	MVG param. l				
keywords of methods	MVG param. Ksat.				
No. of horizons	flag if estim. PSD data				
rating of data quality	comments				
data quality rated by	keywords of methods				

Fig. 19.3 Example of data stored in each of the six tables in HYPRES version 1

It took some time to run quality checks on the considerable amount of data held in HYPRES and some errors were not detected. Some of the quality checks that were undertaken included: checking that the soil had been classified according to an internationally recognised soil classification system and that as many of the descriptive fields as possible were populated. Information that did not conform to the set guidelines was accepted on the assumption that some information was better than none. Additionally, descriptions of the methods and terminology used was standardized and 'keywords' were added to help in data manipulation.

19.3.3 Data Standardization

As the data were collated from many different countries around Europe, there were many different methods used to derive water retention and hydraulic conductivity. Also, and in particular, particle size classes varied widely throughout Europe (Table 19.1). As these were crucial parameters used to derive pedotransfer functions, they had to be standardized.

Table 19.1 Measured particle size class limits reported by the different HYPRES contributors

Country	Particle size limits (μm)
Belgium	2-10-20-50-100-200(or 250)-500-1000-2000
Denmark	2-20-63-125-200-500-2000
England	2-20-60-100-200-600-2000
France	2-50-2000
Germany	2-(6.3)-20-63-(125)-200-630-2000
Greece	2-6-20-60-200-600-2000
The Netherlands	2-16-50-105-150-210-300-2000
Italy	2-50-2000
Northern Ireland	2-60-2000
Portugal	2-20-200-2000
Scotland	2-60-2000
Slovak Republic	1-10-50-250-2000
Spain	2-20-2000
Sweden	2-6-20-60-200-600-2000

Particle Size Distribution

It was decided to standardize the particle size class limits on the FAO/USDA particle size classes (Soil Survey Staff 1951) as this was an internationally recognised system and it was the one in use in the 1:1,000,000 scale Soil Geographical Database (soil map) of Eurasia as described by Jamagne et al. (1994) and by the European Soil Bureau Network (2004). The FAO/USDA classification has the following class limits: clay particles <2 μm, silt-sized particles 2–50 μm and sand-sized particles 50–2,000 μm.

Clearly data from Belgium, France, The Netherlands and Italy were compatible but the data from the remaining countries (>70 % of the profiles) required to be standardized by predicting the cumulative percentage of particles at 50 μm (Table 19.1). Four methods were tested using two test datasets (Nemes et al. 1999).

- *Loglinear distribution function*: this fitting procedure is based on the assumption of a log linear relation between two neighboring points with known particle size classes and measured proportions.
- *Gompertz cumulative distribution function*: although similar to the logistic curve, the Gompertz distribution function allows non-symmetry whereas the logistic curve does not.
- *Spline fitting*: Splines are highly flexible but complicated and are constructed from segments of cubic polynomials between distinct values of a variable and are constrained to be smooth at the junctions.
- *Similarity procedure*: In this procedure, a reference dataset which contained 7 or 8 measured particle-size fractions was used. The procedure involved searching the reference dataset (in this case, a dataset from the Netherlands with over 9,600 particle size distributions) for a particle-size distribution that was the most similar to that of the soil for which data on a particular particle-size limit was missing. Selection of these soils was based on a comparison of the particle-size

fractions within those particle-size limits which were common to both the reference set and the soil under investigation. Once these soils were selected they were used to calculate the particle-size fraction for the missing particle-size class. This was done through an iteration procedure designed to select the ten most similar particle size distributions from the reference dataset for which there were data on the unknown size range. The mean proportion of particles in this size range was calculated from the ten selected distribution curves. In principle, this approach is a non-spatial variant of the k-nearest neighbor technique, adapted to handle the particular data need.

The log linear interpolation gave the poorest results and, as the Gompertz distribution curve required fitting of four parameters, it could not be applied to the all the datasets as those from Northern Ireland, Scotland and Spain only contained three data pairs. Therefore, the methods used to predict the cumulative proportion of particles at 50 μm were fitting the spline function and using the similarity procedure in parallel, following a set of rules described in Nemes et al. (1999).

Data Standardization: Soil Hydrological Data

The water retention and hydraulic conductivity of the soil horizons in HYPRES were derived from many different methods (for example, tension tables, instantaneous profile method, evaporation method, Guelph permeameter, multistep crust method, constant head permeameter etc.). Some of these methods generate considerably more data than others, for example the evaporation method *vs* tension tables to generate θ/h, which would lead to statistical bias when developing pedotransfer functions. Also, there were few commonly applied pressure heads which made calculation of Field Capacity and Permanent Wilting Point difficult for many horizons.

Therefore, it was decided to standardize the water retention and hydraulic conductivity functions by fitting the Mualem-van Genuchten (MVG) model (van Genuchten 1980) to the water retention and hydraulic conductivity curves using the RETC algorithm (van Genuchten et al. 1991) and subsequently using the parameters to back-calculate water retention and hydraulic conductivity at 14 pressure heads (Wösten et al. 1999). This information was then stored in the HYRAULIC_PROPS table linked to all others via the LOCALNGR and, where appropriate, HORIZON fields.

19.4 Lessons Learnt

There were a number of decisions made when compiling the HYPRES database that, in hindsight, were perhaps not the most appropriate.

At the time we were developing the database we decided to follow the pattern of the UNSODA database so that we might be able to merge the two at some point.

Although this never happened, it impacted on the subsequent utility of the database. The water retention, conductivity and particle size data in the 'RAW_...' tables were stored in columns, however, it would be easier to manipulate the data if it were stored in rows.

Although a manual was written explaining the database structure, the data manipulations and the units for each data field, this wasn't always consulted by users of the data and occasionally there was confusion about the units used in the tables. A table with field names and units was subsequently added to the database so that the information is available without the need to consult the manual.

Free text to describe such aspects as site, location, land use and parent material was allowed but while this gave scope to capture as much data as possible, it made it more difficult to use the data in a systematic way. A more hierarchical approach to recording land use and parent material in particular would have been beneficial.

Not all soil physicists are good pedologists and some soils were classified using combinations of Reference Groups and Qualifiers that were not possible in the classification system used. This can cause problems when trying to analyze data by soil type.

A key aspect to developing a soil hydrological database to develop pedotransfer functions is to properly document the methods used to derive the transfer function input data such as particle size and organic carbon content. Confusion between methods and whether data is **organic carbon** or **organic matter** can have major implications for the development of transfer functions (for example, see Nemes et al. 2009). Similarly, determination of particle size by hydrometer or pipette can produce different distributions compared with particle size determination by laser diffraction and the methods used to determine bulk density, in particular the temperature at which the soil was dried can be crucial. Therefore it is vitally important to verify methodology and values.

Another important lesson learnt is the need for simplicity in data fields. In the original version of HYPRES there was a set of rules to record either the presence or absence of a water table and, if present, its depth below the soil surface. However, unless the user read the manual, they were faced with a data field that contained either a number (that is, the depth to the watertable) or the indicators 'ND' (water table present but depth not determined) or 'NA' (no water table present and so 'not applicable'). This caused a degree of confusion.

The primary key used to link the separate HYPRES tables was the georeference of the profile and made use of either latitudinal and longitudinal co-ordinates or a national projection system. This led to two main problems. The first of which was associated with the use of the data after the project ended and the other with the complexity of the linkage between tables.

At the end of the project, and to protect their intellectual property, a number of partners requested that the data they had contributed should be deleted from the database. However, as previously stated, it was felt that the database itself was a major output from the project and efforts were made to retain all the data in the database. The compromise reached was that the data could remain within the database but data could only be used by those outside of the partnership if the georeferences (and any other information that would identify which country the data came from) were removed. Thus having the georeference as part of the primary key meant additional data manipulation to remove it and to add a substitute primary key while retaining linkages between the various tables when sending data to other researchers. Additionally, the georeferences were not in a consistent format throughout the database (as different countries had their own unique systems) and, when combined with horizon symbols (with their mixture of characters and spaces) to link the SOIL_PROPS, HYRAULIC_PROPS and RAW tables, the unique identifier became difficult to work with and lacked consistency in its application.

In retrospect, it would have been more pragmatic to use a unique four figure number for each profile and a unique number for each horizon/sample depth based on its position within the profile, for example, the profile at NJ400500 has three samples from an Ap, a Bg and a Cg horizon (horizon position 1, 2 and 3 from the surface) thus a unique identifier for the profile could be 1234 AND sample id would then be 123401, 123402 & 123403.

19.5 Revisions to HYPRES Version1.1

When a colleague at the Joint Research Centre, Ispra, Italy attempted to extract data from HYPRES to augment another European-scale soil database, he uncovered some errors in HYPRES that had previously been undetected. These included some inconsistencies in data format. This led to some further work on the HYPRES database such as checking and correcting potential errors and further standardization of data formats. Specifically;

- The various georeferences were converted to a common INSPIRE compliant standard
- Terminology was further standardized wherever possible
- Potentially unrealistic values or duplicate data were checked against the original data wherever possible and amended if necessary
- A hierarchical approach to classify soil parent material and land use to facilitate data extraction and analyses at different levels of detail was introduced

A total of 111 records in the BASICDATA table were found to have no soil horizon data and were deleted leaving 1,679 records in the BASICDATA table of HYPRES Version 1.1. This error had probably arisen as checks were originally made to ensure that there were no horizon data without the contextual information such as that contained in the BASICDATA table but insufficient checks were made to ensure that all profiles had associated horizon data. Additionally, two duplicate records were found in the RAWPSD table and were deleted.

Where possible, the georeferences were converted to the INSPIRE compliant ETRS 1989 projection system. This is a new stipulation imposed on spatial data in

Table 19.2 Example of hierarchy in soil parent material

6000	Unconsolidated glacial deposits/glacial drift	6100	Morainic deposits	6110	Glacial till	6111	Boulder clay
				6120	Glacial debris		
		6200	Glaciofluvial deposits	6210	Outwash sand, glacial sand		
					Outwash gravel, glacial gravel		
		6300	Glaciolacustrine deposits	6310	varves		

Table 19.3 Example of a hierarchy in land use for cultivated soils

Cultivated	Cereals	Barley	Spring
			Winter
		Wheat	Spring
			Winter
		etc.. …	etc.. …
	Grass	Recreational	
		Grazed	
	Root crops	Potatoes	
		Beet	

Europe since 2007 (Directive 2007/2/EC of the European Parliament and of the Council) and, by adopting it, it allows for greater integration of HYPRES with other spatial data in the future. Only 1,538 sample sites could be converted to the ETRS 1989 projection system which implies that the georeferences of the remainder were either missing or wrong.

HYPRES had soils classified by three different versions of the FAO soil classification system. It was decided to update this field to the most recent, 1994 revised version of the Soil Map of the World Legend (FAO/UNESCO 1994) although the original classification – as done by the individual partners – was retained. Ideally, the soil classification should be updated to the new WRB classification (IUSS 2006) but there is the potential to introduce errors into the soil classification if this is done in retrospect and not done by the person who originally classified the soil.

The parent material field was updated to take account of a European standard classification used in the European Soil Database (European Commission 2003), which is also hierarchical (Table 19.2), thus allowing the data to be grouped at different levels; for example, we could analyse all soils developed on glacial till (6110) or have more data by including all till soils and all soils developed on glacial debris (6120) by considering all soils with parent materials in the morainic deposits class (6100).

Similarly with land use, HYPRES was updated such that the data could be analyzed by different land uses, for example all cultivated soils or only those growing cereals, etc. (Table 19.3). The land use hierarchy included soils under woodland and under semi-natural vegetation.

19.6 Conclusions

The HYPRES database was developed between 1995 and 1997 during which soil hydrological data comprising over 125,000 K/h and over 200,000 θ/h data pairs were compiled. These data were standardized and subsequently used to derive a suite of pedotransfer functions that have been widely used. During its development, the project team were aware of other soil hydrological databases that were being constructed and had useful discussions with the teams involved in these as well as with those partners directly involved in the development of HYPRES. Thus, the database that emerged was an amalgamation of many different ideas. The decision to host the master copy on an Oracle™ platform meant that there were fewer issues with maintaining the database in a current format and that it could be easily downloaded to other database management systems or spreadsheets when required.

One of the main issues with the database was the restrictions imposed by those partners whose institutions demanded that their intellectual property be protected. While this did not stop the database from being used, it did perhaps limit its use. It seems clear that in any future development of HYPRES or any similar databases it should be a condition of contributing data that it can be used freely by other researchers. Preservation of such a database over a prolonged period of time requires institutional support and this needs to be agreed in advance. This also impacts on who can make updates to the database, how it is preserved, who is responsible for its upkeep and version management as well as issues such as who will be able to access and use the data. Again, these should be made clear at the outset so that those contributing the data are aware of how their data will be maintained and possibly used in the future.

The world has changed since 1995 and increasingly soil data is being sought by non-soil scientists and there is a move towards greater access to soils data in many countries. In this paper, we have tried to indicate some of the potential pitfalls in developing and maintaining a pan-national database. There are likely to be many more than we encountered but such databases become much more than the sum of their component parts and are a vital resource for both the soil science and the wider science communities.

References

European Commission (2003) European soil database (distribution version v2.0). European Commission Joint Research Centre, Italy. URL http://eusoils.jrc.ec.europa.eu/ESDB_Archive/ESDB/index.htm. Accessed 12 Jan 2014

FAO/UNESCO (1994) Soil map of the World. Revised legend, with corrections. ISRIC Technical Paper 20. International Soil Reference and Information Centre, Wageningen

IUSS Working Group WRB (2006) World reference base for soil resources 2006. World Soil Resources Reports No. 103. FAO, Rome

Jamagne M, King D, Le Bas C et al. (1994) Creation and use of a European Soil Geographic Database. 15th International Congress of Soil Science, Transactions, vol. 6a, Commission V, Symposia, Acapulco, Mexico, p 728

Leij FJ, Alves WJ, van Genuchten MTH, Williams JR (1996) The UNSODA Unsaturated Soil Hydraulic Database User's Manual; Version 1.0. EPA/600/R-96/095. United States Office of Research and Development. Environmental Protection Agency, Washington, DC 20460

Nemes A, Wösten JHM, Lilly A et al (1999) Evaluation of different procedures to interpolate the cumulative particle-size distribution to achieve compatibility within a soil database. Geoderma 90:187–202

Nemes A, Timlin DJ, Pachepsky YA et al (2009) Evaluation of the Rawls et al. (1982) Pedotransfer functions for their applicability at the US national scale. Soil Sci Soc Am J 73:1638–1645

Soil Survey Staff (1951) Soil Survey Manual. US Department of Agriculture Handbook No. 18. USDA, Washington, DC

The European Soil Bureau Network (2004) The European Soil Database distribution version 2.0, European Commission and the European Soil Bureau Network, CD-ROM, EUR 19945 EN, 2004 Office for Official Publications of the European Communities, Luxembourg

van Genuchten MTH (1980) A closed-form equation for predicting the hydraulic conductivity of unsaturated soils. Soil Sci Soc Am J 44:892–898

van Genuchten MTH, Leij FJ, Yates SR (1991) The RETC code for quantifying the hydraulic functions of unsaturated soils. USDA, US Salinity Laboratory, Riverside, California. United States EPA document EPA/600/2-91/065

Wösten JHM, Lilly A, Nemes A et al (1998) Using existing soil data to derive hydraulic parameters for simulation models in environmental studies and in land use planning. Final report 156 on the European Union funded project (CHRX-CT94-0639). DLO Winand Staring Centre, Wageningen, The Netherlands

Wösten JHM, Lilly A, Nemes A et al (1999) Development and use of a database of hydraulic properties of European soils. Geoderma 90:169–185

Chapter 20
A Hydrophysical Database to Develop Pedotransfer Functions for Brazilian Soils: Challenges and Perspectives

Marta Vasconcelos Ottoni, Maria Leonor Ribeiro Casimiro Lopes-Assad, Yakov Pachepsky, and Otto Corrêa Rotunno Filho

Abstract Access to soil hydrological data is vital for hydrology projects and for supporting decision-making in issues related to the availability of food and water and the forecasting of phenomena related to geomechanics. Brazil is a country of continental dimensions and has accumulated a significant body of soil information, holding a prominent position in tropical soil science. Nevertheless, a database with hydrophysical information on Brazilian soils has not been compiled so far, whereas much information is registered and analyzed. In this study we discuss the potential for the development of a Brazilian hydrophysical database and pedotransfer functions (PTFs). We present on metadata the measurement methods of soil hydrophysical

M.V. Ottoni (✉)
Department of Hydrology, Companhia de Pesquisa de Recursos Minerais (CPRM), Geological Survey of Brazil, Rio de Janeiro, RJ, Brazil

Laboratório de Recursos Hídricos e Meio Ambiente (LABH2O), Programa de Engenharia Civil, Instituto Alberto Luiz Coimbra de Pós-Graduação e Pesquisa em Engenharia (COPPE), Universidade Federal do Rio de Janeiro (UFRJ), Rio de Janeiro, RJ, Brazil
e-mail: marta.ottoni@cprm.gov.br

M.L.R.C. Lopes-Assad
Department of Natural Resources and Environmental Protection, Federal University of São Carlos (UFSCar), Araras, SP, Brazil
e-mail: assad@cca.ufscar.br

Y. Pachepsky
Environmental Microbial and Food Safety Laboratory, USDA-ARS Beltsville Agricultural Research Center, 10300 Baltimore Ave., Bldg. 173, Beltsville, MD 20705, USA
e-mail: yakov.pachepsky@ars.usda.gov

O.C. Rotunno Filho
Laboratório de Recursos Hídricos e Meio Ambiente (LABH2O), Programa de Engenharia Civil, Instituto Alberto Luiz Coimbra de Pós-Graduação e Pesquisa em Engenharia (COPPE), Universidade Federal do Rio de Janeiro (UFRJ), Rio de Janeiro, RJ, Brazil
e-mail: otto@coc.ufrj.br

W.G. Teixeira et al. (eds.), *Application of Soil Physics in Environmental Analyses: Measuring, Modelling and Data Integration*, Progress in Soil Science, DOI 10.1007/978-3-319-06013-2_20,

attributes and the characteristics of the sites where these soil properties were determined. Statistical analyses were performed to characterize the dataset according to the metadata-based data structure. A total of 8,983 datasets contains soil water retention information associated with physical or chemical properties. Besides these, 1,253 datasets with data on saturated hydraulic conductivity coupled with water retention information are also available. The results of this study suggest that it is possible to develop a representative hydrophysical database for Brazilian soils that covers most of Brazil's federative states, with a substantial volume of data and homogeneous with respect to the methods of measuring soil properties. This creates excellent prospects for PTFs development, especially for estimating water retention, at a national scale. The challenge in the development of the Brazilian soil hydrophysical database is the refinement of the dataset model that can encompass the wide range of available information and can provide answers to queries of interest to different types of users of soil information. Considering the size of the Brazilian territory, it would be of interest that the database development become a joint effort of government agencies, universities and commercial enterprises.

Keywords Water retention • Hydraulic conductivity • Tropical soils • Brazilian soil • Database

20.1 Introduction

The numerical models that simulate physical, chemical and biological processes in the vadose zone are of widely recognized importance to agricultural management, water resources, and climate change- related research and applications. Such modeling also provides technical and scientific support for the development of policies in matters related to food and water supply and earth surface phenomena that can lead to land quality reduction or even disasters, like soil loss or landslides. Typically, these models make a partition of precipitation into infiltration, runoff and evapotranspiration, and utilize soil hydraulic properties as input data to describe water and solute retention and transport in soils. The amount of these input data needed to run the models depends on the spatial and temporal resolution considered in the simulations. The higher this resolution, the greater is the volume of required measurements (Minasny and McBratney 2002). This problem indicates the need for less expensive and rapid tools to estimate soil hydraulic properties (Schaap et al. 2001).

An indirect method of determining soil hydraulic attributes consists in estimating them with pedotransfer functions (PTFs), which are based on pedological variables that can be measured routinely and easily (McBratney et al. 2002). The concept was introduced by Bouma (1989) and has been used worldwide (Wösten et al. 1999; Schaap et al. 2001; Tomasella et al. 2003; Rawls 2004; Stekauerová and Sútor 2004; Reichert et al. 2009; Makó et al. 2010; Wang et al. 2012; Miháliková et al. 2013).

The first PTF for water retention in Brazilian soils was proposed relatively recently (Tomasella et al. 2000). Initiatives for Brazilian PTFs development started much earlier but were limited to a specific soil class or to a certain region (Arruda et al. 1987; Silva et al. 1990; Van Den Berg et al. 1997). In the attempt to better represent the wide variety of soils in Brazil, new PTFs for Brazilian soils were created in 2003, adding to the database of Tomasella et al. (2000) with 630 samples another one with 208 samples (Tomasella et al. 2003). The PTFs presented by Tomasella et al. (2000) were used in very few studies in Brazil like Rossato (2002) and Doyle et al. (2013), but the application of PTFs from temperate regions to Brazilian soils persists, although tropical soils use to have a different mineralogy and texture than soils from temperate regions. Hodnett and Tomasella (2002) observed considerable differences between estimated parameters of the Van Genuchten (1980) equation comparing soils from tropical and temperate regions. Their analysis showed that the application of temperate PTFs for tropical soils is inadequate.

The accuracy of the numerical models depends on the PTF's performance in the simulations (Tomasella et al. 2000; Minasny and McBratney 2002; Jana et al. 2007). The accuracy of these functions requires, among other aspects, databases that present the following characteristics: (a) precise measures of soil attributes (Wösten et al. 2001; Mohanty and Shouse 2002; Perkins and Nimmo 2009; Vereecken et al. 2010); (b) various soil properties and landscape features, such as topographic parameters and terrain vegetation (Pachepsky et al. 2001; Rawls and Pachepsky 2002; Leij et al. 2004; Sharma et al. 2006) that may serve as input variables in these functions; (c) a substantial number of data distributed in relatively homogenous quantities from diverse soils and environments (Perkins and Nimmo 2009; Vereecken et al. 2010); and (d) standardization of measuring methods of soil attributes among data contained in the database (Schaap and Leij 1998; Pachepsky et al. 1999; Perkins and Nimmo 2009; Vereecken et al. 2010).

Brazil has accumulated a vast amount of soil information from pedological surveys, but most of these data is not readily available. One of the initiatives of gathering such information was the study of Cooper et al. (2005) which had its soil profile database updated by Benedetti et al. (2008) considering the second last edition of the Brazilian Soil Classification System (Embrapa 2006). Literature reviews on Brazilian soil surveys and pedological databases can be found in Chagas et al. (2004), Embrapa (2013) and Barros and De Jong Van Lier (2014, this volume).

Generally, soil survey studies in Brazil are focused on taxonomic descriptions, whereas information on hydraulic properties (water retention and hydraulic conductivity) is rarely collected. Such tendency is observed worldwide. ISRIC-WISE (Harmonized Global Soil Profile Dataset, version 3.1) (Batjes 2009), UNSODA (UNsaturated SOil DAtabase) (Leij et al. 1996; Nemes et al. 2001) and HYPRES (HYdraulic PRoperties of European Soils) (Wösten et al. 1999), were the first soil databases in which hydraulic properties associated with soil survey information were available. These datasets have (partially) satisfied the world's demand for hydrophysical information at the time.

Several soil hydrophysical data inventories have been recently developed. The latest effort resulted in the EU-HYDI (EUropean HYdropedological Data Inventory) (European Commission 2013). This initiative assembles information from soil surveys associated with hydraulic measurements from 18 European countries, including around 6,014 soil profiles and 18,682 horizons. MARTHA (Hungarian Detailed Soil Hydrophysical Database) (Makó et al. 2010) and PROPSOLO (Gonçalves et al. 2011) are examples of databases developed for Hungarian and Portuguese soils, respectively, which are included in EU-HYDI. Both were developed for building soil hydraulic PTFs.

A database with hydrophysical information on Brazilian soils has not been compiled so far. At present, diverse data are reported and analyzed in disperse publications. Assad et al. (2001), Tomasella et al. (2003), Urach (2007), Andrade and Stone (2009) and Barros et al. (2013) present examples of studies that used data on soil hydraulic attributes and carried out PTFs development. Assad et al. (2001) employed data from 373 samples from several Brazilian states, Tomasella et al. (2003) used 838 samples representing most of the Brazilian territory, Urach (2007) described 963 samples from the Southern state of Rio Grande do Sul, Andrade and Stone (2009) used more than 2,000 soil datasets, mainly from the Central-Western region of Brazil, and 786 from the Northeastern region of Brazil were collected by Barros et al. (2013). None of these studies included organizing information in a structured database that could be easily accessed, searched, and generalized. The first attempt to organize a hydrophysical database for Brazilian soils was possibly the dissertation of Silva (2005), but it is not available for public access.

Currently, the development of a soil hydrophysical database encompassing the variety of existing data in Brazil is of utmost importance. Many applications of soil hydraulic PTFs can be expected. One of them is the development of a Brazilian soil water availability map, based on the information from pedological survey studies in Brazil (Rossato et al. 1998; Cooper et al. 2005). As Brazil possesses a significant share of terrain affecting the global energy circulation, another important demand for a Brazilian PTF is in the modeling of the global energy balance for climatic forecasts.

In this context, in 2010 the Department of Hydrology of the Geological Survey of Brazil (or simply CPRM, which is the acronym for the company's official name in Portuguese: Companhia de Pesquisa de Recursos Minerais) started the "Hydrophysical Characterization of Soils" project, with the objective of developing a hydrophysical database for Brazilian soils. This project is developed by Brazilian and foreign researchers and is expected to be completed in 2016. The study consists of four stages:

1. to create an inventory of technical and scientific publications that contains hydrophysical attributes of Brazilian soils, and the corresponding data collection;
2. to define the database model;
3. to format and convert data according to the proposed model, and the corresponding data consistence;
4. to develop PTFs for the prediction of the water retention property for Brazilian soils.

In this chapter we will focus on the first of these four stages. We present an inventory of studies developed mostly in the last ten years, which address measuring hydrophysical attributes in Brazilian soils. We intend to evaluate the potential of this information to compose a Brazilian hydrophysical database, with the proposal to develop PTFs to estimate water retention for Brazilian soils.

20.2 Inventory of Publications and Data Survey

Relevant Brazilian publications, such as scientific papers, dissertations, theses, studies presented in soil conferences and symposia, and technical notes from the Brazilian Agricultural Research Corporation (Embrapa) were parsed for information. The study title was used as search reference considering four key words: 'hydraulic conductivity', 'hydrophysical', 'soil database' and 'water retention', as well as their equivalents in Portuguese. The studies selected to be evaluated for availability of hydrophysical data were those that included measurements of soil water retention.

In total, 1,100 studies were evaluated, from which 125 contained, in their titles, 'hydraulic conductivity', 611 'hydrophysical', 185 'soil database', and 170 'water retention'. Out of these studies, 163 contained water retention measurements and were selected for data availability analysis. The approximately 400 authors of these 163 publications were contacted and asked to provide their datasets to the CPRM, including physical, chemical and hydraulic information, and general data of the location where the soil properties measurements were made, with geographical coordinates (latitude and longitude), soil class description, methodological details of the experiments and other possibly relevant information.

For 52 of these studies, the authors made available the complete datasets, in the form of digital spreadsheets or printed material. Great differences were found in presentation format and in data quantity and quality.

Among these 52 studies with complete data availability, 14 refer to MSc dissertations, 7 to PhD theses, 10 to studies published in the Brazilian Journal of Soil Science and 21 to other sources (symposia and Embrapa technical notes). On classifying these studies into the four main topics proposed at the 2nd Brazilian Soil Physics Meeting/2013, the vast majority (41 out of 52) refers to "Applied soil physics: interaction between soil physics, agronomy, ecology and geosciences". Four refer to the section "Database development and pedotransfer functions", and seven others refer to "Modeling flow and transport processes in the vadose zone" and "Development of methods and instrumentation in soil physics" sections. Even though the 52 studies may not be a representative sample of the universe of publications in soil physics in Brazil, this result is an indication that, in Brazil, there seem to be few studies focused on databases, PTFs, modeling and instrumentation, compared to studies focusing on applied soil physics.

20.3 Development of Metadata on Soil Hydrophysical Information

Tables 20.1, 20.2 and 20.3 present the structure of the metadata on soil hydrophysical information with the example data from three publications. The soil attributes shown in Table 20.2 were chosen based on published experience of PTF development for hydraulic properties.

The geographical coordinates (latitude and longitude) of the sites of measurement were also registered in the metadata (Table 20.1). When coordinates of the sampling location were unavailable, they were taken as the center of the respective city or state. Locations with these coordinates, estimated or extracted from the original publications, were plotted on the map of Brazil in order to visually evaluate the distribution of available data by Brazilian state and by biomes.

The distribution of the total number of datasets compiled from the 52 publications across the Brazilian soils was evaluated at the level of soil order, according to the second last edition of the Brazilian Soil Classification System (Embrapa 2006).

20.4 Statistical Analysis

The goal of the statistical analysis was to assess the potential of the hydrophysical datasets from the 52 publications for creation of a hydrophysical database in Brazil, specifically suitable for PTFs development. In order to do so, the following aspects were evaluated: (1) total number of datasets on soil hydraulic properties; (2) scope and geographical distribution; (3) representation of soils, in terms of pedological and textural classes, as well as in terms of depth in the soil profile; (4) the variety of measuring methods of soil attributes; and (5) data characteristics for water retention PTFs development. In the latter, we intended to focus on the description of available physico-chemical attributes that can be used as PTFs predictors. Characterization of the water retention data was also done based on the frequency of measurements at different matric potential levels.

20.5 Summary Statistics of Soil Hydraulic data

The number of datasets with soil water retention measurements (water retention content versus matric potential) is shown in Table 20.4, together with the number of datasets including information on saturated or unsaturated hydraulic conductivity.

Comparing the data of Table 20.4 to the statistics of continental and international databases, we can conclude that Brazil is in a good position when it comes to water retention measurements, but not for the hydraulic conductivity data (Wösten et al. 1999; Nemes et al. 2001). In the European continental database HYPRES,

Table 20.1 Metadata with hydrophysical information on Brazilian soils – part 1 ('-' symbol, indicates that the information was not reported in the study)

		General information of the studied area							
Study	Origin	City	State[b]	Soil class	Latitude	Longitude	Geographic coordinate information	Number of datasets	Depth
Assad et al. (2001)[a]	Soil dataset of the Federal University of São Carlos (UFSCAR)/SP	Many	CE, GO, MG, PR, RJ, RN, SC, SP	Many	Many	Many	Each sample has its own coordinate	1,409	Soil depths varying from 0 m to 9.00 m (with soil horizons description)
Carducci et al. (2011)	Cerrado biome	Rio Verde	GO	Latossolo Vermelho - Amarelo distrófico (1 profile) (Embrapa 2006)	17°47′01″S	50°57′55″W	Each sample has its own coordinate	4	Bw (0.80–1.00 m)
				Latossolo Vermelho distrófico (5 profiles) (Embrapa 2006)				20	
				Latossolo Vermelho distroférrico (4 profiles) (Embrapa 2006)				16	
Ottoni (2005)	Santa Maria and Cambiocó micro-watersheds	São José de Ubá	RJ	Cambissolo Háplico (Embrapa 1999)	21°23′52″S	41°55′03″W	Each profile has its own coordinate	18	Soil depths varying from 0.075 to 0.50 m, approximately (with soil horizons description)
				Argissolo Vermelho-Amarelo (Embrapa 1999)				24	Soil depths varying from 0.10 to 0.60 m, approximately (with soil horizons description)
				Neossolo Litólico (Embrapa 1999)				6	0.085, 0.125, 0.185 m (with soil horizons description)
				Gleissolo Háplico (Embrapa 1999)				16	Soil depths varying from 0.09 to 0.60 m, approximately

(continued)

Table 20.1 (continued)

Study	Origin	General information of the studied area					Geographic coordinate information	Number of datasets	Depth
		City	State[b]	Soil class	Latitude	Longitude			
									(with soil horizons description)
				Planossolo Háplico (Embrapa 1999)				16	Soil depths varying from 0.08 to 0.60 m, approximately (with soil horizons description)
				Luvissolo Háplico (Embrapa 1999)				4	0.10, 0.25, 0.37 m (with soil horizons description)
				Extra profile				8	0.11,0.27, 0.47 e 0.65 (without soil horizons description)

[a]Assad et al. (2001) used 373 datasets with water retention measurements, extracted from the UFSCAR hydrophysical database (not published) which contains 1,409 datasets as shown in Table 20.1

[b]CE – Ceará, GO – Goiás, MG – Minas Gerais, PR – Paraná, RJ – Rio de Janeiro, RN – Rio Grande do Norte, SC – Santa Catarina, SP – São Paulo

Table 20.2 Metadata with hydrophysical information on Brazilian soils – part 2 ('-' symbol, indicates that the information was not reported in the study)

	Soil hydrophysical property methodology								
	Grain size analysis					Bulk density (BD)	Soil penetration resistance (PR)	Water retention curve	
Study	Method	Texture class	Organic carbon (OC)	Organic matter (OM)	Particle density (PD)	*Method; sample type[b]; sample size*	*Method; sample type[b]; sample size*	*Number of different values of matric potential*	*Matric potential values*
Assad et al. (2001)[a]	Embrapa (1997)	Clay (<0.002 mm), silt (0.002–0.05 m-m), sand (0.05–2 mm)	Modified Walkley-Black (Jackson 1982), with wet oxidation and determination by titration	OC × 1.72	Embrapa (1997)	Volumetric ring; UN;100 cm^3	-	3	10, 33, 1,500 kPa
Carducci et al. (2011)	Pipette	Clay (<0.002 mm), silt (0.002–0.05 mm), sand (0.05–2 mm)	-	-	Pycnometer	Volumetric ring; UN; 0.064 m of diameter and 0.025 m of height	Static; UN; 0.064 m of diameter and 0.025 m of height	11	1, 2, 4, 6, 8, 10 kPa 33, 60, 100,500, 1,500 kPa
								19–28	>1,500 kPa (1,500–300,000 kPa)
Ottoni (2005)	Densimeter	Clay (<0.002 mm), silt (0.002–0.05 mm), sand (0.05–2 mm)	Modified Walkley-Black (Jackson 1982), with wet oxidation and determination by titration	OC × 1.724	Volumetric flask	Volumetric ring; UN; 100 cm^3	-	3	0.006, 0.033, 1.5 Mpa

[a]Assad et al. (2001) used 373 datasets with water retention measurements, extracted from the UFSCAR hydrophysical database (not published) which contains 1,409 datasets as shown in Table 20.1

[b]UN – undisturbed sample; D – disturbed sample

Table 20.3 Metadata with hydrophysical information on Brazilian soils – part 3 ('-' symbol, indicates that the information was not reported in the study)

	Soil hydrophysical property methodology				
	Water retention curve			Hydraulic conductivity K	
Study	*Time of measurement*	*Method*	*Sample type*[b]*; sample size*	*K saturated; Method; sample type; sample size*	*K unsaturated; Method; sample type; sample size*
Assad et al. (2001)[a]	-	Pressure chamber	UN; 100 cm^3	-	-
Carducci et al. (2011)	-	Hanging water column	UN; 0.064 m of diameter and 0.025 m of height	-	-
		Pressure chamber			
		Dew point psychrometer – WP4-T	D		
Ottoni (2005)	-	Pressure chamber	UN; 100 cm^3	-	-

[a]Assad et al. (2001) used 373 datasets with water retention measurements, extracted from the UFSCAR hydrophysical database (not published) which contains 1,409 datasets as shown in Table 20.1

[b]UN – undisturbed sample; D – disturbed sample

Table 20.4 Number of datasets with soil hydraulic properties measurements

Hydraulic properties	Number of datasets
Both water retention and saturated hydraulic conductivity	1,253
Both water retention and unsaturated hydraulic conductivity	30
Water retention data[a]	8,983
Water retention data associated with geographic coordinates information	3,715

[a]Including datasets with saturated and unsaturated hydraulic conductivity

for example, there are 2,894 horizons with information on water retention and 1,136 with joint measurements of water retention and hydraulic conductivity.

The low number of datasets containing information on unsaturated hydraulic conductivity may be due to our adopted criteria for selection of publications (see Sect. 20.2). As previously noted, only those studies including water retention data were selected. Thus, publications that focused solely on unsaturated hydraulic conductivity measurements were not considered. Even though, we can affirm that few studies in Brazil determine unsaturated hydraulic conductivity, and those that do so use mostly the instantaneous profile method, the measured tension range being limited to the suction tensiometer range (0–800 cm).

From the total of 8,983 datasets containing information about water retention, 3,715 have geographic coordinates, which is a positive factor for the development of maps on soil hydraulic properties. These numbers as well as other statistics presented in Table 20.4 are, however, subject to revision after evaluating the consistency of all compiled information.

20.6 Scope and Geographical Distribution

The geographical distribution of the locations where hydrophysical measurements were done (52 publications) is shown in Fig. 20.1. The number of datasets containing water retention information for each Brazilian state is also indicated in Fig. 20.1 and, when inexistent, suggests that there are no measurements of this soil attribute.

The distribution of the water retention datasets (8,983) across the states and geographic regions of Brazil, with indication of data density per region (number of datasets per 100,000 km^2) is shown in Table 20.5.

The Southeastern region has a significant amount of datasets (2,328) and is the part of Brazil with the highest spatial density, 251 datasets per 100,000 km^2 (Table 20.5). Among the states in this region, São Paulo is represented by 1,176 datasets or 13 % of all water retention data. Goiás is the only state with more data quantity, with 26 %. Of the other states in the Southeast, Minas Gerais and Rio de Janeiro together present 1,110 datasets, each with relatively similar quantities of water retention measurements. The state of Espirito Santo represents only 42 datasets (~0.5 %). Figure 20.1, from Tomasella et al. (2003), shows a significant amount of datasets concentrated in the Southeastern region and might increase the percentage for this region. Currently the Tomasella et al. (2003) database can be found in the 1,561 datasets classified as 'no information' (Table 20.5).

The reason for the impressive volume of soil hydraulic information in the Southeast, especially in São Paulo, is due to the existence of large centers of soil research in this state. The Southeastern region is the second smallest region in terms of surface area, however, it represents the economic center of Brazil, with high population density, large iron ore deposits, hydro-electric power plants and ports, located between two important biomes, Cerrado and Atlantic Forest (Fig. 20.1).

In contrast to the Southeast, the Northern Brazilian region, the largest of the country and containing the richest biodiversity, surface and subsurface water resources and carbon concentration due to the presence of the Amazon Rain Forest (Fig. 20.1), is represented by only about 10 % of the entire water retention database, predominantly from the state of Amazonas, with 707 datasets (~8 %). The state of Pará, the second largest state in Brazil after Amazonas, is represented by only 16 datasets. The Northern region shows the lowest spatial density, with only 23 datasets per 100,000 km^2. Some of its states (Acre, Roraima and Amapá) have no water retention data at all.

Unlike the humid Northern Brazil, the Northeastern region is sub-humid to semi-arid, characterized by little rainfall. The main biome of this region is the Caatinga (Fig. 20.1), with a xerophytic vegetation and predominantly shallow and stony soils, commonly inappropriate for rainfed agriculture. This region is represented by 13 % of all database (8,983), thus being the territory with the second lowest spatial density, 75 datasets per 100,000 km^2. The states of Alagoas, Rio Grande do Norte and Bahia together are represented by about 1,000 datasets, 11 % of the total amount of water retention measurements. The states of Piauí and Paraíba have a low quantity of datasets (two and nine datasets, respectively). Maranhão has no water retention data at all. The remaining states of the region with hydraulic information are Ceará, Sergipe and Pernambuco, with 92, 37 and 24 datasets, respectively.

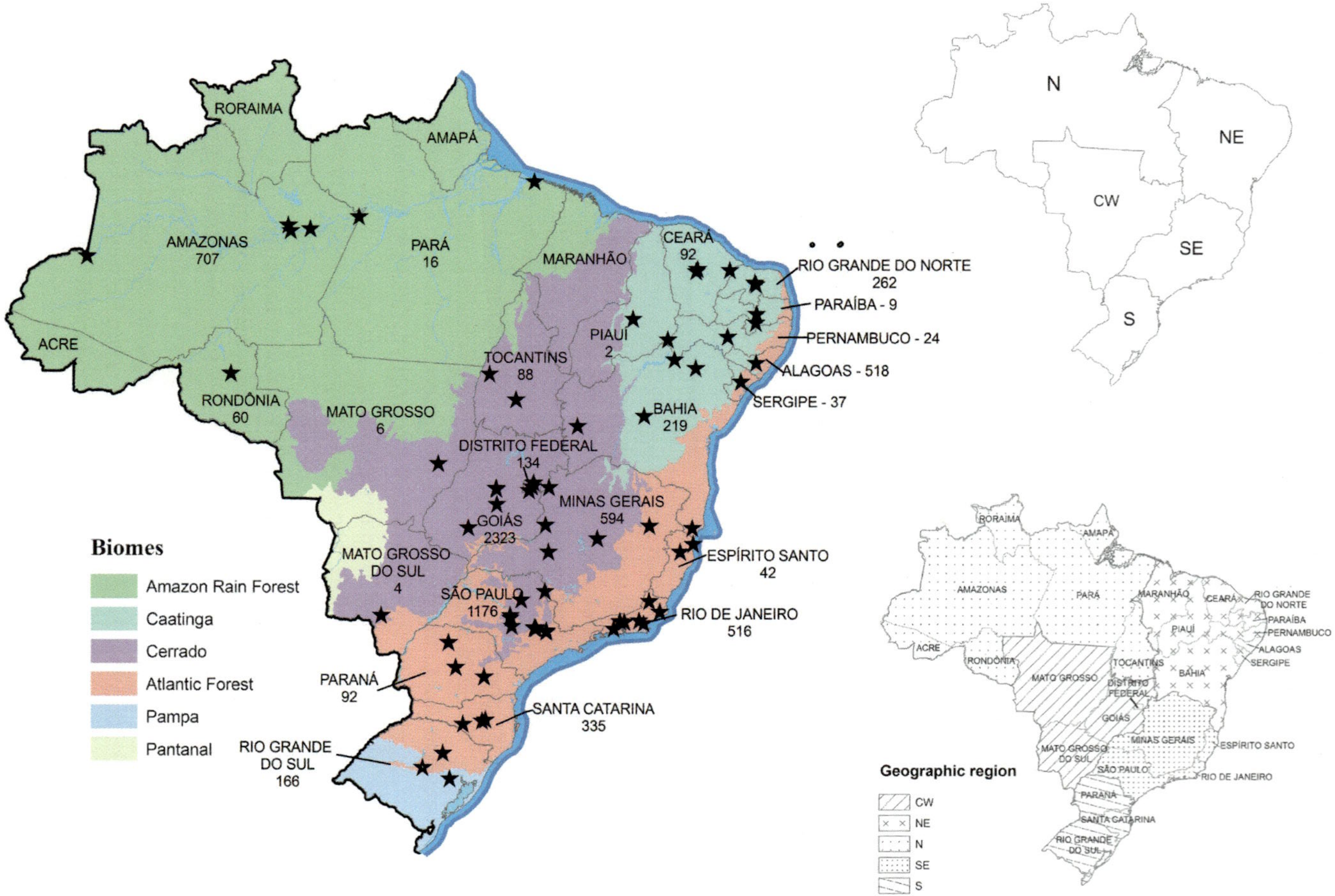

Fig. 20.1 Distribution of used sampling locations with water retention data in Brazil; N – North, NE – Northeast, CW – Central-West, SE – Southeast, S – South. The contours of the Brazilian states and biomes were extracted from http://downloads.ibge.gov.br/downloads_geociencias.htm

Table 20.5 Distribution of the water retention database across Brazil

Geographic region	Data density (number of datasets per 100,000 km^2)	Federal states	Number of datasets	Percentages	
				States	Region
Central-West (CW)	94	Goias	2,323	25.86 %	27.46 %
		Distrito Federal	134	1.49 %	
		Mato Grosso	6	0.07 %	
		Mato Grosso do Sul	4	0.04 %	
Total CW			**2,467**		
Southeast (SE)	251	São Paulo	1,176	13.09 %	25.92 %
		Minas Gerais	594	6.61 %	
		Rio de Janeiro	516	5.74 %	
		Espírito Santo	42	0.47 %	
Total SE			**2,328**		
Northeast (NE)	75	Alagoas	518	5.77 %	12.95 %
		Rio Grande do Norte	262	2.92 %	
		Bahia	219	2.44 %	
		Ceará	92	1.02 %	
		Sergipe	37	0.41 %	
		Pernambuco	24	0.27 %	
		Paraiba	9	0.10 %	
		Piaui	2	0.02 %	
		Maranhão	0	0.00 %	
Total NE			**1,163**		
North (N)	23	Amazonas	707	7.87 %	9.70 %
		Tocantins	88	0.98 %	
		Rondônia	60	0.67 %	
		Pará	16	0.18 %	
		Acre	0	0.00 %	
		Amapá	0	0.00 %	
		Roraima	0	0.00 %	
Total N			**871**		
South (S)	103	Santa Catarina	335	3.73 %	6.60 %
		Rio Grande do Sul	166	1.85 %	
		Paraná	92	1.02 %	
Total S			**593**		
No information			1,561	17.38 %	17.38 %
Total			**8,983**	100 %	100 %

The Central-Western region of Brazil overlies a significant portion of the Guarani aquifer (the second largest groundwater reserve in the world), and is covered primarily by the Cerrado biome (the second largest biome in South America). The Cerrado is characterized by savannah vegetation with grasses, shrubs and sparse trees, and by high biodiversity of plant species. This is the region with the highest concentration of datasets (2,467), although data density is not too high. Goiás alone concentrates 2,323 datasets, whereas Distrito Federal is represented by 134 measurements. Mato Grosso and Mato Grosso do Sul are represented by an insignificant number of datasets compared to the other states in the region, adding up to only 10 datasets containing water retention information. These two states represent a considerable portion of the Brazilian area and are partially covered by a major biome, Pantanal, with shallow and marshy soils. Insufficient representation of hydraulic properties of these states can be due to the unavailability of laboratories with suitable equipment for such determinations. On the other hand, the impressive volume of datasets in Goiás is explained by the presence of an Embrapa center (the Rice and Beans research center).

Finally, Brazil's Southern region, characterized in general by a subtropical climate and fertile soils, has only about 7 % of the total number of datasets, highlighting the state of Santa Catarina which has 4 %, followed by 2 % from Rio Grande do Sul and 1 % from the state of Paraná. Despite these low numbers in comparison to assessments conducted for other regions, its data frequency per 100,000 km^2 is the second highest. The Southern region is the home of important centers of soil research and a considerable amount of water retention data could be expected to have been measured, but much of it is currently inaccessible. Therefore, the available information for this region may not be a good overall representation.

Considering the representativeness of the 8,983 datasets for the Brazilian biomes, we found that for all but the Pantanal biome water retention measurements are reported (Fig. 20.1). According to the distribution of sampling locations shown in Fig. 20.1, there is a greater density of these sites in the Atlantic Forest biome, Cerrado and Caatinga, with little representation of the Amazon Rain forest and Pampa.

Regarding the water retention measurements in combination with saturated hydraulic conductivity data (Ksat), the number of datasets 1,253 is much lower than water retention only (Table 20.4). The geographical distribution of this database across Brazilian states and regions is shown in Table 20.6. For comparison purposes the number of datasets with data for water retention is also shown in the last column of this table.

The geographical distribution of the 1,253 Ksat datasets is quite different from the 8,983 datasets with water retention data. The Northern and Northeastern regions, for example, have in total just about 7 % of all data (Table 20.6), compared to 23 % in Table 20.5. The spatial densities of saturated hydraulic conductivity datasets in these regions are very small, especially for the Northern region, with only one dataset per 100,000 km^2. In both regions, most states are not subject to soil surveys including the combined determination of soil water retention and saturated hydraulic conductivity.

Table 20.6 Distribution of the water retention and saturated hydraulic conductivity datasets across Brazil

Geographic region	Data density (number of datasets per 100,000 km^2)	Federal states	Number of datasets (WR[a] + Ksat[b])	Percentages States	Percentages Region	Number of datasets (only WR[a])
Central-West (CW)	28	Goias	696	55.55 %	57.86 %	2,323
		Distrito Federal	29	2.31 %		134
		Mato Grosso	0	0.00 %		6
		Mato Grosso do Sul	0	0.00 %		4
Total CW			**725**			**2,467**
Southeast (SE)	15	São Paulo	29	2.31 %	10.93 %	1,176
		Minas Gerais	0	0.00 %		594
		Rio de Janeiro	108	8.62 %		516
		Espírito Santo	0	0.00 %		42
Total SE			**137**			**2,328**
Northeast (NE)	3	Alagoas	0	0.00 %	3.83 %	518
		Rio Grande do Norte	0	0.00 %		262
		Bahia	0	0.00 %		219
		Ceará	48	3.83 %		92
		Sergipe	0	0.00 %		37
		Pernambuco	0	0.00 %		24
		Paraiba	0	0.00 %		9
		Piaui	0	0.00 %		2
		Maranhão	0	0.00 %		0
Total NE			**48**			**1,163**
North (N)	1	Amazonas	35	2.79 %	3.43 %	707
		Tocantins	8	0.64 %		88
		Rondônia	0	0.00 %		60
		Pará	0	0.00 %		16
		Acre	0	0.00 %		0
		Amapá	0	0.00 %		0
		Roraima	0	0.00 %		0
Total N			**43**			**871**
South (S)	40	Santa Catarina	90	7.18 %	18.36 %	335
		Rio Grande do Sul	140	11.17 %		166
		Paraná	0	0.00 %		92
Total S			**230**			**593**
No information			70	5.59 %	5.59 %	1,561
Total			**1,253**	100.00 %	100.00 %	**8,983**

[a]WR – water retention data
[b]Ksat – saturated hydraulic conductivity

The Southeastern region concentrates only 11 % of the entire Ksat database (Table 20.6). The state of São Paulo has only 29 datasets, very few if compared to the 1,176 datasets for water retention alone. This percentage of 11 % is even lower than the percentage presented in the Southern region where this percentage is 18 %.

Brazil's Central-Western region represents the largest proportion of dataset concentration, similar to the information from Table 20.5, with about 60 % of the total number of datasets with saturated hydraulic conductivity information (1,253). The state of Goiás alone is represented by 55 %.

Datasets with both water retention and unsaturated hydraulic conductivity measurements (30) are concentrated in the state of São Paulo.

20.7 Soils Diversity Representation

20.7.1 Pedological Class

The Brazilian soil classification system contains 13 main soil classes; the percentage of area occupied by these classes in Brazil, as well as the approximate correspondence of them to those of the FAO classification system are shown in Table 20.7.

Figure 20.2a illustrates the distribution of the number of datasets with soil water retention measurements (8,983) across the Brazilian soil classes and the corresponding data densities. In Fig. 20.2b, the distribution of the number of datasets containing water retention and saturated hydraulic conductivity (1,253) across the 13 pedological classes is also shown.

Table 20.7 Area occupied by the main Brazilian soils and correspondence to the FAO classification system

FAO soil class[a]	Brazilian soil class	Area/area of Brazil (%)[b]
Ferrasols	Latossolos	32.14
Acrisols, Lixisols, Alisols	Argissolos	27.52
Fluvisols, Leptosols, Arenosols, Regosols	Neossolos	13.50
Plinthosols	Plintossolos	7.09
Cambisols	Cambissolos	5.37
Gleysols, Solonchaks	Gleissolos	4.79
Luvisols	Luvissolos	2.98
Planosols, Solonetz	Planossolos	2.74
Podzols	Espodossolos	2.03
Nitosols, Lixisols, Alisols	Nitossolos	1.15
Chernozems, Kastanozems, Phaeozems, Greyzems	Chernossolos	0.45
Vertisols	Vertissolos	0.21
Histosols	Organossolos	0.03

[a]Cited in Embrapa (2013)

[b]Source: Brazilian soil map (scale 1:5,000,000; Santos et al. 2011); in the case of associations, only the predominant soil class was considered

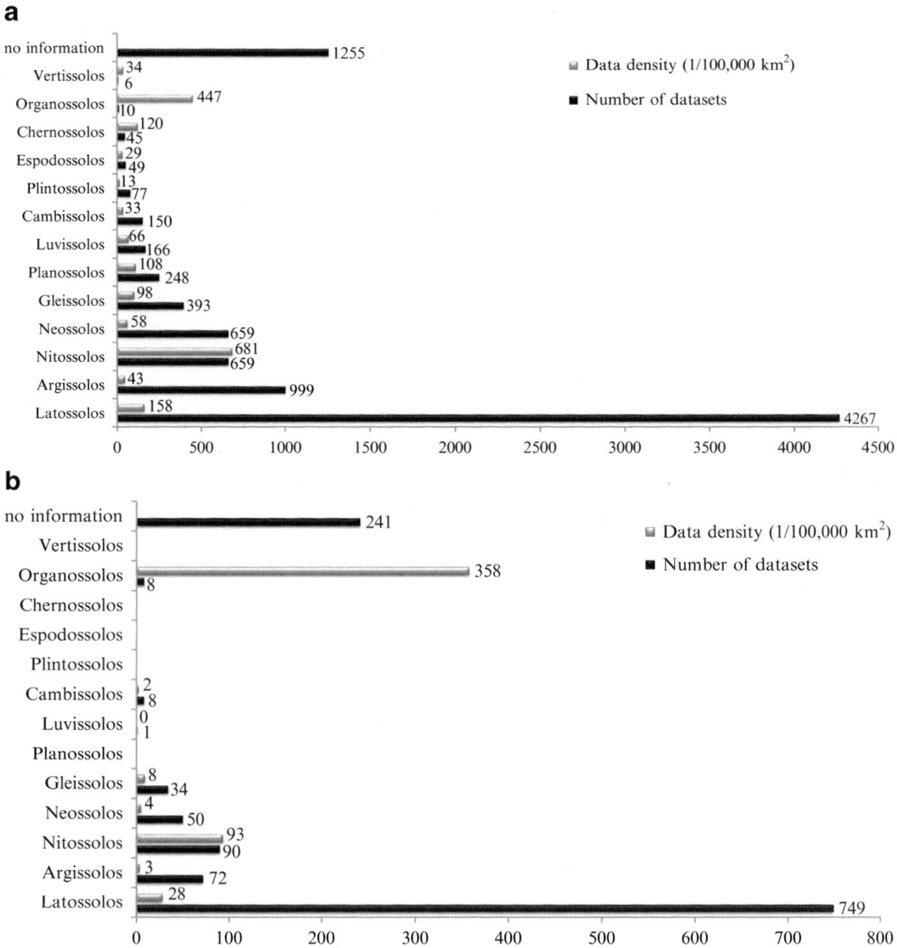

Fig. 20.2 Distribution of the number of datasets across 13 Brazilian soils classes with available data on (**a**) soil water retention; (**b**) saturated hydraulic conductivity and water retention with corresponding datasets frequency per 100,000 km^2

From these figures it can be seen that the predominant soil classes, in terms of dataset concentration, are Latossolos, Argissolos, Nitossolos, Neossolos, and Gleissolos. In the case of Fig. 20.2a, all other classes are characterized. In Fig. 20.2b however, five classes do not have measurements of saturated hydraulic conductivity and water retention, and Cambissolos and Luvissolos have low representation.

In both figures, Latossolos and Argissolos together represent more than 50 % of the total number of datasets, although neither showed the highest data density (Fig. 20.2a, b). These soil classes cover, approximately, 60 % of the Brazilian territory (Table 20.7). In general, when their chemical deficiencies are remediated, they usually are very suitable for agriculture purposes, since they offer good physical and morphological conditions.

The Nitossolos, generally excellent soils for agricultural use because of their fertility and physical conditions, occupy only about 1 % of Brazil, mostly in the states of Southern Brazil. For this soil class 659 datasets are available (Fig. 20.2a), against only 90 in the case of Fig. 20.2b, thus being the class with the highest data density when considering only water retention measurements (Fig. 20.2a).

The Neossolos, which is the third more common soil class in Brazil (Table 20.7) are mostly shallow soils with inherited characteristics of the parental material, occupying mainly the Northeastern region of Brazil. These soils are represented by 659 datasets containing information about water retention. This number of datasets is about three times larger than that for Plintossolos and Cambissolos classes together, even though the latter two soils occupy almost the same area as the Neossolos (Fig. 20.2a). As for the total number of datasets containing saturated hydraulic conductivity and water retention measurements (Fig. 20.2b), 50 of them pertain to the Neossolos class, with only four datasets per 100,000 km^2.

The Gleissolos, which are lowland soils, covering about 5 % of the Brazilian territory, mainly in the Northern region, are represented by 393 datasets with water retention information. This is the fifth predominant soil class in term of dataset concentration (Fig. 20.2a, b). Yet, this soil class has only 34 Ksat measurements (Fig.20.2b).

The Organossolos are soils with a high organic matter content. These show a high data frequency in the two figures if considered their small geographical importance (0.03 %). Contrastingly, Chernossolos and Luvissolos, soils of high fertility and with a small geographical importance, exhibit a lower data density (Fig. 20.2a, b).

Unsaturated hydraulic conductivity data (30 datasets) represent only the Latossolos and Nitossolos classes.

20.7.2 Textural Class

Out of the 8,983 datasets containing water retention information, 7,055 present particle size distribution data, distributed among textural classes as observed in Fig. 20.3a.

The prevalence of soils from the clay (C) and the sandy clay loam (SCL) classes, covering 2,963 and 1,339 datasets, respectively, can be observed in Fig. 20.3a. Classes with higher silt content (>40 %), located in the lower right corner of the textural triangle (silty clay – SiC, silty clay loam – SiCL, silty loam – SiL and silty – Si) include 432 datasets. The silt class shows the lowest percentage of all classes, with only four soils. In contrast, soils containing more than 50 % of sand found in the lower left corner of the triangle (sandy clay – SC , sandy clay loam – SCL, sandy loam – SL , loamy sand – LS and sandy – S), are represented by 3,171 datasets. The intermediate classes clay loam (CL) and loam (L) are represented by 489 datasets.

The distribution of soil samples, as shown in Fig. 20.3b, corresponds approximately to the database of Benedetti et al. (2008), who compiled 11,232

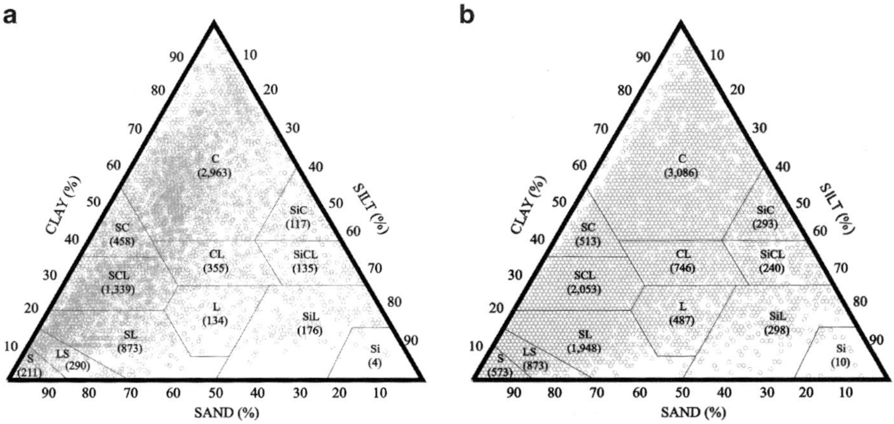

Fig. 20.3 Distribution of the total number of datasets (**a**) compiled in this study and (**b**) from Benedetti et al. (2008) over the 12 conventional textural classes. The number of datasets for each texture class is shown between brackets below the respective abbreviation

soil horizons (Fig. 20.3b) with physical and chemical information for the entire Brazilian territory. Comparison of numbers of datasets in the same classes at Figs. 20.3a, b shows that the hydraulic property database inventoried here for Brazilian soils is representative, when it comes to their texture information.

20.7.3 Depth in Soil Profile

The soil depths defined as topsoil and subsoil are represented here by depth ranges of 0–30 cm and over 30 cm, respectively. These two groups, when characterized in terms of number of datasets with water retention information (8,983) shows that about 33 % of all database are classified as subsoil, whereas approximately 55 % as topsoil and 12 % do not contain soil depth information. For the saturated hydraulic conductivity measurements (1,253 datasets), 11 % of them are classified as subsoil, 26 % as topsoil and for most datasets, about 63 %, the information is missing. As for the soils with unsaturated hydraulic conductivity measurements (30 datasets), 80 % of these datasets are from the subsoil, and only 20 % for the 0–30 cm soil depth range.

20.8 Measurement Methods of Soil Hydrophysical Attributes

Figure 20.4 shows the distribution of different methods for determining soil water retention for various matric potential levels. The pressure chamber and the centrifuge methods are used in about 60 % and 30 % of the water retention database (8,983),

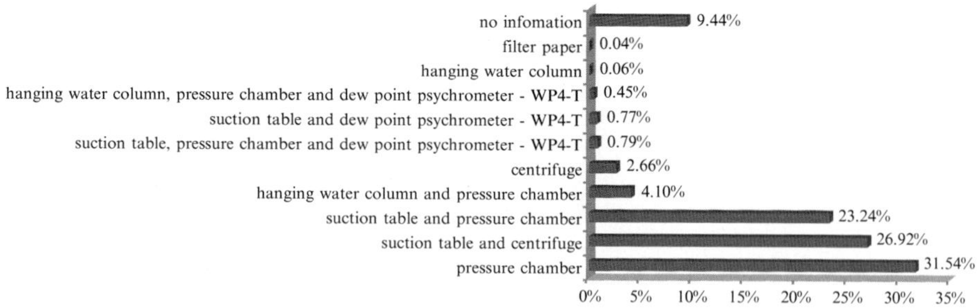

Fig. 20.4 Distribution of different methods for determining soil water retention

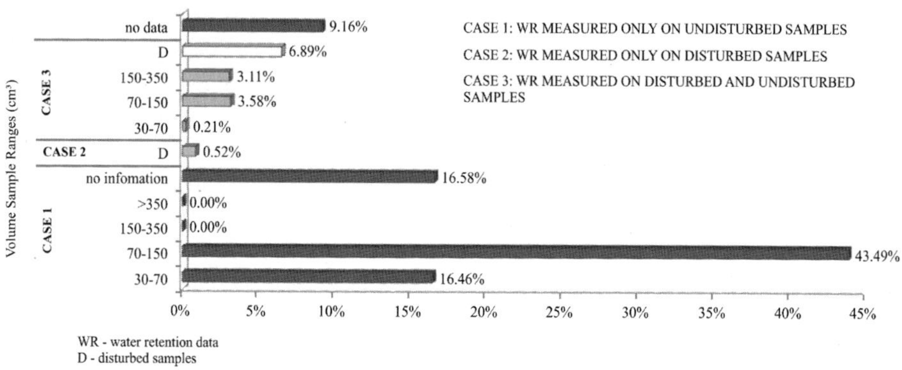

Fig. 20.5 Distribution of sample volume ranges (cm^3) for measuring soil water retention in undisturbed samples (case 1), disturbed samples (case 2), or both disturbed and undisturbed samples (case 3)

respectively. Water retention data determined by the centrifuge method should be carefully interpreted, especially when grouping them together with datasets obtained by other methods for developing water retention pedotransfer functions. Nevertheless, Reatto et al. (2008) stated that for 468 soil samples from the Cerrado biome, there was a good agreement ($R^2 = 0.99$) between the water retention determination by the centrifuge and by the pressure chamber methods. An extension of this study to other types of Brazilian soils would be useful to evaluate the general reliability of this method.

About 56 % of the total number of datasets (8,983) have their water retention data measured with two or three methods (Fig. 20.4). The suction table and hanging water column are used for lower tension ranges (0–100 cm), the pressure chamber for medium (100–500 cm) and high ranges (500–15,000 cm) and the WP4-T device exclusively for high and very high tensions (>15,000 cm).

In Fig. 20.5 the distribution of sample volumes for three cases of measuring water retention content versus matric potential is shown. The first two cases (case 1 and 2) make reference to water content measurements determined only in undisturbed or disturbed samples for the entire range of matric potential, respectively, whereas the third group (case 3) refers to measurements in both sample types together. In the entire water retention database (8,983) 9,648 samples are available. In 665 datasets water retention content was measured on both undisturbed and disturbed samples (case 3) (665 disturbed samples + 665 undisturbed samples); the disturbed ones were usually employed in the highest suction range (dry range) and the undisturbed in the lowest (wet range). Taking into account case 1 and case 2 7,384 and 50 datasets have their water retention data determined exclusively in undisturbed and disturbed samples, respectively.

The most common sample size range is 70–150 cm^3, and 100 cm^3 is the most frequent single value. The range of 30–70 cm^3 includes almost 17 % of the total number of samples, most of them of 50 cm^3. About 7 % of the considered database (9,648 samples) are represented by disturbed samples.

About 74 % of saturated hydraulic conductivity measurements (1,253 datasets) were done with the constant head permeameter, 10 % using a Guelph permeameter, 8 % by the falling head permeameter method and other 8 % by using other methods. About 60 % of all 1,253 datasets have their saturated hydraulic conductivity data determined in undisturbed soil samples in the volume range of 70–150 cm^3, with predominant use of 100 cm^3, as for the water retention data. In approximately 10 % of the saturated hydraulic conductivity database, information on the sample type (disturbed or undisturbed) is unavailable, which is a restrictive factor for the use of these datasets, as hydraulic conductivity is strongly structure dependent. The unsaturated hydraulic conductivity measurements (30 datasets) were all determined in field instruments by the instantaneous profile method.

Table 20.8 summarizes the methods used to measure the particle size distribution, organic carbon content, bulk density and penetration resistance observed in the metadata, as well as the respective number of publications. The datasets compiled for most of these soil properties are not organized for immediate handling. Therefore, the observed distribution in Table 20.8 is based on the number of publications and not on the number of datasets, except for textural composition.

Two traditional methods for determining the particle size distribution stand out: the pipette and densimeter method (Table 20.8). Among these techniques, there is a predominance of measurements with the pipette, concentrating about 43 % of the database with particle size information (7,055). In some publications, the method for determining soil texture is referred to as Embrapa (1997), which, in turn, describes the methodological procedures for both methods (pipette and densimeter). In 4,739 datasets containing soil texture and water retention information (out of the 7,055 datasets), the ranges of particle size used for the clay, silt and sand fractions are described. The same range is used for all dataset (4,739): clay (<0.002 mm), silt (0.002–0.05 mm) and sand (0.05–2 mm).

In relation to organic carbon content, the modified Walkley-Black method (Jackson 1982), with wet oxidation and determination by titration, is the most popular one (Table 20.8), cited in 58 % of the studies. Measurements of the bulk density were mostly acquired by the volumetric ring method.

Table 20.8 Methods for determining the physico-chemical attributes reported in the selected publications

Texture			Organic carbon content			Bulk density			Penetration resistance		
Method	Number of datasets	Percentage	Method	Number of studies	Percentage	Method	Number of studies	Percentage	Method	Number of studies	Percentage
Pipette	3,066	43 %	Modified Walkley-Black (Jackson 1982), with wet oxidation and determination by titration	30	58 %	Volumetric ring	42	81 %	Static	6	12 %
Embrapa (1997)	1,820	26 %	Dry combustion (CHNS/O)	3	6 %	Technique of gamma-ray attenuation	1	2 %	Dynamic	2	4 %
Densimeter	1,336	19 %	Wet oxidation after preheating	1	2 %	–	–	–	–	–	–
–	–	–	Modified Walkley-Black (Jackson 1982), with colorimetric determination	1	2 %	–	–	–	–	–	–
No information	833	12 %	No information	4	8 %	No information	8	15 %	No information	–	–
–	–		No data	13	25 %	No data	1	2 %	No data	44	85 %
Total	7,055	100 %	Total	52	100 %	Total	52	100 %	Total	52	100 %

Finally, only eight studies in our inventory determined penetration resistance. Six of the sources indicated the use of the static method. The other two reported the dynamic method as the technique to determine penetration resistance.

20.9 Characteristics of Soil Hydrophysical Database for Development of PTFs for Water Retention Estimation

Figure 20.6 illustrates the distribution of the compiled database with information of water retention (8,983 datasets) at different levels of matric potential.

About 92 % of the water retention datasets include more than three values of matric potential, six values being the most frequent, in approximately 24 % of all cases (Fig. 20.6). In most cases, the experimental data of soil water content range from the wet range to the very dry range (60 or 100 cm to 15,000 cm suction), which can ensure greater suitability of these measurements to fit equations that represent the water retention curve.

Regarding the availability of predictors variables for PTF development for water retention (Fig. 20.7), about 63 % of the publications present data on water retention associated with particle size distribution, organic matter content (or organic carbon content) and bulk density. This percentage drops to approximately 12 % when penetration resistance is included. Figure 20.8 shows that about 58 % of the datasets that contain information on soil texture and water retention (7,055 datasets), have their particle size distribution represented by more than four size fractions. All the aspects discussed above are beneficial for the derivation of PTFs for water retention prediction.

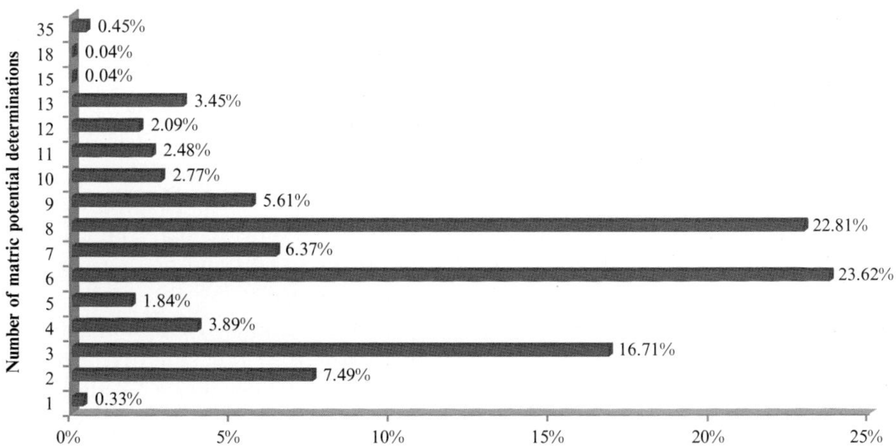

Fig. 20.6 Distribution of the number of different measurements of matric potential per dataset over the water retention database

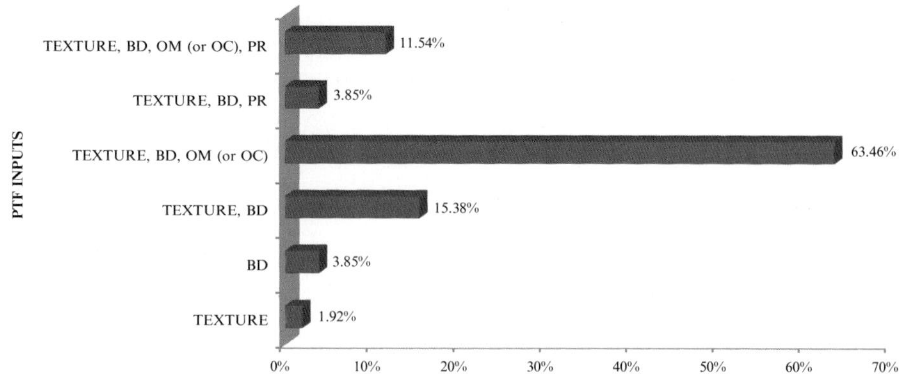

Fig. 20.7 Distribution of publications in sets of associations of PTFs predictors

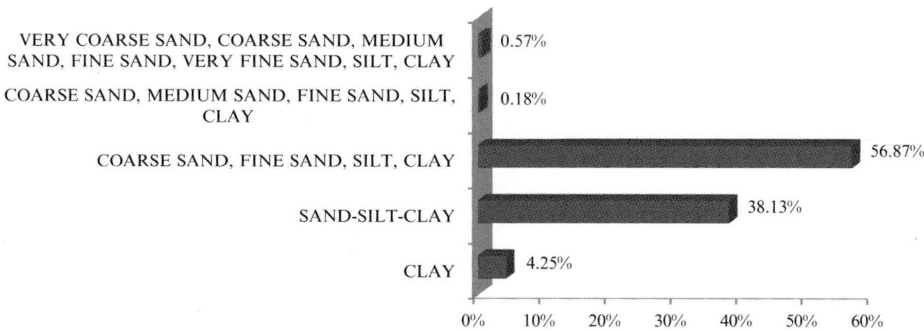

Fig. 20.8 Distribution of the total number of datasets containing water retention and soil textural composition data among different particle size fractions associations

20.10 Summary and Conclusion

Access to soil hydrological data is vital for hydrologic projects and for supporting decision-making in issues related to the availability of food and water, as well as to several land surface phenomena, including natural disasters. This topic has been widely discussed at national and international forums, such as the G8 International Conference on Open Data for Agriculture and other forums related to the Global Soil Map initiative, which aim to develop global soil maps and predict soil properties at fine resolution (http://www.globalsoilmap.net/).

Brazil, a country of continental dimensions, has accumulated a significant body of soil information and holds a prominent position in the studies on tropical soils. In general, the analytical results from the Brazilian pedological survey studies are scattered and many of them do not have information on soil hydraulic properties, which represents a setback for understanding and modeling of physical and hydrodynamic processes.

In this study, the status of hydrophysical database on Brazilian soils for water retention PTFs development was discussed. We presented on metadata the measurement methods of some soil hydrophysical attributes and the characteristics of the sites where these soil properties were estimated. Statistical analyses were performed to characterize the database based on the metadata-based data structure.

A total of 8,983 datasets have soil water retention information (at various matric potential levels) associated with physical or chemical properties, such as organic matter content (or organic carbon content), particle size fractions, bulk density, and penetration resistance. A total of 63 % of the publications present data of water retention related to determinations of organic matter content (or organic carbon content), particle size distribution and bulk density.

The distribution of the water retention datasets across the Brazilian territory shows that some regions are better represented than others, especially the Southeastern and Central-Western region, with 2,328 and 2,467 datasets, respectively. The Northern region of Brazil is not represented proportional to its social and environmental importance. All Brazilian soils classes are characterized. Two dominant soil types are Latossolos (or FAO Ferrasols) and Argissolos (Acrisols, Lixisols or Alisols, according to FAO). Information on soil water retention is relatively well distributed among topsoil and subsoil. All texture classes, except those with high silt content, are well characterized. This reflects the fact that highly weathered Brazilian soils generally contain very small silt contents.

A total of 1,253 datasets containing saturated hydraulic conductivity associated with water retention measurements (at various matric potential levels) are also found. The study showed that this database can not represent all Brazilian soil classes satisfactorily, neither the Brazilian states, in particular the ones in the Northern and Northeastern regions. Only 30 datasets are found for unsaturated hydraulic conductivity data, all determined by the instantaneous profile method, with the disadvantage of making measurements only in the water-filled tensiometer range (0–800 cm).

The methods used to measure water retention versus matric potential levels, saturated hydraulic conductivity and other physico-chemical attributes, presented in the metadata, are relatively homogeneous, but the quality of measurement of these properties has not been evaluated. The homogeneity observed for methods of determining soil water retention became evident not only by the similarities in the measuring methodology itself, but also because the measurements have been performed in undisturbed samples for almost 77 % of the cases.

The inventory of water retention data shows a significant amount of measurements associated with more than three matric potential determinations, most of them characterizing the suction range from 60 (or 100 cm) to 15,000 cm. Most of the publications analyzed in this study presents measurements of water retention related to other physico-chemical properties, usually reported in the literature as potential predictors of soil water retention. Besides this, about 58 % of the database containing information on soil textural composition and water retention (7,055 datasets) have their particle size distribution represented by more than four size fractions. All these factors, along with the homogeneity of the methods for determining the soil water retention curve, and the scope and representation of such data on Brazilian soils, create excellent prospects for studies on pedotransfer functions for water retention, at the Brazilian national scale, in an attempt to obtain sufficient hydrophysical data on different scales and for several applications. This is in contrast to data on saturated and unsaturated hydraulic conductivity.

The challenge in the development of the Brazilian soil hydrophysical database for PTFs application is to define a dataset model that can encompass the wide range of available information and that meets the different kinds of queries of interest to users of soil science information. A difficulty relates to data standardization in a database model. Existing data are available in different formats. Thus, an effort will be required to make the information uniform, including the elimination of some data after data consistence analysis. The need of statistical methods for analysis and processing of the information will exist. A preliminary structure of a database to store Brazilian soil hydrophysical information is in development. Considering the size of the Brazilian territory, it would be of interest that the database development becomes a joint effort of government agencies, universities and commercial enterprises.

Acknowledgment The authors thank the Geological Survey of Brazil (CPRM) for the financial and technical support and all researchers and students for providing the data mentioned here.

References

Andrade RS, Stone LF (2009) Uso do índice S na determinação da condutividade hidráulica não-saturada de solos do cerrado brasileiro. R Bras Agric Ambiental 13:376–381

Arruda FB, Zullo J, Oliveira JB (1987) Parâmetros de solo para cálculo de água disponível com base na textura do solo. R Bras Sci Sol 11:11–15

Assad MLL, Sans LMA, Assad ED et al (2001) Relação entre água retida e conteúdo de areia total em solos brasileiros. Rev Bras Agrometeorologia 9(3):588–596

Barros AHC, De Jong Van Lier Q (2014) Pedotransfer functions for Brazilian soils. In: Teixeira WG, Ceddia MB, Ottoni MV et al (eds) Application of soil physics in environmental analyses: measuring, modelling and data integration, 1st edn. Springer, p ??

Barros AHC, De Jong Van Lier Q, Maia AHN et al (2013) Pedotransfer functions to estimate water retention parameters of soils in Northeastern Brazil. R Bras Ciênc Solo 37:379–391

Batjes NH (2009) Harmonized soil profile data for applications at global and continental scales: update to the WISE database. Soil Use Manage 25(2):124–127

Benedetti MM, Sparovek G, Copper M et al (2008) Representatividade e potencial de utilização de um banco de dados de solos do Brasil. Rev Bras Ciênc Solo 32:2591–2600

Bouma J (1989) Using soil survey data for quantitative land evaluation. Adv Soil Sci 9:177–213

Carducci CE, De Oliveira GC, Severiano EC et al (2011) Modelagem da curva de retenção de água de Latossolos utilizando a equação duplo Van Genuchten. R Bras Ciênc Solo 35:77–86

Chagas CS, Carvalho W Jr, Bhering SB et al (2004) Estrutura e organização do sistema de informações georreferenciadas de solos do Brasil (sigsolos – versão 1.0). R Bras Ciênc Solo 28:865–876

Cooper M, Mendes LMS, Silva WLC et al (2005) A national soil profile database for Brazil available to international scientists. Soil Sci Soc Am J 69:649–652

Doyle ME, Tomasella J, Rodriguez DA et al (2013) Experiments using new initial soil water conditions and soil map in the Eta model over La Plata Basin. Meteorol Atmos Phys 121:119–136

Empresa Brasileira de Pesquisa Agropecuária – Embrapa (1997) Manual de métodos de análises de solo, 2nd edn. Embrapa Solos, Rio de Janeiro

Empresa Brasileira de Pesquisa Agropecuária – Embrapa (1999) Sistema brasileiro de classificação de solos, 6th edn. Embrapa Solos, Rio de Janeiro

Empresa Brasileira de Pesquisa Agropecuária – Embrapa (2006) Sistema brasileiro de classificação de solos, 2nd edn. Embrapa Solos, Rio de Janeiro

Empresa Brasileira de Pesquisa Agropecuária – Embrapa (2013) Sistema brasileiro de classificação de solos, 3rd edn. Embrapa Solos, Rio de Janeiro

European Commission (2013) European Hydropedological Data Inventory (EU-HYDI). European Commission Joint Research Centre. doi:10.2788/5936

Gonçalves MC, Ramos TB, Pires FP (2011) Base de dados georreferenciada das propriedades do solo. In: Coelho PS, Reis P (eds) Agrorrural contributos científicos. Instituto Nacional dos Recursos Biológicos I.P. e Imprensa Nacional – Casa da Moeda, S.A., Oeiras, pp 564–574

Hodnett MG, Tomasella J (2002) Marked differences between Van Genuchten soil water-retention parameters for temperate and tropical soils: a new water-retention pedo-transfer function developed for tropical soils. Geoderma 108:155–180

Jackson ML (1982) Analisis quimico de suelos. Omega, Barcelona, pp 282–309

Jana RB, Mohanty BP, Springer EP (2007) Multiscale pedotransfer functions for soil water retention. Vadose Zone J 6:868–878

Leij FJ, Alves WJ, Van Genuchten MT et al (1996) The UNSODA unsaturated soil hydraulic database. Rep. EPA/600/R-96/095. USEPA Natl. Risk Manage. Lab., Cincinna, Ohio

Leij FJ, Romano N, Palladino M et al (2004) Topographical attributes to predict soil hydraulic properties along a hillslope transect. Water Resour Res 40:1–15

Makó A, Tóth B, Hernádi H et al (2010) Introduction of the Hungarian Detailed Soil Hydrophysical Database (MARTHA) and its use to test external pedotransfer functions. Agrokém Talajtan 59:29–38

McBratney AB, Minasny B, Cattle SR et al (2002) From pedotransfer functions to soil inference systems. Geoderma 109:41–73

Miháliková M, Matula S, Doležal F (2013) HYPRESCZ – Database of soil hydrophysical properties in the Czech Republic. Soil Water Res 8:34–41

Minasny B, McBratney AB (2002) Uncertainty analysis for pedotransfer functions. Eur J Soil Sci Eur J Soil Sci 53(3):417–429

Mohanty BP, Shouse PJ (2002) Soil property database: Southern Great Plains 1997 Hydrology Experiment. Water Resour Res 38(5):1–5. doi:10.1029/2000WR000076

Nemes A, Schaap MG, Leij FJ et al (2001) Description of the unsaturated soil hydraulic database. UNSODA version 2.0. J Hydrol 251:151–162

Ottoni MV (2005) Classificação físico-hídrica de solos e determinação da capacidade de campo *in situ* a partir de testes de infiltração. Dissertação, Programa de Engenharia Civil-COPPE, Universidade Federal do Rio de Janeiro (UFRJ)

Pachepsky YA, Rawls WJ, Timlin DJ (1999) The current status of pedotransfer functions: their accuracy, reliability, and utility in field- and regional-scale modeling. In: Corwin DL, Loague K, Ellsworth TR (eds) Assessment of non-point source pollution in the vadose zone. American Geophysical Union, Washington, DC. doi:10.1029/GM108p0223

Pachepsky YA, Timlin DJ, Rawls WJ (2001) Soil water retention as related to topographic variables. Soil Sci Soc Am J 65:1787–1795

Perkins KS, Nimmo JR (2009) High-quality unsaturated zone hydraulic property data for hydrologic applications. Water Resour Res 45, W07417. doi:10.1029/2008WR007497

Rawls WJ (2004) Pedotransfer Functions for the United States. In: Pachepsky YA, Rawls WJ (eds) Development of pedotransfer functions in soil hydrology, vol 30, Developments in soil science. Elsevier B.V., The Netherlands, pp 437–447

Rawls WJ, Pachepsky YA (2002) Using field topographic descriptors to estimate soil water retention. Soil Sci 167:423–435

Reatto A, Da Silva EM, Bruand A et al (2008) Validity of the centrifuge method for determining the water retention properties of tropical soils. Soil Sci Soc Am J 72:1547–1553

Reichert JM, Albuquerque JA, Kaiser DR et al (2009) Estimation of water retention under availability in soils of Rio Grande do Sul. R Bras Ciênc Solo 33:1547–1560

Rossato L (2002) Estimativa da capacidade de armazenamento de água no solo do Brasil. INPE-(INPE-8915-TDI/809), São José dos Campos

Rossato L, Alvalá RCS, Tomasella J (1998) Distribuição geográfica da capacidade de armazenamento de água e das propriedades físicas do solo no Brasil. In: X Congresso Brasileiro de Meteorologia/VIII Congresso da FLISMET, Brasilia, Artigo HL98009 (Hidrometeorlogia), CD-ROM

Santos HG dos, Carvalho Junior W de, Dart R de O et al (2011) Mapa de solos do Brasil (in shapefiles). Escala 1:5.000.000. Embrapa Solos, Rio de Janeiro

Schaap MG, Leij FJ (1998) Database-related accuracy and uncertainty of pedotransfer functions. Soil Sci 163:765–779

Schaap MG, Leij FJ, Van Genuchten MT (2001) ROSSETA: computer program for estimating soil hydraulic parameters with hierarchical pedotransfer functions. J Hydrol 251:163–176

Sharma SK, Mohanty BP, Zhu J (2006) Including topography and vegetation attributes for developing pedotransfer functions. Soil Sci Soc Am J 70:1430–1440

Silva AM (2005) Banco de dados de curvas de retenção de água de solos brasileiros. Dissertação, Escola de Engenharia de São Carlos

Silva A de SS, Brito LT de L, De Oliveira CAV et al (1990) Parâmetros de solo em função da umidade na capacidade de campo em áreas irrigáveis do trópico semi-árido brasileiro. Pesquisa Agropecuária Brasileira 25:103–116

Stekauerová V, Sútor J (2004) Pedotransfer Funtions for the Rye Island – Southwest Slovakia. In: Pachepsky YA, Rawls WJ (eds) Development of pedotransfer functions in soil hydrology, vol 30, Developments in soil science. Elsevier B.V., The Netherlands, pp 465–473

Tomasella J, Hoodnett MG, Rossato L (2000) Pedotransfer functions for the estimation of soil water retention in Brazilian soils. Soil Sci Soc Am J 64:327–338

Tomasella J, Pachepsky YA, Crestana S et al (2003) Comparison of two techniques to develop pedotransfer functions for water retention. Soil Sci Soc Am J 67:1085–1092

Urach FL (2007) Estimativa da retenção de água em solos para fins de irrigação. Dissertação, Universidade de Santa Maria

Van Den Berg M, Klamt E, Van Reeuwijk LP et al (1997) Pedotransfer functions for the estimation of water retention characteristics of Ferrasols and related soils. Geoderma 78:161–180

Van Genuchten MT (1980) A closed-form equation for predicting the hydraulic conductivity of unsaturated soils. Soil Sci Soc Am J 44:892–898

Vereecken H, Weynants M, Javaux M et al (2010) Using pedotransfer functions to estimate the Van Genuchten-Mualem soil hydraulic properties: a review. Vadose Zone J 9:795–820

Wang Y, Shao M, Liu Z (2012) pedotransfer functions for predicting soil hydraulic properties of the Chinese Loess Plateau. Soil Sci 177(7):424–432. doi:10.1097/SS.0b013e318255a449

Wösten JHM, Lilly A, Nemes A (1999) Development and use of a database of hydraulic properties of European soils. Geoderma 90:169–185

Wösten JHM, Pachepsky YA, Rawls WJ (2001) Pedotransfer functions: bridging the gap between available basic soil data and missing soil hydraulic characteristics. J Hydrol 251:123–150

Index

W.G. Teixeira et al. (eds.), *Application of Soil Physics in Environmental Analyses: Measuring, Modelling and Data Integration*, Progress in Soil Science, DOI 10.1007/978-3-319-06013-2, © Springer International Publishing Switzerland 2014

T

U

V

W

Printed by Printforce, the Netherlands